THE FINITE ELEMENT METHOD
FOR ENGINEERS

THE FINITE ELEMENT METHOD FOR ENGINEERS

Third Edition

KENNETH H. HUEBNER
Ford Motor Company

EARL A. THORNTON
University of Virginia

TED G. BYROM
Consulting Engineer

A WILEY-INTERSCIENCE PUBLICATION

JOHN WILEY & SONS, INC.

New York • Chichester • Brisbane • Toronto • Singapore

Copyright © 1995 by John Wiley & Sons, Inc.

Library of Congress Cataloging in Publication Data:
Huebner, Kenneth H., 1942–
      The finite element method for engineers / Kenneth H. Huebner, Earl
  A. Thornton, Ted G. Byrom.—3rd ed.
          p.   cm.
      Includes bibliographical references and indexes.
      ISBN 0-471-54742-5 (acid-free paper)
      1. Finite element method.   I. Thornton, Earl A. (Earl Arthur),
  1936– . II. Byrom, Ted. G.   III. Title.
  TA347.F5H83   1995
  620'.001'5153--dc20   94-14031

Printed in the United States of America

10 9 8 7 6 5 4 3 2

To Our Wives,
Louise M. Huebner
and
Margaret B. Thornton
and
Anne K. Byrom

PREFACE

In the decade since the second edition of this book was published, finite element analysis has remained the preeminent approach for structural analysis and continues to find acceptance in nonstructural applications, including heat transfer and fluid mechanics. Rapid advancements in computer hardware with substantial increases in computer memory and computational speeds at lower cost have meant that large mainframe computers are no longer required for finite element analysis. Indeed, substantial computational capability with color graphics is now available to individuals at relatively modest costs. The availability of personal computers and workstations with advanced finite element software brings powerful simulation capability to the design process.

Engineering schools now routinely offer courses in finite elements, with an introductory course in the fourth year being a routine requirement in many programs. Virtually all graduate programs contain one or more courses in more advanced theory and applications. This book is intended primarily to serve as an introductory text that presents the finite element method in a clear, easy-to-understand manner. The text is aimed at mechanical, aeronautical, and civil engineers as well as engineering mechanics and engineering science majors. The emphasis remains on establishing a clear understanding of fundamentals to facilitate using the method in research and/or to solve practical, real-world problems. In this edition, as in the two previous editions, the mathematics level remains at the advanced undergraduate level found in U.S. engineering programs. Extensive use of matrices continues, but again tensors are omitted. A basic familiarity with computing is assumed.

In this third edition of the book we updated existing subject matter, introduced new advancements in finite element technology, and deleted obsolete material. In some places the text has been revised to improve clarity and in other places revisions were made to offer a more modern viewpoint. In Part I, the book begins with an introductory chapter (Chapter 1) that discusses the basic concept of the method, without mathematics, and presents a brief history of the method. The next three chapters (Chapters 2–4) develop basic finite element mathematical formulations starting from physical considerations (Chapter 2), proceeding to the well-established variational approach (Chapter 3), and then presenting the more versatile method of weighted residuals (Chapter 4). The principal revisions to these chapters include changes for clarity, additions of more recent references, and additional problems. The method of weighted residuals continues to receive more emphasis than the variational approach in this text because of its importance in nonstructural applications. Chapter 5 continues to give a basic and reasonably complete treatment of elements and interpolation functions.

In Part II, which is devoted to applications, we made more significant revisions to reflect recent developments in the subject. The chapter on elasticity has been revised, but the emphasis remains on linear static and dynamic problems. The mathematical formulation of elasticity problems using the method of weighted residuals is a new addition. Chapter 7 continues to consider equilibrium, eigenvalue, and propagation problems. The section on solving transient equations was expanded to give a more thorough treatment of algorithm stability and accuracy.

A chapter on lubrication problems was deleted to keep the book within a reasonable length as other new material was added. The chapter on heat transfer (Chapter 8) presents a complete discussion of finite element analysis of conduction, conduction combined with radiation, the convective–diffusion equation, and free and forced convection. Theoretical developments are illustrated with several practical examples.

The chapter on fluid mechanics problems (Chapter 9) received a major revision to reflect the major research effort expanded in finite element solution of flow problems in the last decade. The Chapter begins with inviscid incompressible flow, presents an expanded treatment of viscous incompressible flow, and concludes with a major new section on compressible flow. Along the way examples are presented illustrating adaptive mesh refinement. In the chapter on a sample computer code and other practical considerations, a new section describes adaptive refinement methods and presents illustrative examples as well as numerous references for this major new advancement.

The appendices (A–E) have been only slightly modified and should continue to make the book self-contained. Reference lists in all chapters have been updated and enlarged to reflect current finite element trends. Because of the continued very rapid growth in the literature we could no longer provide a list of other finite element books. Instead Chapter 1 identifies a comprehensive survey paper that lists numerous other finite element texts, conference proceedings, and other sources of information.

Throughout the book, the reference lists have been revised to include only widely available archival literature with an emphasis on journal papers.

The revised book is suitable as a textbook for two one-semester courses. For several years the second author has taught an introductory course in finite elements to advanced undergraduates and beginning graduate students covering primarily the chapters that constitute Part I. The introductory course covers most of Chapters 1–5 and selected topics from Chapter 6 and 10. Appendices A–C are used as background material. At the graduate level, a course on finite element thermal-fluid analysis covers Chapters 7–9 and selected topics from Chapter 10. For the thermal-fluid analysis course, Appendices D and E are used as reference material. We would like to hear from others concerning their experiences with the book either as a text or reference book. We welcome comments, corrections, and suggestions on ways to improve the book.

We take this opportunity to thank our colleagues and students for their assistance in the preparation of this third edition. Graduate students at the University of Virginia made many important contributions. Paul Hernan drew a substantial number of new illustrations for the text and new problems. Other students performed numerical computations for new illustrative examples. Lori Frietag performed the computations for the heated circular tube with radiation in Chapter 8. Frank Giraldo wrote the adaptive remeshing program used in Chapters 9 and 10 and with Yun Song performed the computations and prepared the illustrations for the adaptive remeshing solutions presented in these Chapters. Yun Song used the equal-order velocity and pressure formulation to solve the illustrative example of Chapter 9 and prepared the illustrations presented there. Phil Yarrington developed the new algorithm in Chapter 9 for the solution of low-speed flows with variable density. He also solved the illustrative example and prepared the illustrations. Other students, too numerous to mention, have used the book as a text and offered valuable suggestions and found several errors that have persisted for far too long.

Over the years the second and third authors have benefited greatly from their interactions with research colleagues at the Aerothermal Loads Branch at the NASA Langley Research Center. The Branch Head, Allan R. Wieting, has been a staunch supporter of the development of new finite element methodology. For the last ten years, he has provided financial support for several university researchers in the United States and the United Kingdom, and he deserves much credit for recent advancements such as the solution algorithms for compressible flow and adaptive mesh refinement. The consequences of his support of finite element CFD research are far-reaching, and we are most appreciative of his support and encouragement.

At the University of Virginia, Paulette Hughes has steadfastly performed numerous chores in the preparation of the third edition, including typing much of the revised manuscript and performing numerous multimedia communication tasks. All of these activities were performed with patience and skill and are most appreciated by the authors. We are also most appreciative of other unnamed friends, colleagues, students, and former students who have contributed in a variety of ways to this book. Finally and most of all, the authors are greatly appreciative for the support of their wives, to whom this book is sincerely dedicated.

KENNETH H. HUEBNER
EARL A. THORNTON
TED G. BYROM

Dearborn, Michigan
Charlottesville, Virginia
Fort Worth, Texas

PREFACE TO THE SECOND EDITION

Since the first edition of this book, the subject of finite element analysis gained importance and maturity in structural applications and received increasing acceptance in nonstructural areas such as heat transfer and fluids. As a result, most engineering schools now offer courses in this subject, usually at the graduate level but sometimes for advanced undergraduate students as well. This book is intended primarily to serve as an introductory text that presents the finite element method in a clear, easy-to-understand manner. The text is aimed primarily at mechanical, aeronautical, and civil engineers and engineering mechanics or engineering science majors. The emphasis remains on establishing a clear understanding of fundamentals to facilitate using the method in research and/or to solve practical real-world problems. In this edition, the mathematics remains at the same level as in the first edition in order to reach persons with mathematical skills typically learned in undergraduate engineering programs. Extensive use of vectors and matrices continues, but again tensors are omitted. A knowledge of FORTRAN and a basic familiarity with computing is assumed.

In this second edition we updated the subject matter and added homework problems to make the book more suitable for classroom or self-study. In Part One, the book begins with an introductory chapter (Chapter 1) that discusses the basic concept of the method without recourse to mathematics and gives a brief history of the method. The next three chapters (Chapters 2–4) develop basic finite element mathematical formulations starting from physical considerations (Chapter 2), proceeding to the well-established variational approach (Chapter 3), and then giving more generalized mathematical approaches (Chapter 4). The principal revisions to these chapters include changes for clarity and additions of more recent references and problems.

The method of weighted residuals receives increased emphasis in Chapter 4 (and later chapters) because of its important role in problems for which variational principles do not exist. Chapter 5 gives a basic and reasonably complete treatment of elements and interpolation functions. The treatment of isoparametric elements was expanded to reflect the importance of these popular elements.

In Part Two, which is devoted to applications, we have made more significant revisions in order to reflect recent developments in the subject. The chapter on elasticity (Chapter 6) has been enlarged, with increased emphasis on structural dynamics. The mathematical formulation of finite element solutions to free vibrations and transient response is more fully developed. A finite element application to free vibrations of composite panels has been included. The coverage of numerical methods for integrating the second-order structural dynamics equations has been expanded to include both the central difference and Newmark methods. Chapter 7, on general field problems, has been reorganized into sections dealing with equilibrium, eigenvalue, and propagation problems. A new section on solving first-order time-dependent equations has been added which parallels the treatment of the second-order equations given in Chapter 6. Popular methods of numerical integration, lumped versus consistent "mass" matrices, and the oscillation and stability of solutions are now discussed.

Chapter 8, which treats lubrication problems, needed only slight revisions to reflect recent trends. The principal change is a discussion of a method of handling the Reynolds' boundary condition. Chapter 9 on fluid mechanics problems has been more extensively modified to reflect recent progress in this area. The sections on inviscid compressible flow and viscous incompressible flows have been enlarged. Recent methods of formulating and solving these problems are presented and illustrated with examples. Some of the more recent methods included are the concept of upwind weighting functions and the penalty function approach. A new chapter (Chapter 10) on heat transfer problems has been added to reflect a growing interest in this area. The material consists of a complete discussion of finite element analysis of conduction, conduction combined with radiation, the convective-diffusion equation, and free and forced convection. Theoretical developments are illustrated with several practical examples. In the chapter on computer codes and practical applications (Chapter 11), a new computer code has been included. The code is modular and demonstrates features found in production-type computer programs. The code is illustrated with two new sample problems. A section on methods of solving algebraic equations has also been added.

The appendixes (A to D) have been only slightly modified and should continue to make the book self-contained. A new appendix (Appendix E) on basic equations from heat transfer has been added to support Chapter 10.

Reference lists in all chapters have been updated and enlarged to reflect current finite element trends. Chapter 1 in particular lists other finite element texts, recent conference proceedings, and other sources of information on the subject. As in the first edition, but perhaps more so, the reference lists are by no means exhaustive. Because of the continued rapid growth in the literature, we could only cite the more fundamental, widely available references.

The revised book is currently being used as a text for a two-semester sequence of courses at Old Dominion University. The first semester, taken by advanced undergraduates and graduate students, serves as an introduction to the finite element method. In this semester we cover most of Chapters 1–5 and selected topics from Chapters 7 and 11. Appendixes A and B are used as background material. The second semester focuses on advanced topics and is limited to graduate students. Applications of finite elements to elasticity, fluid mechanics, and heat transfer (Chapters 6, 9, and 10) are covered, with the emphasis varying with class interest. Appendixes C to E are used as reference material. We would like to hear from others concerning their experiences with the book as a text or as a reference book and welcome comments, corrections, or suggestions on ways to improve the book.

We take this opportunity to thank our colleagues for their assistance in the preparation of this second edition. We appreciate particularly the assistance of Professor J. Booker of Cornell University, who offered valuable suggestions for the revision to Chapter 8 on lubrication problems. We also would like to thank Professors C. Mei and O. A. Kandil of Old Dominion University, who reviewed chapters and offered important comments. Research colleagues of the second author at the NASA–Langley Research Center, A. R. Wieting and C. P. Shore, also reviewed chapters and made helpful suggestions. Beverly Dorton typed the revisions to the manuscript, and Deborah Miller prepared the new illustrations. For their patience and most skillful efforts we are most appreciative. In addition, the students of the Mechanical Engineering and Mechanics Department at Old Dominion University were quite helpful. By using a draft of the book as a text, they made a significant contribution in "debugging" the book by reviewing the text and working the problems. Finally, the authors are most appreciative of the support of their wives, to whom this book is sincerely dedicated.

KENNETH H. HUEBNER
EARL A. THORNTON

Waterloo, Iowa
Norfolk, Virginia
November 1982

PREFACE TO THE FIRST EDITION

One of the newest and most popular numerical techniques which engineers and scientists everywhere are using is the finite element method. With the help of high-speed digital computers, the finite element method has greatly enlarged the range of engineering problems amenable to numerical analysis.

The method of finite elements originated some 15 years ago in the aircraft industry as an effective means for analyzing complex airframe structures. It began as an extension of matrix methods for structural analysis, but now it is recognized as a powerful and versatile tool which permits a computer solution of almost all previously intractable problems in stress analysis. Many analysts agree that the finite element method represents a true breakthrough in solid mechanics.

An appealing feature of the finite element method is that it is not restricted to solid mechanics. Although this important fact was recognized about 8 years ago, the method has only recently been applied in other areas. Actually, it is applicable to almost all continuum or field problems. Hence it is not surprising that the method is receiving much attention today in engineering, physics, and mathematics.

This book is for persons who are involved in analysis in the physical sciences and want to learn about the nature and capabilities of finite element analysis. It is a "starting-point" text that presents the finite element method at an easy-to-understand, introductory level. Most previous treatments of the method center on structural mechanics problems. This book views the finite element method more generally as a numerical analysis tool for engineering mechanics problems. The approach has been to draw freely from the technical journals and to consolidate scattered information. The exposition, it is hoped, is a balanced mixture of theory and examples.

In this book, I have attempted to meet three objectives: (1) to describe as simply as possible the fundamentals of the finite element method; (2) to give the reader a coherent working familiarity with these fundamentals so that he can, without difficulty, apply the method to his problems; and (3) to offer a useful overall view of the method to establish a convenient point of departure for further study of the special advanced topics continually appearing in the literature.

I have tried to begin the discussion of each topic at a relatively elementary level and to work up gradually to its more complex aspects. An elementary knowledge of engineering mechanics and the associated mathematical and computer skills is assumed. However, special aspects of matrix algebra and variational calculus that are employed in the method are treated in sufficient detail. Tensor notation (although it can be useful in this field) is omitted for simplicity.

The book is divided into two parts. The first part presents basic concepts and the fundamental theory of the method. It begins with a largely historical and discursive discussion which provides an overview and general orientation for the reader. Then, to provide a physical basis, a detailed development of the method (as it originated from structural mechanics) is given. Because matrices and variational calculus are important to an understanding of finite element analysis, these topics are also treated in the first two appendices. Once the physical basis has been established, the mathematical basis for the method is presented. Here the reader is shown how to apply the method to widely diverse problems in engineering mechanics. The fourth chapter in this section presents the most recent generalized finite element concepts. In summary, Part I enables the reader to comprehend the method, to discover how it extends from structures problems to general continuum problems, and to appreciate the limitations of the method.

The second part of the book offers five chapters on the applications of finite element methods in engineering mechanics. Chapter 6 presents finite element formulations for linear elasticity theory, and Appendix C provides a brief review of the relevant basic equations from solid mechanics. The formulations for the elasticity problems are based on the commonly used displacement method of analysis, though mention is also made of some references in which other possible formulations are treated. General field problems such as heat conduction, electromagnetics, and torsion are the subject of Chapter 7. Since there are many practical field problems in which time is an independent parameter, an extension of finite element concepts to the time domain is examined in some detail. An entire chapter is devoted to a particular type of field problem, namely, the fluid-film lubrication problem. Inclusion of this chapter reflects my special interest in this area.

The penultimate chapter of the book studies the application of finite element techniques in fluid dynamics. The basic equations and available variational principles are reviewed in Appendix D. A discussion of the diverse types of fluid mechanics problems amenable to finite element analysis is the central theme. Here the reader will find a treatment that takes him or her up to the latest applications of the method. Solving a problem by the finite element method ultimately reduces to writing a computer program to generate and solve a set of simultaneous equations. To guide the reader through this procedure, a typical program is presented and explained in detail in Chapter 10. Here other practical considerations associated with implementing the finite element method on a digital computer are also discussed.

I have attempted to make the book self-contained so that a person wishing to apply the finite element method to his particular problem need study only one text. Although this volume treats the most important aspects of finite element analysis, space limitations made it necessary to omit or considerably abridge the treatment of some specialized aspects. I believe, however, that the treatment is sufficiently comprehensive to meet the needs of students as well as those of most practicing engineers and scientists.

The references listed at the end of each chapter are those in which readers can find additional information or detailed developments of the more advanced topics. Because of the rapidly expanding literature on this subject, the reference lists are by no means exhaustive. Instead, they were selected to represent some of the more fundamental research works.

The most important contributors to this book were all the researchers—too numerous to mention here—whose works are copiously cited throughout the book. These were the persons who provided the information which I have organized, summarized, and reported. To them, I am most grateful.

My special thanks go also to several of my colleagues. D. F. Hays granted some of the time and helped to create the atmosphere in which completion of the book was possible. A. O. "Butch" De Hart has a large measure of my gratitude for suggesting the writing of this book in the first place and for giving much active encouragement throughout the project. Professor J. F. Booker of Cornell University has been most helpful in reviewing part of the manuscript, providing stimulating ideas, and making valuable suggestions. To Professor O. C. Zienkiewicz, whose friendship and research association were most treasured during this writing project, I express my sincere appreciation.

I am also indebted to a number of persons who helped to prepare the manuscript. In particular, I wish to thank Sallie Ellison and Jim Carter, staff members of the G. M. Research Library, for finding and providing many references, and Sue Moreau for her excellent typing assistance.

Finally, I am especially grateful to my wife Louise for her patience and understanding while enduring the "book widow syndrome" for over a year.

KENNETH H. HUEBNER

Warren, Michigan
July 1974

CONTENTS

PART I

1

MEET THE FINITE
ELEMENT METHOD

1.1 WHAT IS THE FINITE ELEMENT METHOD?

The finite element method is a numerical analysis technique for obtaining approximate solutions to a wide variety of engineering problems. Although originally developed to study stresses in complex airframe structures, it has since been extended and applied to the broad field of continuum mechanics. Because of its diversity and flexibility as an analysis tool, it is receiving much attention in engineering schools and in industry.

Although this brief comment on the finite element method answers the question posed by the section heading, it does not give us the operational definition we need to apply the method to a particular problem. Such an operational definition—along with a description of the fundamentals of the method—requires considerably more than one paragraph to develop. Hence Part I of this book is devoted to basic concepts and fundamental theory. Before discussing more aspects of the finite element method, we should first consider some of the circumstances leading to its inception, and we should briefly contrast it with other numerical schemes.

In more and more engineering situations today, we find that it is necessary to obtain approximate numerical solutions to problems rather than exact closed-form solutions. For example, we may want to find the load capacity of a plate that has several stiffeners and odd-shaped holes, the concentration of pollutants during nonuniform atmospheric conditions, or the rate of fluid flow through a passage of arbitrary shape. Without too much effort, we can write down the governing equations and boundary conditions for these problems, but we see immediately that no simple analytical solution can be found. The difficulty in these three examples lies in the fact that either the geometry or some other feature of the problem is irregular or "arbitrary." Analytical solutions to problems of this type seldom exist; yet these are the kinds of problems that engineers are called upon to solve.

The resourcefulness of the analyst usually comes to the rescue and provides several alternatives to overcome this dilemma. One possibility is to make simplifying assumptions—to ignore the difficulties and reduce the problem to one that can be handled. Sometimes this procedure works; but, more often than not, it leads to serious inaccuracies or wrong answers. Now that computers are widely available, a more viable alternative is to retain the complexities of the problem and find an approximate numerical solution.

Several approximate numerical analysis methods have evolved over the years—a commonly used method is the finite difference [1][1] scheme. The familiar finite difference model of a problem gives a *pointwise* approximation to the governing equations. This model (formed by writing difference equations for an array of grid points) is improved as more points are used. With finite difference techniques we can treat some fairly difficult problems; but, for example, when we encounter irregular geometries or an unusual specification of boundary conditions, we find that finite difference techniques become hard to use.

In addition to the finite difference method, another, more recent numerical method (known as the "finite element method") has emerged. Unlike the finite difference method, which envisions the solution region as an array of grid points, the finite element method envisions the solution region as built up of many small, interconnected subregions or elements. A finite element model of a problem gives a *piecewise* approximation to the governing equations. The basic premise of the finite element method is that a solution region can be analytically modeled or approximated by replacing it with an assemblage of discrete elements. Since these elements can be put together in a variety of ways, they can be used to represent exceedingly complex shapes.

As an example of how a finite difference model and a finite element model might be used to represent a complex geometrical shape, consider the turbine blade cross section in Figure 1.1. For this device we may want to find the distribution of displacements and stresses for a given force loading or the distribution of temperature for a given thermal loading. The interior coolant passage of the blade, along with its exterior shape, gives it a nonsimple geometry.

A uniform finite difference mesh would reasonably cover the blade (the solution region), but the boundaries must be approximated by a series of

---

[1]Numbers in brackets denote references at the end of the chapter.

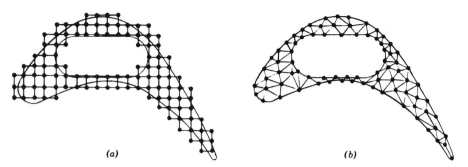

(a)   *(b)*

Figure 1.1 (*a*) Finite difference and (*b*) finite element discretizations of a turbine blade profile.

horizontal and vertical lines (or "stair steps"). On the other hand, the finite element model (using the simplest two-dimensional element—the triangle) gives a better approximation to the region and requires fewer nodes. Also, a better approximation to the boundary shape results because the curved boundary is represented by a series of straight lines. This example is not intended to suggest that finite element models are decidedly better than finite difference models for all problems. The only purpose of the example is to demonstrate that the finite element method is particularly well suited for problems with complex geometries.

1.2 HOW THE METHOD WORKS

We have been alluding to the essence of the finite element method, but now we shall discuss it in greater detail. In a continuum[2] problem of any dimension the field variable (whether it is pressure, temperature, displacement, stress, or some other quantity) possesses infinitely many values because it is a function of each generic point in the body or solution region. Consequently, the problem is one with an infinite number of unknowns. The finite element discretization procedures reduce the problem to one of a finite number of unknowns by dividing the solution region into elements and by expressing the unknown field variable in terms of assumed approximating functions within each element. The approximating functions (sometimes called *interpolation functions*) are defined in terms of the values of the field variables at specified points called *nodes* or *nodal points*. Nodes usually lie on the element boundaries where adjacent elements are connected. In addition to boundary nodes, an element may also have a few interior nodes. The nodal values of the field variable and the interpolation functions for the elements completely define the behavior of the field variable within the elements. For

---

[2]We define a continuum to be a body of matter (solid, liquid, or gas) or simply a region of space in which a particular phenomenon is occurring.

the finite element representation of a problem the nodal values of the field variable become the unknowns. Once these unknowns are found, the interpolation functions define the field variable throughout the assemblage of elements.

Clearly, the nature of the solution and the degree of approximation depend not only on the size and number of the elements used but also on the interpolation functions selected. As one would expect, we cannot choose functions arbitrarily, because certain compatability conditions should be satisfied. Often functions are chosen so that the field variable or its derivatives are continuous across adjoining element boundaries. The essential guidelines for choosing interpolation functions are discussed in Chapters 3 and 5. These are applied to the formulation of different kinds of elements.

Thus far we have briefly discussed the concept of modeling an arbitrarily shaped solution region with an assemblage of discrete elements, and we have pointed out that interpolation functions must be defined for each element. We have not yet mentioned, however, an important feature of the finite element method that sets it apart from other numerical methods. This feature is the ability to formulate solutions for individual elements before putting them together to represent the entire problem. This means, for example, that if we are treating a problem in stress analysis, we find the force–displacement or stiffness characteristics of each individual element and then assemble the elements to find the stiffness of the whole structure. In essence, a complex problem reduces to considering a series of greatly simplified problems.

Another advantage of the finite element method is the variety of ways in which one can formulate the properties of individual elements. There are basically three different approaches. The first approach to obtaining element properties is called the *direct approach* because its origin is traceable to the direct stiffness method of structural analysis. Although the direct approach can be used only for relatively simple problems, it is presented in Chapter 2 because it is the easiest to understand when meeting the finite element method for the first time. The direct approach suggests the need for matrix algebra (Appendix A) in dealing with the finite element equations.

Element properties obtained by the direct approach can also be determined by the *variational approach*. The variational approach relies on the calculus of variations (Appendix B) and involves extremizing a *functional*. For problems in solid mechanics the functional turns out to be the potential energy, the complementary energy, or some variant of these, such as the Reissner variational principle. Knowledge of the variational approach (Chapter 3) is necessary to work beyond the introductory level and to extend the finite element method to a wide variety of engineering problems. Whereas the direct approach can be used to formulate element properties for only the simplest element shapes, the variational approach can be employed for both simple and sophisticated element shapes.

A third and even more versatile approach to deriving element properties has its basis in mathematics and is known as the *weighted residuals approach* (Chapter 4). The weighted residuals approach begins with the governing equations of the problem and proceeds without relying on a variational statement. This approach is advantageous because it thereby becomes possible to extend the finite element method to problems where no functional is available. The method of weighted residuals is widely used to derive element properties for nonstructural applications such as heat transfer and fluid mechanics.

Regardless of the approach used to find the element properties, the solution of a continuum problem by the finite element method always follows an orderly step-by-step process. To summarize in general terms how the finite element method works we will succinctly list these steps now; they will be developed in detail later.

1. *Discretize the continuum.* The first step is to divide the continuum or solution region into elements. In the example of Figure 1.1 the turbine blade has been divided into triangular elements that might be used to find the temperature distribution or stress distribution in the blade. A variety of element shapes (such as those cataloged in Chapter 5) may be used, and different element shapes may be employed in the same solution region. Indeed, when analyzing an elastic structure that has different types of components such as plates and beams, it is not only desirable but also necessary to use different elements in the same solution. Although the number and the type of elements in a given problem are matters of engineering judgment, the analyst can rely on the experience of others for guidelines. The discussion of applications in Chapters 6 to 9 reveals many of these useful guidelines.

2. *Select interpolation functions.* The next step is to assign nodes to each element and then choose the interpolation function to represent the variation of the field variable over the element. The field variable may be a scalar, a vector, or a higher-order tensor. Often, polynomials are selected as interpolation functions for the field variable because they are easy to integrate and differentiate. The degree of the polynomial chosen depends on the number of nodes assigned to the element, the nature and number of unknowns at each node, and certain continuity requirements imposed at the nodes and along the element boundaries. The magnitude of the field variable as well as the magnitude of its derivatives may be the unknowns at the nodes.

3. *Find the element properties.* Once the finite element model has been established (that is, once the elements and their interpolation functions have been selected), we are ready to determine the matrix equations expressing the properties of the individual elements. For this task we may use one of the three approaches just mentioned: the direct approach, the variational approach, or the weighted residual approach.

4. *Assemble the element properties to obtain the system equations.* To find the properties of the overall system modeled by the network of elements we must "assemble" all the element properties. In other words, we combine the matrix equations expressing the behavior of the elements and form the matrix equations expressing the behavior of the entire system. The matrix equations for the system have the same form as the equations for an individual element except that they contain many more terms because they include all nodes.

The basis for the assembly procedure stems from the fact that at a node, where elements are interconnected, the value of the field variable is the same for each element sharing that node. A unique feature of the finite element method is that they system equations are generated by assembly of the individual *element* equations. In contrast, in the finite difference method the system equations are generated by writing nodal equations. In Chapter 2 we demonstrate how the assembly process leads to the system equations.

5. *Impose the boundary conditions.* Before the system equations are ready for solution they must be modified to account for the boundary conditions of the problem. At this stage we impose known nodal values of the dependent variables or nodal loads. In Chapter 2 we will see examples of how nodal boundary conditions are introduced.

6. *Solve the system equations.* The assembly process gives a set of simultaneous equations that we solve to obtain the unknown nodal values of the problem. If the problem describes steady or equilibrium behavior then we must solve a set of linear or nonlinear algebraic equations. In Chapter 10 we briefly discuss standard solution techniques for solving these equations. If the problem is unsteady, the nodal unknowns are a function of time, and we must solve a set of linear or nonlinear ordinary differential equations. We describe techniques for solving time-dependent equations in Part II of the book in Chapters 6, 7, 8, and 9.

7. *Make additional computations if desired.* Many times we use the solution of the system equations to calculate other important parameters. For example, in a structural problem the nodal unknowns are displacement components. From these displacements we calculate element strains and stresses. Similarly, in a heat-conduction problem the nodal unknowns are temperatures, and from these we calculate element heat fluxes.

1.3   A BRIEF HISTORY OF THE METHOD

Although the label *finite element method first* appeared in 1960, when it was used by Clough [2] in a paper on plane elasticity problems, the ideas of finite element analysis date back much further. In fact, the questions Who originated the finite element method? and When did it begin? have three different answers depending on whether one asks an applied mathematician, a physicist, or an engineer. All of these specialists have some justification for claiming the finite element method as their own, because each developed the essential ideas independently at different times and for different reasons. The applied mathematicians were concerned with boundary value problems of continuum mechanics; in particular, they wanted to find approximate upper and lower bounds for eigenvalues. The physicists were also interested in solving continuum problems, but they sought means to obtain piecewise approximate functions to represent their continuous functions. Faced with increasingly complex problems in aerospace structures, engineers were searching for a way in which to find the stiffness influence coefficients of shell-type structures reinforced by ribs and spars. The efforts of these three groups resulted in three sets of papers with distinctly different viewpoints.

The first efforts to use piecewise continuous functions defined over triangular domains appear in the applied mathematics literature with the work of Courant [3] in 1943. Courant used an assemblage of triangular elements and the principle of minimum potential energy to study the St. Venant torsion problem.

In 1959 Greenstadt [4], motivated by a discussion in the book by Morse and Feshback [5], outlined a discretization approach involving "cells" instead of points; that is, he imagined the solution domain to be divided into a set of contiguous subdomains. In his theory he describes a procedure for representing the unknown function by a series of functions, each associated with one cell. After assigning approximating functions and evaluating the appropriate variational principle to each cell he uses continuity requirements to tie together the equations for all the cells. By this means he reduces a continuous problem to a discrete one. Greenstadt's theory allows for irregularly shaped cell meshes and contains many of the essential and fundamental ideas that serve as the mathematical basis for the finite element method as we know it today.

As the popularity of the finite element method began to grow in the engineering and physics communities, more applied mathematicians became interested in giving the method a firm mathematical foundation. As a result, a number of studies were aimed at estimating discretization error, rates of convergence, and stability for different types of finite element approximations. These studies most often focused on the special case of linear elliptic boundary value problems. Since the late 1960s the mathematical literature on the finite element method has grown more than in any previous period. In this book we shall not study the rigorous mathematical basis of the finite element method because such knowledge is unnecessary for most practical applications. Instead we shall call upon pertinent results when they are needed.

While the mathematicians were developing and using finite element concepts, the physicists were also busy with similar ideas. The work of Prager and Synge [6] leading to the development of the hypercircle method is a key example. As a concept in function space, the hypercircle method was originally developed in connection with classical elasticity theory to give its

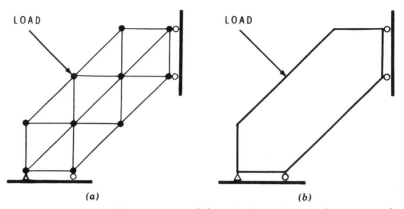

Figure 1.2   Example of (*a*) a truss and (*b*) a similarly shaped plate supporting the same load.

minimum principles a geometric interpretation. Outgrowths of the hypercircle method (such as the one suggested by Synge [7]) can be applied to the solution of continuum problems in much the same way as finite element techniques can be applied.

Physical intuition first brought finite element concepts to the engineering community. In the 1930s when a structural engineer encountered a truss problem such as the one shown in Figure 1.2*a*, he immediately knew how to solve for component stresses and deflections as well as the overall strength of the unit. First, he would recognize that the truss was simply an assembly of rods whose force–deflection characteristics he knew well. Then he would combine these individual characteristics according to the laws of equilibrium and solve the resulting system of equations for the unknown forces and deflections for the overall system.

This procedure worked well whenever the structure in question had a *finite* number of interconnection points, but then the following question arose: What can we do when we encounter an elastic continuum structure such as a plate that has an *infinite* number of interconnection points? For example, in Figure 1.2*b*, if a plate replaces the truss, the problem becomes considerably more difficult. Intuitively, Hrenikoff [8] reasoned that this difficulty could be overcome by assuming the continuum structure to be divided into elements or structural sections (beams) interconnected at only a finite number of node points. Under this assumption the problem reduces to that of a conventional structure which could be handled by the old methods. Attempts to apply Hrenikoff's "framework method" were successful, and thus the seed to finite element techniques began to germinate in the engineering community.

Shortly after Hrenikoff, McHenry [9] and Newmark [10] offered further development of these discretization ideas, while Kron [11, 12] studied topological properties of discrete systems. There followed a 10-year spell of inactivity which was broken in 1954 when Argyris and his collaborators [13–17] began to publish a series of papers extensively covering linear structural analysis and efficient solution techniques well-suited to automatic digital computation.

The actual solution of plane stress problems by means of triangular elements whose properties were determined from the equations of elasticity theory was first given in the now classical paper of Turner, Clough, Martin, and Topp [18]. These investigators were the first to introduce what is now known as the direct stiffness method for determining finite element properties. Their studies, along with the advent of the digital computer at that time, opened the way to the solution of complex plane elasticity problems. After further treatment of the plane elasticity problem by Clough [2] in 1960, engineers began to recognize the efficacy of the finite element method. In a 1980 paper Clough [19] gives his personal account of the origins of the method, describing the sequence of events from the original efforts at Boeing that led to reference 18 to the paper [2] in which introduced the label of *the finite element method*.

In 1965 the finite element method received an even broader interpretation when Zienkiewicz and Cheung [20] reported that it is applicable to all field problems that can be cast into variational form. During the late 1960s and early 1970s (while mathematicians were working on establishing errors, bounds, and convergence criteria for finite element approximations) engineers and other appliers of the finite element method were also studying similar concepts for various problems in the area of solid mechanics.

In the years since 1960 the finite element method has received widespread acceptance in engineering. Thousands of papers, hundreds of conferences, and many books have been published on the subject. The number of books published over this period illustrates the exponential growth. The first edition of this book in 1974 lists fewer than 10 finite element books. In the second edition in 1982 we list fewer than 40 finite element books. As we write this third edition in the early 1990s, finite element books are now so numerous that we are no longer able to list them. A recent bibliography [21] lists nearly 400 finite element books in English and other languages. The bibliography also identifies over 200 international finite element symposia, conferences and short courses that took place between 1964 and 1991. Clearly, these trends show the amazingly rapid worldwide acceptance of the method.

1.4   RANGE OF APPLICATIONS

Applications of the finite element method can be divided into three categories, depending on the nature of the problem to be solved. In the first category are all the problems known as *equilibrium problems* or time-independent problems. The majority of applications of the finite element method fall into this category. For the solution of equilibrium problems in the solid mechanics area we need to find the displacement distribution and the stress distribution for a given mechanical or thermal loading. Similarly, for the solution of equilibrium problems in fluid mechanics, we need to find pressure, velocity, temperature, and density distributions under steady-state conditions.

In the second category are the so-called *eigenvalue problems* of solid and fluid mechanics. These are steady-state problems whose solution often requires the determination of natural frequencies and modes of vibration of solids and fluids. Examples of eigenvalue problems involving both solid and fluid mechanics appear in civil engineering when the interaction of lakes and dams is considered, and in aerospace engineering when the sloshing of liquid fuels in flexible tanks is involved. Another class of eigenvalue problems includes the stability of structures and the stability of laminar flows.

In the third category is the multitude of time-dependent or *propagation problems* of continuum mechanics. This category is composed of the problems that result when the time dimension is added to the problems of the first two categories.

Just about every branch of engineering is a potential user of the finite element method. But the mere fact that this method can be used to solve a particular problem does not mean that it is the most practical solution technique. Often several attractive techniques are available to solve a given problem. Each technique has its relative merits, and no technique enjoys the lofty distinction of being "the best" for all problems. Consequently, when a designer or analyst has a continuum problem to solve, his first major step is to decide which method to use. This involves a study of the alternative methods of solution, the availability of computer facilities and computer packages, and, most important of all, the amount of time and money that can be spent to obtain a solution. These important aspects of the finite element method are considered further throughout this book.

The range of possible applications of the finite element method extends to all engineering disciplines, but civil, mechanical, and aerospace engineers are the most frequent users of the method. In addition to structural analysis other areas of applications include heat transfer, fluid mechanics, electromagnetism, biomechanics, geomechanics, and acoustics. The method is also finding acceptance in multidisciplinary problems where there is a coupling between one or more of the disciplines. Examples include thermal structures where there is a natural coupling between heat transfer and structures as well as aeroelasticity where there is a strong coupling between external flow and the structural response of flight vehicles.

1.5   THE FUTURE OF THE FINITE ELEMENT METHOD

Our brief look at the history of the finite element method shows us that its early development was sporadic. The applied mathematicians, physicists, and engineers all dabbled with finite element concepts, but they did not recognize at first the diversity and the multitude of potential applications. After 1960 this situation changed and the tempo of development increased markedly. By 1972 the finite element method had become the most active field of interest in the numerical solution of continuum problems. As an analysis technique the finite element method has reached the point where no additional dramatic developments or breakthroughs can be expected. Instead, future growth will involve broader applications to practical problems, increased understanding of special important aspects, and further refinement of the basic techniques.

Although in solid mechanics the finite element method can be used to solve a very large number of complex problems, there are still some areas where more work needs to be done. Some examples are the treatment of problems involving material and geometric nonlinearities. Attention must also be given to the micro- and macroanalyses of advanced composite materials. Solid mechanics problems involving multidisciplinary interactions will also require further work.

Outside of the field of solid mechanics many advances of the finite element method will continue to occur. Mathematicians will doubtless work to put the method on a broader theoretical foundation and to provide insight into problems of determining error bounds and rates of convergence for both linear and nonlinear problems. In the general area of continuum mechanics, effort will continue on developing more effective approaches for nonlinear and time-dependent problems. In computational fluid dynamics (CFD), three-dimensional flows remain a significant challenge, especially for compressible fluids. More research is needed for improved solution algorithms, mesh generation, and adaptive refinement strategies for complex flow phenomena. There is also a growing realization of the importance of validation of numerical solutions with experimental data. This conclusion is particularly valid for nonlinear problems where few, if any, comparative analytical solutions are available.

From a practitioner's viewpoint, the finite element method, like any other numerical analysis technique, can always be made more efficient and easier to use. As the method is applied to larger and more complex problems, it becomes increasingly important that the solution process remain economical. The rapid growth in engineering usage of computer technology will undoubtedly continue to have a significant effect on the advancement of the finite element method. Improved efficiency achieved by computer technology advancements such as parallel processing will surely occur. Since the mid 1970s interactive finite element programs on small but powerful personal computers and workstations have played a major role in the remarkable growth of computer-aided design. With continuing economic pressures to improve engineering productivity, this decade will see an accelerated role of the finite element method in the design process. In design offices the method enjoys routine use for structures, and we expect more routine usage in nonstructural applications such as CFD.

REFERENCES

1. D. A. Anderson, J. C. Tannehill, and R. H. Pletcher, *Computational Fluid Mechanics and Heat Transfer*, Hemisphere, Washington, DC, 1984.

2. R. W. Clough, "The Finite Element Method in Plane Stress Analysis," *Proceedings of 2nd ASCE Conference on Electronic Computation*, Pittsburgh, PA, September 8–9, 1960.

3. R. Courant, "Variational Methods for the Solutions of Problems of Equilibrium and Vibrations," *Bull. Am. Math. Soc.*, Vol. 49, 1943, pp. 1–23.

4. J. Greenstadt, "On the Reduction of Continuous Problems to Discrete Form," *IBM J. Res. Dev.*, Vol. 3, 1959, pp. 355–363.

5. P. M. Morse and H. Feshback, *Methods of Theoretical Physics*, McGraw-Hill, New York, 1953, Section 9.4.

6. W. Prager and J. L. Synge, "Approximation in Elasticity Based on the Concept of Function Space," *Q. Appl. Math.*, Vol. 5, 1947, pp. 241–269.

7. J. L. Synge, "Triangulation in the Hypercircle Method for Plane Problems," *Proc. R. Irish Acad.*, Vol. 54A21, 1952.

8. A. Hrenikoff, "Solution of Problems in Elasticity by the Framework Method," *J. Appl. Mech.*, Vol. 8, 1941, pp. 169–175.

9. D. McHenry, "A Lattice Analogy for the Solution of Plane Stress Problems," *J. Inst. Civ. Eng.*, Vol. 21, 1943, pp. 59–82.

10. N. M. Newmark, in *Numerical Methods of Analysis in Engineering*, L. E. Grinter (ed.), Macmillan, New York, 1949.

11. G. Kron, "Tensorial Analysis and Equivalent Circuits of Elastic Structures," *J. Franklin Inst.*, Vol. 238, No. 6, Dec. 1944, pp. 400–442.

12. G. Kron, "Equivalent Circuits of the Elastic Field," *J. Appl. Mech.*, Vol. 66, 1944, pp. A-149 to A-161.

13. J. H. Argyris, "Energy Theorems and Structural Analysis," *Aircraft Eng.*, Vol. 26, October–November 1954, pp. 347–356, 383–387, 394.

14. J. H. Argyris, "Energy Theorems and Structural Analysis," *Aircraft Eng.*, Vol. 27, February–March–April–May 1955, pp. 42–58, 80–94, 125–134, 145–158.

15. J. H. Argyris, "The Matrix Theory of Statics (In German)," *Ingenieur Archiv*, Vol. 25, 1957, pp. 174–192.

16. J. H. Argyris, "The Analysis of Fuselages of Arbitrary Cross-Section and Taper," *Aircraft Eng.*, Vol. 31, 1959, pp. 62–74, 101–112, 133–143, 169–180, 192–203, 244–256, 272–283.

17. J. H. Argyris and S. Kelsey, *Energy Theorems and Structural Analysis*, Butterworth, London, 1960.

18. M. J. Turner, R. W. Clough, H. C. Martin, and L. C. Topp, "Stiffness and Deflection Analysis of Complex Structures," *J. Aeronaut. Sci.*, Vol. 23, No. 9, 1956, pp. 805–823, 854.

19. R. W. Clough, "The Finite Element Method after Twenty-Five Years: A Personal View," *Comput. Structures*, Vol. 12, 1980, pp. 361–370.

20. O. C. Zienkiewicz and Y. K. Cheung, "Finite Elements in the Solution of Field Problems," *Engineer*, Vol. 220, 1965, pp. 507–510.

21. A. K. Noor, "Bibliography of Books and Monographs on Finite Element Technology," *Appl. Mech. Rev.*, Vol. 44, No. 8, June 1991, pp. 307–317.

2

THE DIRECT APPROACH:
A PHYSICAL INTERPRETATION

2.1 INTRODUCTION

The finite element method offers a way to solve a complex continuum problem by allowing us to subdivide it into a series of simpler interrelated problems. Essentially, it gives a consistent technique for modeling the whole as an assemblage of discrete parts or finite elements. The "whole" may be a body of matter or a region of space in which some phenomenon of interest is occurring. The degree to which the assemblage of elements represents the whole usually depends on the number, size, and type of elements chosen for the representation. Sometimes it is possible to choose the elements in a way that leads to an exact representation, but this occurs only in special cases. Most often the choice of elements is a matter of engineering judgment based on accumulated experience. We will continue to elaborate on this important point throughout the book.

Since the fundamental ideas of discretization in the finite element method stem from the physical procedures used in network analysis and structural framework analysis, we shall begin our discussion of finite element concepts by considering simple examples from these areas. The advantage of this approach is that an understanding of the techniques and essential concepts is gained without much mathematical manipulation. Accordingly, we develop an intuitive "feel" for the process before going on to more advanced topics. As we have seen, in any finite element analysis the first step is to replace a complex system by an equivalent idealized system consisting of individual elements connected to each other at specified points or nodes. Implicit in this procedure is the problem of defining or identifying the elements and then determining their properties. For some problems the part that should be chosen as an element immediately suggests itself, whereas in other cases the choice is not so obvious.

In this chapter we consider several example problems for which we can easily identify discrete elements. Once the elements have been selected we use direct physical reasoning to establish the element equations in terms of pertinent variables. Then we combine the element equations to form the governing equations for the complete system. By this means we illustrate the so-called direct approach for determining element properties, and we establish the general assembly procedures common to all finite element analyses. In the concluding section of this chapter we point out that the early direct approaches to the finite element method lead to far broader interpretations of the method and open the way for applying it to many different kinds of problems in engineering.

2.2   DEFINING ELEMENTS AND THEIR PROPERTIES

2.2.1   Linear Spring Systems

One of the most elementary systems that we can examine from a finite element viewpoint is the linear spring system shown in Figure 2.1. In this system we have two springs connected in series in, say, the x-coordinate direction. One end of the spring, on the left-hand side, is rigidly attached to a wall, while the spring on the right is free to move. We assume that both springs can experience either tension or compression. Forces, displacements, and spring stiffnesses are the only parameters in this system.

The way to subdivide this system into discrete elements is immediately obvious. If we define each spring to be an element, the system consists of two elements and three nodes (points of connection where forces can be transmitted and displacements can exist).

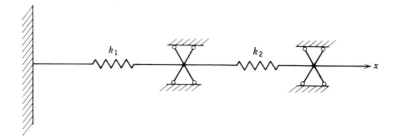

Figure 2.1   A simple linear spring system.

To determine the properties of an element, in this case its force–displace-ment equations, we focus our attention on an isolated element shown in the free-body diagram of Figure 2.2. A force and a displacement are defined at each node, and for convenience we take these forces and displacements in the positive x direction. The field variable for this example is displacement. We do not need to choose an interpolation function to represent the variation of the field variable over the element, because an exact representa-tion is already available. Hooke's law in its simplest form provides the means to relate the nodal displacements and the applied nodal forces.

According to Hooke's law, when an elastic spring experiences an axial load F, it deflects an amount δ, given by

$$\delta = \frac{1}{k}F = aF \tag{2.1}$$

where k is the spring stiffness (spring constant), and a is the spring flexibility. From equation 2.1 we can interpret k as the force required to produce a unit deflection and a as the deflection caused by a unit force. Using equation 2.1, we can write the force–deflection relationship for the spring in terms of the nodal force F_1 and the nodal displacements,

$$F_1 = k\delta_1 - k\delta_2 \tag{2.2}$$

and since equilibrium of forces requires $F_2 = -F_1$, we also have

$$F_2 = -k\delta_1 + k\delta_2 \tag{2.3}$$

Figure 2.2   Free-body diagram of a simple linear spring system.

If we use matrix notation,[1] equations 2.2 and 2.3 may be written as one equation expressing the force–displacement properties of an element:

$$
\begin{bmatrix} k & -k \\ -k & k \end{bmatrix} \begin{Bmatrix} \delta_1 \\ \delta_2 \end{Bmatrix} = \begin{Bmatrix} F_1 \\ F_2 \end{Bmatrix} \tag{2.4a}
$$

or

$$
[K]\{\delta\} = \{F\} \tag{2.4b}
$$

The components of $[K]$ are usually subscripted as k_{ij} to denote their location in the ith row and jth column of $[K]$. The square matrix $[K]$ is known as the element stiffness matrix, the column vector $\{\delta\}$ is the nodal displacement vector, and the column vector $\{F\}$ is the nodal force vector for the element.

Although equation 2.4 was derived for one of the simplest types of finite elements, namely, a linear spring, it possesses many of the characteristics of the equations expressing the properties of more complex elements. For instance, the *form*[2] of equation 2.4 remains the same regardless of the type of problem, the complexity of the element, or the way in which the element properties were derived. In this simple example, Hooke's law enabled us to determine exact values for the stiffness coefficients in the matrix $[K]$; but for the more complex situations that we will encounter later the stiffness coefficients will be determined approximately by using assumed displacement functions. Whether the stiffness coefficients of $[K]$ are determined exactly or approximately, their interpretation is the same—a typical stiffness coefficient of $[K]$, k_{ij}, is defined for this example as the force required at node i to produce a unit deflection at node j. This definition holds because only one force and one displacement exist at each node. We also note that our simple stiffness matrix obeys the Maxwell–Betti reciprocal theorem, which states that all stiffness matrices for linear structures referred to orthogonal coordinate systems must be symmetric.

The element properties given by equation 2.4 apply to either the right- or the left-hand element, depending on which value of spring stiffness we substitute into $[K]$. The fact that the left-hand element is constrained to have zero displacement at one node does not influence the derivation of the element properties. Constraint conditions are taken into account only after the element equations are assembled to form the system equations. We discuss the assembly procedure common to all finite element analysis later in

---

[1] The matrix notation and matrix manipulations used in this book are summarized in Appendix A.
[2] We will call the form of equation 2.4 the "standard form" because element equations and system equations usually appear in this way. Even the matrix equations for nonlinear problems can be written in the standard form, but the coefficients in $[K]$ for this case can be variables instead of constants.

this chapter after we consider several other examples showing how element properties are established by the direct physical approach.

2.2.2 Flow Systems

One-Dimensional Heat Flow   Another simple system that we can conveniently study using finite element concepts is shown in Figure 2.3. Here we have a section of layered material through which heat is flowing in only the x direction. In this heat conduction problem we assume that there is no internal heat generation, that the left-hand side of the wall is held at a uniform temperature higher than that of the right-hand side, and that each layer is a homogeneous solid whose thermal conductivity is known in the direction of heat flow. Heat flux, temperature, thermal conductivity, and layer thickness are the pertinent parameters.

This problem splits into a series of simpler ones if we consider each layer of the material as a finite element whose characteristics can be determined by the basic law of heat conduction. The field variable for this problem is the temperature.

The "nodes" for a typical element are now the bounding *planes* of the layer, and each node is characterized by a temperature that is uniform over the plane. Our system then consists of four elements and five nodes. A typical isolated element is also shown in Figure 2.3. We can find the exact heat flow behavior of an element by using the basic law between heat flow and temperature gradient established by the French mathematician J. B. J. Fourier. Again, we do not need an assumed interpolation function. According to Fourier's law, the quantity of heat crossing a unit area per unit time in

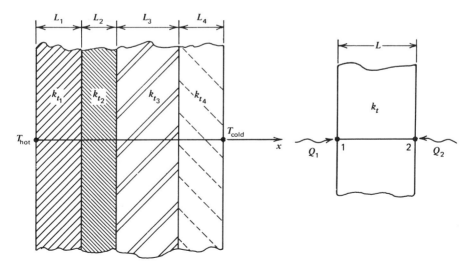

Figure 2.3   One-dimensional heat flow through a composite material.

the x direction is given by

$$q = -k_t A \frac{dT}{dx} \tag{2.5}$$

where k_t is the thermal conductivity of the material, and A is the area normal to the heat flow. For constant thermal conductivity, Fourier's law for a typical layer can be written as

$$q = k_t A \frac{\Delta T}{L}$$

where ΔT is the temperature drop across the layer, and L is the layer thickness. Then we are able to express the nodal heat flow entering a typical node in terms of the element nodal temperatures,

$$Q_1 = \frac{k_t A}{L}(T_1 - T_2) \tag{2.6}$$

and since conservation of energy requires $Q_2 = -Q_1$, we also have

$$Q_2 = -\frac{k_t A}{L}(T_1 - T_2) \tag{2.7}$$

In matrix notation equations 2.6 and 2.7 can be written as

$$\frac{k_t A}{L} \begin{bmatrix} 1 & -1 \\ -1 & 1 \end{bmatrix} \begin{Bmatrix} T_1 \\ T_2 \end{Bmatrix} = \begin{Bmatrix} Q_1 \\ Q_2 \end{Bmatrix}$$

or, more concisely, as

$$[K_t]\{T\} = \{Q\} \tag{2.8}$$

where $[K_t]$ is the matrix of thermal conductance coefficients, $\{T\}$ is the column vector of nodal temperatures, and $\{Q\}$ is the column vector of nodal heat fluxes.

We recognize that equations 2.8 has the standard form and that it completely defines the heat conduction properties of our simple thermal element. The thermal "stiffness" matrix is analogous to the structural stiffness matrix. Later, in Chapter 8 when we consider the general heat conduction problem, we see that element properties are again expressed in the form of equation 2.8. The only difference is in the dimension of $[K_t]$ and the complexity of its terms.

Fluid and Electrical Networks   Figure 2.4 shows a simple fluid flow network that could represent the water distribution system in a small building. The system is composed of many individual flow paths, and the problem is to find how the pressure and flow emanating from a given source, such as a pump, are distributed among the various paths.

We can imagine this system to be a collection of finite elements if we define each flow path between two junctions to be an element. Using an arbitrary numbering scheme, we find that this system has 16 elements and 14 nodes. The element characteristics, that is, the pressure loss—flow rate relations, may then be derived for some cases from first principles of fluid mechanics and no assumed interpolation functions are needed. For example,

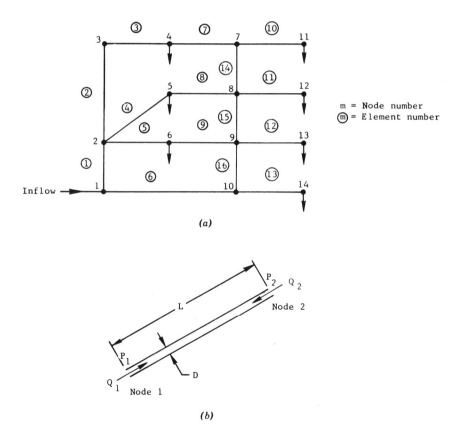

Figure 2.4   A fluid network: (*a*) a schematic representation of a flow system with numbered nodes and elements; (*b*) a typical element.

if we assume that the flow paths are circular pipes of constant cross-sectional area and that fully developed laminar flow exists in each pipe, then the volume flow rate Q is related to the pressure gradient [1] by

$$Q = -\frac{\pi D^4}{128\mu}\frac{dP}{dx} \tag{2.9}$$

where μ is the dynamic viscosity of the fluid, and D is the pipe diameter. For a typical element of length L (Figure 2.4b) we may write

$$Q = \frac{\pi D^4}{128\mu}\frac{\Delta P}{L}$$

where ΔP is the pressure drop. Then we are able to express the nodal volume flow rate entering a typical node in terms of the element nodal pressures

$$Q_1 = \frac{\pi D^4}{128\mu}\frac{P_1 - P_2}{L}$$

and since conservation of mass requires $Q_2 = -Q_1$, we also have

$$Q_2 = -\frac{\pi D^4}{128\mu}\frac{P_1 - P_2}{L}$$

In matrix notation these equations become

$$\frac{\pi D^4}{128L\mu}\begin{bmatrix} 1 & -1 \\ -1 & 1 \end{bmatrix}\begin{Bmatrix} P_1 \\ P_2 \end{Bmatrix} = \begin{Bmatrix} Q_1 \\ Q_2 \end{Bmatrix} \tag{2.10a}$$

or

$$[K_p]\{P\} = \{Q\} \tag{2.10b}$$

where $[K_p]$ is the coefficient matrix, $\{P\}$ is the column vector of nodal pressures, and $\{Q\}$ is the column vector of nodal flow rates.

Again we see that equation 2.10 has the standard form. Our development of the element equations for the special case of laminar pipe flow does not include the pressure losses associated with the fittings and valves that are part of most fluid networks. These pressure losses are normally taken into account by introducing an equivalent length of pipe that can be lumped with the actual pipe length.

If the fluid flow in the network is turbulent, it is still possible to define an element as a length of fluid-carrying conduit, but the element equations are no longer linear. This can be seen by examining the Fanning equation, an empirical relation governing fully developed turbulent pipe flow. According to Fanning's equation [1],

$$P_1 - P_2 = \frac{8f_M L}{\pi D^5} Q^2$$

where L is the length of the pipe, D is its diameter, and f_M is the Moody friction factor, which is a function of the Reynolds number and the pipe roughness. Although it would be possible to use Fanning's equation to develop element equations similar to equation 2.10, the resulting $[K_p]$ matrix would contain known functions of the flow rate Q instead of constants. The same nonlinear character would prevail in the final equations assembled from the individual element equations, and hence special solution techniques would be required.

The idea of defining a flow path as a finite element of a fluid flow network also applies to direct-current electrical networks. A current-carrying member of the electrical network can be taken as a finite element, and Ohm's law then provides the means for establishing the element characteristics. Procedures directly analogous to those used for the fluid flow networks may be used; voltages V_1 and V_2 play the same role as the nodal pressures and a current I replaces the flow rate Q.

2.2.3   Simple Elements From Structural Mechanics

In the engineering community the idea of modeling a structure as a series of finite elements began as an extension of the traditional methods used for analyzing framed structures such as trusses and bridges. These structures consist of bars interconnected only at certain points where forces can be transmitted. Hence it seemed natural to an engineer to regard these structures as assemblages of individual components or elements. In the earliest approaches to find the force–deflection characteristics, or stiffness, of a complete structure, engineers determined the stiffness of its individual components. The method used closely followed the approach we illustrate with the linear spring example in Section 2.2.1. For pin-connected truss members or beam members of rigid-jointed frames, engineers were able to derive the stiffness relationships by the direct physical approach. From these member stiffness matrices, the stiffness characteristics of the complete structure were synthesized. This approach is known as the *direct stiffness method*. When the need arose in the aerospace industry in the early 1950s to model more complex structures built from ribs, spars, and thin-walled skins, the direct stiffness method was extended by Turner et al. [2] for other structural components, including thin triangular elements in plane stress.

The approach developed in the 1950s for deriving stiffness matrices of structural elements contains basic steps that remain valid today. Of course,

over the years, the approach has been generalized to become more sophisticated and broader in scope. We present here the basic steps and illustrate
the approach for a truss element and a triangular element for plane stress
problems. In Chapter 6, we generalize the approach for an elastic continuum.

There are six essential steps in the derivation for a structural element
stiffness matrix:

1. Assume the functional form for the displacement field within an element.
2. Express the displacement field in terms of the displacements defined at
   the nodes.
3. Introduce the strain–displacement equations and thereby determine
   the state of element strain corresponding to the assumed displacement
   field.
4. Write the constitutive equations relating stress to strain and thereby
   introduce the influence of the material properties of the element.
5. Write the equilibrium equations for the element to establish the relationship between the nodal displacements and nodal forces.
6. Combine the results from steps 1 through 5 and algebraically derive the
   element stiffness matrix.

A Truss Element   A pin-connected bar, the simplest finite element, provides
a good starting point to illustrate how the approach is used to derive the
stiffness matrix of a structural element.

Step 1   Figure 2.5 shows a pin-connected truss element and the *local*
element coordinate axes denoted by the subscript L. At each end of the
element we have a displacement component and nodal force directed
along the local axis x_L. The local axis x_L makes an angle θ with the *global*

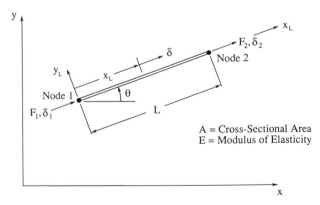

Figure 2.5   Pin-connected truss element with local and global coordinate systems.

x axis. At an interior point the axial displacement along the element is δ, which varies along the element; thus $\delta = \delta(x_L)$. At node 1, we know that δ has the value δ_1, and at node 2 we know that δ has the value δ_2. With these constraints we wish to find an equation for the displacement within the element. With two constraints, we choose the displacement to be a linear function of x_L,

$$\delta = \alpha_1 + \alpha_2 x_L \tag{2.11}$$

where α_1 and α_2 are constants to be determined. For the truss element, the assumed linear displacement variation given in equation 2.11 is actually the exact displacement function. We know this from mechanics of materials [3] because the elongation of a bar of length L fixed at the left end and subjected to an axial force P is given by $\delta = Px_L/EA$, where A is the cross-sectional area, and E is the modulus of elasticity. The only difference here is that for generality we permit the left end, node 1, of our truss member to have a displacement δ_1.

Step 2 Since equation 2.11 defines the displacement within the element, we are now ready to find the constants α_1 and α_2. Evaluating equation 2.11 at the nodes, we write two equations:

For node 1 $\delta(0) = \delta_1 = \alpha_1$

For node 2 $\delta(L) = \delta_2 = \alpha_1 + \alpha_2 L$

We solve these equations for α_1 and α_2 and substitute the results into equation 2.11:

$$\delta = \delta_1 + \frac{\delta_2 - \delta_1}{L} x_L \tag{2.12}$$

A quick check of equation 2.12 shows us that at the nodes the displacement δ takes on the proper values. Between nodes, of course, the displacement varies linearly within the element.

Step 3 With the element displacement function known, we proceed to find the element strain. The only nonzero strain component for the truss element is ϵ, and the strain–displacement relationship is $\epsilon = \partial \delta / \partial x_L$. Thus

$$\epsilon = \frac{\delta_2 - \delta_1}{L} \tag{2.13}$$

The strain is simply the ratio of the change in length of the element to the length. The strain is independent of x_L, and thus it is constant within the element.

Step 4   For our truss member the only nonzero stress is σ, and Hooke's law takes the simple form $\sigma = E\epsilon$. Thus

$$\sigma = E\epsilon = \frac{E}{L}(\delta_2 - \delta_1) \tag{2.14}$$

The stress σ is also constant within the element.

Step 5   The key step in our derivation of the truss element stiffness matrix is to write equilibrium equations that relate the nodal displacements to nodal forces. An approach that we may use for structural elements is the principle of minimum potential energy (see Appendix C). For an elastic structure the total potential energy Π is the sum of the internal strain energy U and the negative of the work V of the external applied forces. The strain energy is obtained by integration over the volume of the element:

$$U = \frac{1}{2} \iiint_{\text{Vol}} \sigma\epsilon \, d(\text{Vol})$$

Substituting equations 2.13 and 2.14 and noting that the integrand is constant, we get

$$U = \frac{1}{2} \frac{E}{L^2}(\delta_2 - \delta_1)^2 \iiint_{\text{Vol}} d(\text{Vol})$$

and since the volume of the element is AL, we obtain

$$U = \frac{1}{2} \frac{EA}{L}(\delta_2 - \delta_1)^2 \tag{2.15}$$

The work of the external applied forces, the nodal forces for our element, is

$$V = F_1\delta_1 + F_2\delta_2 \tag{2.16}$$

Combining the last two results gives the total potential energy for the element:

$$\Pi^{(e)} = \frac{1}{2} \frac{EA}{L}(\delta_2 - \delta_1)^2 - (F_1\delta_1 + F_2\delta_2) \tag{2.17}$$

where (e) denotes the *element* total potential energy. The principle of minimum potential energy states that for equilibrium,

$$\frac{\partial \Pi}{\partial \delta_1} = \frac{\partial \Pi}{\partial \delta_2} = 0$$

Performing the indicated derivatives of equation 2.17 yields the element equilibrium equations:

$$-\frac{EA}{L}(\delta_2 - \delta_1) - F_1 = 0 \qquad (2.18a)$$

$$\frac{EA}{L}(\delta_2 - \delta_1) - F_2 = 0 \qquad (2.18b)$$

Step 6 For the truss element, the last step consists simply of writing the equilibrium equations in the standard matrix form,

$$\frac{EA}{L}\begin{bmatrix} 1 & -1 \\ -1 & 1 \end{bmatrix}\begin{Bmatrix} \delta_1 \\ \delta_2 \end{Bmatrix} = \begin{Bmatrix} F_1 \\ F_2 \end{Bmatrix} \qquad (2.19)$$

and we see immediately that the element stiffness matrix is

$$[K]^{(e)} = \frac{EA}{L}\begin{bmatrix} 1 & -1 \\ -1 & 1 \end{bmatrix} \qquad (2.20)$$

Note that the truss element stiffness matrix given in equation 2.20 is defined in the local coordinate system, (x_L, y_L). In Section 2.2.4 we show how to transform the stiffness matrix to the global coordinate system, (x, y). We note that the truss element stiffness matrix in local coordinates is the same as for the linear spring given in equation 2.4 if we define the spring constant $k = EA/L$.

A Triangular Element Historically, the triangular element was the first two-dimensional structural finite element. As we mentioned earlier, engineers had successfully used the direct stiffness method to derive stiffness characteristics for the truss and beam element, but these elements have only one spatial dimension. The triangular element was first developed by Turner et al. in 1956 [2], although the label *finite element* was not used in this paper. In fact not until four years later was the method called finite elements by Clough in another paper [4]. The original three-node triangular element is the simplest plane stress element and is still widely used.

Step 1 Figure 2.6 shows a typical triangular element of a thin plate experiencing plane stress. The element is referenced to a global x, y coordinate system, and the three nodes are numbered in the counterclockwise direction. If we define two components of displacement at each node, this element has six degrees of freedom, given by the column vector of

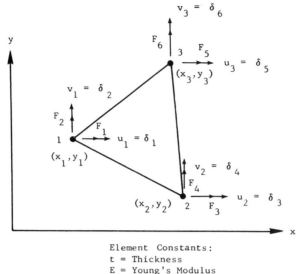

Figure 2.6   A triangular element in plane stress.

displacements:

$$\{\delta\} = \begin{Bmatrix} \delta_1 \\ \delta_2 \\ \delta_3 \\ \delta_4 \\ \delta_5 \\ \delta_6 \end{Bmatrix} = \begin{Bmatrix} u_1 \\ v_1 \\ u_2 \\ v_2 \\ u_3 \\ v_3 \end{Bmatrix} \qquad (2.21)$$

The displacement distribution within the element clearly must be a function of x and y, and it must be *uniquely* determined by the six nodal displacements. With this constraint a linear displacement function is the only choice, that is,

$$\begin{aligned} u &= \alpha_1 + \alpha_2 x + \alpha_3 y \\ v &= \alpha_4 + \alpha_5 x + \alpha_6 y \end{aligned} \qquad (2.22)$$

where α_1 through α_6 are constants yet to be determined. Equation 2.22 states that the displacements vary linearly within the element and along the element boundaries. Such behavior of the displacement functions ensures continuity of the displacement field throughout the structure modeled by the mesh of adjacent elements. This desirable feature derives from the fact that the displacement along a boundary between two adjacent elements is uniquely specified by the displacements at the two nodes defining the ends of the boundary.

Step 2   Having defined the displacement field within an element, we are now ready to find the constants α_1 through α_6 in terms of the coordinates of the node points and the nodal displacements. Evaluating equation 2.22 at each of the nodes gives two sets of three equations each:

$$\text{For node 1} \qquad u_1 = \alpha_1 + \alpha_2 x_1 + \alpha_3 y_1$$

$$\text{For node 2} \qquad u_2 = \alpha_1 + \alpha_2 x_2 + \alpha_3 y_2 \qquad (2.23)$$

$$\text{For node 3} \qquad u_3 = \alpha_1 + \alpha_2 x_3 + \alpha_3 y_3$$

Similarly, for the y component of displacement we have

$$v_1 = \alpha_4 + \alpha_5 x_1 + \alpha_6 y_1$$

$$v_2 = \alpha_4 + \alpha_5 x_2 + \alpha_6 y_2 \qquad (2.24)$$

$$v_3 = \alpha_4 + \alpha_5 x_3 + \alpha_6 y_3$$

With a little algebraic manipulation we can easily solve equations 2.23 for α_1, α_2, and α_3 in terms of u_1, u_2, and u_3 and the nodal coordinates. When the result is substituted into the first of equations 2.22 we obtain the following expression for u:

$$u = \frac{a_1 u_1 + a_2 u_2 + a_3 u_3}{2\Delta} + \frac{b_1 u_1 + b_2 u_2 + b_3 u_3}{2\Delta} x$$
$$+ \frac{c_1 u_1 + c_2 u_2 + c_3 u_3}{2\Delta} y \qquad (2.25)$$

where

$$\begin{array}{lll}
a_1 = x_2 y_3 - x_3 y_2 & a_2 = x_3 y_1 - x_1 y_3 & a_3 = x_1 y_2 - x_2 y_1 \\
b_1 = y_2 - y_3 & b_2 = y_3 - y_1 & b_3 = y_1 - y_2 \qquad (2.26) \\
c_1 = x_3 - x_2 & c_2 = x_1 - x_3 & c_3 = x_2 - x_1
\end{array}$$

and

$$\Delta = \text{area of the triangular element} = \frac{1}{2} \begin{vmatrix} 1 & x_1 & y_1 \\ 1 & x_2 & y_2 \\ 1 & x_3 & y_3 \end{vmatrix} \qquad (2.27)$$

when the nodes of the element are numbered in the counterclockwise direction.

Following the same procedure for equations 2,24, we finally obtain

$$v = \frac{a_1 v_1 + a_2 v_2 + a_3 v_3}{2\Delta} + \frac{b_1 v_1 + b_2 v_2 + b_3 v_3}{2\Delta} x$$
$$+ \frac{c_1 v_1 + c_2 v_2 + c_3 v_3}{2\Delta} y \tag{2.28}$$

Step 3   Now that the displacement field is specified, we can find the element strains corresponding to the assumed displacements. For the plane stress case we are considering, the only strains that exist are those in the x–y plane ($\epsilon_x, \epsilon_y, \gamma_{xy}$), which for small displacements are given by[3]

$$\epsilon_x = \frac{\partial u}{\partial x} \qquad \epsilon_y = \frac{\partial v}{\partial y} \qquad \gamma_{xy} = \frac{\partial v}{\partial x} + \frac{\partial u}{\partial y} \tag{2.29}$$

An expression for the element strain can now be obtained by substituting equations 2.25 and 2.28 into equation 2.29 and differentiating. Using matrix notation, we may write

$$\epsilon_x = \frac{1}{2\Delta} \lfloor b_1 \quad 0 \quad b_2 \quad 0 \quad b_3 \quad 0 \rfloor \begin{Bmatrix} u_1 \\ v_1 \\ u_2 \\ v_2 \\ u_3 \\ v_3 \end{Bmatrix}$$

$$\epsilon_y = \frac{1}{2\Delta} \lfloor 0 \quad c_1 \quad 0 \quad c_2 \quad 0 \quad c_3 \rfloor \begin{Bmatrix} u_1 \\ v_1 \\ u_2 \\ v_2 \\ u_3 \\ v_3 \end{Bmatrix}$$

$$\gamma_{xy} = \frac{1}{2\Delta} \lfloor c_1 \quad b_1 \quad c_2 \quad b_2 \quad c_3 \quad b_3 \rfloor \begin{Bmatrix} u_1 \\ v_1 \\ u_2 \\ v_2 \\ u_3 \\ v_3 \end{Bmatrix}$$

[3]The reader who is unfamiliar with these relations may want to refer to the review of the basic relations of solid mechanics given in Appendix C.

or

$$\{\epsilon\} = \begin{Bmatrix} \epsilon_x \\ \epsilon_y \\ \gamma_{xy} \end{Bmatrix} = \frac{1}{2\Delta} \begin{bmatrix} b_1 & 0 & b_2 & 0_3 & 0 & b_3 \\ 0 & c_1 & 0 & c_2 & 0 & c_3 \\ c_1 & b_1 & c_2 & b_2 & c_3 & b_3 \end{bmatrix} \begin{Bmatrix} u_1 \\ v_1 \\ u_2 \\ v_2 \\ u_3 \\ v_3 \end{Bmatrix} \qquad (2.30a)$$

If we let

$$[B] = \frac{1}{2\Delta} \begin{bmatrix} b_1 & 0 & b_2 & 0 & b_3 & 0 \\ 0 & c_1 & 0 & c_2 & 0 & c_3 \\ c_1 & b_1 & c_2 & b_2 & c_3 & b_3 \end{bmatrix} \qquad (2.30b)$$

then

$$\{\epsilon\}^{(e)} = [B]^{(e)}\{\delta\}^{(e)} \qquad (2.30c)$$

where the superscript (e) designates that the equation holds for one element. Equation 2.30 are the important relations that enable us to calculate the strain of an element once the displacements of its nodes are known. We see that the choice of a linear displacement field leads to a constant strain state in the element.

Step 4   Hooke's law provides the linear constitutive equations we need to relate stress to strain. In contrast to the one-dimensional case, we see from Appendix C that the three-dimensional version of Hooke's law has six components of stress related to six components of strain through a 6×6 proportionality matrix $[C]$:

$$\begin{Bmatrix} \sigma_x \\ \sigma_y \\ \sigma_z \\ \tau_{xy} \\ \tau_{yz} \\ \tau_{xz} \end{Bmatrix} = \begin{bmatrix} C_{11} & C_{12} & \cdots & C_{16} \\ \vdots & & & \vdots \\ C_{61} & \cdots & \cdots & C_{66} \end{bmatrix} \begin{Bmatrix} \epsilon_x \\ \epsilon_y \\ \epsilon_z \\ \gamma_{xy} \\ \gamma_{yz} \\ \gamma_{xz} \end{Bmatrix} \qquad (2.31a)$$

or

$$\{\sigma\} = [C]\{\epsilon\} \qquad (2.31b)$$

The matrix $[C]$ in equation 2.31a contains 36 constants, but because of its symmetry there are actually only 21 distinct elastic constants. Equation 2.31a simplifies considerably for the special case of homogeneous and isotropic solids experiencing *plane stress*. For this case $[C]$ contains only 2 independent constants instead of 21. Two such constants are Young's modulus, E, and Poisson's ratio ν. In terms of these constants the stress–strain equations may be written as follows:

$$\left\{\begin{array}{c} \sigma_x \\ \sigma_y \\ \tau_{xy} \end{array}\right\} = \frac{E}{1-\nu^2} \left[\begin{array}{ccc} 1 & \nu & 0 \\ \nu & 1 & 0 \\ 0 & 0 & \dfrac{1-\nu}{2} \end{array}\right] \left\{\begin{array}{c} \epsilon_x \\ \epsilon_y \\ \gamma_{xy} \end{array}\right\} \tag{2.32a}$$

or

$$\{\sigma\}^{(e)} = [C]^{(e)}\{\epsilon\}^{(e)} \tag{2.32b}$$

where, again, the superscript (e) signifies that the equation holds for a specific element.

Although equation 2.32 applies only when the material of the element is isotropic, we may use the element formulation we are deriving here to model inhomogeneous materials in plane stress. We can do this by assigning different E and ν values to each element. Equation 2.32 reveals another important consequence of assuming a linear displacement field within the element: Not only is the state of strain constant within the element, but the state of stress is constant also.

Step 5   Once again to find the element equilibrium equations, we use the principle of minimum potential energy. The total potential energy for the element is

$$\Pi^{(e)} = U^{(e)} - V^{(e)}$$

where $U^{(e)}$ is the strain energy, and $V^{(e)}$ is the work done by the external applied forces. From Appendix C,

$$U^{(e)} = \tfrac{1}{2} \iiint\limits_{\text{Vol}} \lfloor\epsilon\rfloor[C]\{\epsilon\} \, d(\text{Vol})$$

In steps 4 and 5 we learned that the strains $\{\epsilon\}$ and elasticity matrix $[C]$ are constant within an element. Thus the integrand is constant, and we may write

$$U^{(e)} = \tfrac{1}{2}\lfloor\epsilon\rfloor[C]\{\epsilon\} \iiint\limits_{\text{Vol}} d(\text{Vol})$$

and since the element volume is $t\Delta$,

$$U^{(e)} = \tfrac{1}{2} t \Delta \lfloor \epsilon \rfloor [C] \{\epsilon\} \tag{2.33}$$

where t is the element thickness, and Δ is the element area given in equation 2.27. Equation 2.30 relates the element strains to the nodal displacements. Substituting for the strain components in equation 2.33, we may express the element strain energy as

$$U^{(e)} = \tfrac{1}{2} t \Delta \lfloor \delta \rfloor [B]^T [C][B]\{\delta\} \tag{2.34}$$

where the strain interpolation matrix $[B]$ is defined in equation 2.30, and the superscript T denotes the transpose of $[B]$. Equation 2.34 shows that the element strain energy is a function of the six nodal displacements, the geometric terms that appear in the $[B]$ matrix, and the elastic constants that define the $[C]$ matrix. If we were to multiply the matrix products we would obtain the strain energy as a quadratic function of the six displacement components. In mathematics, such a function is called a quadratic form (see Appendix A).

The work of the applied nodal forces is

$$V^{(e)} = F_1 \delta_1 + F_2 \delta_2 + \cdots + F_6 \delta_6$$

or in matrix notation,

$$V^{(e)} = \lfloor \delta \rfloor \{F\} \tag{2.35}$$

Combining equations 2.34 and 2.35, we obtain the element potential energy:

$$\Pi^{(e)} = \tfrac{1}{2} t \Delta \lfloor \delta \rfloor [B]^T [C][B]\{\delta\} - \lfloor \delta \rfloor \{F\} \tag{2.36}$$

For equilibrium,

$$\frac{\partial \Pi^{(e)}}{\partial \delta_1} = \frac{\partial \Pi^{(e)}}{\partial \delta_2} = \cdots = \frac{\partial \Pi^{(e)}}{\partial \delta_6} = 0$$

or more compactly we denote the derivatives with respect to the nodal displacements as

$$\left\{ \frac{\partial \Pi^{(e)}}{\partial \delta} \right\} = 0$$

To differentiate equation 2.36 explicitly would be very tedious. However,

Appendix A shows for a quadratic function

$$I = \tfrac{1}{2}\lfloor x \rfloor [A]\{x\} - \lfloor x \rfloor\{B\}$$

that

$$\left\{ \frac{\partial I}{\partial x} \right\} = [A]\{x\} - \{B\}$$

Using this result we see that minimizing the element potential energy in equation 2.36 gives

$$t\Delta[B]^T[C][B]\{\delta\} - \{F\} = 0 \tag{2.37}$$

Equation 2.37 represents six element equilibrium equations.

Step 6   By inspection we recognize that equation 2.37 has the standard form (equation 2.4b) if we define the element stiffness matrix as

$$[K]^{(e)} = t\Delta[B]^T[C][B] \tag{2.38}$$

Note that by transposing equation 2.38 we may show that the element stiffness matrix is symmetric. Symmetry is a general characteristic of element stiffness matrices as we have seen for the linear spring and the truss element. Performing the matrix multiplications indicated by equation 2.38 leads to the following expression for the element stiffness matrix:

$$[K] = \frac{Et}{4\Delta(1 - v^2)}$$

$$\times
\begin{bmatrix}
b_1^2 + \lambda c_1^2 & (v + \lambda)b_1 c_1 & b_1 b_2 + \lambda c_1 c_2 & vb_1 c_2 + \lambda b_2 c_1 & b_1 b_3 + \lambda c_1 c_3 & vb_1 c_3 + \lambda b_3 c_1 \\
 & c_1^2 + \lambda b_1^2 & vb_2 c_1 + \lambda b_1 c_2 & c_1 c_2 + \lambda b_1 b_2 & vb_3 c_1 + \lambda b_1 c_3 & c_1 c_3 + \lambda b_1 b_3 \\
 & & b_2^2 + \lambda c_2^2 & (v + \lambda)b_2 c_2 & b_2 b_3 + \lambda c_2 c_3 & vb_2 c_3 + \lambda b_3 c_2 \\
 & & & c_2^2 + \lambda b_2^2 & vb_3 c_2 + \lambda b_2 c_3 & c_2 c_3 + \lambda b_2 b_3 \\
 & \text{Symmetric} & & & b_3^2 + \lambda c_3^2 & (v + \lambda)b_3 c_3 \\
 & & & & & c_3^2 + \lambda b_3^2
\end{bmatrix}$$

$$\tag{2.39}$$

where $\lambda = (1 - v)/2$. We write this stiffness matrix explicitly to demonstrate the details. For computer implementation, we often use equation 2.38 and perform the matrix multiplications directly in the program.

The stiffness matrix for a triangular element in plane strain may be derived using this same procedure with one exception: Equation 2.32 must be changed to give the stress–strain relations for the plane strain case instead of the plane stress case.

In later chapters, other examples of deriving stiffness matrices are presented. For structural elements we typically follow the steps demonstrated here with some changes to make the approach more general. For nonstructural applications, we modify the approach because the principle of minimum potential energy no longer applies. For many applications no comparable physical principle exists, and we rely on mathematics to derive element equations.

2.2.4 Coordinate Transformations

In the preceding section we discuss an approach for establishing the stiffness matrices for simple elements from structural mechanics. For a truss element, we derive the element stiffness matrix in a local coordinate system. Before going on to discuss how element characteristics are assembled to obtain the characteristics of the entire system of elements, we need to consider coordinate transformations.

Regardless of the means used to find the characteristics of elements, we sometimes find it more convenient or easier to derive the element matrices in a local coordinate system—a coordinate system associated with the element. This is exactly the approach we follow for the truss element. In the actual structure, because of different orientations of the members, the local coordinate system may be different for each element in the assembly. If local coordinate systems are used, it is necessary, before the individual element matrices are assembled, to transform the element matrices so that all element characteristics are referred to a common global coordinate system. Coordinate transformations are required when the nodal unknowns are components of a vector. Then the vector components vary depending on the coordinate system we select. Displacement components in structural applications are the most common example encountered, and we will demonstrate coordinate transformations used in structural problems.

In general, element matrix equations will have the standard form

$$[K_L]\{\delta_L\} = \{F_L\} \qquad (2.40)$$

where the subscript L denotes local coordinates. We wish to transform the element stiffness matrix $[K_L]$ and force vector $\{F_L\}$ to a global coordinate system. The key is the transformation equation between the displacement components in the local coordinates $\{\delta_L\}$, and the displacement components in the global system, which we denote simply as $\{\delta\}$. From geometry we can write the transformation equation as

$$\{\delta_L\} = [R]\{\delta\} \qquad (2.41)$$

where $[R]$ is the transformation matrix. Sometimes $[R]$ is called the rotation matrix because the transformation represents a coordinate system rotation.

Note that, in general, the transformation matrix is not a square matrix because the number of displacement components in the local coordinate system may be different from the number of displacement components in the global coordinate system.

To develop the transformation equations for an element's stiffness matrix and load vector we use conservation of potential energy. The total potential of an element $\Pi^{(e)}$ is a scalar independent of the coordinate system. We may, in general, write this energy for an element in a local coordinate system as

$$\Pi^{(e)} = \tfrac{1}{2}\lfloor \delta_L \rfloor [K_L]\{\delta_L\} - \lfloor \delta_L \rfloor \{F_L\} \tag{2.42}$$

The preceding section illustrates equation 2.42 for the truss element. If we substitute the transformation equation 2.41 into equation 2.42, then

$$\Pi^{(e)} = \tfrac{1}{2}\lfloor \delta \rfloor [R]^T [K_L][R]\{\delta\} - \lfloor \delta \rfloor [R]^T \{F_L\} \tag{2.43}$$

If we represent the total energy in global coordinates as

$$\Pi^{(e)} = \tfrac{1}{2}\lfloor \delta \rfloor [K]\{\delta\} - \lfloor \delta \rfloor \{F\} \tag{2.44}$$

then we recognize the element transformation equations to be

$$[K]^{(e)} = [R]^T [K_L]^{(e)}[R] \tag{5.45a}$$

and

$$\{F\}^{(e)} = [R]^T \{F_L\}^{(e)} \tag{2.45b}$$

Equations 2.45 are often used in structural mechanics for two- and three-dimensional problems. Although derived for structural elements, the equations apply to other problems where the nodal unknowns are vectors that may be resolved into components that differ in local or global systems. If we are solving a problem where the nodal unknown is a scalar, e.g., temperature, then we have no need for transformation equations.

Example   To illustrate the transformation equations, let us reconsider a truss element. In the preceding section we give the truss element stiffness matrix in local coordinates in equation 2.20. Element displacements for the element in the local and global coordinate systems are shown in Figure 2.7. The stiffness matrix given in equation 2.20 is associated with displacements δ_1 and δ_2 along the x_L axis. From the figure we see that in global coordinates each local nodal displacement can be represented by a horizontal and vertical displacement component along the global x and y axes. Thus from the figure

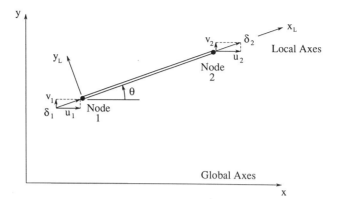

Figure 2.7   Truss element displacements in local and global coordinate systems.

we may write

$$\delta_1 = u_1 \cos \theta + v_1 \sin \theta$$

$$\delta_2 = u_2 \cos \theta + v_2 \sin \theta$$

or in the matrix notation of equation 2.41,

$$\begin{Bmatrix} \delta_1 \\ \delta_2 \end{Bmatrix} = \begin{bmatrix} \cos \theta & \sin \theta & 0 & 0 \\ 0 & 0 & \cos \theta & \sin \theta \end{bmatrix} \begin{Bmatrix} u_1 \\ v_1 \\ u_2 \\ v_2 \end{Bmatrix} \qquad (2.46a)$$

so that

$$[R] = \begin{bmatrix} \cos \theta & \sin \theta & 0 & 0 \\ 0 & 0 & \cos \theta & \sin \theta \end{bmatrix} \qquad (2.46b)$$

This last equation is the required transformation matrix that we may use to derive the element stiffness matrix in global coordinates. Substituting 2.20 and 2.46b into equation 2.45a yields

$$[K]^{(e)} = \frac{EA}{L} \begin{bmatrix} \cos \theta & 0 \\ \sin \theta & 0 \\ 0 & \cos \theta \\ 0 & \sin \theta \end{bmatrix} \begin{bmatrix} 1 & -1 \\ -1 & 1 \end{bmatrix} \begin{bmatrix} \cos \theta & \sin \theta & 0 & 0 \\ 0 & 0 & \cos \theta & \sin \theta \end{bmatrix}$$

or after multiplication,

$$[K]^{(e)} = \frac{EA}{L} \begin{bmatrix} \cos^2\theta & \cos\theta\sin\theta & -\cos^2\theta & -\cos\theta\sin\theta \\ \cos\theta\sin\theta & \sin^2\theta & -\cos\theta\sin\theta & -\sin^2\theta \\ -\cos^2\theta & -\cos\theta\sin\theta & \cos^2\theta & \cos\theta\sin\theta \\ -\cos\theta\sin\theta & -\sin^2\theta & \cos\theta\sin\theta & \sin^2\theta \end{bmatrix}$$

$$(2.47)$$

Equation 2.47 is a truss element's stiffness matrix in global coordinates; its rows and columns are identified with the global displacements u_1, v_1, u_2, v_2. Note that the matrix is symmetric, as we should expect. We will use the equation for analyses of truss problems. We could also use the rotation matrix and equation 2.45b to transform any element force components given in the local coordinate system to components in global coordinates.

2.3 ASSEMBLING THE PARTS

Assuming that by some means we have found the necessary algebraic equations describing the characteristics of each element of our system, the next step in the finite element analysis of the system is to combine all these equations to form a complete set governing the composite of elements. The procedure for constructing the system equations from the element equations is the same regardless of the type of problem being considered or the complexity of the system of elements. Even if the system is modeled with a mixture of several different kinds of elements, the system equations are assembled from the element equations in the same way.

The system assembly procedure is based on our insistence of *compatibility* at the element nodes. By this we mean that at nodes where elements are connected the value (values) of the unknown nodal variable (or variables if more than one exists at the node) is (are) the same for all elements connecting at that node. The consequence of this rule is the basis for the assembly process. For example, in elasticity problems the nodal variables are usually generalized displacements that can be translations, rotations, or spatial derivatives of the translations. When these generalized displacements are matched at the nodes, the nodal stiffnesses and nodal loads for each of the elements sharing the node are added to obtain the net stiffness and the net load at the node. These assembly concepts, which originated in matrix structural analysis and network analysis, are an essential part of every finite element solution.

2.3.1 Assembly Rules Derived From an Example

Before we explicitly state the steps in the general assembly procedure and discuss an algorithm for its execution, we introduce and illustrate its essential

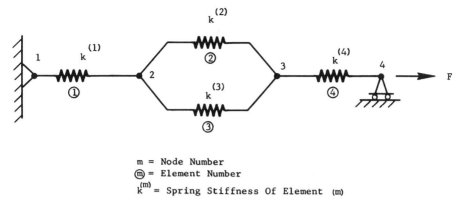

m = Node Number
ⓜ = Element Number
$k^{(m)}$ = Spring Stiffness Of Element (m)

Figure 2.8   Linear spring system consisting of four elements.

features with an elementary example. Consider finding the force–displacement behavior of the system of linear one-dimensional springs shown in Figure 2.8. We assume that the spring elements in this system can experience only tension and compression. By definition, this system has four elements and four nodes.

An arbitrary global numbering scheme, as indicated in Figure 2.8, is established to identify these nodes and elements. In general, the systematic numbering of all the nodes and elements in a finite element discretization is an essential part of the solution process. Once the numbering scheme[4] has been established for a finite element mesh, we create the system's topology—the record of which nodes belong to each of the elements. This topology, given as input to the computer program or generated internally by the program, defines the connectivity of the element mesh; in other words, it tells how the elements are joined together. On the element level the topology is simply the ordered numbering of the nodes. Table 2.1 illustrates the system topology that we have established for our four-element spring system. For example, Table 2.1 tells us that element 4 has nodes 3 and 4, and that node 1 of the *element* actually is node 3 of the *system*, while node 2 of the *element* is node 4 of the *system*.

Having specified the topology, we proceed to the element equations. From our first derivation in Section 2.2 we recall that for a typical spring element (see Figure 2.2) the relations expressing its stiffness are

$$\begin{bmatrix} k_{11} & -k_{12} \\ -k_{21} & k_{22} \end{bmatrix} \begin{Bmatrix} \delta_1 \\ \delta_2 \end{Bmatrix} = \begin{Bmatrix} F_1 \\ F_2 \end{Bmatrix} \qquad (2.4)$$

[4]Node and element numbering may be done by hand or by a computer operating from a programmed algorithm.

TABLE 2.1. System Topology: The Correspondence Between Local and Global Numbering Schemes

Element	Scheme	
	Local	Global
1	1	1
	2	2
2	1	2
	2	3
3	1	2
	2	3
4	1	3
	2	4

where

$$k_{11} = k_{12} = k_{21} = k_{22} = k$$

Since these element equations were derived in a local coordinate reference, which is the same as the global coordinates for this example, we do not need to transform the element stiffness and loads.

We recognize that under a given loading condition each element as well as the system of elements must be in equilibrium. If we impose this equilibrium condition at a particular node i, we find that

$$\sum_e F_i^{(e)} = F_i^{(1)} + F_i^{(2)} + F_i^{(3)} + \cdots = R_i \qquad (2.48)$$

which states that the sum of all the nodal forces in one direction at node i equals the resultant external load applied at node i.

Before evaluating the various $F_i^{(e)}$ and applying equation 2.48 to find the system equations, we review the standard subscript notation for stiffness coefficients. Each coefficient in a stiffness matrix is assigned a double subscript, say ij; the number i is the subscript designating the force F_i produced by a unit value of the displacement whose subscript is j. The force F_i is that which exists when $\delta_j = 1$ and all the other displacements are fixed. A displacement and a resultant force in the direction of the displacement carry the same subscript.

Consider evaluating equation 2.48 at each node in our linear spring system. We shall use the subscript notation based on the global numbering scheme. Since we insist upon the condition of displacement compatibility, that is, the displacement at each node is the same for all elements sharing that node, we can write the following force balance:

At node 1:

$$k_{11}^{(1)}\delta_1 + k_{12}^{(1)}\delta_2 = R_1$$

At node 2:

$$k_{21}^{(1)}\delta_1 + \left(k_{22}^{(1)} + k_{22}^{(2)} + k_{22}^{(3)}\right)\delta_2 + \left(k_{23}^{(2)} + k_{23}^{(3)}\right)\delta_3 = 0$$

At node 3:

$$\left(k_{32}^{(2)} + k_{32}^{(3)}\right)\delta_2 + \left(k_{33}^{(2)} + k_{33}^{(3)} + k_{33}^{(4)}\right)\delta_3 + k_{34}^{(4)}\delta_4 = 0$$

At node 4:

$$k_{43}^{(4)}\delta_3 + k_{44}^{(4)}\delta_4 = F \tag{2.49}$$

Using matrix notation, we can write these system equilibrium equations as

$$\begin{bmatrix} k_{11}^{(1)} & k_{12}^{(1)} & 0 & 0 \\ k_{21}^{(1)} & \left(k_{22}^{(1)} + k_{22}^{(2)} + k_{22}^{(3)}\right) & \left(k_{23}^{(2)} + k_{23}^{(3)}\right) & 0 \\ 0 & \left(k_{32}^{(2)} + k_{32}^{(3)}\right) & \left(k_{33}^{(2)} + k_{33}^{(3)} + k_{33}^{(4)}\right) & k_{34}^{(4)} \\ 0 & 0 & k_{43}^{(4)} & k_{44}^{(4)} \end{bmatrix} \begin{Bmatrix} \delta_1 \\ \delta_2 \\ \delta_3 \\ \delta_4 \end{Bmatrix} = \begin{Bmatrix} R_1 \\ 0 \\ 0 \\ F \end{Bmatrix}$$

or

$$[K]\{\delta\} = \{R\} \tag{2.50}$$

Equations 2.50 are the assembled force–displacement characteristics for the complete system, and $[K]$ is the assembled stiffness matrix. These equations cannot be solved for the nodal displacements until they have been modified to account for the boundary conditions. Element stiffness matrices and the assembled stiffness matrix are always singular; that is, their inverse cannot be found because their determinants are zero. In our linear spring system, as well as in other general elasticity problems, the displacement field or the set of nodal displacements cannot be found unless enough nodal displacements are fixed to prevent the structure from moving as a rigid body when external loads are applied. As we see in Figure 2.8, this requirement is satisfied by the condition $\delta_1 = 0$, node 1 being rigidly fixed to some foundation. Since we are concerned here only with assembly procedures, we shall delay discussion of the procedures for treating boundary conditions until the next section.

An alternative approach to deriving these system equilibrium equations from first principles is to employ the principle of minimum potential energy. Hence if we wrote the expression for the potential energy of our spring system in terms of spring stiffnesses, nodal displacements, and the external load F and then minimized this expression with respect to $\{\delta\}$, equation 2.50 would result.

Inspection of equation 2.50 reveals that the assembled stiffness matrix contains stiffness coefficients obtained by directly adding the individual element stiffness coefficients in the appropriate locations in the global stiffness matrix. The resultant load vector for the system is also obtained by adding individual element loads at the appropriate locations in the column matrix of resultant nodal loads. This result suggests that the element matrices may be thought of as submatrices for the entire system and that the system matrices may be obtained by simple addition of the element submatrices. This observation is indeed valid, and it is the essence of the general assembly procedure for all finite element analyses. The element matrices are, of course, not added in a random fashion. Matrix addition is defined only for matrices of the same size. Thus before we add the various element matrices we first expand them to the dimension of the system matrix. The system matrix allows for the possibility that each nodal unknown (degree of freedom) can be related to each nodal "action"; thus if the system has n nodal unknowns (degrees of freedom), the system matrix $[K]$ will be a square matrix of dimension $n \times n$. Our spring system has four nodes and only one nodal unknown (displacement) per node; hence, as we have seen, the system matrix $[K]$ has the dimensions 4×4. Expanded element matrices are constructed from the original element matrices by inserting the known stiffness coefficients into their proper locations in the enlarged matrices. It is helpful to think of the expanded element matrices initially as sets of $n \times n$ null matrices (all zero entries) that are then partially filled in by the insertion of the element submatrices.[5]

Consider developing the expanded element submatrices for our linear spring system. For element (1) the local and global numbering schemes are, by coincidence, the same (see Table 2.1), so that the subscripts of the element stiffness coefficients remain unchanged and we have

$$[\bar{K}]^{(1)} = \begin{bmatrix} k_{11}^{(1)} & k_{12}^{(1)} & 0 & 0 \\ k_{21}^{(1)} & k_{22}^{(1)} & 0 & 0 \\ 0 & 0 & 0 & 0 \\ 0 & 0 & 0 & 0 \end{bmatrix} \qquad (2.51)$$

where $[\bar{K}]^{(1)}$ designates the expanded stiffness matrix for element (1).

For element (2) the correspondence between local and global numbering schemes indicates that the following holds:

$$\begin{array}{cc} \text{Local} & \text{Global} \\ k_{11}^{(2)} \rightarrow & k_{22}^{(2)} \\ k_{12}^{(2)} \rightarrow & k_{23}^{(2)} \\ k_{21}^{(2)} \rightarrow & k_{32}^{(2)} \\ k_{22}^{(2)} \rightarrow & k_{33}^{(2)} \end{array}$$

---

[5]This is only a way to visualize the process. As we will see, we do not actually do this in practice.

Hence when these coefficients are inserted into the expanded matrix, we have

$$
[\bar{K}]^{(2)} = \begin{bmatrix} 0 & 0 & 0 & 0 \\ 0 & k_{22}^{(2)} & k_{23}^{(2)} & 0 \\ 0 & k_{32}^{(2)} & k_{33}^{(2)} & 0 \\ 0 & 0 & 0 & 0 \end{bmatrix} \tag{2.52}
$$

Similarly, for the remaining two elements the expanded element matrices are

$$
[\bar{K}]^{(3)} = \begin{bmatrix} 0 & 0 & 0 & 0 \\ 0 & k_{22}^{(3)} & k_{23}^{(3)} & 0 \\ 0 & k_{32}^{(3)} & k_{33}^{(3)} & 0 \\ 0 & 0 & 0 & 0 \end{bmatrix} \tag{2.53}
$$

and

$$
[\bar{K}]^{(4)} = \begin{bmatrix} 0 & 0 & 0 & 0 \\ 0 & 0 & 0 & 0 \\ 0 & 0 & k_{33}^{(4)} & k_{34}^{(4)} \\ 0 & 0 & k_{43}^{(4)} & k_{44}^{(4)} \end{bmatrix} \tag{2.54}
$$

Now we observe that the master stiffness matrix of equation 2.50 can be obtained by simply adding equations 2.51–2.54, representing the contribution from each element. The mathematical statement of this assembly procedure is

$$
[\bar{K}] = \sum_{e=1}^{M} [\bar{K}]^{(e)} = [\bar{K}]^{(1)} + [\bar{K}]^{(2)} + [\bar{K}]^{(3)} + \cdots \tag{2.55}
$$

where M is the total number of elements in the assemblage. In our example $M = 4$, but in an actual problem there might be several hundred elements. Even if the assemblage contains many different kinds of elements, equation 2.55 still holds—each individual element matrix is expanded (according to the global numbering scheme) to the dimension of the system matrix, and then these matrices are added.

The same expansion and summation principle applies for finding the column vectors of resultant external nodal actions[6] (forces in our example) from the element subvectors:

$$
\{R\} = \sum_{e=1}^{M} \{\bar{R}\}^{(e)} \tag{2.56}
$$

---

[6]The nodal actions are the forcing mechanisms of the problem. These may be generalized forces (applied loads, torques, stresses, pressures, etc.) or generalized fluxes (velocities, heat flow, concentration, etc.).

where $\{\overline{R}\}^{(e)}$ is the expanded column vector for element (e), and M is the total number of elements.

2.3.2 General Assembly Procedure

The preceding assembly procedure, although developed from a simple example, is in principle the general procedure that applies to all finite element systems. We have assembled systems matrices by hand for a simple problem with four elements, but for real problems involving hundreds of elements the process is done by computer. The general procedure is summarized in the following steps[7] (for clarity we omit special considerations that sometimes improve computing efficiency):

1. Set up $n \times n$ and $n \times 1$ null matrices (all zero entries), where $n =$ number of system nodal variables.
2. Starting with one element, transform the element equations from local to global coordinates if these two coordinates systems are not coincident.
3. Perform any necessary matrix operations on the element matrices.[8]
4. Using the established correspondence between local and global numbering schemes, change to the global indices (a) the subscript indices of the coefficients in the square matrix and (b) the single subscript index of the terms in the column matrix.
5. Insert these terms into the corresponding $n \times n$ and $n \times 1$ master matrices in the locations designated by their indices. Each time a term is placed in a location where another term has already been placed, it is added to whatever value is there.
6. Return to step 2 and repeat this procedure for one element after another until all elements have been treated. The result will be an $n \times n$ master matrix $[K]$ of stiffness coefficients and an $n \times 1$ column matrix $\{R\}$ of resultant nodal actions. The complete system equations are then

$$\overset{n \times n \ \ n \times 1}{[K] \, \{x\}} = \overset{n \times 1}{\{R\}} \tag{2.57}$$

where $\{x\}$ is the column matrix of nodal unknowns for the assemblage.

The generality of this assembly process for the finite element method offers a definite advantage: Once a computer program for the assembly process has been developed for the solution of one particular class of problems by the

---

[7]We assume that the element matrices have already been determined or can be determined as they are needed in the assembly process.
[8]Sometimes elements have one or more nodes that have no connectivity. When this occurs, it is necessary to eliminate the nodal unknowns or degrees of freedom associated with these nodes. This concept is discussed in Chapter 5, where we treat different types of elements.

finite element method, it may be used again for the finite element solution of other classes of problems.

2.3.3   Features of the Assembled Matrix

Before discussing how equations 2.57 may be modified to account for boundary conditions, we will point out several important and useful features of the overall system matrix. The simple matrix of equations 2.50, for example, displays these features. First we note that the system matrix has its nonzero terms clustered about its main diagonal, while locations distant from the diagonal contain zero terms. Essentially, the nonzero terms are contained in a band centered on the diagonal, and outside the band all terms are zero. Consequently, the matrix is said to be banded as well as *sparse*, because it is not fully populated. Bandedness and sparseness reflect the connectivity of a finite element mesh. In a general finite element mesh the system matrices are sparse because each element has relatively few nodes compared to all the system nodes and only a few elements share each node. Numbering of the nodes causes the system matrices to be banded. If an efficient nodal numbering scheme is used, the bandwidth can be minimized. Figure 2.9 shows

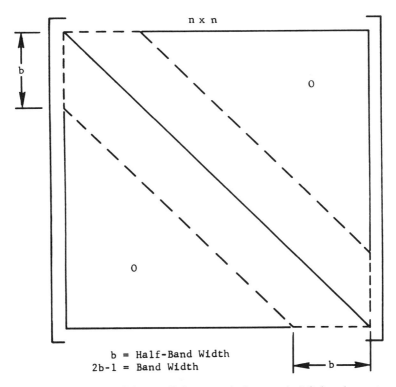

Figure 2.9   The structure of the coefficient matrix for a typical finite element system.

schematically a typical coefficient matrix for a finite element system. The region of the matrix enclosed by dotted lines is the band that contains mostly nonzero coefficients. The width of this band is directly related to the maximum difference between any two global node numbers for an element. Another important feature is that system coefficient matrices are usually symmetric—a characteristic that can often be used to advantage in storing the matrices.

Many of the practical aspects of handling system matrices and numbering node prints to obtain well-conditioned matrices are discussed in Chapter 10 on computational considerations.

2.3.4   Introducing Boundary Conditions

Regardless of the type of problem for which a set of system equations has been assembled, the final equations will have the form

$$
\overset{n \times n}{[K]} \, \overset{n \times 1}{\{x\}} = \overset{n \times 1}{\{R\}}
\tag{2.57}
$$

These equations already take into account the boundary conditions involving specified external or applied nodal "actions," because these are included in the resultant column vector $\{R\}$ during assembly. However, for a unique solution of equations 2.57, at least one and sometimes more than one nodal variable must be specified, and $[K]$ must be modified to render it nonsingular. The required number of specified nodal variables is dictated by the physics of the problem. Nodal variables may be specified for either interior or boundary nodes, but for nodes i whose coordinates are fixed, it is physically impossible to specify both x_i and R_i. However, at every node i we know either x_i or R_i.

There are a number of ways to apply the boundary conditions to equation 2.57, and when they are applied, the number of nodal unknowns and the number of equations to be solved are effectively reduced. However, it is most convenient to introduce the known nodal variables in a way that leaves the original number of equations unchanged and avoids major restructuring of computer storage. Two straightforward means to accomplish this are described in the following discussion.

One way to include prescribed nodal variables in equation 2.57 while retaining an $n \times n$ system of equations is to modify the matrices $[K]$ and $\{R\}$ as follows. If i is the subscript of a prescribed nodal variable, the ith row and the ith column of $[K]$ are set equal to zero and k_{ii} is set equal to unity. The term R_i of the column vector $\{R\}$ is replaced by the known value of x_i. Each of the $n - 1$ remaining terms of $\{R\}$ is modified by subtracting from it the value of the prescribed nodal variable multiplied by the appropriate column term from the original $[K]$ matrix. This procedure is repeated for each prescribed x_i until all of them have been included.

To illustrate this procedure for entering boundary conditions we consider a simple example with only four system equations. Equation 2.57 expands to the form

$$
\begin{bmatrix}
k_{11} & k_{12} & k_{13} & k_{14} \\
k_{21} & k_{22} & k_{23} & k_{24} \\
k_{31} & k_{32} & k_{33} & k_{34} \\
k_{41} & k_{42} & k_{43} & k_{44}
\end{bmatrix}
\begin{Bmatrix}
x_1 \\ x_2 \\ x_3 \\ x_4
\end{Bmatrix}
=
\begin{Bmatrix}
R_1 \\ R_2 \\ R_3 \\ R_4
\end{Bmatrix}
$$

Suppose that for this hypothetical system nodal variables x_1 and x_3 are specified as

$$
x_1 = \beta_1 \qquad x_3 = \beta_3
$$

When these boundary conditions are inserted, the equations become

$$
\begin{bmatrix}
1 & 0 & 0 & 0 \\
0 & k_{22} & 0 & k_{24} \\
0 & 0 & 1 & 0 \\
0 & k_{42} & 0 & k_{44}
\end{bmatrix}
\begin{Bmatrix}
x_1 \\ x_2 \\ x_3 \\ x_4
\end{Bmatrix}
=
\begin{Bmatrix}
\beta_1 \\
R_2 - k_{21}\beta_1 - k_{23}\beta_3 \\
\beta_3 \\
R_4 - k_{41}\beta_1 - k_{43}\beta_3
\end{Bmatrix}
$$

This set of equations, unaltered in dimension, is now ready to be solved for all the nodal variables. The solution, of course, yields $x_1 = \beta_1$ and $x_3 = \beta_3$ along with the actual unknowns, x_2 and x_4.

Another way to insert prescribed nodal variables into the system equations is to modify certain diagonal terms of $[K]$. The diagonal term of $[K]$ associated with a specified nodal variable is multiplied by a large number, say 1×10^{15}, while the corresponding term in $\{R\}$ is replaced by the specified nodal variable multiplied by the same large factor times the corresponding diagonal term. This procedure is repeated until all prescribed nodal variables have been treated. Effectively, this procedure makes the unmodified terms of $[K]$ very small compared to the modified terms (those associated with the specified nodal variables). After these modifications have been made, we proceed with the simultaneous solution of the complete set of n equations.

If we use this procedure to modify the original equations of our previous example, we obtain

$$
\begin{bmatrix}
k_{11} \times 10^{15} & k_{12} & k_{13} & k_{14} \\
k_{21} & k_{22} & k_{23} & k_{24} \\
k_{31} & k_{32} & k_{33} \times 10^{15} & k_{34} \\
k_{41} & k_{42} & k_{43} & k_{44}
\end{bmatrix}
\begin{Bmatrix}
x_1 \\ x_2 \\ x_3 \\ x_4
\end{Bmatrix}
=
\begin{Bmatrix}
\beta_1 k_{11} \times 10^{15} \\
R_2 \\
\beta_3 k_{33} \times 10^{15} \\
R_4
\end{Bmatrix}
$$

To see that this procedure gives the desired result consider the first equation of the set,

$$k_{11} \times 10^{15}x_1 + k_{12}x_2 + k_{13}x_3 + k_{14}x_4 = \beta_1 \times k_{11} \times 10^{15}$$

For all practical purposes this equation expresses the fact that

$$x_1 = \beta_1$$

since

$$k_{11} \times 10^{15} \gg k_{1j}, \qquad j = 2, 3, 4$$

Computer coding is a little easier for the second method than for the first one. Both methods preserve the sparse, banded, and usually symmetric properties of the original master matrix.

After the modified equations have been solved for the unknown nodal variables, we may return to the original set of equations and find the unknown nodal reactions if desired. An orderly, though impractical, way to do this is as follows. Starting with the original equations, we may rearrange and partition them to obtain

$$\begin{matrix} n \times n \\ \begin{bmatrix} [K_{11}] & [K_{12}] \\ [K_{12}]^T & [K_{22}] \end{bmatrix} \end{matrix} \begin{matrix} n \times 1 \\ \begin{Bmatrix} \{\tilde{x}_1\} \\ \{\tilde{x}_2\} \end{Bmatrix} \end{matrix} = \begin{matrix} n \times 1 \\ \begin{Bmatrix} \{\tilde{R}_1\} \\ \{\tilde{R}_2\} \end{Bmatrix} \end{matrix} \qquad (2.58)$$

where $\{\tilde{R}_1\}$ is the column vector of known nodal reactions, and $\{\tilde{x}_2\}$ is the column vector of *known* nodal variables. The matrix $[K_{11}]$ now possesses an inverse, so we may find the unknown $\{\tilde{x}_1\}$ by solving

$$[K_{11}]\{\tilde{x}_1\} = \{\tilde{R}_1\} - [K_{12}]\{\tilde{x}_2\} \qquad (2.59)$$

With $\{\tilde{x}_1\}$ known from equation 2.59, the vector $\{\tilde{R}_2\}$ can be found from

$$\{\tilde{R}_2\} = [K_{12}]^T\{\tilde{x}_1\} + [K_{22}]\{\tilde{x}_2\} \qquad (2.60)$$

The procedure represented symbolically by equations 2.58–2.60 is usually impractical because the reordering process requires much additional programming effort and tends to destroy the bandedness property of the original. Once the full set of equations has been assembled and the prescribed boundary conditions introduced, the next step is to solve the equations for the unknowns. Many schemes are available for this purpose. The most efficient schemes take advantage of the possible symmetry, sparseness, and bandedness of $[K]$. Although a detailed treatment of equation-solving routines is beyond the scope of this book, the more popular solution

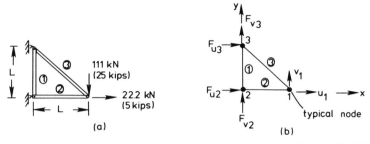

Figure 2.10   A truss and its finite element model: (*a*) truss with applied loads; (*b*) finite element model with unknown displacements and reactions.

techniques are mentioned, and references are cited in Chapter 10.

Numerical Example   To help solidify the fundamental ideas presented in this chapter we consider a two-dimensional truss example (Figure 2.10*a*). Given the applied forces, the problem is to determine the joint displacements, the joint reactions, element forces, and element stresses. We use the finite element method as presented in this chapter to solve the problem in a systematic fashion. Although we use a simple problem to illustrate the basic steps, the approach is general and is followed throughout the remainder of this book.

The steps of the solution procedure are as follows.

1. *Discretize the structure.* The objective of the finite element procedure is to divide the problem into elements whose properties can be easily found. If we choose each truss member as a single finite element, then we have a problem with three elements and three nodes, as shown in Figure 2.10*b*. A truss element exactly represents a truss member, hence there is no need to refine further the discretization as might be required, for example, in modeling a continuum with triangular elements.

2. *Find element properties.* Now that we have identified the individual elements of the structure, we prepare the input data required to compute the truss element matrices. We use the truss element shown in Figure 2.5 and prepare the pertinent element data given in Table 2.2.

3. *Compute element matrices.* The nodal unknowns are the x and y global displacement components u and v, hence each node has two degrees of freedom and each two-noded truss element has four degrees of freedom. An element stiffness matrix depends on its orientation angle θ (see Figure 2.5) tabulated in Table 2.2. Note that the global connections column gives the local numbering system with local node 1 first. Observe that in applications with scalar nodal unknowns (e.g., the flow systems in Section 2.2.2) a node

TABLE 2.2. Element Property Data for Numerical Example

Element	A (cm²)	A (in.²)	E (N/m²)	E (lb/in.²)	L (m)	L (in.)	Global Node Connections	θ (deg)
1	32.3	5	$6.9E + 10$	$10E + 6$	2.54	100	2 to 3	90
2	38.7	6	$20.7E + 10$	$30E + 6$	2.54	100	2 to 1	0
3	25.8	4	$20.7E + 10$	$30E + 6$	3.59	141	1 to 3	135

has only one scalar unknown, and the element matrices can be computed in local coordinates without considering the element orientation. However, here we compute the truss element stiffness matrices in global coordinates using equation 2.47 and the data in Table 2.2:

$$[K]^{(1)} = 87.7 \begin{bmatrix} 0 & 0 & 0 & 0 \\ 0 & 1 & 0 & -1 \\ 0 & 0 & 0 & 0 \\ 0 & -1 & 0 & 1 \end{bmatrix} \begin{matrix} u_2 \\ v_2 \\ u_3 \\ v_3 \end{matrix} \; \text{MN/m}$$

$$[K]^{(2)} = 315 \begin{bmatrix} 1 & 0 & -1 & 0 \\ 0 & 0 & 0 & 0 \\ -1 & 0 & 1 & 0 \\ 0 & 0 & 0 & 0 \end{bmatrix} \begin{matrix} u_2 \\ v_2 \\ u_1 \\ v_1 \end{matrix} \; \text{MN/m}$$

$$[K]^{(3)} = 74.4 \begin{bmatrix} 1 & -1 & -1 & 1 \\ -1 & 1 & 1 & -1 \\ -1 & 1 & 1 & -1 \\ 1 & -1 & -1 & 1 \end{bmatrix} \begin{matrix} u_1 \\ v_1 \\ u_3 \\ v_3 \end{matrix} \; \text{MN/m}$$

The relevant global displacement components are written to the right of each matrix to aid us in assembling the global stiffness matrix.

4. *Assemble the element equations.* We form the global stiffness matrix by summing each of the element stiffness matrices above, taking care to place matrix elements in the proper rows and columns:

$$\begin{bmatrix} 315 + 74.4 & -74.4 & -315 & 0 & -74.4 & 74.4 \\ -74.4 & 74.4 & 0 & 0 & 74.4 & -74.4 \\ -315 & 0 & 315 & 0 & 0 & 0 \\ 0 & 0 & 0 & 87.7 & 0 & -87.7 \\ -74.4 & 74.4 & 0 & 0 & 74.4 & -74.4 \\ 74.4 & -74.4 & 0 & -87.7 & -74.4 & 87.7 + 74.4 \end{bmatrix} \begin{Bmatrix} u_1 \\ v_1 \\ u_2 \\ v_2 \\ u_3 \\ v_3 \end{Bmatrix} = \begin{Bmatrix} F_{u_1} \\ F_{v_1} \\ F_{u_2} \\ F_{v_2} \\ F_{u_3} \\ F_{v_3} \end{Bmatrix}$$

For clarity we retain the u–v notation for nodal displacements and denote the nodal forces with similar subscripts. In problems with element load vectors (e.g., member weights), we would assemble the system load vector taking these contributions into account. Here the only nodal forces are external forces, which we consider as boundary conditions in the next step.

5. *Impose the boundary conditions.* At each node we know either a displacement component or the corresponding force component. If we know the displacement, the corresponding force component is an unknown reaction; if we know the force component, the corresponding displacement is unknown. Referring to Figure 2.10b, we list the knowns and unknowns:

$$\lfloor u \rfloor = \left| u_1 = ? \quad v_1 = ? \quad u_2 = 0 \quad v_2 = 0 \quad u_3 = 0 \quad v_3 = 0 \right|$$

$$\lfloor F \rfloor = \left| \begin{matrix} F_{u_1} = 0.0222 & F_{v_1} = -0.111 \\ F_{u_2} = ? & F_{v_2} = ? & F_{u_3} = ? & F_{v_3} = ? \end{matrix} \right| \text{MN}$$

Thus we can write the system equation as

$$\begin{bmatrix} 389 & -74.4 & -315 & 0 & -74.4 & 74.4 \\ -74.4 & 74.4 & 0 & 0 & 74.4 & -74.4 \\ -315 & 0 & 315 & 0 & 0 & 0 \\ 0 & 0 & 0 & 87.7 & 0 & -87.7 \\ -74.4 & 74.4 & 0 & 0 & 74.4 & -74.4 \\ 74.4 & -74.4 & 0 & -87.7 & -74.4 & 162 \end{bmatrix} \begin{Bmatrix} u_1 \\ v_1 \\ 0 \\ 0 \\ 0 \\ 0 \end{Bmatrix} = \begin{Bmatrix} 0.0222 \\ -0.111 \\ F_{u_2} \\ F_{v_2} \\ F_{u_3} \\ F_{v_3} \end{Bmatrix}$$

Next we segregate the equations to permit the solution for the nodal unknowns. Note that this problem was set up so that known and unknown displacements are naturally segregated; usually more effort is required because reordering of the equations is needed to segregate the known and unknown displacements. Following equations 2.59 and 2.60, we write

$$\begin{bmatrix} 389 & -74.4 \\ -74.4 & 74.4 \end{bmatrix} \begin{Bmatrix} u_1 \\ v_1 \end{Bmatrix} = \begin{Bmatrix} 0.0222 \\ -0.111 \end{Bmatrix}$$

$$\begin{Bmatrix} F_{u_2} \\ F_{v_2} \\ F_{u_3} \\ F_{v_3} \end{Bmatrix} = \begin{bmatrix} -315 & 0 \\ 0 & 0 \\ -74.4 & 74.4 \\ 74.4 & -74.4 \end{bmatrix} \begin{Bmatrix} u_1 \\ v_1 \end{Bmatrix}$$

where the computations are simplified because of the specified *zero* displacements.

6. *Solve the system equations.* We have two equations in two unknown displacements u_1 and v_1 that we solve simultaneously to obtain

$$\begin{Bmatrix} u_1 \\ v_1 \end{Bmatrix} = \begin{Bmatrix} -0.282 \\ -1.77 \end{Bmatrix} \text{ mm or } \begin{Bmatrix} -0.0111 \\ -0.070 \end{Bmatrix} \text{ in.}$$

With these displacements we compute the reactions by direct substitution:

$$\begin{Bmatrix} F_{u_2} \\ F_{v_2} \\ F_{u_3} \\ F_{v_3} \end{Bmatrix} = \begin{bmatrix} -315 & 0 \\ 0 & 0 \\ -74.4 & 74.4 \\ 74.4 & -74.4 \end{bmatrix} \begin{Bmatrix} -0.282 \\ -1.77 \end{Bmatrix}$$

$$= \begin{Bmatrix} 88.8 \\ 0 \\ -111 \\ 111 \end{Bmatrix} \text{ kN or } \begin{Bmatrix} 20 \\ 0 \\ -25 \\ 25 \end{Bmatrix} \text{ kips}$$

Next, we check our solution by drawing a free-body diagram of the truss with the applied forces and the computed reactions and ask the question: Is the truss in equilibrium? The reader should do this check as an exercise.

7. *Solve for element forces and stresses.* To compute element forces and stresses we first compute the element displacements in a local element coordinate system. For a typical truss member we may compute the local nodal displacements (Figure 2.11) Δ_1 and Δ_2 from the global displacements δ_i, $i = 1, 4$, by equation 2.46a

$$\begin{Bmatrix} \Delta_1 \\ \Delta_2 \end{Bmatrix}^{(e)} = \begin{bmatrix} \cos\theta & \sin\theta & 0 & 0 \\ 0 & 0 & \cos\theta & \sin\theta \end{bmatrix} \begin{Bmatrix} \delta_1 \\ \delta_2 \\ \delta_3 \\ \delta_4 \end{Bmatrix}^{(e)}$$

Then, using the data in Table 2.2 and the global displacements, we compute

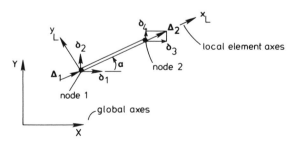

Figure 2.11   Truss element displacements in global and local coordinate systems.

the local element displacements:

$$\left\{ \begin{matrix} \Delta_1 \\ \Delta_2 \end{matrix} \right\}^{(1)} = \begin{bmatrix} 0 & 1 & 0 & 0 \\ 0 & 0 & 0 & 1 \end{bmatrix} \left\{ \begin{matrix} 0 \\ 0 \\ 0 \\ 0 \end{matrix} \right\}^{(1)} = \left\{ \begin{matrix} 0 \\ 0 \end{matrix} \right\}$$

$$\left\{ \begin{matrix} \Delta_1 \\ \Delta_2 \end{matrix} \right\}^{(2)} = \begin{bmatrix} 1 & 0 & 0 & 0 \\ 0 & 0 & 1 & 0 \end{bmatrix} \left\{ \begin{matrix} 0 \\ 0 \\ -0.282 \\ -1.77 \end{matrix} \right\}^{(2)} = \left\{ \begin{matrix} 0 \\ -0.282 \end{matrix} \right\} \text{mm}$$

$$\left\{ \begin{matrix} \Delta_1 \\ \Delta_2 \end{matrix} \right\}^{(3)} = \begin{bmatrix} -0.707 & 0.707 & 0 & 0 \\ 0 & 0 & -0.707 & 0.707 \end{bmatrix} \left\{ \begin{matrix} -0.282 \\ -1.77 \\ 0 \\ 0 \end{matrix} \right\}^{(3)}$$

$$= \left\{ \begin{matrix} -1.05 \\ 0 \end{matrix} \right\} \text{mm}$$

Finally, we compute the element forces for a typical member by

$$F = \frac{AE}{L}(\Delta_2 - \Delta_1)$$

and the element stress by

$$\sigma = \frac{F}{A}$$

For example, for member 3

$$F_3 = \frac{25.8 \times 10^{-4}(20.7 \times 10^{10})}{3.59}[0 - (-1.05)] = 156 \text{ kN or 35 kips}$$

$$\sigma_3 = \frac{156 \times 10^3}{25.8 \times 10^{-4}} = 60 \text{ MN/m}^2 \text{ or 8.8 ksi}$$

You may want to compute the stresses in the other elements as an exercise.

2.4   CLOSURE

In this chapter we introduced the concepts of the finite element method for the elementary cases. We used the so-called direct approach to derive equations expressing the behavior of a few simple elements. By this means several important features of the finite element method were exposed, and the physical interpretations of modeling a real system by a network of finite elements were introduced. As we progressed, we introduced the basic steps for deriving characteristics of simple elements from structural mechanics.

The utility of the approach described in this chapter is severely limited because it cannot be generalized to study the behavior of continuum problems for nonstructural applications. In succeeding chapters we consider more mathematical approaches that permit us to develop the method for a broad range of engineering applications. But, with the element equations we did derive, we proceeded to develop the mathematical procedure by which element equations are assembled to form system equations. The assembly concepts we established are general and apply to all finite element analyses regardless of the means used to derive element equations.

We concluded the chapter with a discussion of how the assembled system equations should be modified before solution to account for boundary conditions. Having treated the assembly procedures and the methods for handling boundary conditions, we may now focus our attention in succeeding chapters on the formulation of element equations for many types of problems.

The direct approach is very helpful in explaining some of the finite element concepts, but to work beyond the elementary concepts we have to consider the mathematical foundations of the method. As we noted in Section 1.3, engineers began to recognize the mathematical foundations of the finite element method in 1963. For finite elements of an elastic continuum civil engineers realized that the matrix equations they had previously derived by direct physical reasoning or by virtual work concepts could also be derived by directly minimizing an energy functional with respect to the nodal values of the field variable. If the field variable was the displacement field, the appropriate functional for the derivation was the integral expression for the total potential energy of the system. The discovery that the physically formulated finite element method was actually a mathematical minimization process and that element properties could be derived from variational principles governing the particular problem of interest, opened the way to far broader applications of the method. In addition, this mathematical interpretation of the method helped to establish continuity requirements for the assumed interpolation functions and facilitated the study of convergence properties and associated error estimates (bounds).

In the next chapter we examine the variational basis of the finite element method and show how the method can be applied to almost all problems governed by an appropriate variational principle.

REFERENCES

1. R. W. Fox and A. T. McDonald, *Introduction to Fluid Mechanics*, 2d ed., Wiley, New York, 1978.
2. M. J. Turner, R. W. Clough, H. C. Martin, and L. C. Topp, "Stiffness and Deflection Analysis of Complex Structures," *J. Aeronaut. Sci.*, Vol. 23, No. 9, September 1956, pp. 807–823, 854.

3. E. P. Popov, *Mechanics of Materials*, 2d ed., Prentice Hall, Englewood Cliffs, NJ, 1972.

4. R. W. Clough, "The Finite Element Method in Plane Stress Analysis," *Proceedings of 2nd ASCE Conference on Electronic Computation*, Pittsburgh, PA, September 1960.

PROBLEMS

1. Read the paper "Stiffness and Deflection Analysis of Complex Structures" by Turner et al., [2]. Write a summary of the paper and discuss its principal contributions.

2. Use the stiffness matrix transformation, equation 2.45, to derive a truss element stiffness matrix for an element in a general three-dimensional orientation. Express the stiffness matrix in terms of the direction cosines (l, m, n) of the angles between the element axes and the global coordinate axes.

3. Ohm's law states that the voltage V across a resistor is directly proportional to the current I flowing through the resistor. The constant of proportionality is the resistance R of the resistor, that is, $V = IR$. Following the procedures of Section 2.2.2, derive a finite element matrix equation describing a resistive component of a direct-current electrical network.

4. Use the direct stiffness approach to develop the element stiffness matrix for a uniform bar in torsion. T is the applied torque, ϕ is the angle of twist, G is the shearing modulus, and J is the cross-sectional torsional constant.

5. Each node in the truss shown in Figure P2.5 has two degrees of freedom.
   a. Compute the half-bandwidth of the stiffness matrix for the node-numbering scheme shown.
   b. Renumber the nodes to achieve a minimum half-bandwidth.
   c. Show the stiffness matrices schematically (nonzero values with x's) before and after renumbering the nodes.

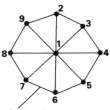

Typical truss member   **Figure P2.5**

6. Each node in the plane-stress finite element model shown in Figure P2.6 has two degrees of freedom.

   a. Compute the half-bandwidth of the stiffness matrix for the node-numbering scheme shown.

   b. Remember the nodes to achieve a minimum half-bandwidth.

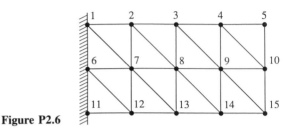

Figure P2.6

7. Element matrices for the two-dimensional problem shown in Figure P2.7 are

$$[K]^{(e)} = \begin{bmatrix} 1 & -1 & 0 \\ -1 & 2 & -1 \\ 0 & -1 & 1 \end{bmatrix} \qquad \{F\}^{(e)} = \begin{Bmatrix} 4 \\ 4 \\ 4 \end{Bmatrix}, \qquad e = 1, 2, 3$$

   a. Assemble the system matrix and system load vector assuming the nodal unknowns are ϕ_i, $i = 1, 2, 3, 4, 5$.

   b. What is the half-bandwidth of the assembled $[K]$ matrix?

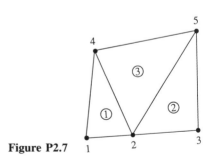

Figure P2.7

8. Figure P2.8 shows a system of three rigid carts connected by six linear springs. The carts may experience only horizontal translation.

   a. Develop the system stiffness matrix and system load vector by writing equilibrium equations for each rigid cart. Write the resulting equations in matrix form.

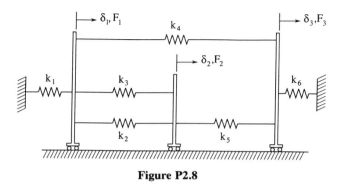

Figure P2.8

b. Develop the system stiffness matrix and system load vector by the energy approach. Write the total potential energy of the system, and then use the principle of minimum potential energy to derive the system equilibrium equations. Write the resulting equations in matrix form.

c. Use the finite element concept to assemble the system stiffness matrix from the six linear spring element matrices. Write the system load vector. Impose the boundary conditions and derive the system equilibrium equations.

9. A uniform bar (Figure P2.9) is loaded by concentrated nodal forces as shown. Use a two-element finite element model to compute nodal displacements, element forces, and stresses.

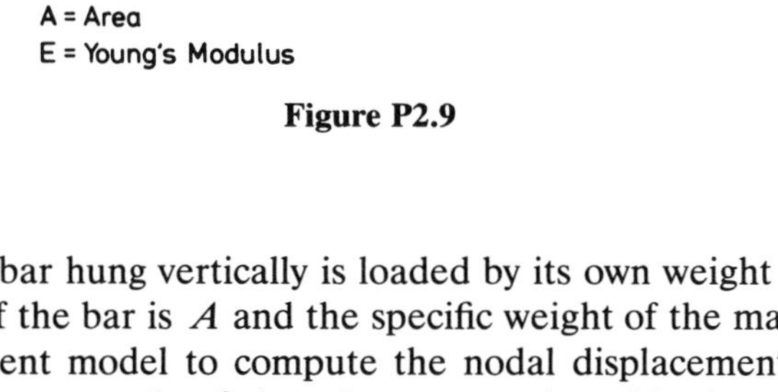

1.52 m   3.05 m
|← (60") →|← (120") ————→|
           4.45 kN (1kip)
⌐1   2→   → 2.23 kN (0.5 kip)
                      3
A = 12.9 cm² (2.0 in²)   A = 8.06 cm² (1.25 in²)

$E = 6.9 \times 10^{10} \frac{N}{m^2}$ (10^7 psi) $E = 3.5 \times 10^{10} \frac{N}{m^2}$ (5×10^6 psi)

A = Area
E = Young's Modulus

Figure P2.9

10. A uniform bar hung vertically is loaded by its own weight (Figure P2.10). The area of the bar is A and the specific weight of the material is γ. Use a two-element model to compute the nodal displacements and element stresses. Compare the finite element results with the exact analytical solution to the problem.

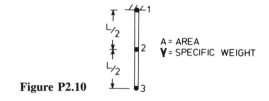

Figure P2.10

11. Derive the stiffness matrix for the assembly of three springs shown in Figure P2.11 considering boundary conditions of zero displacement at nodes 2, 3 and 4. Assume that all springs have equal stiffness k.

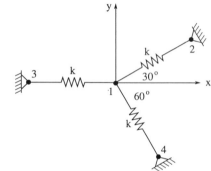

Figure P2.11

12. All members of the three-element truss shown in Figure P2.12 have the same L and EA. The unknown nodal displacements are the vertical displacement v_1 at node 1 and the horizontal displacement u_2 at node 2. A finite element solution of the global equilibrium equations shows that $v_1 = 0$ and $u_2 = -4PL/5EA$. Using these results, determine the extension (or contraction), strain, stress, and force for each element.

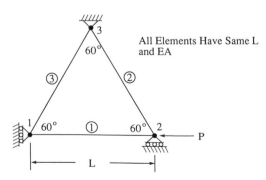

Figure P2.12

13. Consider the assembly of four springs with equal stiffness k shown in Figure P2.13.

   a. Derive the stiffness matrix for the system in the x–y coordinate system, taking into consideration the zero displacement boundary conditions at nodes 2–4.

   b. Using a matrix transformation find the stiffness matrix in the $\bar{x}$–$\bar{y}$ coordinate system.

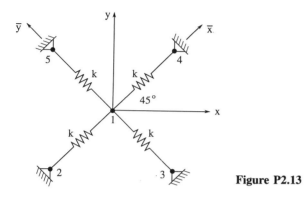

Figure P2.13

14. For the truss shown in Figure P2.14 find the displacements of node 1 and all member forces, assuming that the members have equal cross-section area A and modulus of elasticity E.

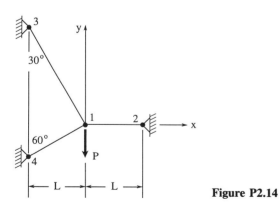

Figure P2.14

15. A truss (Figure P2.15) is made up of a vertical bar of length L and cross-sectional area A_1 and two equal inclined bars of length L and cross-sectional areas A_2 and supports a vertical load P.

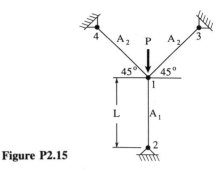

Figure P2.15

a. Using a finite element model, determine the displacement of node 1 and the forces in the bars.

b. Find the ratio of the areas A_2/A_1 that will make each force in an inclined member equal two times the force in the vertical member.

16. For the three-member truss shown in Figure P2.16 compute the following:

a. the nodal displacements

b. the reactions

c. member forces

d. member stresses

Take $E = 6.9 \times 10^{10}$ N/m^2 (10^7 psi).

$$A_1 = A_3 = 12.9 \text{ cm}^2 \ (2 \text{ in.}^2)$$

$$A_2 = 25.8 \text{ cm}^2 \ (4 \text{ in.}^2)$$

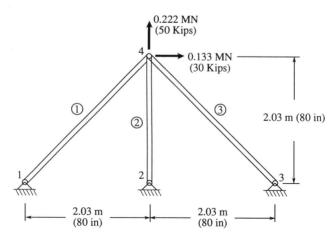

Figure P2.16

17. Determine the nodal displacements and reactions for the truss shown in Figure P2.17. The members are made of steel with $E = 20.7 \times 10^{10}$ N/m² (3×10^7 psi).

	Area	
Member	(cm²)	(in.²)
1, 4	30	4.65
2, 5	20	3.10
3	25	3.87

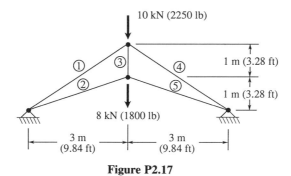

Figure P2.17

18. The plane truss shown (Figure P2.18) has six members and four joints. The material is aluminum with $E = 6.9 \times 10^{10}$ N/m² (10^7 psi), and the cross-sectional area is 6.5 cm² (1.0 in.²) for all members except for the diagonals, which have an area of 4.6 cm² (0.707 in.²). Compute the joint displacements, reactions, and member stresses.

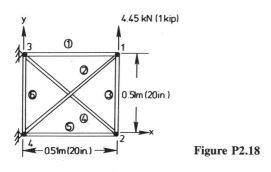

Figure P2.18

19. A composite wall (Figure P2.19) consists of layers of aluminum, copper, and steel. The surface temperatures are 264 K (475° R) and 311 K (560° R). Use a three-element model to determine the interfacial temperatures and the heat flow per unit area.

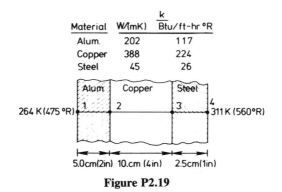

Material	W/(mK)	$\dfrac{k}{\text{Btu/ft-hr °R}}$
Alum.	202	117
Copper	388	224
Steel	45	26

Figure P2.19

20. The ceiling of many homes consists of a sheet of plaster board supported by ceiling joists, with the space between the joists filled by insulation (Figure P2.20). Neglecting the effect of the wooden joists, determine the heat transfer rate per unit area for a ceiling surface temperature of 297 K (75° F) and an insulation surface temperature of 317 K (110° F). Use a finite element model with three nodes. For the insulation $k = 0.0346$ W/m K (0.020 Btu/h ft °F), and for the plaster board $k = 0.0519$ w/m K (0.030 Btu/h ft °F).

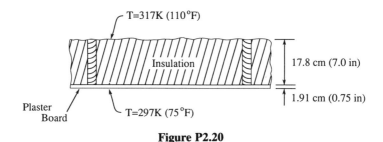

Figure P2.20

21. The radial heat flux for one-dimensional heat conduction in the hollow circular cylinder shown in Figure P2.21 is given by

$$q = \frac{2\pi L k}{\ln(r_2/r_1)}(T_1 - T_2)$$

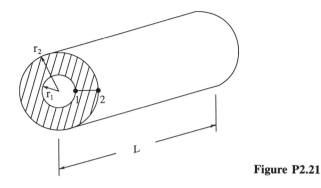

Figure P2.21

Following the procedures of Section 2.2.2, derive a finite element matrix equation describing a one-dimensional radial heat conduction element.

22. Refrigerant flows in a copper tube have an outside diameter of 4.83 cm (1.9 in.) and 0.714 cm (0.281 in.) wall thickness. The tube is covered with a 10.2-cm (4-in.) layer of insulation. The inside surface temperature of the tube is 258 K (5 °F), and the outside surface temperature of the tube covering is 294 K (70 °F). Use a finite element model based on the matrix equation developed in Problem 2.21, to determine the following:

    a. the temperature on the outside surface of the copper tube

    b. the radial heat flux in the tube

	Thermal Conductivity k
Pipe	381 W/M K (220 Btu/h ft °F)
Insulation	0.152 W/m K (0.088 Btu/h ft °F)

23. Laminar incompressible flow occurs in the branched network of circular pipes shown in Figure P2.23. If 0.085 m³/s (3 ft³/s) (Q) of fluid enters and leaves the piping network, use a three-node, four-element model to compute the fluid nodal pressures and the volume flow rate in each pipe.

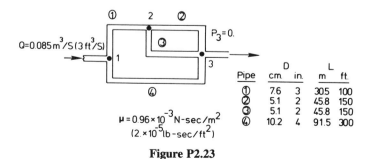

Figure P2.23

24. The voltages at the output terminals of the direct-current circuit (Figure P.2.24) have the values shown. Use a four-node finite element model to compute the voltages at each node and the current in each resistor.

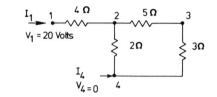

Figure P2.24

3

THE MATHEMATICAL APPROACH:
A VARIATIONAL INTERPRETATION

3.1 INTRODUCTION

In Chapter 2 we began our discussion of finite element analysis with one-dimensional applications where we were able to develop basic element equations from conservation laws. This approach is helpful in gaining an initial understanding of the method, but it is limited to relatively simple problems. Later in Chapter 2 we described a six-step process for deriving element equations in structural mechanics problems. The first step in the process is to assume a functional form for the displacement field within an element. This step is fundamental to the finite element method, and we continue to apply and expand the idea through the remainder of the book. Another essential step in the development of the finite element method, step 5 in Chapter 2, is the writing of the element equilibrium equations. There, for structural elements, we use the principle of minimum potential energy. In this chapter and the succeeding chapter we begin to generalize these basic concepts by relying less on direct physical interpretations and physical principles and more on a mathematical approach. In this chapter we take a broader view than in Chapter 2 and interpret the finite element method as an approximate means for solving variational problems.[1] At this point, however, we cannot fruitfully discuss the many specific techniques that are useful for particular types of problems. These specialized aspects of the finite element method will be introduced in the chapters of Part II, where applications are discussed.

To set the stage for the introduction of the mathematical concepts and to give them a place in the general collection of solution techniques, we begin with a general discussion of the continuum problems of mathematical physics. After briefly mentioning some more popular solution techniques for different classes of problems, we establish the necessary terminology and definitions to show how variational problems and the finite element method are related. The variational basis of the finite element method dictates the criteria to be satisfied by the functions that approximate the solution within the element and enables us to make definitive statements about convergence of the results as we use an ever-increasing number of smaller and smaller elements. A treatment of the rigorous mathematical considerations on error bounds and convergence is beyond the scope of this book, but some work in this area was mentioned in Chapter 1. After a discussion of the variational approach to the formulation of element equations we consider a detailed example.

3.2   CONTINUUM PROBLEMS

3.2.1   Introduction

Problems in engineering and science fall into two fundamentally different categories, depending on which point of view we adopt. One point of view is that all matter consists of discrete particles that retain their identity and nature as they move through space. Their position at any instant is given by coordinates in some reference system, and these coordinates are functions of time—the only independent variable for any process. This viewpoint, known as the *rigid body* viewpoint, is the basis for Newtonian particle mechanics.

---

[1]We shall assume in this chapter that the reader is familiar with the calculus of variational problems. To review the aspects of this subject relevant to the finite element method the reader may refer to Appendix B.

The other viewpoint, the one that we shall use, stems from the *continuum* rather than the discrete molecular or particle approach to nature. In the continuum viewpoint we say that all bodies of interest are continuous at all points in space to the extent that all field quantities describing the state of the body are sufficiently differentiable in the independent variables of the continuum, space and time. This allows us to focus our attention on one point in space and time and observe the phenomena occurring there.

Continuum problems are concerned with the fields of temperature, stress, mass concentration, displacement, and potentials (e.g., electromagnetic and acoustic), to name just a few examples. These problems arise from phenomena in nature that are approximately characterized by partial differential equations and their boundary conditions.

We shall briefly discuss the nature of continuum problems and some possible means of solution; then we will turn to the topic of solving these problems by the finite element method. Continuum problems of mathematical physics are often called *boundary value problems* because their solution is sought in some domain defined by a given boundary on which certain conditions, called *boundary conditions*, are specified. Except for free-boundary problems,[2] the shape of the boundary and its location are always known. The boundary may be defined by a curve or a surface in n-dimensional space, and the domain it defines may be finite or infinite, depending on the extremities of the boundary. The boundary is said to be *closed* if conditions affecting the solution of the problem are specified everywhere on the boundary (though part of the boundary may extend to infinity), and *open* if part of the boundary extends to infinity and no boundary conditions are specified at the part at infinity [1]. It is important to note that our definition of a boundary problem departs from the usual one. The usual definition distinguishes between boundary value problems and *initial value problems*, where time is an independent variable. Because of our definition of the boundary of the domain, we describe all partial differential equations and their boundary conditions as boundary value problems. Therefore, our definition groups the usual boundary value problems and initial value problems into one category.

3.2.2 Problem Statement

The kinds of continuum problems that we wish to solve are usually formulated in general terms as follows. Consider some domain Ω bounded by the surface Γ. Let ϕ be a scalar function[3] defined in the interior of Ω such that

---

[2] In free-boundary problems part of the problem is to determine the shape and location of the boundary.

[3] Without loss of generality we restrict our attention here to scalar functions. Were we to consider vector functions, we would simply have a scalar equation like equation 3.1 for each component of the vector.

the behavior of ϕ at any point in Ω is given by

$$A(\phi) - f = 0 \qquad (3.1)$$

where f is a known scalar function of the independent variables and A is some linear or nonlinear differential operator. We assume that the physical parameters of the differential operator are known constants or known functions. A simple example of a linear differential operator is

$$A(\ ) = \frac{\partial(\ )}{\partial x}$$

Two other examples of linear differential operators are the Laplace operator (in two dimensions),

$$A(\ ) = \frac{\partial^2(\ )}{\partial x^2} + \frac{\partial^2(\ )}{\partial y^2}$$

and the biharmonic operator (also in two dimensions),

$$A(\ ) = \frac{\partial^4(\ )}{\partial x^4} + \frac{\partial^4(\ )}{\partial x^2 \partial y^2} + \frac{\partial^4(\ )}{\partial y^4}$$

In n dimensions second-order differential operators can usually be reduced, by a suitable transformation, to the form

$$A(\ ) = \sum_{i=1}^{n} a_i \frac{\partial^2(\ )}{\partial x_i^2} + \sum_{i=1}^{n} b_i \frac{\partial(\ )}{\partial x_i} + c(\ ) + d \qquad (3.2)$$

where coefficients a_i, b_i, and c and the term d may be functions. The operator as given in equation 3.2 is *linear* if a_i, b_i, c, and d are functions only of the independent variables $(x_1, x_2, x_3, \ldots, x_n)$, and *quasilinear* if a_i, b_i, c, and d are functions of x_i and the dependent parameter, as well as the first derivatives of the dependent parameter. An operator is linear if and only if

$$A(\alpha f + \beta g) = \alpha A(f) + \beta A(g)$$

where α and β are scalars. The general definition we have given the operator $A(\ )$ in equation 3.1 precludes a discussion of appropriate boundary conditions. However, without boundary conditions, equation 3.1 does not describe a specific problem. For some cases it may be possible to integrate equation 3.1, but if this is done, the result will always contain arbitrary constants if $A(\ )$ is an *ordinary* differential operator, or arbitrary functions if $A(\ )$ is a *partial* differential operator. These constants or functions can be found only if the "proper" boundary conditions are specified.

3.2.3 Classification of Differential Equations

The important question of what kinds of conditions constitute "proper" boundary conditions for a given differential operator, or, equally important, what kinds of conditions are improper and cannot be satisfied, can be answered by investigating the classification of the given operator. All partial differential equations of the form of equation 3.1 can be classified as either elliptic, parabolic, or hyperbolic, or some combination of these three categories, such as ultrahyperbolic, elliptically parabolic, or hyperbolically parabolic. For example, we consider the following general partial differential equation in two independent variables:

$$a\frac{\partial^2 \phi}{\partial x^2} + 2b\frac{\partial^2 \phi}{\partial x\,\partial y} + c\frac{\partial^2 \phi}{\partial y^2} = d\left(x, y, \phi, \frac{\partial \phi}{\partial x}, \frac{\partial \phi}{\partial y}\right) \tag{3.3}$$

where a, b, and c are functions of x and y only.

Equation 3.3 can be linear or nonlinear, depending on the form of the function d:

If $b^2 - ac < 0$, the equation is elliptic.
If $b^2 - ac = 0$, the equation is parabolic.
If $b^2 - ac > 0$, the equation is hyperbolic.

For parabolic and hyperbolic equations the solution domains are usually open, whereas for elliptic equations the solution domains are defined by a closed boundary. Because a, b, and c may be functions of each generic point in the solution region, the classification of equation 3.3 may change from one point in the solution region to another.

To determine the classification of more complex partial differential equations, it is necessary to study their *characteristics*. At the risk of oversimplifying the concepts, we can say that the characteristics of the equations are the lines (perhaps in *n*-dimensional space) along which the highest-order derivatives in the equation may be discontinuous. Once the characteristics of an equation are known, we have the following rules:

• Elliptic equations have no real characteristics.
• Parabolic equations have mixed real and imaginary characteristics.
• Hyperbolic equations have all real and distinct characteristics.

The classification of partial differential equations and the role played by the characteristics in determining proper boundary conditions for differential equations are beyond the scope of our central theme. Refer to the books by Petrovsky [2] and Crandall [3] for further details on these subjects.

Our purpose in mentioning the classification of partial differential equations is to point out that the application of finite element techniques to the solution of differential equations and their boundary conditions depend on the classification of a given equation only insofar as the boundary conditions are concerned. If the proper boundary conditions are specified, the classification of the equation does not enter explicitly into consideration. When proper boundary conditions are given,[4] in principle the finite element method is applicable to linear and partial nonlinear differential equations valid over domains of *any* geometrical shape.

3.3   SOME METHODS FOR SOLVING CONTINUUM PROBLEMS

3.3.1   An Overview

Returning to equation 3.1, we see that our general problem is to find the unknown function ϕ that satisfies equation 3.1 in the domain Ω and the associated boundary conditions specified on Γ. There are a number of approaches to the solution of linear and nonlinear boundary value problems, and they range from completely analytical to completely numerical. Of these, the following deserve attention:

    I. Direct Integration (exact solutions)
       A. Separation of variables
       B. Similarity solutions
       C. Fourier and Laplace transforms
   II. Approximate Solutions
       A. Perturbation
       B. Power series
       C. Probability schemes
       D. Method of weighted residuals (MWR)
       E. Ritz method
       F. Finite difference techniques
       G. Finite element method

For a few problems it is possible to obtain an exact solution by direct integration of the differential equation. This is accomplished sometimes by an obvious separation of variables or by applying a transformation that makes the variables separable and leads to a similarity solution. Occasionally a Fourier or Laplace transformation of the differential equation leads to an

---

[4]In this book we shall assume that we are dealing with well-posed problems whose statements contain the proper boundary conditions.

exact solution. However, the number of problems with exact solutions is severely limited, and most of these have already been solved. They are the classical problems.

Because regular and singular perturbation methods are applicable primarily when the nonlinear terms in the equation are small in relation to the linear terms, their usefulness is limited. The power series method is robust and has been employed with success, but since the method requires the generation of a coefficient for each term in the series, it is relatively tedious. Also, it is difficult, if not impossible, to demonstrate that the series converges.

The probability schemes, usually classified under the heading of Monte Carlo methods [4], are used for obtaining a statistical estimate of a desired quantity by random sampling. These methods work best when the desired quantity is a statistical parameter and sampling is done from a selective population.

With the advent of computers the currently outstanding methods for obtaining approximate solutions of high accuracy are the method of weighted residuals, the Ritz method, the finite difference method, and the finite element method. These methods are related, as we shall see in this chapter and the next. More than one volume would be needed to do justice to all of the important aspects of these methods, but since this book deals exclusively with the finite element method we will only touch on the essentials of the Ritz method and method of weighted residuals as they apply to the finite element method. The interested reader is referred to references 5–7 for more thorough treatment of the finite difference method and method of weighted residuals.

3.3.2 The Variational Approach

Often continuum problems have a different, but equivalent formulation—a differential formulation and a variational formulation. In the differential formulation, as we have seen, the problem is to integrate a differential equation or a system of differential equations subject to specified boundary conditions. In the *classical variational formulation*[5] the problem is to find the unknown function or functions that extremize (maximize, minimize) or make stationary a functional or system of functionals subject to the same specified boundary conditions. The two formulations are equivalent because the functions that satisfy the differential equation system and its boundary conditions also extremize or make stationary the functionals. This equivalence is apparent from the calculus of variations, which shows that the functionals are extremized or made stationary when one or more of the Euler equations and their boundary conditions are satisfied. And these equations are precisely the governing differential equations of the problem. The classical variational

---

[5]Generally, we shall be concerned only with the classical variational formulation in this book.

formulation of a continuum problem often has advantages over the differential formulation from the standpoint of obtaining an approximate solution. First, the functional, which may actually represent some physical quantity in the problem, contains derivatives of lower order than that of the differential operator, and consequently an approximation solution can be sought from among a larger class of functions. Second, the problem may possess reciprocal variational formulations, that is, one functional to be minimized and another functional of a different form to be maximized. In such cases we have means for finding upper and lower bounds on the functional and this capability may have significant engineering value. Third, the variational formulation allows us to treat very complicated boundary conditions as natural boundary conditions.[6] And, fourth, from a purely mathematical standpoint the variational formulation is helpful because with the calculus of variations it can sometimes be used to prove the existence of a solution.

In the past when engineers first used the finite element method to solve particular continuum problems, they most often relied on variational methods to derive finite element equations. This approach is especially convenient when it is applicable; but before it can be used, a variational statement of the continuum problem must be obtained, that is, we must pose the problem in variational form. Though the topic of obtaining variational formulations is not discussed in this text, we will assume that the variational formulation for a given problem is known.

Historically, variational methods are among the oldest means of obtaining solutions to problems in physics and engineering. One general method for obtaining approximate solutions to problems expressed in variational form is known as the *Ritz* or *Rayleigh–Ritz method*. Since this method is basically a mathematical forerunner of the finite element method, we introduce some of its features before discussing its incorporation into the finite element approach. Actually, the finite element method is a special case of the Ritz method when the interpolation functions obey certain continuity requirements.

3.3.3 The Ritz Method

The Ritz begins by assuming the form of the unknown solution in terms of known functions (*trial functions*) with unknown adjustable parameters. From the class of trial functions we select the function that renders the functional stationary. The procedure consists of substituting the trial functions into the functional and thereby expressing the functional in terms of the adjustable parameters which are the unknowns. The functional is then differentiated with respect to each parameter, and the resulting equation is set to zero. If

---

[6]As noted in Appendix B, boundary conditions may be either *essential* or *natural*. In the finite element method we explicitly impose essential boundary conditions via the procedure in Chapter 2, whereas the variational statement implicitly imposes the natural boundary conditions.

there are n unknown parameters, then there will be n simultaneous equations to be solved for these parameters. By this means the approximate solution is chosen from the class of assumed solutions.

The procedure does nothing more than give us the "best" solution from the class of assumed solutions. Clearly then, the accuracy of the approximate solution depends on the choice of trial functions. We require that the trial functions be defined over the whole solution domain and that they satisfy at least some and usually all of the boundary conditions. Sometimes, if we know the general nature of the desired solution, we can improve the approximation by choosing the trial functions to reflect this nature. If, by chance, the exact solution is contained in the class of trial solutions, the Ritz procedure gives the exact solution. Generally, the approximation improves as the size of the class of trial functions and the number of adjustable parameters increase. If the trial functions are part of an infinite set of functions that are capable of representing the unknown function to any degree of accuracy, the process of including more and more terms of that set in the trial functions leads to a series of approximate solutions that converges to the true solution. Often a class of trial functions is constructed from polynomials of successively increasing degree, but in certain cases other kinds of functions may offer advantages [8].

Example: The Ritz Method   To illustrate the Ritz method we shall consider a simple one-dimensional example. Suppose that we want to find the function $\phi(x)$ satisfying

$$\frac{d^2\phi}{dx^2} = -f(x), \qquad a < x < b$$

with boundary conditions

$$\phi(a) = A \qquad \phi(b) = B$$

Note that these are essential boundary conditions, that is, the value of the dependent variable ϕ is specified on both boundaries. We assume $f(x)$ is a smooth function in the closed interval $[a, b]$. This problem is equivalent to finding the function $\phi(x)$ that minimizes the quadratic functional

$$J(\phi) = \int_a^b \left[\frac{1}{2} \left(\frac{d\phi}{dx} \right)^2 - f(x)\phi(x) \right] dx$$

We shall ignore the fact that this problem has an exact solution and proceed to find an approximate solution. According to the Ritz method we assume that the desired solution can be approximated in $[a, b]$ by a combination of selected trial functions of the form

$$\phi(x) \approx \tilde{\phi}_n(x) = \psi_0 + C_1\psi_1 + C_2\psi_2 + \cdots + C_n\psi_n \qquad a \leq x \leq b$$

where the n constants C_i are the adjustable parameters to be determined. The trial functions must be selected so that the essential boundary conditions are satisfied regardless of the choice of the constants C_i. To clarify an often confusing point, if ψ_1 and ψ_2 are trial functions in a particular class and C_1 and C_2 are scalars, then $\tilde{\phi} = C_1\psi_1 + C_2\psi_2$ (a linear combination of ψ_1 and ψ_2) is also in that class and is also referred to as a trial function. Because the essential boundary conditions must be satisfied we must select our trial functions so that

$$\psi_0(a) = A \qquad \psi_0(b) = B$$

and

$$\psi_i(a) = \psi_i(b) = 0, \qquad i = 1, \ldots, n$$

Using polynomials is a simple and convenient way to construct trial functions, so now we illustrate the procedure with a specific example:

$$\frac{d^2\phi}{dx^2} = -kx, \qquad 0 < x < 1$$

$$\phi(0) = 0 \qquad \phi(1) = 1$$

where k is a constant. Thus we can write

$$\psi_0 = x$$

$$\psi_i(x) = (1 - x)x^i$$

so that

$$\phi(x) \approx \tilde{\phi}_n(x) = \psi_0 + \sum_{i=1}^{n} C_i\psi_i(x)$$

is a possible trial function formed from a series of trial functions. Notice that the essential boundary conditions are satisfied by ψ_0 and that ψ_i is zero on the boundaries. When we substitute this trial function into the quadratic functional to be minimized we obtain, after carrying out the integration,

$$J(\tilde{\phi}) = J(C_1, C_2, \ldots, C_n)$$

Now we require $C_1, C_2, \ldots, C_n$ be chosen to minimize J. Hence, from

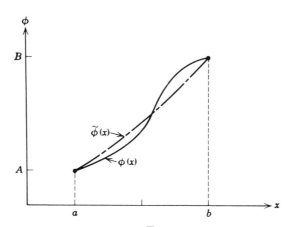

Figure 3.1 Ritz approximation $\overline{\phi}(x)$ of the exact solution $\phi(x)$.

differential calculus, we have

$$\frac{\partial J}{\partial C_1} = 0, \quad \frac{\partial J}{\partial C_2} = 0, \quad \ldots, \quad \frac{\partial J}{\partial C_n} = 0$$

When these n equations are solved for $C_1, \ldots, C_n$, the approximate solution to $\phi'' = -kx$ is obtained, as shown in Figure 3.1. The accuracy of the approximation depends, among other things, on the number of terms used in the trial function. Generally, as n increases the accuracy improves. To assess the improvement in accuracy as n increases, we can solve the problem repeatedly by taking successively more terms in the trial function; that is, we can use

$$\phi(x) \approx \tilde{\phi}_1(x) = x + (1 - x)xC_1$$

$$\phi(x) \approx \tilde{\phi}_2(x) = x + (1 - x)xC_1 + (1 - x)x^2C_2$$

$$\phi(x) \approx \tilde{\phi}_3(x) = x + (1 - x)xC_1 + (1 - x)x^2C_2 + (1 - x)x^3C_3$$

and so on. By comparing the results at the end of each calculation we can estimate the effect of adding more terms on accuracy. For instance, in the above problem for $n = 1$, the procedure outlined leads to $C_1 = k/4$ and

$$\phi(x) \approx \tilde{\phi}_1(x) = x + \frac{k}{4}(x - x^2)$$

For $n = 2$, then, $C_1 = C_2 = k/6$ and

$$\phi(x) \approx \tilde{\phi}_2(x) = x + \frac{k}{6}(x - x^2) + \frac{k}{6}(x^2 - x^3)$$

or

$$\phi(x) \approx \tilde{\phi}_2(x) = x + \frac{k}{6}(x - x^3)$$

which happens to be the *exact solution* for this particular problem.

There are several important points that may be subtle enough to go unnoticed in this example. We cannot simply use any trial function we wish, because the problem imposes certain constraints. The trial functions $\tilde{\phi}_1$, $\tilde{\phi}_2, \tilde{\phi}_3$ are second-, third-, and fourth-degree polynomials, respectively. Though it is almost a trivial point, for this second-order problem the trial function must be sufficiently differentiable to substitute into this quadratic functional, hence there is a smoothness constraint on the trial function. Next, to evaluate the integral for $J(\phi)$ the first derivative of the trial function must be *square integrable*, i.e.,

$$\int_a^b \left(\frac{d\tilde{\phi}}{dx}\right)^2 dx < \infty$$

Since we are solving for the constants in a system of linear equations, the rows (columns) of the coefficient matrix must be linearly independent. We have already stated that the trial function must satisfy the essential boundary conditions. We see that we cannot have our trial function just any way we please, but instead, the selection process can become quite restrictive [9, 10].

We have seen briefly how the Ritz method works and its potential for approximating solutions to certain problems, but it does have limitations. First, it is a global scheme in that it approximates a solution for the entire domain and satisfies all of the essential boundary conditions and often many of the natural boundary conditions. For this reason if we wish to do several analyses of a particular problem with different types of boundary conditions (a typical engineering application) the problem would have to be reformulated for each change in boundary conditions. Another disadvantage is that for a complex geometry it may be difficult or impossible to find a trial function that will satisfy the essential boundary conditions. Further, there is no systematic approach to finding trial functions for given problems. We shall see in the next section how some of these limitations are overcome with the finite element method, but the reader interested in more information on global variational methods is referred to references 9 and 10.

3.4   THE FINITE ELEMENT METHOD

3.4.1   Relation to the Ritz Method

The finite element method and the Ritz method are essentially equivalent. Each method uses trial functions as a starting point for obtaining an approximate solution; both methods take linear combinations of the class of functions that make up the trial functions; and both methods seek the trial functions that makes a given functional stationary. The major difference in the methods is that trial functions in the finite element method are not defined over the whole solution domain, and they have to satisfy no boundary conditions but only certain continuity conditions. While we saw in the last section that the Ritz method by itself is a global method, the finite element method is in essence the Ritz method applied to discrete local domains which are usually quite simple in contrast to the global domain. By imposing certain continuity requirements at the boundaries of the discrete local domains they may be assembled into an exceedingly more complex global domain. These discrete domains are known as elements, hence the name of the method. From a strict mathematical standpoint the finite element method is a special case of the Ritz method only when the piecewise trial functions obey certain continuity and completeness conditions to be discussed later.

3.4.2   Generalizing the Definition of an Element

The basic idea of the finite element method is to divide the solution domain into a finite number of subdomains (elements). These elements are connected only at node points in the domain and on the element boundaries. In this way the solution domain is discretized and represented as a patchwork of elements. Frequently the boundaries of the finite elements are straight lines or planes, so if the solution domain has curved boundaries, these are approximated by a series of straight or flat segments.[7] In Chapter 2, when we considered a physical interpretation of the finite element method, we imagined the elements to be individual segments or parts of the actual system, for example, a spring, a fluid- or current-carrying conduit, a bar, or a triangular plate.

   The nodes for these elements were part of the elements, and in cases where displacement was the unknown field variable the nodes could move as the element deformed under load. In our example of a fluid-carrying conduit the nodes were simply places or locations in the system where pressure and flow were defined—the finite element and its nodes represented not a part of the flowing fluid, but only a region through which the flow passed.

---

[7]An exception to this statement occurs when isoparametric elements are used. We shall introduce the concept of isoparametric elements in Chapter 5.

The mathematical interpretation of the finite element method requires us to generalize our definition of an element and to think of elements in less physical terms. Instead of viewing an element as a physical part of the system, we view the element as a part of the solution domain where the phenomena of interest are occurring. We imagine the solution domain to be sectioned by lines (or general planes in n dimensions) that define the boundaries of the elements. The elements are interconnected only at imaginary node points on the boundaries or surfaces of the elements. For solid mechanics problems we no longer need to imagine that the elements deform or change shape; rather, we define them as regions of space where a displacement field exists. The nodes of an element are then simply locations in space where the displacement and possibly its derivatives are known or sought. Similarly, for fluid mechanics problems the elements are regions over which a pressure field exists and through which fluid is flowing. The mathematical interpretation of a finite element mesh is that it is a *spatial subdivision* rather than a material subdivision. This broader interpretation of an element allows us to carry over many of the basic ideas from one problem area to another.

In the finite element procedure once the element mesh for the solution domain has been decided, the behavior of the unknown field variable over each element is approximated by continuous functions expressed in terms of the nodal values of the field variable and sometimes the nodal values of its derivatives up to a certain order. The functions defined over each finite element are called *interpolation functions*, *shape functions*, or field variable models. The collection of interpolation functions for the whole solution domain provides a piecewise approximation to the field variable.

3.4.3 Example of a Piecewise Approximation

To illustrate the nature of this piecewise approximation we consider the representation of a two-dimensional field variable, $\phi(x, y)$. We will show how the nodal values of ϕ can uniquely and continuously define $\phi(x, y)$ throughout the domain of interest in the x–y plane, and we will introduce the notation for an interpolation function.

Suppose that we have the domain shown in Figure 3.2, and we section it into triangular elements with nodes at the vertices of the triangles. With this type of domain discretization we can allow ϕ to vary linearly over each element (see Figure 3.3). The plane passing through the three nodal values of ϕ associated with element (e) is described by

$$\phi^{(e)}(x, y) = \beta_1^{(e)} + \beta_2^{(e)}x + \beta_3^{(e)}y \tag{3.4}$$

We can express the constants $\beta_1^{(e)}$, $\beta_2^{(e)}$, and $\beta_3^{(e)}$ in terms of the coordinates of the element's nodes and the nodal values of ϕ by evaluating equation 3.4

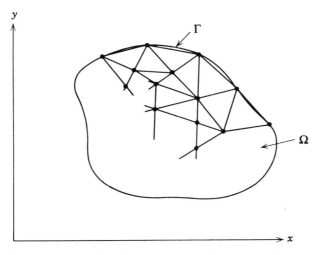

Figure 3.2 Two-dimensional domain divided into triangular elements.

at each node. Hence

$$\phi_i = \beta_1^{(e)} + \beta_2^{(e)}x_i + \beta_3^{(e)}y_i$$
$$\phi_j = \beta_1^{(e)} + \beta_2^{(e)}x_j + \beta_3^{(e)}y_j \qquad (3.5)$$
$$\phi_k = \beta_1^{(e)} + \beta_2^{(e)}x_k + \beta_3^{(e)}y_k$$

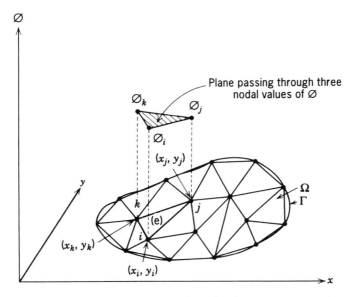

Figure 3.3 Subdivided domain Ω and a piecewise linear solution surface $\phi(x, y)$.

Solving yields

$$\beta_1^{(e)} = \frac{\phi_i(x_j y_k - y_j x_k) + \phi_j(x_k y_i - x_i y_k) + \phi_k(x_i y_j - x_j y_i)}{2\Delta}$$

$$\beta_2^{(e)} = \frac{\phi_i(y_j - y_k) + \phi_j(y_k - y_i) + \phi_k(y_i - y_j)}{2\Delta} \tag{3.6}$$

$$\beta_3^{(e)} = \frac{\phi_i(x_k - x_j) + \phi_j(x_i - x_k) + \phi_k(x_j - x_i)}{2\Delta}$$

where (as in equation 2.27)

$$2\Delta = \begin{vmatrix} 1 & x_i & y_i \\ 1 & x_j & y_j \\ 1 & x_k & y_k \end{vmatrix} = 2\begin{bmatrix} \text{area of triangle whose} \\ \text{vertices are } i, j, k \end{bmatrix}$$

Substituting equation 3.6 into equation 3.4 and rearranging terms, we have

$$\phi^{(e)}(x, y) = \frac{a_i + b_i x + c_i y}{2\Delta} \phi_i + \frac{a_j + b_j x + c_j y}{2\Delta} \phi_j + \frac{a_k + b_k x + c_k y}{2\Delta} \phi_k \tag{3.7}$$

where

$$a_i = x_j y_k - x_k y_j \qquad b_i = y_j - y_k \qquad c_i = x_k - x_j$$

and the other coefficients are obtained through a cyclic permutation of the subscripts i, j, and k. If $i = 1$, $j = 2$, and $k = 3$, the coefficients are given explicitly by equations 2.26.

Now we define

$$N_n^{(e)} = \frac{a_n + b_n x + c_n y}{2\Delta} \qquad n = i, j, k$$

and let

$$\{\phi^{(e)}\} = \begin{Bmatrix} \phi_i \\ \phi_j \\ \phi_k \end{Bmatrix} \qquad \lfloor N^{(e)} \rfloor = \lfloor N_i^{(e)}, N_j^{(e)}, N_k^{(e)} \rfloor$$

In general, the functions $N^{(e)}$ are called shape functions or interpolation functions, and they play a most important role in all finite element analyses. (Here the $N^{(e)}$ are linear interpolation functions for the three-node triangular element.) In matrix notation we can write equation 3.7 as

$$\phi^{(e)}(x, y) = \lfloor N^{(e)} \rfloor \{\phi^{(e)}\} = N_i \phi_i + N_j \phi_j + N_k \phi_k \qquad (3.8)$$

If the domain contains M elements, the complete representation of the field variable over the whole domain is given by

$$\phi(x, y) = \sum_{e=1}^{M} \phi^{(e)}(x, y) = \sum_{e=1}^{M} \lfloor N^{(e)} \rfloor \{\phi^{(e)}\} \qquad (3.9)$$

From equation 3.9 we see that if the nodal values of ϕ are known, we can represent the complete solution surface $\phi(x, y)$ as a series of interconnected, triangular planes. This many-faceted surface has no discontinuities or "gaps" at interelement boundaries because the values of ϕ at any two nodes defining an element boundary uniquely determine the linear variation of ϕ along that boundary. Figure 3.4 shows a wire model of the piecewise representation of the field variable expressed by equation 3.9.

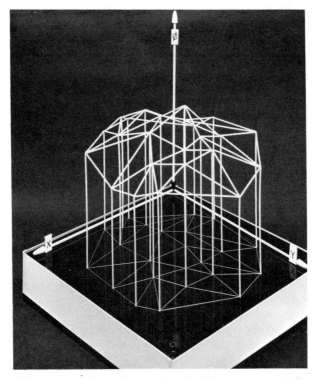

Figure 3.4   Wire model illustrating a piecewise linear representation of $\phi(x, y)$.

Although we obtained equations 3.8 and 3.9 for a particular interpolation function (linear) and a particular element type (three-node triangle), these equations have general validity. For more complex interpolation functions and other element types, such as those discussed in Chapter 5, the form of equations 3.8 and 3.9 remains the same; only the numbers of terms in the row and column matrices are different. Hence if a solution domain is subdivided into elements, we can represent the unknown field variable in each element as

$$\phi^{(e)} = \overset{1 \times r}{\lfloor N^{(e)} \rfloor} \overset{r \times 1}{\{\phi\}} \tag{3.10}$$

where $\lfloor N^{(e)} \rfloor$ is the row vector of interpolation functions that are functions of the coordinates of the nodes, and $\{\phi\}^{(e)}$ is the column vector, which is the collection of r discrete values consisting of nodal values of ϕ associated with the element *and* perhaps some other parameters that characterize the element and which are not identified with any node. Such "nodeless" [11] variables may appear when constraints are imposed on the field variable or when parameters are assigned to an element via modes of the interpolation functions that vanish on the element boundaries. These nodeless variables appear as *Lagrange multipliers* in the augmented functional. The nature of Lagrange multipliers and their relation to variational principles are discussed in Appendix B, but their application in finite element analysis is beyond the scope of this book.

We shall now assume that we have the field variable ϕ completely represented in the solution domain in terms of a collection of nodal values of ϕ. Under this assumption if these discrete values were known, the problem of finding an approximation to ϕ would be solved.

3.4.4   Element Equations From a Variational Principle

The finite element solution to the problem involves picking the nodal values of ϕ so as to make the functional $I(\phi)$ stationary. To make $I(\phi)$ stationary with respect to the nodal values of ϕ we require that

$$\delta I(\phi) = \sum_{i=1}^{n} \frac{\partial I}{\partial \phi_i} \delta \phi_i = 0 \tag{3.11}$$

where n is the total number of discrete values of ϕ assigned to the solution domain. Since the $\delta \phi_i$'s are independent, equation 3.11 can hold only if

$$\frac{\partial I}{\partial \phi_i} = 0, \quad i = 1, 2, \ldots, n \tag{3.12}$$

If the interpolation functions giving our piecewise representation of ϕ obey certain continuity and compatibility conditions, which we discuss shortly, then the functional $I(\phi)$ may be written as a sum of individual functionals defined for all elements of the assemblage, that is,

$$I(\phi) = \sum_{e=1}^{M} I^{(e)}(\phi^{(e)}) \qquad (3.13)$$

where M is the total number of elements and the superscript (e) denotes an element. Hence, instead of working with the functional defined over the whole solution region, we may focus our attention on the functionals defined for the individual elements. From equation 3.13 we have

$$\delta I = \sum_{e=1}^{M} \delta I^{(e)} = 0 \qquad (3.14)$$

where the variation of $I^{(e)}$ is taken only with respect to the nodal values associated with elements (e). Equation 3.14 implies that

$$\left\{ \frac{\partial I^{(e)}}{\partial \phi} \right\} = \frac{\partial I^{(e)}}{\partial \phi_j} = 0, \qquad j = 1, 2, \ldots, r \qquad (3.15)$$

where r is the number of nodes assigned to element (e). Equations 3.15 comprise a set of r equations that characterizes the behavior of element (e). The fact that we can represent the functional for the assemblage of elements as the sum of the functionals for all individual elements provides the key to formulating individual element equations from a variational principle. If, for example, the governing differential equations and boundary conditions for a problem are linear and self-adjoint,[8] the corresponding variational statement of the problem involves a *quadratic functional*. When $I(\phi)$ is quadratic, $I^{(e)}$ is also quadratic, and equation 3.15 for element (e) can always be written as

$$\overset{r \times 1}{\left\{ \frac{\partial I^{(e)}}{\partial \phi} \right\}} = \overset{r \times r}{[K]^{(e)}} \overset{r \times 1}{\{\phi\}^{(e)}} - \overset{r \times 1}{\{F\}^{(e)}} = \overset{r \times 1}{\{0\}} \qquad (3.16)$$

where $[K]^{(e)}$ is a square matrix of constant "stiffness" coefficients, $\{\phi\}^{(e)}$ is the column vector of nodal values, and $\{F\}$ is the vector of resultant nodal

---

[8]A differential operator is said to be self-adjoint in domain Ω if

$$\int_{\Omega} uA(v) \, d\Omega = \int_{\Omega} vA(u) \, d\Omega$$

A detailed mathematical definition of an adjoint for A can be found in reference 1.

actions. The complete set of finite element equations for the problem is assembled by adding all the derivatives of I as given by equations 3.15 for all the elements. The assembly procedure is identical to that discussed in Chapter 2. Symbolically, we can write the complete set of equations as

$$\frac{\partial I}{\partial \phi_i} = \sum_{e=1}^{M} \frac{\partial I^{(e)}}{\partial \phi_i} = 0, \qquad i = 1, 2, \ldots, n \qquad (3.17)$$

or

$$\left\{ \frac{\partial I}{\partial \phi} \right\} = \{0\}$$

Our problem is solved when the set of n equations 3.17 is solved simultaneously for the n nodal values of ϕ. If there are q nodes in the solution domain where ϕ is specified by boundary conditions, there will be $n - q$ equations to be solved for the $n - q$ unknowns. Note that the summation indicated in equation 3.17 contains many zero terms, because only elements sharing node i will contribute to $\partial I/\partial \phi_i$. If node i does not belong to element e, then $\partial I/\partial \phi_i = 0$. This fact is manifested in the "narrow-band" and "sparseness" properties of the resulting matrix of "stiffness" coefficients.

We have now established the means for formulating individual finite element equations from a variational principle. The procedure can be summarized as follows.

If the functional for a given problem can be expressed as the sum of functionals evaluated for all elements, we may focus our attention on an isolated element without regard for its eventual location in the assemblage. To derive the equations governing the element's behavior we first use interpolation functions to define the unknown field variable ϕ in terms of its nodal values associated with the element; then we evaluate the functional $I^{(e)}$ by substituting the assumed form for $\phi^{(e)}$ and its derivatives and carry out the integration over the domain defined by the element boundaries. Finally, we perform the differentiations indicated by equation 3.15, and the result is the set of equations defining the element behavior. Since the differentiations are with respect to the discrete nodal values, we employ only calculus, not calculus of variations.

3.4.5   Requirements for Interpolation Functions

Our procedure for formulating the individual element equations from a variational principle and our ability to assemble these equations to obtain the system equations rely on the assumption that the interpolation functions satisfy certain requirements. The requirements we place on the choice of the interpolation functions stem from the need to ensure that equation 3.13 holds and that our approximate solution converges to the correct solution

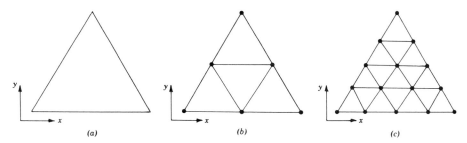

Figure 3.5 Example of successive mesh refinements: (*a*) Original solution domain; (*b*) Discretization with 4 triangular elements; (*c*) Discretization with 16 triangular elements.

when we use an increasing number of smaller elements, that is, when we refine the element mesh. Mathematical proofs of convergence assume that the process of mesh refinement occurs in a regular fashion defined by three conditions [12]: (1) the elements must be made smaller in such a way that every point of the solution domain can always be within an element, regardless of how small the element may be; (2) all previous meshes must be contained in the refined meshes; and (3) the form of the interpolation functions must remain unchanged during the process of mesh refinement.

These three conditions are illustrated in Figure 3.5, where a simple two-dimensional solution domain in the form of an equilateral triangle is discretized with an increasing number of three-node triangles. We note that when elements with straight boundaries are used to model solution domains with curved boundaries, the first two conditions are not satisfied and rigorous mathematical proofs of convergence may not be obtainable. In spite of this limitation, many applications of the finite element method to problems with nonpolygonal solution domains yield acceptable engineering solutions.

To guarantee monotonic convergence in the sense just described and to make the assembly of the individual element equations meaningful we require that the interpolation functions $N^{(e)}$ in the expressions

$$\phi^{(e)} = \lfloor N^{(e)} \rfloor \{\phi\}^{(e)}, \qquad e = 1, 2, \ldots, M$$

be chosen so that the following general requirements are met:

1. At element interfaces (boundaries) the field variable ϕ and any of its partial derivatives up to one order less than the highest-order derivative appearing in $I(\phi)$ must be continuous.
2. All uniform states of ϕ and its partial derivatives up to the highest order appearing in $I(\phi)$ should have representation in $\phi^{(e)}$ when, in the limit, the element size shrinks to zero.

These requirements were given in Felippa and Clough [12] and justified by Oliveira [13]. The first one is known as the *compatibility* requirement, and the second as the *completeness* requirement. Elements whose interpolation functions satisfy the first requirement are called *compatible* or *conforming elements*, those satisfying the second requirement are called *complete elements*.

The definition of an *incompatible element* is obvious. When, for example, our field variable ϕ is defined as the displacement field in a body, these requirements on the interpolation functions have particular meanings. For instance, the compatibility requirement ensures the continuity of displacement between adjacent elements. Clearly, we could not expect a good approximation to reality if our displacement field representation were discontinuous across element boundaries. The second requirement, the completeness requirement, ensures that our displacement field representation allows for the possibility of rigid-body displacements and constant-strain states within the element. These conditions are discussed more fully in Part II of this book.

The compatibility requirement helps to ensure that the integral in $I(\phi)$ is well defined. Without interelement continuity we cannot be sure that the integral in $I(\phi)$ is unique. Uncertain contributions may arise from the "gaps" between elements. If the compatibility condition is violated, it is sometimes possible to add special boundary integrals for compensation [14]. It is always desirable when carrying out a finite element analysis to be sure that mesh refinement will lead to answers that are converging to the correct solution. For some problems, however, choosing interpolation functions that meet all the requirements may be difficult and may involve excessive numerical computation. For this reason some investigators have ventured to formulate interpolation functions for elements that do not meet all the compatibility and completeness requirements. In some instances acceptable convergence has been obtained, whereas in others no convergence or convergence to an incorrect solution has occurred. In solid mechanics applications we shall encounter a number of cases where complete but incompatible elements have been successfully used. Some applications [15] have shown that the compatibility condition does not always lead to the most rapid convergence. However, when extending the finite element method for use in other areas where far less experience is available, the safest approach is to pick functions and formulate elements that satisfy all the requirements. In Chapter 5 we shall discuss the formulation of many types of elements using polynomial interpolation functions.

Example   To show how the element equations can be derived from a variational principle we consider the same simple example we used to illustrate the Ritz method in Section 3.3. The problem was to find the function $\phi(x)$ satisfying

$$\frac{d^2\phi}{dx^2} = -f(x) \qquad \text{with boundary conditions } \phi(a) = A, \phi(b) = B$$

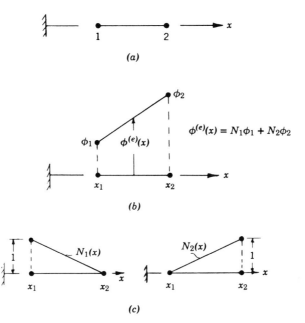

Figure 3.6   Linear representation of a field variable over a one-dimensional element; (*a*) One-dimensional line element; (*b*) Linear variation of $\phi(x)$ over element (*e*); (*c*) Linear interpolation functions for $\phi^{(e)}(x)$.

Equivalently, we seek the function $\phi(x)$ that minimizes

$$J(\phi) = \int_a^b \left[\frac{1}{2} \left(\frac{d\phi}{dx} \right)^2 - f(x)\phi(x) \right] dx$$

Since this is a one-dimensional problem (one independent variable), our elements will be line segments along the x axis (line elements). The number of nodes we assign to each element (and hence the type of interpolation function we can use for the element) is open to choice. Since the highest-order derivative appearing in $J(\phi)$ is a first-order derivative, the simplest interpolation functions satisfying our compatibility and completeness requirements are linear functions. Two nodal values of ϕ are required to define uniquely a linear variation of ϕ over an element; hence a typical element must have two nodes (Figure 3.6*a*). A linear variation of ϕ over element (*e*) can be written in terms of its nodal values and the coordinates of the nodes (Figure 3.6*b*) as

$$\phi^{(e)}(x) = \left(\frac{x_2 - x}{x_2 - x_1} \right)\phi_1 + \left(\frac{x - x_1}{x_2 - x_1} \right)\phi_2$$

$$\phi^{(e)} = \lfloor N_1, N_2 \rfloor \begin{Bmatrix} \phi_1 \\ \phi_2 \end{Bmatrix} = \lfloor N \rfloor \{\phi\}^{(e)}$$

where

$$N_1(x) = \frac{x_2 - x}{x_2 - x_1} \qquad N_2(x) = \frac{x - x_1}{x_2 - x_1}$$

The interpolation functions N_1 and N_2 are shown in Figure 3.6c. Since our interpolation functions for modeling the behavior of ϕ satisfy the compatibility and completeness requirements, we may write

$$J(\phi) = \sum_{e=1}^{M} J^{(e)}(\phi^{(e)})$$

where M is the number of elements with which we choose to model the domain $a \le x \le b$. We also have the mathematical assurance that as M increases, our approximate solutions will converge to the correct solution. To find $J^{(e)}(\phi^{(e)})$ we substitute the equation for $\phi^{(e)}$ into the expression for $J(\phi)$ and obtain (integrating only over the domain of the element)

$$J^{(e)}(\phi^{(e)}) = \int_{x_1}^{x_2} \left[\frac{1}{2} \left(\lfloor N_1' N_2' \rfloor \begin{Bmatrix} \phi_1 \\ \phi_2 \end{Bmatrix} \right)^2 - f(x) \lfloor N_1 N_2 \rfloor \begin{Bmatrix} \phi_1 \\ \phi_2 \end{Bmatrix} \right] dx$$

where the prime denotes differentiation with respect to x. Carrying out the minimization of $J^{(e)}(\phi^{(e)})$ with respect to nodal values of ϕ gives the equations for an element:

$$\frac{\partial J^{(e)}}{\partial \phi_1} = \int_{x_1}^{x_2} \left[N_1' \lfloor N_1' N_2' \rfloor \begin{Bmatrix} \phi_1 \\ \phi_2 \end{Bmatrix} - f N_1 \right] dx = 0$$

and

$$\frac{\partial J^{(e)}}{\partial \phi_2} = \int_{x_1}^{x_2} \left[N_2' \lfloor N_1' N_2' \rfloor \begin{Bmatrix} \phi_1 \\ \phi_2 \end{Bmatrix} - f N_2 \right] dx = 0$$

These equations may be written concisely as

$$[K]^{(e)} \{\phi\}^{(e)} = \{F\}^{(e)}$$

where[9]

$$[K]^{(e)} = \int_{x_1}^{x_2} \begin{bmatrix} N_1' N_1' & N_1' N_2' \\ N_2' N_1' & N_2' N_2' \end{bmatrix}^{(e)} dx$$

$$\{\phi\}^{(e)} = \begin{Bmatrix} \phi_1 \\ \phi_2 \end{Bmatrix} \qquad \{F\}^{(e)} = \int_{x_1}^{x_2} \begin{Bmatrix} f N_1 \\ f N_2 \end{Bmatrix}^{(e)} dx$$

[9]Integration of each term of the matrix is implied in this expression. We also note that $[K]^{(e)}$ is symmetric.

The components of $\{F\}^{(e)}$ may be viewed as the nodal forcing functions. With these definitions we see that the functional for one element can be written in quadratic form as

$$J^{(e)}(\phi^{(e)}) = \tfrac{1}{2}\lfloor \phi \rfloor^{(e)}[K]^{(e)}\{\phi\}^{(e)} - \lfloor F \rfloor^{(e)}\{\phi\}^{(e)}$$

The equations $[K]^{(e)}\{\phi\}^{(e)} = \{F\}^{(e)}$ give the characteristics of a particular finite element of our one-dimensional continuum. The same procedure we used to derive these equations can be employed to obtain the characteristics of more complicated elements. For instance, in this example the first *higher-order element* would be formulated as a line element with three nodes (two exterior nodes and one interior node) and with quadratic interpolation functions such as those shown in Figure 3.7. It is easy to see how we can formulate still higher-order elements with more nodes and subsequently higher-order interpolation functions. Regardless of the type of element we choose to formulate for this problem, the element equations will have the form $[K]^{(e)}\{\phi\}^{(e)} = \{F\}^{(e)}$.

To complete the finite element solution of this problem it is necessary to derive the equations for all the elements in the assemblage and then to assemble these algebraic equations according to the prescriptions in Chapter 2. After introducing the boundary conditions we solve the resulting set of equations for the nodal values of ϕ. The result of the solution will be a piecewise representation of ϕ over $[a, b]$, such as that shown in Figure 3.8. By comparing Figures 3.1 and 3.8 we can see that primary difference between the Ritz method and the finite element method.

In view of the preceding example several general comments are in order. The task of deriving the equations for the individual elements for a problem

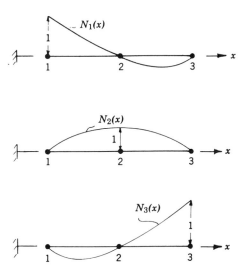

Figure 3.7 Quadratic interpolation functions for three-node line elements.

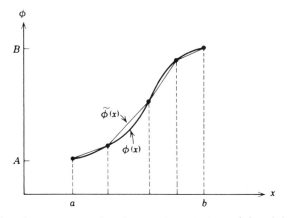

Figure 3.8   Finite element approximation to the solution $\phi(x)$. $\tilde{\phi}(x)$ is an approximate finite element solution.

may appear to be tedious, but actually the process is routine. If the same type of element is used throughout the solution domain, the element equations are given by one general expression. The expression contains nodal coordinates and other known parameters; hence it can serve as the computational algorithm for generating all the equations by computer. If several different types of elements are used, we simply have several different algorithms. Usually, however, for a given problem we use only one type of element. As we saw in Chapter 2, the assembly procedure is automatically done within the computer also. Usually the element matrices are inserted in the master matrix as they are computed.

Before leaving our simple example we note an interesting fact about one-dimensional finite element solutions. A rigorous proof based on a variational approach [16] exists for one-dimensional equations to show that if the problem $\phi'' = -f$, $\phi(a) = A$, $\phi(b) = B$ is solved by the finite element method using *linear* interpolation functions over two-node elements, and if at least two elements are used in the solution, then, regardless of the number of elements, the discrete nodal values of ϕ from the solution lie on the exact solution $\phi(x)$. Thornton et al. [17] use the method of weighted residuals to show that for general second-order differential equations finite elements can be formulated that yield a nodeless parameter and interpolation functions based on the homogeneous and particular solutions of the governing differential equation.

3.4.6 Domain Discretization

The first task in a finite element solution consists of discretizing the continuum by dividing it into a series of elements. Underlying the discretization process is the goal of achieving a good representation of the phenomena under study. There are no set rules for reaching this goal because of the vastly different circumstances encountered from one problem to another, but some helpful guidelines emerge from the large amount of available experience in finite element analysis.

Frequently, the first questions an analyst asks are, What type of element should I use, and Should I mix several different types of elements? The answers to these questions depend on the problem being considered. Often only one type of element is used to represent the continuum, unless circumstances dictate otherwise. It is easy to imagine a problem for which several different types of elements would be necessary. An example from solid mechanics would be an elastic body supported by pin-connected bars. In this case the elastic body would be represented by three-dimensional solid elements such as "bricks," while the bars would be approximated by one-dimensional elements like those we considered in Chapter 2. The most popular and versatile elements, because of the ease with which they can be assembled to fit complex geometries, are triangular elements in two dimensions. These elements can have any number of exterior and interior nodes, depending on the type of interpolation functions defined for them. In Chapter 5 we discuss many different types of elements and their usefulness in particular problems.

A uniform element mesh is easy to construct, but it may not always provide a good representation of the continuum. In regions of the solution domain where the gradient of the field variable is expected to vary relatively quickly a finer element mesh should be used. Also, it is most convenient to place nodes and element boundaries at locations where point external actions (such as forces in structural problems) are applied and where there are abrupt changes in the continuum. If the boundary of the region has any corners, nodes should also be placed at these corners. More elements should be used in regions where the boundary is irregular than in regions where it is smooth.

Another guideline is that the continuum should be discretized so as to give the element a well-proportioned shape, or aspect ratio; in other words, the ratio of an element's smallest dimension to its largest dimension should be near unity. Long, narrow elements should be avoided because they lead to a solution with directional bias that may not be correct. For a given element in a given location in the solution domain the best or optimum ratio of extreme dimensions depends on the local gradient of the field variable, and this is unknown a priori. Hence the conservative procedure is to use elements whose aspect ratios are near one, although this is not always possible.

Provided that the elements obey the requirements for a convergent solution, we may expect that the more elements we use to model the solution domain, the better the accuracy of our results. Since increasing the number of elements leads to higher computational expenses, analysts must base their decisions on rational compromise as usual. When solving a particular type of problem for the first time, it is good practice to obtain several solutions with different numbers of elements. By comparing the results it is then possible to see whether enough elements are being used in the solution. A newly emerging approach is the concept of adaptive mesh refinement. Adaptive mesh refinement automates mesh refinement by using a solution for a given mesh and the idea of *error indicators* to generate a new mesh. The approach minimizes errors due to the mesh discretization by successively refining the mesh in the regions of steep solution gradients and derefining the mesh in regions of small gradients. We discuss and illustrate adaptive mesh refinement in Chapter 10.

Some problems have domains that are so large in relation to the area of specific interest that they are mathematically represented as *infinite domains*. Representing such a domain with a finite number of elements can present problems when the boundary conditions on the infinite boundaries are imposed at boundaries that are a finite distance from the area of interest. The usual approach to such a problem is a trial and error procedure in which a finite boundary is moved farther away from the area of interest in successive approximations until it is deemed to have a minimally acceptable effect on the solution. This normally implies the use of larger and larger elements toward the boundary, but for some problems the boundary effects are so significant that a large number of elements must be employed to reduce the effects of using a finite boundary in place of an infinite boundary. In recent years one successful approach to these types of problems has been developed using special *infinite elements* to model the infinite boundary [18].

3.4.7   Example of a Complete Finite Element Solution

In Chapter 1 we discuss in general terms the seven-step finite element solution procedure. Now that we have seen how the finite element equations for a continuum problem can be derived from variational principles, we return to these basic steps and illustrate them in some detail with an example. Recall from Chapter 1 the seven basic steps of the method:

1. Discretize the continuum.
2. Select interpolation functions.
3. Find element properties.
4. Assemble the element properties.
5. Impose the boundary conditions.
6. Solve the system of equations.
7. Make additional computations if desired.

Using each of the steps, we shall now carry out a complete finite element solution of a specific problem.

Suppose we wish to solve the following one-dimensional boundary value problem:

$$\frac{d}{dx}\left(a\frac{du}{dx}\right) - ku + b = 0, \qquad 0 < x < L$$

$$-a\frac{du}{dx}(0) = P_1 \qquad a\frac{du}{dx}(L) = -P_2 = -\hat{k}u(L) \qquad (a)$$

This governing equation physically represents a piling or caisson in the ground subjected to some axial load P at the surface and whose base at L is supported by an elastic resisting force of the soil proportional to the axial displacement at the base (see Figure 3.9). Then u is an axial displacement (length), $a = EA$ is the product of the elastic modulus (force/length^2) and the transverse cross sectional area (length^2) of the piling, k is an elastic resistance constant (force/length^2) of the soil (similar to a spring constant), $\hat{k}$ is an elastic resistance at the foot of the piling (force/length), and b is the body force or weight per unit length (force/length) of the piling. For simplicity we assume here that a, k, $\hat{k}$, and b are constants. This problem is equivalent to finding the displacement field that satisfies the boundary conditions and minimizes

$$J(u) = \frac{1}{2}\int_0^L\left[a\left(\frac{du}{dx}\right)^2 - ku^2 + 2bu\right]dx \qquad (b)$$

Consider now the application of the finite element method to approximate a solution. Our first step is to discretize the domain into elements. For a

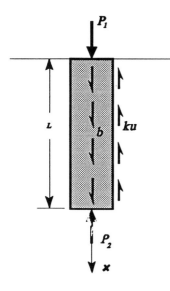

Figure 3.9 A piling with an axial loading buried in a resistive soil.

one-dimensional domain this is a simple accomplishment, and we use linear elements as developed in the example of Section 3.4.5. By choosing linear interpolation functions for each element we can uniquely and continuously represent the displacement field variable $u(x)$ in the solution domain. For one element we can write

$$u^{(e)} = \lfloor N^{(e)} \rfloor \{u^{(e)}\} = N_1 u_1 + N_2 u_2 \tag{c}$$

where $\{u^{(e)}\}$ is the column vector of nodal displacements for element (e) and the interpolation functions are those given in Section 3.4.5. Since these interpolation functions satisfy the interelement compatibility requirement (i.e., the displacement field is continuous at the element boundaries) we may focus our attention on a single element.

To derive the element equations we substitute equation c into equation b and perform the differentiation:

$$\frac{\partial J^{(e)}}{\partial u_i} = \frac{\partial J(u^{(e)})}{\partial u_i}$$

$$= \int_{L^{(e)}} \left[a \frac{\partial u^{(e)}}{\partial x} \frac{\partial}{\partial u_i} \left(\frac{\partial u^{(e)}}{\partial x} \right) - k u^{(e)} \frac{\partial u^{(e)}}{\partial u_i} + b \frac{\partial u^{(e)}}{\partial u_i} \right] dx = 0$$

$$= \int_{L^{(e)}} \left[a \left(\frac{\partial N_1}{\partial x} u_1 + \frac{\partial N_2}{\partial x} u_2 \right) \frac{\partial N_i}{\partial x} - k (N_1 u_1 + N_2 u_2) N_i + b N_i \right] dx = 0 \tag{d}$$

We may write a similar equation for $\partial J^{(e)}/\partial u_j$. The resulting two equations may be written in the matrix form

$$\begin{Bmatrix} \dfrac{\partial J^{(e)}}{\partial u_i} \\[2mm] \dfrac{\partial J^{(e)}}{\partial u_j} \end{Bmatrix} = [K^{(e)}]\{u^{(e)}\} + \{R^{(e)}\} = 0 \tag{e}$$

where the terms of the matrices $[K^{(e)}]$ and $\{R^{(e)}\}$ are given by

$$k_{ij}^{(e)} = \int_{L^{(e)}} \left(a \frac{\partial N_i}{\partial x} \frac{\partial N_j}{\partial x} - k N_i N_j \right) dx, \qquad i = 1, 2, \quad j = 1, 2$$

$$\tag{f}$$

$$r_i^{(e)} = \int_{L^{(e)}} b N_i \, dx, \qquad i = 1, 2$$

On carrying out the element integration for an element we arrive at

$$[K^{(e)}] = \begin{bmatrix} -\dfrac{3a + kL^2}{3L} & \dfrac{6a - kL^2}{6L} \\[2ex] \dfrac{6a - kL^2}{6L} & -\dfrac{3a + kL^2}{3L} \end{bmatrix} \tag{g}$$

and

$$\{r^{(e)}\} = \begin{Bmatrix} \dfrac{bL}{2} \\[2ex] \dfrac{bL}{2} \end{Bmatrix} \tag{h}$$

Now that we have determined the element there is the next step of assembling the element matrices into the global stiffness and forcing vector. This is easily accomplished with a nodal connectivity matrix whereby each local element node is shown to correspond to some a global node number. Assuming that we employ three elements in our solution scheme (see Figure 3.10), the connectivity matrix is

$$C = \begin{bmatrix} 1 & 2 \\ 2 & 3 \\ 3 & 4 \end{bmatrix} \tag{i}$$

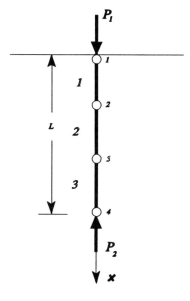

Figure 3.10 An element representation of the boundary value problem.

The rows of this connectivity matrix correspond to the element numbers, the columns correspond to the local element node numbers, and the actual values appearing in the matrix are the global node numbers. For instance, node 1 of element 2 appears at c_{21} where the c_{21} value is 2, indicating that node 1 of element 2 is global node number 2. This matrix is quite simple for our one-dimensional example and may seem unnecessary once the assembly scheme is understood. However, to instruct a computer on how to assemble the global matrices it is necessary, as will be seen in a not so trivial example in the problem set. Now the global stiffness matrix for this problem with three elements is given by

$$[K] = \begin{bmatrix} k_{11}^{(1)} & k_{12}^{(1)} & 0 & 0 \\ k_{21}^{(1)} & k_{22}^{(1)} + k_{11}^{(2)} & k_{12}^{(2)} & 0 \\ 0 & k_{21}^{(2)} & k_{22}^{(2)} + k_{11}^{(3)} & k_{12}^{(3)} \\ 0 & 0 & k_{21}^{(3)} & k_{22}^{(3)} \end{bmatrix} \tag{j}$$

and the global forcing vector by

$$\{R\} = \begin{Bmatrix} r_1^{(1)} \\ r_1^{(1)} + r_2^{(2)} \\ r_2^{(2)} + r_2^{(3)} \\ r_2^{(3)} \end{Bmatrix} \tag{k}$$

This approach to the assembly of the stiffness matrix and force vector is slightly different in notation from that in Chapter 2 in that the subscript indices here refer to the position in the original element stiffness matrices as opposed to the global stiffness matrix as before. This is a common method and is employed here to acquaint the reader with an alternative.

The matrix equations once assembled are now ready for the imposition of the boundary conditions. At global node 1 we have a positive force P_1 which we add to the force vector at that respective node, and at global node 4 we have a negative force $P_2 = \hat{k}u_4$ whose magnitude is not known but depends on the solution, i.e., the displacement at global node 4. We add that force to the force vector at global node 4 giving us a new force vector:

$$\{R\} = \begin{Bmatrix} r_1^{(1)} + P_1 \\ r_2^{(1)} + r_1^{(2)} \\ r_2^{(2)} + r_1^{(3)} \\ r_2^{(3)} - \hat{k}u_4 \end{Bmatrix} \tag{l}$$

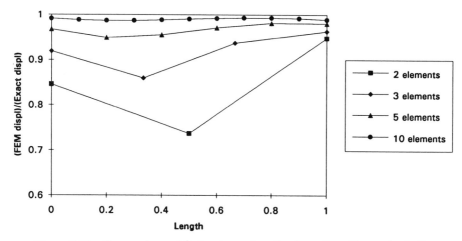

Figure 3.11   Comparison of finite element method results with exact solution.

We notice that one of our unknowns appears on the right-hand side of the system of equations, so we merely move it to the left-hand side:

$$
\begin{bmatrix}
k_{11}^{(1)} & k_{12}^{(1)} & 0 & 0 \\
k_{21}^{(1)} & k_{22}^{(1)} + k_{11}^{(2)} & k_{12}^{(2)} & 0 \\
0 & k_{21}^{(2)} & k_{22}^{(2)} + k_{11}^{(3)} & k_{12}^{(3)} \\
0 & 0 & k_{21}^{(3)} & k_{22}^{(3)} + \hat{k}
\end{bmatrix}
\begin{Bmatrix}
u_1 \\ u_2 \\ u_3 \\ u_4
\end{Bmatrix}
=
\begin{Bmatrix}
r_1^{(1)} + P_1 \\
r_2^{(1)} + r_1^{(2)} \\
r_2^{(2)} + r_1^{(3)} \\
r_2^{(3)}
\end{Bmatrix}
\quad (m)
$$

Once we solve equation m we may compute the element forces and stresses much the same as we did in the example in Chapter 2, except that here we have chosen our coordinate system so that it is not necessary to transform between global and local coordinates. We may also improve on our approximation by refining our mesh (adding more elements). Figure 3.11 shows a comparison between our finite element approximation with various meshes and the exact solution to this problem. It can readily be seen from the figure that as we increase the number of elements, the finite element approximation is converging to the exact solution.

Before leaving this boundary value problem there is a particular aspect that is worth noting: The global stiffness matrix is nonsingular and hence will result in a unique solution even before we apply any boundary conditions. The physical significance in this example is that the weight of the piling is supported by the distributed axial force due to the soil in the term ku, i.e., the piling is in equilibrium without any loading at the boundaries. This is not normally the case with most physical problems to be solved with the finite element method. If there were no soil resistance in this particular problem ($k = \hat{k} = 0$), then the stiffness matrix would be singular and an infinite number of solutions would be possible, all differing by a constant, before applying any boundary conditions. In most applications the stiffness matrix will be singular, and it will be necessary to apply at least one essential boundary condition (for one-dimensional problems) to a node to give a unique solution to the equations. Note that, unlike this example, in many problems natural boundary conditions (such as forces) only modify the right-hand side of the equation and do not affect the stiffness matrix.

3.5   CLOSURE

In this chapter we laid the foundation for deriving element equations from variational principles. We presented (1) the nature of continuum problems, (2) the variational approach to their solution, and (3) the relation of a well-known variational method, the Ritz method, to the finite element method. Taking a variational approach to the finite element method allowed us to extend the method to the broad class of problems governed by classical variational principles. We saw that the interpolation functions defined over each element give a piecewise representation of the unknown field variable, and as examples we used one- and two-dimensional interpolation functions for two simple elements. In Chapter 5 we consider a wide variety of elements and interpolation functions. From the variational approach, if general requirements are satisfied, we can be sure that refining the element mesh leads to a series of approximate solutions that approach the exact solution.

Although the variational approach allows us to treat many problems of practical interest, some problems do not have equivalent variational statements and cannot be treated in this way. In the next chapter we develop a more general approach to the formulation of finite element equations. This approach can be used when the variational approach cannot, and it opens the way for the application of finite element techniques in many areas not envisioned in the earlier development of the method.

REFERENCES

1. P. M. Morse and H. Feshback, *Methods of Theoretical Physics*, Vol. 1, McGraw-Hill, New York, 1953.
2. I. G. Petrovsky, *Lectures on Partial Differential Equations*, Wiley-Interscience, New York, 1954.
3. S. H. Crandall, *Engineering Analysis*, McGraw-Hill, New York, 1956.
4. E. F. Beckenbach (ed.), *Modern Mathematics for the Engineer*, McGraw-Hill, New York, 1956.
5. G. E. Forsythe and W. R. Wasow, *Finite Difference Methods for Partial Differential Equations*, Wiley, New York, 1960.
6. B. A. Finlayson and L. E. Scriven, "The Method of Weighted Residuals—A Review," *Appl. Mech. Rev.*, Vol. 19, No. 9, September 1966, pp. 735–748.

7. B. A. Finlayson, *The Method of Weighted Residuals and Variational Principles*, Academic, New York, 1972.

8. F. B. Hildebrand, *Methods of Applied Mathematics*, Prentice-Hall, Englewood Cliffs, NJ, 1965.

9. T. Mura and T. Koya, *Variational Methods in Mechanics*, Oxford University Press, New York, 1992.

10. J. N. Reddy, *Applied Functional Analysis and Variational Methods in Engineering*, McGraw-Hill, New York, 1986.

11. O. C. Zienkiewicz, *The Finite Element Method*, 3d ed., McGraw-Hill, New York, 1977.

12. C. A. Felippa and R. W. Clough, "The Finite Element Method in Solid Mechanics," in *SIAM–AMS Proceedings*, Vol. 2, American Mathematical Society, Providence, RI, 1970, pp. 210–252.

13. E. R. A. Oliveira, "Theoretical Foundations of the Finite Element Method," *Int. J. Solids Struct.*, Vol. 4, 1968, pp. 929–952.

14. T. H. H. Pian and P. Tong, "Basis of Finite Element Methods for Solid Continua," *Int. J. Numer. Methods Eng.*, Vol. 1, 1969, pp. 3–28.

15. G. P. Bazeley, Y. K. Cheung, B. M. Irons, and O. C. Zienkiewicz, "Triangular Elements in Plate Bending: Conforming and Nonconforming Solution," *Proceedings of the 1st Conference on Matrix Methods in Structural Mechanics*, Wright-Patterson Air Force Base, Dayton, OH, 1965.

16. P. Tong, "Exact Solution of Certain Problems by the Finite Element Method," *AIAA J.*, Vol. 7, No. 1, 1969, pp. 179–180.

17. E. A. Thornton, P. Dechaumphai, and K. K. Tamma, "Exact Finite Element Solutions for Linear Steady-State Thermal Problems," Chapter 2 in Numerical *Methods in Heat Transfer*, Vol. II, edited by R. W. Lewis, K. Morgan, and B. A. Schrefler, Wiley, New York, 1983.

18. P. Bettess, *Infinite Elements*, Penshaw Press, Sunderland, UK, 1992.

PROBLEMS

1. Use the method of Section 3.2.3 to classify the following partial differential equations:

a. $\dfrac{\partial^2 \phi}{\partial x^2} = \alpha \dfrac{\partial^2 \phi}{\partial y^2}$ (α = constant)

b. $\dfrac{\partial^2 \phi}{\partial x^2} + \dfrac{\partial^2 \phi}{\partial y^2} = 0$

c. $\dfrac{\partial^2 \phi}{\partial x^2} = \beta \dfrac{\partial \phi}{\partial y}$ (β = constant)

d. $\dfrac{\partial^2 \phi}{\partial x^2} - 2\dfrac{\partial^2 \phi}{\partial x \, \partial y} + \dfrac{\partial^2 \phi}{\partial y^2} = 0$

e. $\dfrac{\partial^2 \phi}{\partial x^2} + 2x\dfrac{\partial^2 \phi}{\partial x \, \partial y} + (1 - y^2)\dfrac{\partial^2 \phi}{\partial y^2} = 0$

2. A tapered bar (Figure P3.2) is subjected to an axial load P. The potential energy of the system is

$$\Pi = \frac{1}{2} \int_0^L EA(x) \left(\frac{du}{dx}\right)^2 dx - Pu(L)$$

where E is the modulus of elasticity, $A(x)$ is the cross-sectional area, L is the bar length, and $u(x)$ is the axial displacement. Use the principal of minimum potential energy (Appendix C) and calculus of variations (Appendix B) to derive the governing differential equation and the essential and natural boundary conditions for the bar.

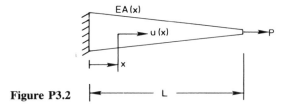

Figure P3.2

3. Use the Ritz method (Section 3.3.3) to obtain a solution to

$$\frac{d^2\phi}{dx^2} = -k, \qquad 0 < x < L$$

$$\phi(0) = 0 \qquad \phi(L) = 1$$

where k is a constant. Assume a trial function

$$\tilde{\phi}_2 = \frac{x}{L} + x(x - L)C_1 + x^2(x - L)C_2$$

Compare the Ritz solution to the exact solution.

4. Find finite element solutions to Problem 3 using
   a. One linear element
   b. Two linear elements
   c. Three linear elements

Compare these solutions with the exact solutions. Also compare $d\phi/dx$ from the finite element approximations in part c at the two interior nodes with the exact values. Note that you will have two values $d\phi/dx$ at each of the two interior nodes. Why?

5. Three nonuniform bars (Figure P3.5) are in equilibrium under the action of the forces shown. The potential energy of the system is

$$\Pi = \tfrac{1}{2}k_1 u_1^2 + \tfrac{1}{2}k_2(u_1 - u_1)^2 + \tfrac{1}{2}k_3 u_2^2 - F_1 u_1 - F_2 u_2$$

where for $i = 1, 2, 3$, and k_i are stiffnesses of each bar. The nodal displacements are u_1 and u_2. Minimize this functional with respect to the nodal displacements and write the system equilibrium equations. Identify the system stiffness matrix $[K]$ and load vector $\{F\}$.

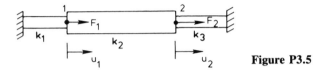

Figure P3.5

6. Use the variational statement for the tapered bar in Problem 2 to derive an equation for the bar stiffness matrix $[K]^{(e)}$ for linear representation of the displacement over a two-node element.

7. Follow the approach used in the example of Section 3.4.5 to derive the equations for $[K]^{(e)}$ and $\{F\}^{(e)}$ when quadratic interpolation functions (Figure 3.7) are used for a three-node line element. Assume that the element is of length L and that node 2 is located at the element centroid.

8. Bending of a beam is governed by the fourth-order equation

$$EI\frac{d^4 w}{dx^4} = p(x)$$

in the region $a \leq x \leq b$, where E is the modulus of elasticity, I is the moment of inertia of the beam cross-sectional area, and $p(x)$ is the transverse loading per unit length. The associated variational formulation is

$$I(w) = \int_a^b \left[\frac{EI}{2}\left(\frac{d^2 w}{dx^2}\right)^2 - pw \right] dx$$

Discuss the selection of an interpolation function for a beam element and identify the nodal degrees of freedom.

9. For a beam element (Figure P3.9) we assume that the displacement interpolation is

$$w(x) = w_1 N_1(x) + \frac{dw_1}{dx}N_2(x) + w_2 N_3(x) + \frac{dw_2}{dx}N_4(x)$$

where

$$N_1(x) = 1 - 3\left(\frac{x}{L}\right)^2 + 2\left(\frac{x}{L}\right)^3 \qquad N_3(x) = \left(\frac{x}{L}\right)^2\left(3 - 2\frac{x}{L}\right)$$

$$N_2(x) = x\left(\frac{x}{L} - 1\right)^2 \qquad\qquad N_4(x) = \frac{x^2}{L}\left(\frac{x}{L} - 1\right)$$

Use the variational formulation presented in Problem 8 to show that the beam element stiffness matrix is

$$[K]^{(e)} = \frac{EI}{L^2}\begin{bmatrix} 12 & 6L & -12 & 6L \\ 6L & 4L^2 & -6L & 2L^2 \\ -12 & -6L & 12 & -6L \\ 6L & 2L^2 & -6L & 4L^2 \end{bmatrix}$$

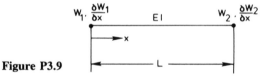

Figure P3.9

10. For the beam element of Problem 9 assume a uniform element loading $p(x) = p$ and derive the element load vector $\{F\}^{(e)}$.

4

THE MATHEMATICAL APPROACH: A GENERALIZED INTERPRETATION

4.1 INTRODUCTION

In chapter 3 we discussed the well-known Ritz Method as a global method and how it may be applied locally on discrete portions of the global domain to derive the equations of the finite element method. This enabled us to view the finite element discretization procedure as simply another means of finding approximate solutions to variational problems. In fact, we saw how the finite element equations are derived by requiring that a given functional be stationary. This broad variational interpretation is widely used to derive element equations, and it is a most convenient approach whenever a classical variational statement exists for a given problem. However, applied scientists and engineers encounter practical problems for which classical variational principles are unknown. In these cases finite element techniques are still applicable, but more generalized procedures must be used to derive the element equations.

In this chapter we introduce these generalizations and show how finite element equations may be derived directly from the governing differential equations of the problem without reliance on any classical, quasivariational, or restricted variational principles. We discuss the method of weighted residuals, which is a global method but which may be employed locally at a discrete elemental level so that the finite element method may be applied to almost all practical problems of mathematical physics.

4.2 DERIVING FINITE ELEMENT EQUATIONS FROM THE METHOD OF WEIGHTED RESIDUALS

The method of weighted residuals (MWR) is a global technique for obtaining approximate solutions to linear and nonlinear partial differential equations. It has nothing to do with the finite element method other than offering another means with which to formulate the element equations. In this section we review the general method of weighted residuals and show how one particular technique can be used to derive finite element equations. References 3 and 7 of Chapter 3 are good sources of more detailed information on general weighted residual techniques.

Applying the method of weighted residuals involves basically two steps. The first step is to assume the general functional behavior of the dependent field variable in some way as to satisfy the given differential equation and boundary conditions. Substitution of this approximation function into the differential equation and boundary conditions then results in some error called a *residual*. This residual is required to vanish in some average sense over the entire solution domain.

The second step is to solve the equation (or equations) resulting from the first step and thereby specialize the general functional form to a particular function, which then becomes the approximate solution sought.

To be more specific we consider a typical problem. Suppose that we want to find an approximate functional representation for a field variable ϕ governed by the differential equation

$$A(\phi) - f = 0 \qquad (4.1)$$

in the domain Ω bounded by the surface Γ. The function f is a known function of the independent variables, and we assume that proper boundary conditions are prescribed on Γ.

The method of weighted residuals is applied in two steps as follows. First, the unknown exact solution ϕ is approximated by $\tilde{\phi}$, where either the functional behavior of $\tilde{\phi}$ is completely specified in terms of the unknown parameters, or the functional dependence on all but one of the independent variables is specified while the remaining independent variable is left unspecified. Thus the dependent variable is approximated by

$$\phi \approx \tilde{\phi} = \sum_{i=1}^{m} N_i C_i \qquad (4.2)$$

where the N_i are the assumed functions, and the C_i are either the unknown parameters or unknown functions of one of the independent variables. Typically for steady-state problems the C_i are constants, and for unsteady problems the C_i are functions of time. The upper limit of the summation, m, is the number of unknowns C_i. The m functions N_i are usually chosen to satisfy the global boundary conditions.

When $\tilde{\phi}$ is substituted into equation 4.1, it is unlikely that the equations will be satisfied, that is,

$$A(\tilde{\phi}) - f \neq 0$$

In fact,

$$A(\tilde{\phi}) - f = R$$

where R is the residual or error that results from approximating ϕ by $\tilde{\phi}$. The method of weighted residuals seeks to determine the m unknown C_i in such a way that the residual R over the entire solution domain is small. This is accomplished by forming a weighted average of the error and specifying that this weighted average vanish over the solution domain. Hence we choose m linearly independent weighting functions W_i and then insist that if

$$\int_\phi \left[A(\tilde{\phi}) - f \right] W_i \, d\Omega = \int_\Omega R W_i \, d\Omega = 0, \qquad i = 1, 2, \ldots, m \qquad (4.3)$$

then $R \approx 0$ in some sense.

The form of the error distribution principle expressed in equation 4.3 depends on our choice for the weighting functions. Once we specify the weighting functions, equation 4.3 represents a set of m equations, either algebraic equations or ordinary differential equations to be solved for the C_i.[1] The second step, then, is to solve equation 4.3 for the C_i and hence obtain an approximate representation of the unknown field variable ϕ via equation 4.2. There are many linear problems and even some nonlinear problems for which it can be shown that as $m \to \infty$, $\tilde{\phi} \to \phi$.

We have a variety of weighted residual techniques because of the broad choice of weighting functions or error distribution principles that we can use [1, 2]. The error distribution principle most often used to derive finite element equations is known as the *Galerkin method*. According to the *Bubnov–Galerkin* method, the weighting functions are chosen to be the same as the approximating functions used to represent ϕ, that is $W_i = N_i$ for $i = 1, 2, \ldots, m$. There are many other choices for weighting functions in which $W_i \neq N_i$. These choices of weighting functions are collectively referred

---

[1]Ordinary differential equations result when the dependent variable is a function of spatial variables and time. In this case the C_i are functions of time, $C_i(t)$.

to as the *Petrov–Galerkin* method, but a number of specific choices of weighting functions under this general heading have proven popular enough to bear their own names. For example, for $W_i = \partial R / \partial C_i$ where R is the residual, the method is called the *least-squares* method, and for $W_i = \delta(x - x_i)$, where $\delta(\ )$ is the Dirac delta function, the method is called the *collocation method*. Applications of the Petrov–Galerkin method where W_i denote *upwind weighting functions* appear in Chapter 8. Here we employ the Bubnov–Galerkin method, which we call simply the Galerkin method. Thus the Galerkin method requires that

$$\int_\Omega \left[A(\tilde{\phi}) - f \right] N_i \, d\Omega = 0, \qquad i = 1, 2, \ldots, m \tag{4.4}$$

In the preceding discussion we assumed that we were dealing with the entire solution domain. However, because equation 4.1 holds for any point in the solution domain, it also holds for any collection of points defining an arbitrary subdomain or element of the whole domain. For this reason we may focus our attention on an individual element and define a local approximation analogous to equation 4.2 and valid for only one element at a time. Now the familiar finite element representations of a field variable become available. The functions N_i are recognized as the interpolation functions $N_i^{(e)}$ defined over the element, and the C_i are the undetermined parameters, which may be the nodal values of the field variable or its derivatives. Then from Galerkin's method we can write the equations governing the behavior of an element as

$$\int_{\Omega^{(e)}} \left[A(\phi^{(e)}) - f^{(e)} \right] N_i^{(e)} \, d\Omega = 0, \qquad i = 1, 2, \ldots, r \tag{4.5}$$

where, as before, the superscript (e) restricts the range to one element, and

$\phi^{(e)} = \lfloor N^{(e)} \rfloor \{ \phi \}^{(e)}$

$f^{(e)} = $ forcing function defined over element (e)

$r = $ number of unknown parameters assigned to the element

We have a set of equations like equation 4.5 for each element of the whole assemblage. Before we can assemble the system equations from the element equations we require that our choice of approximating functions N_i guarantee the interelement continuity necessary for the assembly process. Recall that the assembly process does not allow any spurious contributions if we choose interpolation functions that ensure continuity of ϕ at the element boundaries, as well as continuity of its partial derivatives up to one order less than the highest-order derivative appearing in the expression to be integrated. Since the differential operator A in the integrand usually has derivatives of higher order than occur in the integrand resulting from a variational principle (when one exists), we can see that the Galerkin method can lead to a more stringent choice of interpolation functions than we have previously encountered in the variational approach. As we shall see in Chapter 5, the higher the order of continuity we require of the interpolations, the narrower our choice of functions becomes. Many interpolation functions provide continuity of value, fewer provide continuity of slope, and only several can ensure continuity of curvature.

Often there is a way to escape this dilemma by changing the form of equation 4.5. By applying integration by parts [3] to the integral expression of equation 4.5, we can obtain expressions containing lower-order derivatives, and hence we can use approximating functions with lower-order interelement continuity. When integration by parts is possible, it also offers a convenient way to introduce the natural boundary conditions that must be satisfied on some portion of the boundary. Although the boundary terms containing the natural boundary conditions appear in the equations for each element, in the assembly of the element equations only the boundary elements give nonvanishing contributions. After the assembly process the fixed boundary conditions (for example, fixed displacement or specified temperature) are introduced in the manner described in Chapter 2.

For convenience of reference, we summarize the integration by parts equations used in finite element formulations with Galerkin's method. In one dimension, where the domain of interest is $a \leq x \leq b$, integration by parts is

$$\int_a^b u \, dv = uv \big|_a^b - \int_a^b v \, du \qquad (4.6)$$

where in equation 4.3 we identify u as the weighting function W_i, and dv with derivatives in the differential operator A. Integration by parts in two dimensions is known as Green's theorem; integration by parts in three dimensions is known as Gauss's theorem [3]. For a two- or three-dimensional domain Ω with boundary Γ, integration by parts is

$$\int_\Omega u(\nabla \cdot \mathbf{v}) \, d\Omega = \int_\Gamma u(\mathbf{v} \cdot \hat{n}) \, d\Gamma - \int_\Omega \mathbf{v} \cdot \nabla u \, d\Omega \qquad (4.7)$$

where, as in one dimension, we identify u as the weighting function W_i and $\nabla \cdot \mathbf{v}$ with derivatives in the differential operator A. In equation 4.7 ∇ is the gradient operator,

$$\nabla = \frac{\partial}{\partial x} \hat{i} + \frac{\partial}{\partial y} \hat{j} + \frac{\partial}{\partial z} \hat{k}$$

and $\hat{n}$ is a unit normal vector to the boundary,

$$\hat{n} = n_x \hat{i} + n_y \hat{j} + n_z \hat{k}$$

The immediate result of integration by parts is that the order of differentiation on $\tilde{\phi}$ is reduced by one in the resulting equation. This means that the continuity restrictions on the approximating function $\tilde{\phi}$ have been reduced or *weakened*. The resulting integral equation is referred to as the *weak form* of the boundary value problem. Other benefits are also gained from the integration by parts, as will be seen in several examples given to illustrate and clarify these general ideas.

4.2.1 Example: One-Dimensional Poisson Equation

To illustrate the method of weighted residuals we consider the one-dimensional Poisson equation previously considered by the variational approach in Section 3.4.5. The problem is to find a function $\phi(x)$ satisfying

$$\frac{d^2\phi}{dx^2} = -f(x) \tag{4.8a}$$

with boundary conditions $\phi(a) = A$, $\phi(b) = B$. In Section 3.4.5 the element matrices $[K]^{(e)}$ and $\{F\}^{(e)}$ were derived for a linear representation of $\phi(x)$ over a one-dimensional element with two nodes. Here we derive the same element matrices but use Galerkin's method with weighting functions $W_i = N_i$.

We first approximate the unknown exact solution $\phi(x)$ by $\tilde{\phi}(x)$, which has the form

$$\tilde{\phi} = \sum_{i=1}^{m} N_i(x)\phi_i$$

where the element interpolation functions N_i appear in Figure 3.6 and for an element with two nodes, $m = 2$. The unknown nodal values are ϕ, $i = 1, 2$. We do not consider the fixed boundary conditions at the element level; instead, these are included, as before, after the assembly process. As we shall see, the natural boundary conditions emerge automatically in the formulation of the element equations.

Applying a weighting function to equation 4.8a and casting it in integral form gives

$$\int_{x_1}^{x_2} \left[\frac{d^2\tilde{\phi}}{dx^2} + f(x) \right] N_i(x) \, dx = 0, \qquad i = 1, 2 \tag{4.8b}$$

where x_1 and x_2 are the coordinates of the end nodes of the line element. If we apply integration by parts, equation 4.6, to the term with the derivatives of $\tilde{\phi}$, we obtain

$$N_i \frac{d\tilde{\phi}}{dx}\Big|_{x_1}^{x_2} - \int_{x_1}^{x_2} \frac{d\tilde{\phi}}{dx} \frac{dN_i}{dx} dx + \int_{x_1}^{x_2} f(x) N_i(x) \, dx = 0, \qquad i = 1, 2 \quad (4.8c)$$

This equation is the weak form of equation 4.8a. Notice especially that ϕ must have C^1 continuity in the original boundary value problem as must $\tilde{\phi}$ in the Galerkin integral of equation 4.8b, but in the weak form the continuity requirement on $\tilde{\phi}$ has been reduced to C^0. We also note that integration by parts has brought the natural boundary conditions into the formulation (the first term of equation 4.8c). Further, for this particular problem, the integral term containing the derivatives of $\tilde{\phi}$ and N_i is now symmetric in $\tilde{\phi}$ and N_i, which will greatly reduce the numerical effort required to solve the resulting equations. We make these points now because of their immense importance, which will become much clearer later.

Utilizing our definition of $\tilde{\phi}$, we have

$$\frac{d\tilde{\phi}}{dx} = \sum_{i=1}^{m} \frac{dN_i}{dx} \phi_i = \left\lfloor \frac{dN}{dx} \right\rfloor \{\phi\}^{(e)}$$

where $\{\phi\}^{(e)}$ is the column vector of nodal unknowns for the element. Hence the Galerkin criterion becomes

$$\int_{x_1}^{x_2} \left\lfloor \frac{dN}{dx} \right\rfloor \frac{dN_i}{dx} dx \{\phi\}^{(e)} = N_i \frac{d\tilde{\phi}}{dx}\Big|_{x_1}^{x_2} + \int_{x_1}^{x_2} f(x) N_i(x) \, dx, \qquad i = 1, 2 \quad (4.8d)$$

The first term on the right-hand side of the equation represents natural boundary conditions for the element. We evaluate these as

$$i = 1, \qquad N_i \frac{d\tilde{\phi}}{dx}\Big|_{x_1}^{x_2} = N_1(x_2)^0 \frac{d\tilde{\phi}}{dx}(x_2) - N_1(x_1)^1 \frac{d\tilde{\phi}}{dx}(x_1) = -\frac{d\tilde{\phi}}{dx}(x_1)$$

$$i = 2, \qquad N_i \frac{d\tilde{\phi}}{dx}\Big|_{x_1}^{x_2} = N_2(x_2)^1 \frac{d\tilde{\phi}}{dx}(x_2) - N_2(x_1)^0 \frac{d\tilde{\phi}}{dx}(x_1) = \frac{d\tilde{\phi}}{dx}(x_2)$$

where we use the end-point values of N_i shown in Figure 3.6. The element equations are then

$$\begin{bmatrix} K_{11} & K_{12} \\ K_{21} & K_{22} \end{bmatrix}^{(e)} \begin{Bmatrix} \phi_1 \\ \phi_2 \end{Bmatrix}^{(e)} = \begin{Bmatrix} -\dfrac{d\tilde{\phi}}{dx}(x_1) \\[2mm] \dfrac{d\tilde{\phi}}{dx}(x_2) \end{Bmatrix}^{(e)} + \begin{Bmatrix} F_1 \\ F_2 \end{Bmatrix}^{(e)} \qquad (4.8e)$$

where

$$[K]^{(e)} = \int_{x_1}^{x_2} \begin{bmatrix} \dfrac{dN_1}{dx} \dfrac{dN_1}{dx} & \dfrac{dN_1}{dx} \dfrac{dN_2}{dx} \\[2mm] \dfrac{dN_2}{dx} \dfrac{dN_1}{dx} & \dfrac{dN_2}{dx} \dfrac{dN_2}{dx} \end{bmatrix} dx$$

$$\{F\}^{(e)} = \int_{x_1}^{x_2} \begin{Bmatrix} fN_1 \\ fN_2 \end{Bmatrix} dx$$

This set of equations expresses the behavior of a line element of two nodes. Although we derived equation 4.8e for an element with two nodes, the extension to an element with m nodes follows the same steps, but with $i = 1, 2, \ldots, m$. The matrices for an element with m nodes contain terms similar to equation 4.8e, but with additional rows and columns to account for m element equations.

The element matrices $[K]^{(e)}$ and $\{F\}^{(e)}$ are identical to the element matrices derived in Section 3.4.5 by the variational method. Explicit evaluation of these equations is possible after we specify the interpolation functions N_i. The natural boundary conditions are taken into account when we assemble the element matrices. During assembly of the matrices the natural boundary condition terms $d\tilde{\phi}/dx$ will cancel at all interior nodes of the solution domain, leaving only the natural boundary conditions to be evaluated at points a and b. If the value of ϕ is specified at an end point, for example, $\phi(a) = A$, then $d\phi(a)/dx$ is unknown; if the value of $d\phi(a)/dx$ is specified, then $\phi(a)$ is unknown. In a numerical solution of a problem we consider the boundary conditions by one of the methods described in Section 2.3.4.

In this example we saw the three important roles played by integration by parts: First, it brought into effect the boundary information; second, it lowered to one the order of the highest-order derivative appearing in the integrands of the matrix $[K]^{(e)}$, and, third, for this even-ordered differential equation, it made the matrix $[K]^{(e)}$ symmetric.

Another important observation can also be made. In this example we derived element matrices for a two-node line element and found that their forms were identical to matrices from a variational principle. This special result follows from a far more general relationship. It can be shown that for linear self-adjoint operators A, application of a classical variational principle or Galerkin's method leads to identical calculations. A thorough discussion of this equivalence can be found in Finlayson and Scriven [1] and in the references cited therein.

The equivalence of the classical variational approach and the Galerkin approach to deriving finite element equations has an advantageous interpretation: By means of the Galerkin approach we can derive finite element equations for a given problem from the governing equations and boundary conditions, regardless of whether we are familiar with variational calculus. When a classical variational principle exists for a given problem, we obtain the same results. When a classical variational principle is unknown or does not exist, we may still proceed to derive a finite element model.

4.2.2 Example: Two-Dimensional Heat Conduction

As a further illustration of the use of Galerkin's method in deriving element equations, we shall consider the problem of steady two-dimensional heat conduction. The governing differential equation may be expressed in the general form as

$$\frac{\partial}{\partial x}\left(k_x\frac{\partial T}{\partial x}\right) + \frac{\partial}{\partial y}\left(k_y\frac{\partial T}{\partial y}\right) + \tilde{Q} = 0 \tag{4.9a}$$

where k_x and k_y are the thermal conductivities in the x and y directions, respectively, and $\tilde{Q}$ is the internal heat generation; these are known, specified functions of x and y. On the boundary of the domain Ω we have the following general boundary conditions (see Figure 4.1):

$$T = T(x, y) \qquad \text{on } S_1$$

$$k_x\frac{\partial T}{\partial x}n_x + k_y\frac{\partial T}{\partial y}n_y + q + h(T - T_\infty) = 0 \qquad \text{on } S_2 \tag{4.9b}$$

where $T(x, y)$ is a specified boundary temperature distribution, n_x and n_y are the direction cosines of the outward normal vector $\hat{n}$ to the bounding curve Γ, q is the heat loss at the boundary due to conduction, and $h(T - T_\infty)$ is the heat loss at the boundary due to convection to ambient temperature T_∞ with convection heat transfer coefficient h. In this example we compare the

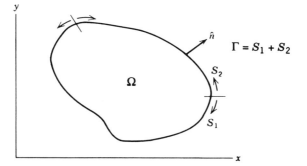

Figure 4.1 Two-dimensional domain for the heat conduction equation.

procedures of the variational method and the Galerkin method by deriving finite element equations in both ways.

Variational Method   From the variational techniques we may show that the function $T(x, y)$ that satisfies equations 4.9a and 4.9b also minimizes the functional

$$I(T) = \iint_{\Omega} \left[\frac{k_x}{2} \left(\frac{\partial T}{\partial x} \right)^2 + \frac{k_y}{2} \left(\frac{\partial T}{\partial y} \right)^2 - \tilde{Q}T \right] dx \, dy$$

$$+ \int_{S_2} \left(qT + \frac{1}{2} hT^2 - hTT_{\infty} \right) d\Gamma \tag{4.9c}$$

Suppose that the domain Ω is divided into polygonal elements and that the distribution of T within each element is assumed to be

$$T^{(e)}(x, y) = \sum_{i=1}^{r} N_i(x, y) T_i = \lfloor N \rfloor \{T\}^{(e)} \tag{4.9d}$$

where r is the number of nodes assigned to element (e), and the T_i are the discrete nodal temperatures. By substituting equation 4.9d into equation 4.9c we obtain for one element the discretized functional $I(T^{(e)})$ expressed in terms of the discrete nodal temperatures. Requiring that $I(T^{(e)})$ be a minimum is equivalent to requiring that

$$\frac{\partial I(T^{(e)})}{\partial T_i} = 0, \qquad i = 1, 2, \ldots, r$$

Hence for a typical node i we have

$$\frac{\partial I(T^{(e)})}{\partial T_i} = 0 = \iint_{\Omega^{(e)}} \left[k_x \frac{\partial T^{(e)}}{\partial x} \frac{\partial}{\partial T_i} \left(\frac{\partial T^{(e)}}{\partial x} \right) \right.$$

$$\left. + k_y \frac{\partial T^{(e)}}{\partial y} \frac{\partial}{\partial T_i} \left(\frac{\partial T^{(e)}}{\partial y} \right) - \tilde{Q} \frac{\partial T^{(e)}}{\partial T_i} \right] dx \, dy$$

$$+ \int_{S_2^{(e)}} \left(q \frac{\partial T^{(e)}}{\partial T_i} + hT^{(e)} \frac{\partial T^{(e)}}{\partial T_i} - hT_{\infty} \frac{\partial T^{(e)}}{\partial T_i} \right) d\Gamma \tag{4.9e}$$

The surface integral in equation 4.9e has nonzero values only for elements whose boundaries are part of the domain boundary segment S_2 or for elements that have external nodal heat fluxes specified via q.

Utilizing equation 4.9d we may rewrite equation 4.9e as

$$0 = \iint_{\Omega^{(e)}} \left(k_x \left\lfloor \frac{\partial N}{\partial x} \right\rfloor \{T\}^{(e)} \frac{\partial N_i}{\partial x} + k_y \left\lfloor \frac{\partial N}{\partial y} \right\rfloor \{T\}^{(e)} \frac{\partial N_i}{\partial y} - \tilde{Q} N_i \right) dx\,dy$$

$$+ \int_{S_2^{(e)}} \left(q N_i + h \lfloor N \rfloor \{T\}^{(e)} N_i - h T_\infty N_i \right) d\Gamma, \qquad i = 1, 2, \ldots, r$$

or, in matrix notation, we have for the entire element

$$[K_c]^{(e)} \{T\}^{(e)} = \{Q\}^{(e)} - \{q\}^{(e)} - [K_h]^{(e)} \{T\}^{(e)} + \{q_{T_\infty}\}^{(e)} \qquad (4.9f)$$

where

$$K_{c_{ij}} = \iint_{\Omega^{(e)}} \left(k_x \frac{\partial N_i}{\partial x} \frac{\partial N_j}{\partial x} + k_y \frac{\partial N_i}{\partial y} \frac{\partial N_j}{\partial y} \right) dx\,dy$$

$$Q_i = \iint_{\Omega^{(e)}} \tilde{Q} N_i \, dx\,dy$$

$$q_i = \int_{S_2^{(e)}} q N_i \, d\Gamma$$

$$K_{h_{ij}} = \int_{S_2^{(e)}} h N_i N_j \, d\Gamma, \qquad q_{T_{\infty_i}} = \int_{S_2^{(e)}} h T_\infty N_i \, d\Gamma$$

Equation 4.9f expresses the general behavior of a two-dimensional thermal element. After we select the element shape and an appropriate set of interpolation functions we can explicitly evaluate these element equations and then routinely assemble them for an aggregate of elements representing the whole solution domain. Nodal temperatures result from the simultaneous solution of the system equations.

We note that since the functional contains only first-order derivatives of the temperature, our approximating functions need only guarantee continuity of temperature across interelement boundaries, and they must be able to represent constant values of both first derivatives within any element.

Galerkin's Method   To derive element equations for equations 4.9a and 4.9b by Galerkin's method we first express the approximate behavior of the temperature within each element according to equation 4.9d. Then, applying Galerkin's method, we may write

$$\iint_{\Omega^{(e)}} N_i \left[\frac{\partial}{\partial x} \left(k_x \frac{\partial T^{(e)}}{\partial x} \right) + \frac{\partial}{\partial y} \left(k_y \frac{\partial T^{(e)}}{\partial y} \right) + \tilde{Q} \right] dx\,dy = 0, \qquad i = 1, 2, \ldots, r$$

$$(4.10a)$$

Equation 4.10a expresses the desired averaging of the error or residual within the element boundaries, but it does not admit the influence of the boundary. We use integration by parts to reduce the order of the derivatives in equation 4.10a and to introduce the influence of the natural boundary conditions.[2]

Focusing our attention on the first two terms in equation 4.10a, we integrate by parts using equation 4.7 with $u = N_i$ and

$$\mathbf{v} = k_x \frac{\partial T^{(e)}}{\partial x} \hat{i} + k_y \frac{\partial T^{(e)}}{\partial y} \hat{j}$$

The result is the symmetric weak form

$$-\iint_{\Omega^{(e)}} \left(k_x \frac{\partial T^{(e)}}{\partial x} \frac{\partial N_i}{\partial x} + k_y \frac{\partial T^{(e)}}{\partial y} \frac{\partial N_i}{\partial y} \right) dx\,dy + \iint_{\Omega^{(e)}} N_i \tilde{Q}\, dx\,dy$$

$$+ \int_{S_2^{(e)}} \left(k_x \frac{\partial T^{(e)}}{\partial x} n_x + k_y \frac{\partial T^{(e)}}{\partial y} n_y \right) N_i\, d\Gamma = 0 \qquad (4.10b)$$

The surface integral (boundary residual) in equation 4.10b now enables us to introduce the natural boundary conditions of equation 4.9b because, for elements on the boundary of Ω,

$$k_x \frac{\partial T^{(e)}}{\partial x} n_x + k_y \frac{\partial T^{(e)}}{\partial y} n_y = -q^{(e)} - h \left(T^{(e)} - T_\infty \right) \quad \text{on } S_2$$

Hence, noting that $T^{(e)} = \lfloor N \rfloor \{T\}^{(e)}$, we may write equation 4.10b as

$$\iint_{\Omega^{(e)}} \left(k_x \left\lfloor \frac{\partial N}{\partial x} \right\rfloor \{T\}^{(e)} \frac{\partial N_i}{\partial x} + k_y \left\lfloor \frac{\partial N}{\partial y} \right\rfloor \{T\}^{(e)} \frac{\partial N_i}{\partial y} \right) dx\,dy$$

$$- \iint_{\Omega^{(e)}} N_i \tilde{Q}\, dx\,dy + \int_{S_2^{(e)}} \left(q N_i + h \lfloor N \rfloor \{T\}^{(e)} N_i - h T_\infty N_i \right) d\Gamma = 0$$

which we recognize to be identical to equation 4.9f, derived by the variational method.

These examples illustrate the use of Galerkin's method in deriving finite element equations from a governing differential equation and its boundary conditions. The resulting element equations are linear algebraic equations identical to those derived from a variational principle, because each of the differential operators we considered was linear and self-adjoint. In each case we had to apply integration by parts to invoke the boundary conditions. It is

---

[2] Fixed boundary conditions are introduced in the usual manner after assembly of the element equations.

important to note that the procedures used in these examples of linear problems apply equally well to nonlinear problems and to problems containing more than one dependent variable. When the independent variable is a vector, such as the displacement field in an elasticity problem, we treat each component of the vector in the same way as we treated a scalar function in the foregoing.

4.2.3   Example: Time-Dependent Heat Conduction

The steady-state equations in the two preceding examples are elliptic differential equations. When time is considered, the governing differential equations become parabolic, and no classical variational principle can be applied.

The differential equation governing two-dimensional transient heat conduction may be expressed in the general form as

$$\frac{\partial}{\partial x}\left(k_x \frac{\partial T}{\partial x}\right) + \frac{\partial}{\partial y}\left(k_y \frac{\partial T}{\partial y}\right) + \tilde{Q} - \rho c \frac{\partial T}{\partial t} = 0 \qquad (4.11a)$$

with boundary and initial conditions

$$T = T(x, y, t) \quad \text{on } S_1, \qquad t > 0$$

$$k_x \frac{\partial T}{\partial x}n_x + k_y \frac{\partial T}{\partial y}n_y + q + h(T - T_\infty) = 0 \quad \text{on } S_2, \qquad t > 0 \quad (4.11b)$$

$$T = T_0(x, y) \quad \text{in } \Omega, \qquad t = 0$$

where the nomenclature of equation 4.9a applies, and T_0 is the initial condition of the system. In equation 4.11a, ρ is the material density and c is the specific heat.

Since the temperature distribution is a function of both space and time, we shall assume that the distribution of T within each element has the form

$$T^{(e)}(x, y, t) = \sum_{i=1}^{r} N_i(x, y)T_i(t) \qquad (4.11c)$$

If we carry out the steps of the preceding example, first introducing Galerkin's method and then integrating by parts, we obtain identical matrix equations, except that the new term

$$-\int_{\Omega^{(e)}} N_i \rho c \frac{\partial T^{(e)}}{\partial t} dx\, dy$$

arises in equation 4.10b. When equation 4.11c is substituted, the resulting equation takes the form

$$[K_c]^{(e)}\{T\}^{(e)} + [C]\left\{\frac{dT}{dt}\right\}^{(e)} = \{Q\}^{(e)} - \{q\}^{(e)} - [K_h]^{(e)}\{T\}^{(e)} + \{q_{T_\infty}\}^{(e)}$$

$$(4.11d)$$

where

$$C_{ij} = \int_{\Omega^{(e)}} \rho c N_i N_j \, dx \, dy$$

and the remaining terms are the same as those in equation 4.9f. Equation 4.11d, a linear first-order differential equation in time, expresses the transient behavior of a typical element. Methods for solving transient heat conduction equations of this type appear in Chapter 8.

As we saw in Chapter 3, the use of classical variational principles to derive finite element equations considerably enlarges the range of problems amenable to finite element techniques. Now the use of Galerkin's method extends even further the applicability of finite element analysis. Galerkin's method offers an alternative approach, and it not only encompasses the variational approach but also goes far beyond, because it can be applied to *any* well-posed system of differential equations and their boundary conditions.

Some of the early applications of Galerkin's method appear in references 4–10. Many other applications of the method have appeared in the literature as analysts have recognized the usefulness and generality of the method. We see in later chapters some specific examples in solid mechanics, fluid mechanics, and heat transfer of finite element equations formulated by Galerkin's method.

4.3   CLOSURE

In this chapter we have been concerned with a more general approach to the derivation of finite element models for continuum problems. Including the procedures discussed in Chapters 2 and 3, we now have at hand three distinct approaches to deriving element equations. For the simplest problems the *direct method*, employing physical reasoning, can be employed and is useful for clarifying some of the concepts in finite element analysis. A more general approach, however, is the *variational method*, which employs classical variational principles. When only the governing differential equations and their boundary conditions are available, *Galerkin's method* is convenient; this approach surpasses the variational method in generality and further broadens the range of applicability of the finite element method.

Before considering how these methods apply to a multitude of practical problems, we examine in the next chapter the important ideas pertaining to the establishment of element shapes and appropriate interpolation functions. When the element shape and the interpolation functions N_i are specified, we can explicitly evaluate the element equations and proceed with the solution in a routine manner.

REFERENCES

1. B. A. Finlayson and L. E. Scriven, "The Method of Weighted Residuals—A Review," *Appl. Mech. Rev.*, Vol. 19, No. 9, September 1966, pp. 735–748.
2. L. Collatz, *The Numerical Treatment of Differential Equations*, Springer, Berlin, 1966.
3. W. Kaplan, *Advanced Calculus*, Addison-Wesley, Reading, MA, 1952.
4. M. M. Aral, P. G. Mayer, and C. V. Smith, Jr., "Finite Element Galerkin Method Solutions to Selected Elliptic and Parabolic Differential Equations," *Proceedings of the 3rd Conference on Matrix Methods in Structural Mechanics*, Wright-Patterson Air Force Base, Dayton, OH, October 1971.
5. B. A. Szabo and G. C. Lee, "Stiffness Matrix for Plates by Galerkin's Method," *Proc. ASCE J. Eng. Mech. Div.*, Vol. 95, No. EM3, June 1969, pp. 571–585.
6. B. A. Szabo and G. C. Lee, "Derivation of Stiffness Matrices for Problems on Plane Elasticity by Galerkin Method," *Int. J. Numer. Methods Eng.*, Vol. 1, No. 3, July 1969, pp. 301–310.
7. J. W. Leonard and T. T. Bramlette, "Finite Element Solutions to Differential Equations," *Proc. ASCE J. Eng. Mech. Div.*, Vol. 96, EM6, December 1970, pp. 1277–1283.
8. O. C. Zienkiewicz and C. J. Parekh, "Transient Field Problems: Two Dimensional and Three Dimensional Analysis by Isoparametric Finite Elements," *Int. J. Numer. Methods Eng.*, Vol. 2, No. 1, January 1970, pp. 61–71.
9. Y. Tada and G. C. Lee, "Finite Element Solution to an Elastic Problem of Beams," *Int. J. Numer. Methods Eng.*, Vol. 2, No. 2, April 1970, pp. 229–241.
10. O. C. Zienkiewicz and C. Taylor, "Weighted Residual Processes in Finite Element with Particular Reference to Some Transient and Coupled Problems," in *Lectures on Finite Element Methods in Continuum Mechanics*, J. T. Oden and E. R. A. Oliveria (eds.), Lecture for NATO Advanced Study Institute on Finite Element Methods in Continuum Mechanics, Lisbon 1971, University of Alabama Press, Huntsville, 1973, pp. 415–458.

PROBLEMS

1. Consider the one-dimensional ordinary differential equation

$$-\frac{d^2\phi}{dx^2} + \alpha\phi = \beta$$

for $a < x < b$, where α and β are constants. The boundary conditions consist of specifying either ϕ or $d\phi/dx$ at the end points a and b of the domain. Use the Galerkin method with $W_i = N_i$ to formulate general equations for the element matrices $[K]^{(e)}$ and $\{F\}^{(e)}$, assuming an element with two nodes. Discuss how the boundary conditions are treated in the finite element solution.

2. Repeat the finite element formulation for the differential equation of Problem 1, but for the boundary conditions

$$\frac{d\phi}{dx}(a) = 0 \qquad \frac{d\phi}{dx}(b) = c_1\phi + c_2$$

where c_1 and c_2 are constants.

3. A differential operator A is said to be self-adjoint in domain Ω if

$$\int_\Omega uA(v)\,d\Omega = \int_\Omega vA(u)\,d\Omega$$

where u and v are any two arbitrary functions that satisfy the essential and natural boundary conditions. Whether a system is self-adjoint can be established by integration by parts. Determine if the following operators are self-adjoint for the boundary conditions $\phi(a) = \phi(b) = 0$.

a. $A(\phi) = \dfrac{d}{dx}\left(\dfrac{d\phi}{dx}\right)$

b. $A(\phi) = \dfrac{d}{dx}\left(x\dfrac{d\phi}{dx}\right)$

c. $A(\phi) = \dfrac{d^2\phi}{dx^2} + \dfrac{1}{x}\dfrac{d\phi}{dx}$

4. Galerkin's method with $W_i = N_i$ always leads to a symmetric coefficient matrix $[K]^{(e)}$ for self-adjoint differential equations, but for nonself-adjoint differential equations an unsymmetric coefficient matrix results. Verify this statement for the differential operators of Problem 3 when $A(\phi) = 0$.

5. A second-order differential equation

$$a_0(x)\frac{d^2\phi}{dx^2} + a_1(x)\frac{d\phi}{dx} + a_2(x)\phi = 0$$

can be written in self-adjoint form by multiplying by the integrating factor $P(x) = \exp[\int(a_1/a_0)\,dx]$ to yield

$$\frac{d}{dx}\left[P(x)\frac{d\phi}{dx}\right] + Q(x)\phi = 0$$

where $Q(x) = a_2 P/a_0$. Determine the integrating factors for $A(\phi) = 0$, where the $A(\phi)$ are given in Problem 3, and write the differential operators in self-adjoint form.

6. Consider the one-dimensional ordinary differential equation

$$-\frac{d}{dx}\left(x^2\frac{d\phi}{dx}\right) = \alpha x^2$$

with the boundary condition $\phi(a) = A, \phi(b) = B$, where α, A, and B are constants. Use Galerkin's method with $W_i = N_i$ to formulate general equations for the element matrices $[K]^{(e)}$ and $\{F\}^{(e)}$ for an element with m nodes.

7. Use the Galerkin method with $W_i = N_i$ to formulate the element matrices for the Reynolds–Poisson equation

$$\frac{\partial^2 P}{\partial x^2} + \frac{\partial^2 P}{\partial y^2} = G = \text{constant}$$

in domain Ω with the condition $P = 0$ on boundary Γ.

8. Verify equation 4.11d, which defines the finite element matrices for transient heat conduction.

9. Consider the linear one-dimensional equation

$$-\frac{\partial^2 \phi}{\partial x^2} + \alpha\frac{\partial \phi}{\partial x} + \beta\frac{\partial \phi}{\partial t} = 0$$

in domain $a < x < b$, where α and β are constants. The boundary conditions consist of specifying either ϕ or $d\phi/dx$ at the end points of the domain, and we have an initial condition $\phi(x, 0) = f(x)$. Use the Petrov–Galerkin method with $W_i \neq N_i$ to derive element matrices $[C]^{(e)}$ and $[K]^{(e)}$. Specialize $W_i = N_i$ (the Bubnov–Galerkin method) and discuss the symmetry or asymmetry of the element matrices that result from the two formulations.

10. Develop a finite element formulation for the nonlinear Burger's equation

$$-\frac{\partial^2 \phi}{\partial x^2} + \phi\frac{\partial \phi}{\partial x} + \frac{\partial \phi}{\partial t} = 0$$

for the boundary conditions $\phi(0, t) = 0, \phi(1, t) = 0$ and initial condition $\phi(x, 0) = f(x)$. Use Galerkin's method with $W_i = N_i$ and linearize the equation by approximating $\phi\, \partial\phi/\partial x$ by $\phi_n\, \partial\phi/\partial x$, where $\phi_n(x)$ is an approximate solution to the problem from a previous interation in a nonlinear solution scheme.

11. Consider the pair of simultaneous partial differential equations

$$\frac{\partial}{\partial x}\left(\frac{\partial \phi}{\partial x}\right) + a_1(\phi - \psi) + a_2\frac{\partial \phi}{\partial t} = 0$$

$$-\frac{\partial}{\partial x}\left(\frac{\partial \psi}{\partial x}\right) + a_1(-\phi + \psi) + a_3\frac{\partial \psi}{\partial t} = 0$$

where a_i, $i = 1, 2, 3$ are constants, and we seek $\phi(x, t)$ and $\psi(x, t)$ in the domain $a < x < b$. The boundary conditions specify either ϕ or $d\phi/dx$ and ψ or $d\psi/dx$ at the end points of the domain; initial distributions $\phi(x, 0)$ and $\psi(x, 0)$ are known. Use Galerkin's method with $W_i = N_i$ to develop a finite element formulation for the problem.

5

ELEMENTS AND INTERPOLATION FUNCTIONS

5.1   INTRODUCTION

A subject of utmost importance in finite element analysis is the selection of particular finite elements and the definition of the appropriate interpolation functions within each element. As we saw in Chapters 3 and 4, after the unknown field variable has been expressed in each element in terms of appropriate nodal parameters and interpolation functions, the derivation of the element equations according to one of the mathematical approaches follows a well-established routine.

Except for two special cases in Chapter 2, where we used the two-node line element and the three-node triangle element, our derivations of the element equations were left in terms of the general interpolation function N_i and general element types. Before these general element equations and others similarly derived can be used for actual problem solving we must specify the particular type of element and the particular interpolation functions; that is, we must choose the functions N_i in the expression

$$\phi = \sum N_i \phi_i = \lfloor N \rfloor \{\phi\}$$

In Chapter 3 we saw that the interpolation functions cannot be chosen arbitrarily; rather, certain continuity requirements must be met to ensure that the convergence criteria, which are usually different from one problem to another, are satisfied.

It is helpful at this point to introduce a standard definition and notation to express the degree of continuity of a field variable at element interfaces. If the field variable is continuous at element interfaces, we say that we have C^0 continuity; if, in addition, first derivatives are continuous, we have C^1 continuity; if second derivatives are also continuous, we have C^2 continuity; and so on. With this definition we may now restate the compatibility and completeness requirements for the interpolation functions representing the behavior of a field variable.

Suppose that the functions appearing under the integrals in the element equations contain derivatives up to the $(r + 1)$th order. Then, to have rigorous assurance of convergence as element size decreases, we must satisfy the following requirements.

Compatibility requirement: At element interfaces we must have C^r continuity.

Completeness requirement: Within an element we must have C^{r+1} continuity.

These requirements hold regardless of whether the element equations (integral expressions) were derived using the variational method, the Galerkin method, or some other method yet to be devised.

We will see in this chapter that constructing elements and interpolation functions to achieve C^0 continuity is not especially difficult, but that the difficulty increases rapidly when higher-order continuity is desired. For problems requiring C^0 continuity we can construct an infinite number of suitable elements, but from this wide variety we usually use only the simplest types to avoid excessive computational labor.

Construction of suitable finite elements to give a specified continuity of order $C^0, C^1, C^2, \ldots$, requires skill and ingenuity stemming from much experience. Fortunately, analysts have developed a variety of elements applicable to many different types of problems, and we may turn to the results of their work for help in formulating new elements, or we may borrow directly from their existing catalog of elements.

Each element in the catalog is characterized by several features. When one practitioner of the finite element method asks another what type of element he is using in his problem, the questioner is really asking for four distinct pieces of information: the shape of the element, the number and type of the nodes, the type of nodal variables, and the type of interpolation function. If any one of these characterizing features is lacking, the description of an element is incomplete.

5.2 BASIC ELEMENT SHAPES

Since the fundamental premise of the finite element method is that a continuum or solution domain of arbitrary shape can be accurately modeled by an assemblage of simple shapes, most finite elements are geometrically simple. For one-dimensional problems, that is, problems with only one independent variable, the elements are line segments (Figure 5.1). The number of nodes assigned to a particular element, as we shall see, depends on the type of nodal variables, the type of interpolation function, and the

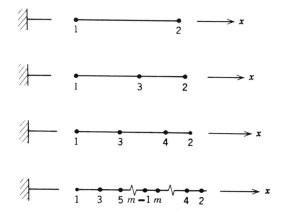

Figure 5.1   A family of one-dimensional line elements.

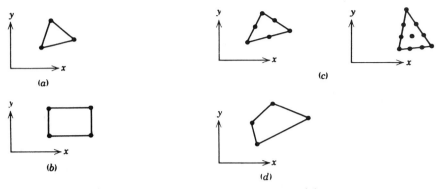

Figure 5.2 Examples of two-dimensional elements: (a) Three-node triangle; (b) Rectangle; (c) Triangles with 6 and 10 nodes; (d) General quadrilateral.

degree of continuity required.[1] Often there is no reason to apply finite element methods to solve one-dimensional problems, because these are governed by linear or nonlinear *ordinary* differential equations that can often be solved by other standard analytical and numerical techniques [1, 2]. However, for some one-dimensional problems the finite element method is the most rational approach. For example, when we are dealing with one-dimensional domains that have abrupt or step changes in properties such as occur in the simple heat conduction problem in Section 2.2.2, each portion of the domain with continuously varying properties can be defined as an element. Frame analyses in solid mechanics and flow network analyses in fluid mechanics offer additional examples in which one-dimensional elements are employed. In elasticity problems where beams are used as stiffeners, one-dimensional elements can often represent the beams while being connected to other two- or three-dimensional elements that represent the rest of the elastic solid.

Common two-dimensional element shapes are shown in Figure 5.2. The three-node flat triangle element (Figure 5.2a) is the simplest two-dimensional element, and it enjoys the distinction of being the first and most often used basic finite element. The reason for this is that an assemblage of triangles can always represent a two-dimensional domain of any shape. A simple but less useful two-dimensional element is the four-node rectangle (Figure 5.2b) whose sides are parallel to the global coordinate system. This type of element is easy to construct because of its regular shape, but it is not well suited for approximating curved boundaries.

In addition to the simple triangle and the rectangle, other common two-dimensional elements are the six-node triangle (Figure 5.2c) and the general quadrilateral (Figure 5.2d). Quadrilateral elements may be formed

---

[1]This statement holds regardless of the dimension of the element.

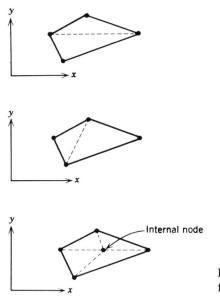

Figure 5.3 The quadrilateral element formed by combining triangles.

directly or they may be developed by combining two or four basic triangles elements, as shown in Figure 5.3.

Other types of elements that are actually three-dimensional but are described by only one or two independent variables are axisymmetric or ring-type elements (Figure 5.4). These elements are useful when treating problems that possess axial symmetry in cylindrical coordinates. Many practical engineering problems are axisymmetric. Solids such as storage tanks,

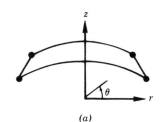

(a)

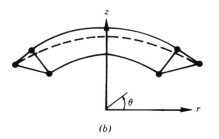

(b)

Figure 5.4 Examples of axisymmetric ring elements: (a) One-dimensional ring element; (b) Two-dimensional triangular ring element.

pistons, valves, shafts, rocket nozzles, and reentry vehicle heat shields fall into this category. Before we can classify a problem as axisymmetric and define elements for it we must be sure not only that the solution domain is axisymmetric, but also that the forcing functions (loads) are axisymmetric. We have axisymmetry only when all the parameters in the problem are the same in any plane passing through the symmetry axis of the solution domain. Since we may construct axisymmetric elements from any one- or two-dimensional element, the variety of these types of elements is virtually unlimited. We shall not discuss ring elements further in this chapter, but we shall consider them again in the following chapters on applications.

The four-node tetrahedron element in three dimensions (Figure 5.5a), is the three-dimensional counterpart of the three-node triangle element in two dimensions. Another simple three-dimensional element is the right prism shown in Figure 5.5b. A general hexahedron (Figure 5.5c), analogous to the general quadrilateral in two dimensions, may be constructed from five tetrahedra. Figure 5.6 is a photograph of a wire model formed by assembling five tetrahedra.

In addition to the endless variety of straight-edged elements that can be constructed, it is also possible, as we shall see, to construct elements with curved boundaries—the so-called, isoparametric element families and others. These elements, some examples of which are illustrated in Figure 5.7, are

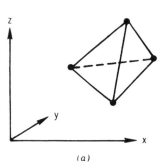

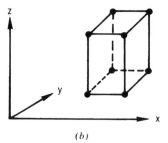

(a)

(b)

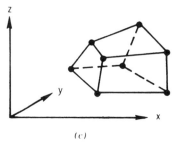

(c)

Figure 5.5 Three-dimensional elements; (a) Tetrahedron; (b) Right prism; (c) General hexahedron.

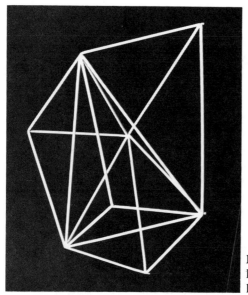

Figure 5.6 Wire model of a general hexahedron constructed from five tetrahedra.

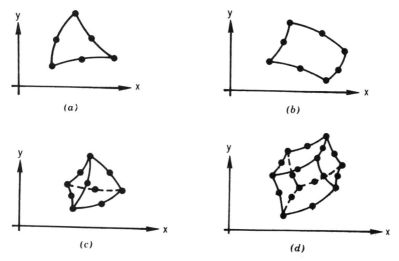

Figure 5.7 Common isoparametric elements: (*a*) Triangle; (*b*) Quadrilateral; (*c*) Tetrahedron; (*d*) Hexahedron.

most useful when it is desirable to approximate curved boundaries with only a few elements. They have been especially helpful in the solution of three-dimensional problems, where it is often necessary to reduce the cost of computation by using fewer elements.

5.3   TERMINOLOGY AND PRELIMINARY CONSIDERATIONS

5.3.1   Types of Nodes

In Figures 5.1–5.5 we illustrated a number of basic element shapes and typical locations of nodes assigned to these elements. We alluded to exterior and interior nodes, but now we can be more specific. Nodes are classified as either *exterior* or *interior*, depending on their location relative to the geometry of an element. *Exterior nodes* lie on the boundary of an element, and they represent the points of connection between bordering elements. Nodes positioned at the corners of elements, along the edges, or on the surfaces are all exterior nodes. For one-dimensional elements such as that in Figure 5.1 there are only two exterior nodes, because only the ends of the element connect to other one-dimensional elements. If one-dimensional elements are connected to two-dimensional elements along a side, all the nodes of a one-dimensional element can be exterior nodes. In contrast to exterior nodes, *interior nodes* are those that do not connect with neighboring elements. The 10-node triangle in Figure 5.2c has one such interior node.

5.3.2   Degrees of Freedom

Two other features, in addition to shape, characterize a particular element type: (1) the number of nodes assigned to the element and (2) the number and type of nodal variables chosen for it. Often the nodal variables or the parameters assigned to an element are called the *degrees of freedom* of the element. This terminology, which we shall adopt, is a spin-off from the solid mechanics field, where the nodal variables are usually nodal displacements and sometimes derivatives of displacements. Nodal degrees of freedom can be interior or exterior in relation to element boundaries, depending on whether they are assigned to interior or exterior nodes.

5.3.3   Interpolation Functions—Polynomials

In the finite element literature the functions used to represent the behavior of a field variable within an element are called *interpolation functions*, *shape functions*, or *approximating functions*. We have used and will continue to use only the first term in this text. Although it is conceivable that many types of functions could serve as interpolation functions, only polynomials have received widespread use. The reason is that polynomials are relatively easy to manipulate mathematically—in other words, they can be integrated or differentiated without difficulty. Trigonometric functions also possess this property, but they are seldom used as interpolation functions. References 3 and 4 offer examples of the use of trigonometric functions in finite element analyses. Here we employ only polynomials to generate suitable interpolation functions. We begin by discussing polynomials in one, two, and three dimensions.

One Independent Variable   In one dimension a general complete nth-order polynomial may be written as

$$P_n(x) = \sum_{k=1}^{T_n^{(1)}} \alpha_k x^i, \qquad i \leq n \tag{5.1}$$

where the number of terms in the polynomial is $T_n^{(1)} = n + 1$. For $n = 1$, $T_1^{(1)} = 2$ and $P_1(x) = \alpha_0 + \alpha_1 x$; for $n = 2$, $T_2^{(1)} = 3$ and $P_2(x) = \alpha_0 + \alpha_1 x + \alpha_2 x^2$; and so on.

Two Independent Variables   In two dimensions a complete nth-order polynomial may be written as

$$P_n(x, y) = \sum_{k=1}^{T_n^{(2)}} \alpha_k x^i y^j, \qquad i + j \leq n \tag{5.2}$$

where the number of terms in the polynomial is $T_n^{(2)} = (n + 1)(n + 2)/2$. For $n = 1$, $T_1^{(2)} = 3$ and $P_1(x, y) = \alpha_1 + \alpha_2 x + \alpha_3 y$; for $n = 2$, $T_2^{(2)} = 6$ and $P_2(x, y) = \alpha_1 + \alpha_2 x + \alpha_3 y + \alpha_4 xy + \alpha_5 x^2 + \alpha_6 y^2$; and so on. If the terms are placed in a triangular array in ascending order, we obtain an arrangement similar to the Pascal triangle (Figure 5.8). We note that the sum of the

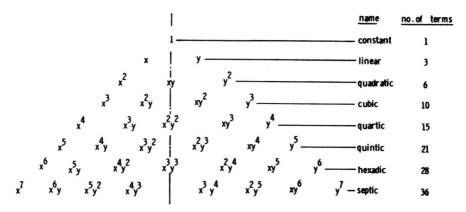

Figure 5.8   Array of terms in a complete polynomial in two dimensions.

exponents of any term in this triangular array is the corresponding number in the well-known Pascal triangle of binomial coefficients.

Three Independent Variables In three dimensions a complete nth-order polynomial may be written as

$$P_n(x, y, z) = \sum_{l=1}^{T_n^{(3)}} \alpha_l x^i y^j z^k, \qquad i + j + k \leq n \tag{5.3}$$

where the number of terms in the polynomial is

$$T_n^{(3)} = \frac{(n + 1)(n + 2)(n + 3)}{6}$$

For $n = 1$, $T_1^{(3)} = 4$ and $P_1(x, y, z) = \alpha_1 + \alpha_2 x + \alpha_3 y + \alpha_4 z$; for $n = 2$, $T_2^{(3)} = 10$ and $P_2(x, y, z) = \alpha_1 + \alpha_2 x + \alpha_3 y + \alpha_4 z + \alpha_5 xy + \alpha_6 xz + \alpha_7 yz + \alpha_8 x^2 + \alpha_9 y^2 + \alpha_{10} z^2$; and so on.

The terms in a complete three-dimensional polynomial may also be arrayed in a manner analogous to the triangular array in two dimensions. The array becomes a tetrahedron, with the various terms placed at different planar levels, as in Figure 5.9.

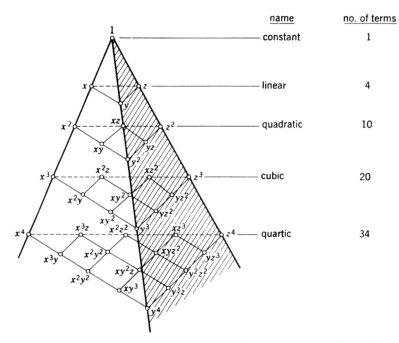

Figure 5.9 Array of terms in a complete polynomial in three dimensions.

5.4   GENERALIZED COORDINATES AND THE ORDER OF THE POLYNOMIAL

5.4.1   Generalized Coordinates

Ignoring for the moment any interelement continuity considerations, we can say that the order of the polynomial we use to represent the field variable within an element depends on the number of degrees of freedom we assign to the element. In other words, the number of coefficients in the polynomial should equal the number of nodal variables available to evaluate these coefficients. In section 3.4.3, where we obtained a piecewise linear representation for a field variable over a triangle, we wrote the linear polynomial series as

$$\phi(x, y) = \alpha_1 + \alpha_2 x + \alpha_3 y$$

The three nodal values of ϕ were then sufficient to find the three coefficients α_1, α_2, and α_3; that is, we evaluated the expression at the three nodes and thereby obtained three equations to be solved for the α_i in terms of the coordinates of the nodes and the nodal values of ϕ. If we select a complete or six-term polynomial to represent ϕ, we need six nodal variables (six degrees of freedom) to find the six α_i.

The coefficients α_i in a polynomial series representation of a field variable are called the *generalized coordinates* of the element. The generalized coordinates are independent parameters that specify or fix the magnitude of the prescribed distribution for ϕ, while the shape of the prescribed distribution is determined by the order of the polynomial we select. As illustrated in equation 3.6, the generalized coordinates (denoted as β_i in this case) have no direct physical interpretation but, rather, are linear combinations of the physical nodal degrees of freedom; also, they are not identified with particular nodes.

5.4.2   Geometric Isotropy

When we choose polynomial expansions as interpolation functions for an element, we usually try to satisfy the compatibility and completeness requirements discussed in Section 3.4.5. As we have seen, these requirements are important because they ensure (1) continuity of the field variable and (2) convergence to the correct solution as the element mesh size is made smaller and smaller. In addition to satisfying these stipulations, we require that the field variable representation within an element, and hence the polynomial expansion for the element, remain unchanged under a linear transformation from one Cartesian coordinate system to another [5]. Polynomials that exhibit this invariance property are said to possess *geometric isotropy*. Clearly, we could not expect a good approximation to reality if our field variable representation changed with a change in origin or in the orientation of the coordinate system. Hence the need to ensure geometric isotropy in our polynomial interpolation functions is apparent.

Fortunately, we have two axioms that allow us to construct polynomial series with the desired property:

1. Polynomials of order n that are complete (contain all their terms) have geometric isotropy.
2. Polynomials of order n that are incomplete, yet contain the appropriate terms to preserve "symmetry," have geometric isotropy.

According to the first axiom, we know that the complete polynomials[2]

$$P_n(x, y) = \sum_{k=1}^{T_n^{(2)}} \alpha_k x^i y^j, \qquad i + j \leq n \tag{5.2}$$

and

$$P_n(x, y, z) = \sum_{l=1}^{T_n^{(3)}} \alpha_l x^i y^j z^k, \qquad i + j + k \leq n \tag{5.3}$$

when used as interpolation functions will remain invariant under linear coordinate transformations. We shall not be concerned with other types of transformations.

It is easy to illustrate what we mean by symmetry in the second axiom by considering a specific example. Suppose that we wish to construct a cubic polynomial expansion for an element that has eight nodal variables assigned to it. In this situation, since a complete cubic polynomial contains 10 terms and hence 10 undetermined coefficients, we may desire to truncate the polynomial by dropping two terms. But then the question arises, How should the polynomial series be truncated so that geometric isotropy is preserved? The answer is that we may drop only terms that occur in symmetric pairs, that is, (x^3, y^3) or $(x^2 y, xy^2)$. Thus, acceptable eight-term cubic polynomial interpolation functions exhibiting geometric isotropy would be

$$P(x, y) = \alpha_1 + \alpha_2 x + \alpha_3 y + \alpha_4 x^2 + \alpha_5 xy + \alpha_6 y^2 + \alpha_7 x^3 + \alpha_{10} y^3$$

or

$$P(x, y) = \alpha_1 + \alpha_2 x + \alpha_3 y + \alpha_4 x^2 + \alpha_5 xy + \alpha_6 y^2 + \alpha_8 x^2 y + \alpha_9 xy^2$$

---

[2]Polynomials in one independent variable inherently possess geometric isotropy regardless of whether they are complete.

Extending this idea to construct other incomplete polynomials is straightforward if we refer to the arrays of terms in Figure 5.8 and 5.9. As Dunne [5] points out, meeting the criterion of geometric isotropy assures us that when we evaluate the interpolation function along any straight edge of the element, it becomes a complete polynomial in the linear coordinate along that edge, and its order is the same as the parent two- or three-dimensional polynomial for the element.

Taylor [6] examines the completeness property of a number of commonly used interpolation functions for C^0 problems and presents guidelines for constructing computationally efficient functions for rectangular elements.

5.4.3 Deriving Interpolation Functions

Thus far we have seen how a field variable can be represented within an element as a polynomial series whose coefficients are the generalized coordinates. In this section we see how the interpolation functions for the physical degrees of freedom are derived. These interpolation functions emerge from the basic procedure for expressing the generalized coordinates in terms of the nodal degrees of freedom. In Section 5.4.1 we alluded to this procedure, but now we consider the details.

The basic idea can be illustrated by a simple example in two dimensions. Suppose that we wish to construct a rectangular element with nodes positioned at the corners of the element (Figure 5.10). If we assign one value of ϕ to each node, the element then has four degrees of freedom, and we may select as an interpolation model a four-term polynomial such as

$$\phi(x, y) = \alpha_1 + \alpha_2 x + \alpha_3 y + \alpha_4 xy \tag{5.4}$$

The generalized coordinates may now be found by evaluating this model at

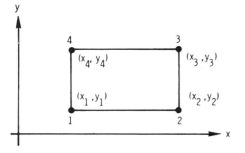

Figure 5.10 A rectangular element with sides parallel to the axes of the global coordinate system.

each of the nodes and then solving the resulting set of simultaneous equations. Thus we may write

$$\phi_1 = \alpha_1 + \alpha_2 x_1 + \alpha_3 y_1 + \alpha_4 x_1 y_1$$

$$\phi_2 = \alpha_1 + \alpha_2 x_2 + \alpha_3 y_2 + \alpha_4 x_2 y_2$$

$$\phi_3 = \alpha_1 + \alpha_2 x_3 + \alpha_3 y_3 + \alpha_4 x_3 y_3$$

$$\phi_4 = \alpha_1 + \alpha_2 x_4 + \alpha_3 y_4 + \alpha_4 x_4 y_4$$

or, in matrix notation,

$$\{\phi\} = [G]\{\alpha\} \tag{5.5}$$

where

$$\{\phi\} = \begin{Bmatrix} \phi_1 \\ \phi_2 \\ \phi_3 \\ \phi_4 \end{Bmatrix}$$

$$[G] = \begin{bmatrix} 1 & x_1 & y_1 & x_1 y_1 \\ 1 & x_2 & y_2 & x_2 y_2 \\ 1 & x_3 & y_3 & x_3 y_3 \\ 1 & x_4 & y_4 & x_4 y_4 \end{bmatrix}$$

$$\{\alpha\} = \begin{Bmatrix} \alpha_1 \\ \alpha_2 \\ \alpha_3 \\ \alpha_4 \end{Bmatrix}$$

In principle, then, we can express the generalized coordinates as the solution of equations 5.5 for $\{\alpha\}$, that is,

$$\{\alpha\} = [G]^{-1}\{\phi\} \tag{5.6}$$

Expressing the terms of the interpolation polynomial equation 5.4 as a product of a row vector and a column vector, we can write

$$\phi = \lfloor P \rfloor \{\alpha\} \tag{5.7}$$

where

$$\lfloor P \rfloor = \lfloor 1 \quad x \quad y \quad xy \rfloor$$

Thus, by substituting equations 5.6 into equation 5.7, we have

$$\phi = \lfloor P \rfloor [G]^{-1}\{\phi\} = \lfloor N \rfloor\{\phi\} \qquad (5.8a)$$

with

$$\lfloor N \rfloor = \lfloor P \rfloor [G]^{-1} \qquad (5.8b)$$

Equations 5.6–5.8, though obtained for one case, are generally applicable to all straight-sided elements. The original interpolation polynomial $\lfloor P \rfloor\{\alpha\}$ should not be confused with the interpolation functions N_i associated with the nodal degrees of freedom. The distinction to note here is that $\lfloor P \rfloor\{\alpha\}$ is an interpolation function that applies to the whole element and expresses the field variable behavior in terms of the generalized coordinates, whereas the interpolation functions N_i refer to individual nodes and individual nodal degrees of freedom; collectively, they represent the field variable behavior. It is easy to see from equation 5.8 that the function N_i referring to node i takes on unit value at node i and zero value at all the other nodes of the element.

The procedure described in equations 5.5–5.8 is a symbolic method for developing element interpolation functions from assumed polynomials. Equation 5.6 represents the key step where we solve for the generalized coordinates $\{\alpha\}$ in terms of the nodal coordinates contained in $[G]$ and the nodal degrees of freedom $\{\phi\}$. We should avoid the actual matrix inversion. Instead, equation 5.5 should be solved directly if possible. Modern symbolic manipulation software can perform the algebra if the number of equations is not too large. An alternative is to perform the inversion numerically as part of element stiffness matrix computations. This approach has been used, but it has drawbacks [7], including the computational effort required to obtain $[G]^{-1}$. For a model with a large number of elements with many degrees of freedom, the cost of the inversion computations may be prohibitive.

These reasons have motivated researchers to develop other methods for deriving interpolation functions. Indeed, these methods are used more frequently than the procedure described by equations 5.5–5.8. The alternative methods often rely on the use of special coordinate systems called *natural coordinates*. In the following section we derive various natural coordinate systems in one, two, and three dimensions.

5.5   NATURAL COORDINATES

A local coordinate system that relies on the element geometry for its definition and whose coordinates range between zero and unity within the element is known as a *natural coordinate system*. Such systems have the property that one particular coordinate has unit value at one node of the element and zero value at the other node(s); its variation between nodes is linear. We may construct natural coordinate systems for two-node line elements, three-node triangular elements, four-node quadrilateral elements, four-node tetrahedral elements, and so on. Since only the simplest coordinates up to three dimensions are commonly used, we shall not consider those in higher dimensions.

The use of natural coordinates in deriving interpolation functions is particularly advantageous because special closed-form integration formulas can often be used to evaluate the integrals in the element equations. Natural coordinates also play a crucial role in the development of curve-sided elements, which we discuss later in this chapter.

The basic purpose of a natural coordinate system is to describe the location of a point inside an element in terms of coordinates associated with the nodes of the element. We denote the natural coordinates as L_i ($i = 1, 2, \ldots, n$), where n is the number of external nodes of the element. One coordinate is associated with node i and has unit value there. It will become evident that the natural coordinates are functions of the global Cartesian coordinate system in which the element is defined.

5.5.1 Natural Coordinates in One Dimension

Figure 5.11 shows a line element in which we desire to define a natural coordinate system. If we select L_1 and L_2 as the natural coordinates, the location of the point x_p may be expressed as a linear combination of the nodal coordinates x_1 and x_2, that is,

$$x_p = L_1 x_1 + L_2 x_2 \tag{5.9}$$

Since x_p can be any point on the line element, we can drop the subscript p for convenience. The coordinates L_1 and L_2 may be interpreted as weighting functions relating the coordinates at the end nodes to the coordinate of any interior point. Clearly, the weighting functions are not independent, since we must have

$$L_1 + L_2 = 1 \tag{5.10}$$

Equations 5.9 and 5.10 may be solved simultaneously for the functions L_1

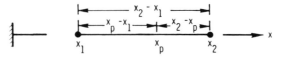

Figure 5.11 Two-node line element with the global coordinate x_p defining some point within the element.

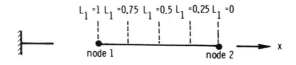

Figure 5.12   Variation of a length coordinate within an element.

and L_2 with the following result:

$$L_1(x) = \frac{x_2 - x}{x_2 - x_1} \qquad L_2(x) = \frac{x - x_1}{x_2 - x_1} \tag{5.11}$$

The functions L_1 and L_2 are seen to be simply ratios of lengths and are often called *length coordinates*. The variation of L_1 is shown in Figure 5.12. We recognize that the natural coordinates for the line elements are precisely the same as the linear interpolation functions we used in the example of Section 3.4.5. Hence the linear interpolation used for the field variable ϕ in that example can be written directly as

$$\phi(x) = \phi_1 L_1 + \phi_2 L_2$$

If ϕ is taken to be a function of L_1 and L_2, differentiation of ϕ follows the chain rule formula

$$\frac{d\phi}{dx} = \frac{\partial \phi}{\partial L_1} \frac{\partial L_1}{\partial x} + \frac{\partial \phi}{\partial L_2} \frac{\partial L_2}{\partial x} \tag{5.12}$$

where

$$\frac{\partial L_1}{\partial x} = \frac{-1}{x_2 - x_1} \qquad \frac{\partial L_2}{\partial x} = \frac{1}{x_2 - x_1} \tag{5.13}$$

Integration of length coordinates over the length of an element is simple with the aid of the following convenient formula:

$$\int_{x_1}^{x_2} L_1^\alpha L_2^\beta \, dx = \frac{\alpha! \beta! (x_2 - x_1)}{(\alpha + \beta + 1)!} \tag{5.14}$$

where α and β are integer exponents and, for instance, $\alpha!$ is the factorial of α.

5.5.2   Natural Coordinates in Two Dimensions

The development of natural coordinates for triangular elements follows the same procedure we used for the one-dimensional case. Again, the goal is to choose coordinates L_1, L_2, and L_3 to describe the location of any point x_p

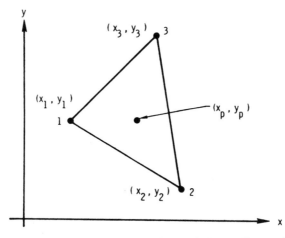

Figure 5.13 Three-node triangle element with global coordinates (x_p, y_p) defining some point within the element.

within the element or on its boundary (Figure 5.13). The original Cartesian coordinates of a point in the element should be linearly related to the new coordinates by the following equations:

$$x = L_1 x_1 + L_2 x_2 + L_3 x_3$$
$$y = L_1 y_1 + L_2 y_2 + L_3 y_3$$

(5.15)

In addition to these equations we impose a third condition requiring that the weighting functions sum to unity, that is,

$$L_1 + L_2 + L_3 = 1$$

(5.16)

From equation 5.16 it is clear that only two of the natural coordinates can be independent, just as in the original coordinate system, where there are only two independent coordinates.

Simultaneous solution of equations 5.15 and 5.16 for L_1, L_2, L_3 gives the natural coordinates in terms of the Cartesian coordinates. Thus

$$L_1(x, y) = \frac{1}{2\Delta}(a_1 + b_1 x + c_1 y)$$

$$L_2(x, y) = \frac{1}{2\Delta}(a_2 + b_2 x + c_2 y)$$

(5.17)

$$L_3(x, y) = \frac{1}{2\Delta}(a_3 + b_3 x + c_3 y)$$

where

$$2\Delta = \begin{vmatrix} 1 & x_1 & y_1 \\ 1 & x_2 & y_2 \\ 1 & x_3 & y_3 \end{vmatrix} = 2(\text{area of triangle } 1\text{-}2\text{-}3) \qquad (5.18)$$

$$a_1 = x_2 y_3 - x_3 y_2 \qquad b_1 = y_2 - y_3 \qquad c_1 = x_3 - x_2$$
$$a_2 = x_3 y_1 - x_1 y_3 \qquad b_2 = y_3 - y_1 \qquad c_2 = x_1 - x_3$$
$$a_3 = x_1 y_2 - x_2 y_1 \qquad b_3 = y_1 - y_2 \qquad c_3 = x_2 - x_3 \qquad (5.19)$$

Recalling our example of piecewise linear interpolation in Section 3.4.3, we see that the natural coordinates L_1, L_2, and L_3 are precisely the interpolation functions for linear interpolation over a triangle, that is, $N_i = L_i$ for the linear triangle. A little algebraic manipulation will reveal that the natural coordinates for a triangle have an interpretation analogous to that of length coordinates for a line. Just as $L_1(x)$ for the line element is a ratio of lengths, $L_1(x, y)$ for the triangle element is a ratio of areas. Figure 5.14 shows how the natural coordinates, often called *area coordinates*, are related to areas. As shown in Figure 5.14, when the point (x_p, y_p) is located on the boundary of the element, one of the area segments vanishes and hence the appropriate area coordinate along that boundary is identically zero. For example, if (x_p, y_p) is on line 1–3, then

$$L_2 = \frac{A_2}{\Delta} = 0, \qquad \text{since } A_2 = 0$$

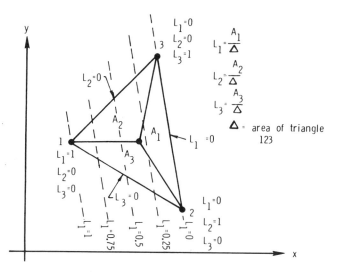

Figure 5.14   Area coordinates for a triangle.

If we interpret the field variable ϕ as a function of L_1, L_2, and L_3 instead of x, y, differentiation becomes

$$\frac{\partial \phi}{\partial x} = \frac{\partial \phi}{\partial L_1}\frac{\partial L_1}{\partial x} + \frac{\partial \phi}{\partial L_2}\frac{\partial L_2}{\partial x} + \frac{\partial \phi}{\partial L_3}\frac{\partial L_3}{\partial x}$$

$$\frac{\partial \phi}{\partial y} = \frac{\partial \phi}{\partial L_1}\frac{\partial L_1}{\partial y} + \frac{\partial \phi}{\partial L_2}\frac{\partial L_2}{\partial y} + \frac{\partial \phi}{\partial L_3}\frac{\partial L_3}{\partial y}$$

$$(5.20)$$

where

$$\frac{\partial L_i}{\partial x} = \frac{b_i}{2\Delta} \qquad \frac{\partial L_i}{\partial y} = \frac{c_i}{2\Delta}, \qquad i = 1, 2, 3 \qquad (5.21)$$

There is also a convenient formula for integrating area coordinates over the area of a triangular element:

$$\int_{A^{(e)}} L_1^\alpha L_2^\beta L_3^\gamma \, dA^{(e)} = \frac{\alpha!\beta!\gamma!}{(\alpha + \beta + \gamma + 2)!}2\Delta \qquad (5.22)$$

The derivation of equation 5.22 appears in reference 8.

Another type of natural coordinate system can be established for a four-node quadrilateral element in two dimensions. Figure 5.15 shows a general quadrilateral element in the global Cartesian coordinate system and a local natural coordinate system. In the nature coordinate system whose origin is at the centroid the quadrilateral element is a square with sides extending to $\xi = \pm 1$, $\eta = \pm 1$. The local and global coordinates are related

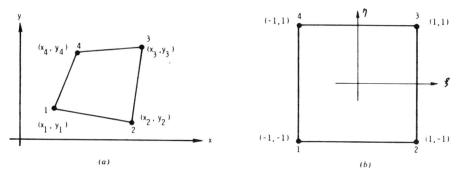

Figure 5.15 Natural coordinates for a general quadrilateral: (*a*) Cartesian coordinates; (*b*) Natural coordinates.

by the following equations:

$$x = \tfrac{1}{4}\big[(1 - \xi)(1 - \eta)x_1 + (1 + \xi)(1 - \eta)x_2$$
$$+ (1 + \xi)(1 + \eta)x_3 + (1 - \xi)(1 + \eta)x_4\big]$$
$$y = \tfrac{1}{4}\big[(1 - \xi)(1 - \eta)y_1 + (1 + \xi)(1 - \eta)y_2$$
$$+ (1 + \xi)(1 + \eta)y_3 + (1 - \xi)(1 + \eta)y_4\big]$$

(5.23a)

We can write equation 5.23a more compactly as

$$x = \sum_{i=1}^{4} L_i(\xi, \eta)x_i$$

$$y = \sum_{i=1}^{4} L_i(\xi, \eta)y_i$$

(5.23b)

where

$$L_i = \tfrac{1}{4}(1 + \xi_i\xi)(1 + \eta_i\eta)$$

(5.23c)

In the last equation ξ_i, η_i represent the coordinates of the nodes in the ξ, η natural coordinates. For example, with reference to Figure 5.15, we write for $i = 1$, $\xi_1 = -1$, $\eta_1 = -1$; for $i = 2$ we write $\xi_2 = 1$, $\eta_2 = -1$; and so on. Equations 5.23 represents a transformation, or mapping, between the Cartesian (x, y) plane and the natural coordinate (ξ, η) plane. It is an important and widely used concept, and we present similar equations for other elements later in this chapter. Earlier in Section 5.4.3 we described a formal procedure for deriving element interpolation functions and illustrated the procedure for the rectangular element in Figure 5.10. Equations 5.23 represent interpolation functions for a general quadrilateral element and include the rectangle as a special case. We make use of equations 5.23 in later chapters and employ the equations in the computer program of Chapter 10.

5.5.3 Natural Coordinates in Three Dimensions

We can define natural coordinates for the four-node tetrahedron in a manner analogous to the procedures for the three-node triangle. The result, as the reader may expect, is a set of *volume coordinates* whose physical interpretation turns out to be a ratio of volumes in the tetrahedron. Figure 5.16 shows a typical element and defines the node-numbering scheme.

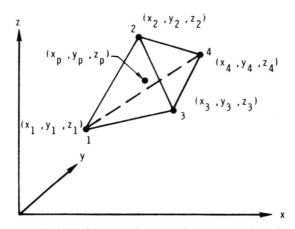

Figure 5.16 A tetrahedral element whose nodes are numbered according to the right-hand rule. The point (x_p, y_p, z_p) is some point *within* the element.

The global Cartesian coordinates and the local natural coordinates are related by

$$x = L_1 x_1 + L_2 x_2 + L_3 x_3 + L_4 x_4$$

$$y = L_1 y_1 + L_2 y_2 + L_3 y_3 + L_4 y_4 \qquad (5.24)$$

$$z = L_1 z_1 + L_2 z_2 + L_3 z_3 + L_4 z_4$$

$$L_1 + L_2 + L_3 + L_4 = 1$$

Equations 5.24 can be solved to give

$$L_i = \frac{1}{6V}(a_i + b_i x + c_i y + d_i z), \qquad i = 1, 2, 3, 4 \qquad (5.25)$$

where

$$6V = \begin{vmatrix} 1 & x_1 & y_1 & z_1 \\ 1 & x_2 & y_2 & z_2 \\ 1 & x_3 & y_3 & z_3 \\ 1 & x_4 & y_4 & z_4 \end{vmatrix} = 6\left(\begin{array}{l} \text{volume of the tetrahedron} \\ \text{defined by nodes } 1, 2, 3, 4 \end{array}\right) \qquad (5.26)$$

and, for instance,

$$a_1 = \begin{vmatrix} x_2 y_2 z_2 \\ x_3 y_3 z_3 \\ x_4 y_4 z_4 \end{vmatrix} \qquad c_1 = - \begin{vmatrix} x_2 & 1 & z_2 \\ x_3 & 1 & z_3 \\ x_4 & 1 & z_4 \end{vmatrix}$$

$$b_1 = - \begin{vmatrix} 1 & y_2 & z_2 \\ 1 & y_3 & z_3 \\ 1 & y_4 & z_4 \end{vmatrix} \qquad d_1 = - \begin{vmatrix} x_2 & y_2 & 1 \\ x_3 & y_3 & 1 \\ x_4 & y_4 & 1 \end{vmatrix} \qquad (5.27)$$

The other constants are obtained through a cyclic permutation of subscripts 1, 2, 3, and 4. Since the constants are the cofactors of the determinant in equation 5.26, attention must be given to the appropriate sign. If the tetrahedron is defined in a right-handed Cartesian coordinate system, equations 5.26 and 5.27 are valid only when the nodes are numbered so that nodes 1, 2, and 3 are ordered counterclockwise when viewed from node 4.

Figures 5.17 illustrates the physical interpretation of natural coordinates for a tetrahedron. The appropriate differentiation formulas are as follows:

$$\frac{\partial \phi}{\partial x} = \sum_{i=1}^{4} \frac{\partial \phi}{\partial L_i} \frac{\partial L_i}{\partial x}$$

$$\frac{\partial \phi}{\partial y} = \sum_{i=1}^{4} \frac{\partial \phi}{\partial L_i} \frac{\partial L_i}{\partial y} \qquad (5.28)$$

$$\frac{\partial \phi}{\partial z} = \sum_{i=1}^{4} \frac{\partial \phi}{\partial L_i} \frac{\partial L_i}{\partial z}$$

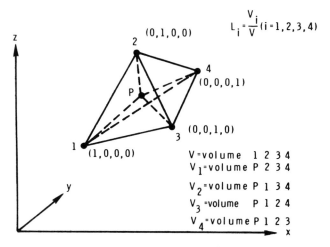

Figure 5.17 Volume coordinates.

where

$$\frac{\partial L_i}{\partial x} = \frac{b_i}{6V} \qquad \frac{\partial L_i}{\partial y} = \frac{c_i}{6V} \qquad \frac{\partial L_i}{\partial z} = \frac{d_i}{6V} \tag{5.29}$$

and the integration formula is

$$\int_{V^{(e)}} L_1^\alpha L_2^\beta L_3^\gamma L_4^\delta \, dV^{(e)} = \frac{\alpha!\beta!\gamma!\delta!}{(\alpha + \beta + \gamma + 3)!} 6V \tag{5.30}$$

Natural coordinates can also be established for general hexahedral elements in three dimensions. For the eight-node hexahedron shown in Figure 5.18 the equations relating the Cartesian coordinates and the natural coordinates are

$$x = \sum_{i=1}^{8} x_i L_i$$

$$y = \sum_{i=1}^{8} y_i L_i \tag{5.31}$$

$$z = \sum_{i=1}^{8} z_i L_i$$

where

$$L_i = \tfrac{1}{8}(1 + \xi\xi_i)(1 + \eta\eta_i)(1 + \zeta\zeta_i), \qquad i = 1, 2, \ldots, 8 \tag{5.32}$$

In equation 5.32, ξ_i, η_i, and ζ_i, $i = 1, 2, \ldots, 8$, are the coordinates of the

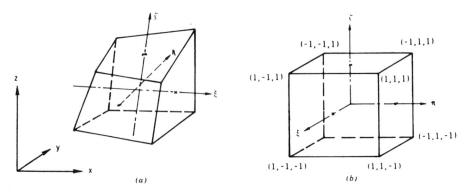

Figure 5.18 Hexahedral coordinates: (a) Cartesian coordinates; (b) Natural coordinates.

nodes in the ξ, η, ζ natural coordinate system As with the similar equations for the quadrilateral, equation 5.32 is widely used for hexahedral elements.

5.6   INTERPOLATION CONCEPTS IN ONE DIMENSION

For a number of different elements it is possible to write the nodal interpolation functions directly by employing natural coordinates *or* classical interpolation polynomials. In this section we review the features of two classical interpolation functions that we shall use later to construct the functions N_i for several types of elements.

5.6.1   Lagrange Polynomials

One type of useful interpolation function is the Lagrange polynomial, defined as

$$L_k(x) = \prod_{\substack{m=0 \\ m \neq k}}^{n} \frac{x - x_m}{x_k - x_m}$$

$$= \frac{(x - x_0) \cdots (x - x_{k-1})(x - x_{k+1}) \cdots (x - x_n)}{(x_k - x_0) \cdots (x_k - x_{k-1})(x_k - x_{k+1}) \cdots (x_k - x_n)} \quad (5.33)$$

Since $L_k(x)$ is a product of n linear factors, it is clearly a polynomial of degree n. We note that when $x = x_k$, the numerator and denominator of $L_k(x_k)$ are identical, and the polynomial has unit value. But when $x = x_m$ and $m \neq k$, the polynomial vanishes. This fact can be used to represent an arbitrary function $\phi(x)$ over an interval on the x axis. For example, suppose that $\phi(x)$ is given by discrete values at four points in the closed interval $[x_0, x_3]$ (Figure 5.19a). A polynomial of degree 3 passing through the four discrete values $\phi_i = \phi(x_i)$ $(j = 0, 1, 2, 3)$ and approximating the function $\phi(x)$ in the interval may be written at once as

$$\phi(x) \simeq \tilde{\phi}(x) = \sum_{i=0}^{3} \phi_i L_i(x) = \lfloor L \rfloor \{\phi\}$$

and we recognize that $L_i(x)$ plays the role $N_i(x)$.

The Lagrange polynomials L_i in the expression for $\tilde{\phi}(x)$ are sometimes called Lagrangian interpolation coefficients. We obtain the Lagrangian interpolation coefficients for the function shown in Figure 5.19a by substitution of $k = 0, 1, 2, 3$ in equation 5.33. After evaluating the indicated product of terms

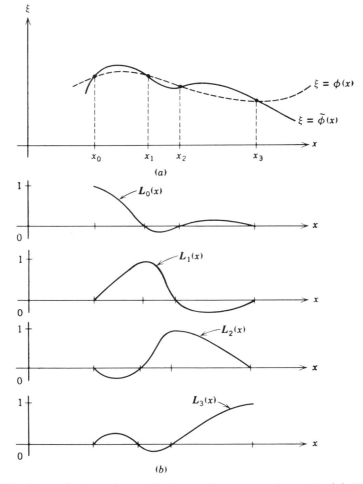

Figure 5.19 Interpolation using a third-order Lagrange polynomial: (*a*) The given function and its approximate representation; (*b*) Lagrange interpolation coefficients.

we obtain

$$L_0(x) = \frac{(x - x_1)(x - x_2)(x - x_3)}{(x_0 - x_1)(x_0 - x_2)(x_0 - x_3)}$$

$$L_1(x) = \frac{(x - x_0)(x - x_2)(x - x_3)}{(x_1 - x_0)(x_1 - x_2)(x_1 - x_3)}$$

$$L_2(x) = \frac{(x - x_0)(x - x_1)(x - x_3)}{(x_2 - x_0)(x_2 - x_1)(x_2 - x_3)}$$

$$L_3(x) = \frac{(x - x_0)(x - x_1)(x - x_2)}{(x_3 - x_0)(x_3 - x_1)(x_3 - x_2)}$$

Figure 5.19b shows how these polynomials take on the values of zero or one as required.

Since the Lagrangian coefficients possess the desired properties of the nodal interpolation functions, we may write immediately for any line element with only ϕ_i specified at the nodes (not derivatives)

$$N_i(x) = L_i(x)$$

with the order of the interpolation polynomials depending on the number of nodes we assign to the element. Since Lagrangian interpolation functions guarantee continuity of ϕ at the connecting nodes, they are suitable for elements used in problems requiring C^0 continuity.

Later in this chapter we see how these concepts can be generalized to two- and three-dimensional elements.

5.6.2 Hermite Polynomials

Just as Lagrange polynomials enable us to write at once various orders of interpolation functions when values of the field variable ϕ are specified at the nodes, Hermite polynomials enables us to construct interpolation functions when ϕ, as well as its derivatives, is specified at the nodes. An nth-order Hermite polynomial in x is denoted as $H^n(x)$ and is a polynomial of order $2n + 1$. Thus, for example, $H^1(x)$ is a first-order Hermite polynomial that is cubic in x. Hermite polynomials are useful as interpolation functions because their value and the value of their derivatives up to order n are unity or zero at the end points of the closed interval $[0, 1]$. This property can be represented symbolically if we assign two subscripts and write $H_{mi}^n(x)$, where m is the order of the derivative, and i designates either node 1 or node 2, the end nodes of a line element. The notation will be clarified with an example.

To illustrate a simple application of Hermite polynomials, suppose that we wish to construct interpolation functions for a two-node line element with ϕ and $\partial\phi/\partial x = \phi_x$ specified as degrees of freedom at each node (Figure 5.20). Since the element has four degrees of freedom (enough to determine uniquely a cubic expansion), we may select first-order or cubic Hermite

Figure 5.20   Higher-order two-node line element with four degrees of freedom.

polynomials as interpolation functions. These are defined as follows:

Node 1

$$H_{01}^1(s) = 1 - 3s^2 + 2s^3$$

$$H_{11}^1(s) = (x_2 - x_1)s(s - 1)^2 \qquad (5.34a)$$

Node 2

$$H_{02}^1(s) = s^2(3 - 2s)$$

$$H_{12}^1(s) = (x_2 - x_1)s^2(s - 1) \qquad (5.34b)$$

with

$$s = \frac{x - x_1}{x_2 - x_1}, \qquad 0 \le s \le 1, \qquad \frac{\partial}{\partial x} = \frac{1}{x_2 - x_1}\frac{\partial}{\partial s}$$

Figure 5.21 shows the behavior of the four Hermite polynomials. The superscript 1 indicates that the polynomials are first-order or cubic polynomials; the right-hand subscript, 1 or 2, identifies the node; and the left-hand subscript indicates the derivative that takes on the value of unity. For instance, H_{01}^1 is a cubic polynomial where the zeroth derivative, the function itself, takes on the value of one at node 1. For the polynomial H_{11}^1, the first derivative is one at node 1. Similar definitions apply to the polynomials for node 2. The first-order cubic polynomial is also known as a *cubic spline*.

Using the first-order Hermite polynomial we may write the interpolation function for the two-node line element in Figure 5.20 as

$$\phi = H_{01}^1\phi_1 + H_{11}^1\phi_{x1} + H_{02}^1\phi_2 + H_{12}^1\phi_{x2} \qquad (5.35)$$

or

$$\phi = \lfloor N \rfloor\{\phi\} \qquad (5.36a)$$

where

$$\lfloor N \rfloor = \lfloor H_{01}^1 \quad H_{11}^1 \quad H_{02}^1 \quad H_{12}^1 \rfloor \qquad (5.36b)$$

and

$$\{\phi\} = \begin{Bmatrix} \phi_1 \\ \phi_{x1} \\ \phi_2 \\ \phi_{x2} \end{Bmatrix} \qquad (5.36c)$$

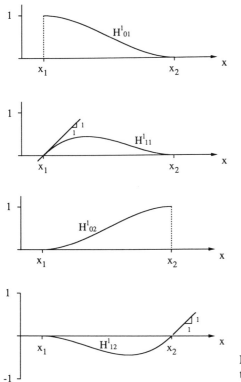

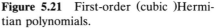

Figure 5.21 First-order (cubic)Hermitian polynomials.

We note that this interpolation model guarantees continuity of both ϕ and ϕ_x at the nodes, and hence it is suitable for one-dimensional problems requiring C^1 continuity. The cubic Hermite polynomial defined in equations 5.34–5.36 finds frequent application to structural beam elements where displacement and slope are the nodal degrees of freedom. Later in this chapter we discuss the use of Hermite polynomials for two- and three-dimensional elements and elements giving continuity of higher derivatives.

5.7 INTERNAL NODES—CONDENSATION / SUBSTRUCTURING

Before considering many different types of elements we need to discuss the problem of how to handle elements that have internal nodes. As we discussed in Section 5.3.1, these nodes, by definition, do not connect with the nodes of other elements during the assembly process, and consequently the degrees of freedom associated with internal nodes do not affect interelement continuity. Though interelement continuity is unaffected, extra degrees of freedom and internal nodes are sometimes used to improve the field variable representation *within* an element [9]. To reduce the overall size of the assembled system matrices and save equation-solving expense we may eliminate internal nodal degrees of freedom at the element level before assembly. This is done by a process called *condensation*.

As we have seen, element equations have the standard form $[K]\{x\} = \{R\}$, where $\{x\}$ is a column vector of all the degrees of freedom of the element. If the element has internal nodal degrees of freedom, we may rearrange and partition $[K]$ as follows:

$$\left[\begin{array}{c|c} [K_{11}] & [k_{12}] \\ \hline [k_{21}] & [k_{22}] \end{array}\right] \left\{\begin{array}{c} \{x_1\} \\ \hline \{x_2\} \end{array}\right\} = \left\{\begin{array}{c} \{R_1\} \\ \hline \{R_2\} \end{array}\right\} \qquad (5.37a)$$

where $\{x_2\}$ is the column vector of internal nodal degrees of freedom, and $\{R_2\}$ is the associated vector of resultant nodal actions or forcing functions. In expanded form these equations become

$$[k_{11}]\{x_1\} + [k_{12}]\{x_2\} = \{R_1\} \qquad (5.37b)$$

$$[k_{21}]\{x_1\} + [k_{22}]\{x_2\} = \{R_2\} \qquad (5.37c)$$

When equation 5.37c is solved for $\{x_2\}$ and the result is substituted into equation 5.37b, we obtain

$$\left[[k_{11}] - [k_{12}][k_{22}]^{-1}[k_{21}]\right]\{x_1\} = \{R_1\} - [k_{12}][k_{22}]^{-1}\{R_2\}$$

or

$$[\bar{K}]\{x_1\} = \{\bar{R}\} \qquad (5.38a)$$

where

$$[\bar{K}] = [K_{11}] - [K_{12}][K_{22}]^{-1}[K_{21}] \qquad (5.38b)$$

$$[\bar{R}] = [R_1] - [K_{12}][K_{22}]^{-1}[R_2] \qquad (5.38c)$$

Equation 5.38a is the condensed form that contains only the degrees of freedom associated with external nodes as unknowns. Thus in the assembly process the matrices $[\bar{K}]$ and $[\bar{R}]$ are used instead of $[K]$ and $[R]$. The internal degrees of freedom, having served their purpose in the element formulation, are then eliminated from further consideration. In addition to reducing the size of the assembled system matrix, the condensation process considerably reduces its bandwidth.

In practice, the procedure symbolized by equations 5.38a and 5.38b is actually carried out by performing a symmetric backward Gaussian elimination. This procedure, which is described in detail and programmed in FORTRAN by Wilson [10], leads to far more efficient digital computation.

The condensation process we have applied to one element can also be applied to groups of elements to eliminate nodal degrees of freedom not lying on the boundary of the assembled group. One simple example of a case where this would be desirable is shown in Figure 5.3c. The internal node resulting when four triangular elements are assembled to form a quadrilateral element would usually be eliminated. By the same procedure it is possible to devise very complex elements from assemblies of simple elements. After the individual element equations are assembled, the internal degrees of freedom are eliminated, leaving only boundary nodes and their degrees of freedom.

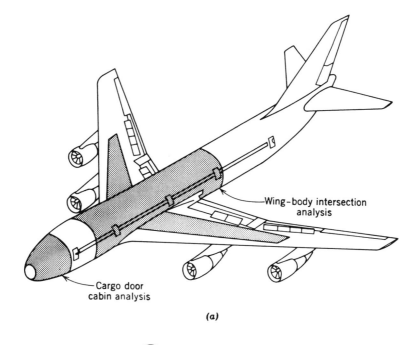

(a)

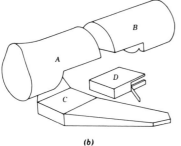

(b)

Figure 5.22 Portions of an aircraft structure treated by substructuring [11]: (*a*) Section of the 747 analyzed by the finite element method; (*b*) Schematic of the individual substructures (complex elements).

A procedure called *substructuring* in structural mechanics problems offers a physical interpretation of the process of eliminating internal nodal variables. Substructuring is a method whereby a large complex structure such as a bridge, aircraft frame, ship, or automobile body is viewed as an assemblage of a small number of very complex elements. There are several reasons for employing substructuring. Sometimes we encounter complex structures requiring hundreds or thousands of elements, nodes, and degrees of freedom. If available computers cannot solve the resulting large-order matrix equations, we must divide the structure into smaller parts that can be handled. For example, in the case of an aircraft structure we could choose to analyze separately the wings, fuselage, and tail sections. Each section would be analyzed in the usual way by discretizing it with simple finite elements, then

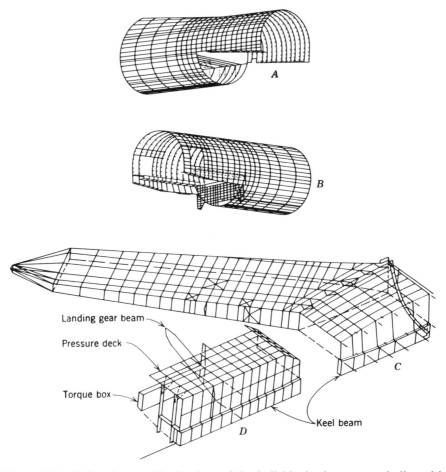

Figure 5.23 Finite element idealizations of the individual substructures indicated in Figure 5.22*b*.

TABLE 5.1.   Substructure Statistics for the 747 Problem [11]

Substructure	Nodes	Elements	Simultaneous Equations
A	1009	2897	1003
B	1014	2728	1017
C	1060	3546	~ 6000
D	894	2526	~ 5000

the individual element equations would be assembled to form the coefficient matrix $[K]$ for the whole section. Since only the external nodes of the section interconnect with the external nodes of the other sections, we use the process of condensation to eliminate the internal nodal degrees of freedom and to form a matrix $[\bar{K}]$ for each section.

Figure 5.22 shows how substructuring was used on a practical structural design problem—the analysis of portions of the 747 aircraft [11]. Four substructures where chosen for detailed discretization, as shown in Figure 5.23. Substructuring was necessary for this problem because the number of nodes and total number of degrees of freedom posed a problem far too large for computers in 1968. As Table 5.1 indicates, even the substructures themselves contain many degrees of freedom. But after the stiffness matrices for the substructures had been assembled, the internal degrees of freedom not interconnecting with the other substructures were eliminated by condensation and the problem became more tractable.

The substructures themselves may be viewed as complex elements with many boundary nodes. Such elements and their matrix equations may then be stored for later use in similar problems. It is important to note that the concept of substructuring and the formulation of complex elements, though originally devised for structural problems, can be used in the finite element analysis of any continuum problem.

5.8   TWO-DIMENSIONAL ELEMENTS

5.8.1   Elements for C^0 Problems

In this section we discuss some families of two-dimensional elements that can be used for problems requiring only continuity of the field variable ϕ at element interfaces. For such problems we usually choose the nodal values of ϕ to be the degrees of freedom of the element. To ensure interelement continuity we require that the number of nodes along a side of the element and hence the number of nodal values of ϕ be sufficient to determine uniquely the variation of ϕ along that side. For example, if ϕ is assumed to have a quadratic variation within the element and retains its quadratic behavior along the element sides, then three values of ϕ or three external nodes must lie along each side.

Although we consider only triangular and rectangular element families, the number of elements capable of satisfying C^0 continuity is infinite— infinite because we can continue to add nodes and degrees of freedom to the elements to form ever-increasing higher-order elements. In general, as the complexity of the elements is increased by adding more nodes and more degrees of freedom and using high-order polynomials, the number of elements and total number of degrees of freedom needed to achieve a given accuracy in a given problem are less than would be required if simpler elements were used. But this does not suggest that we should always favor higher-order over lower-order elements. The computational effort saved by having fewer degrees of freedom for the assembled matrices may be overshadowed by the increased effort required to formulate and compute the individual element equations.

Triangular Elements   The original three-node triangle is only the first of an infinite series of triangular elements that can be specified. Figure 5.24 shows a portion of the family of higher-order elements obtained by assigning additional external and interior nodes to the triangles. Note that each element in this series has a sufficient number of nodes (degrees of freedom in this case) to uniquely specify a complete polynomial of the order necessary to give C^0 continuity. Hence the compatibility, completeness, and geometric isotropy requirements are satisfied. For example, in Figure 5.24c the element contains 10 nodes and each side has 4 degrees of freedom. This is enough to specify a cubic variation of ϕ within the element and along the element boundaries. In general, for triangles with nodes arrayed as in Figure 5.24 a complete nth-order polynomial requires $\frac{1}{2}(n + 1)(n + 2)$ nodes for its specification.

For the three-node triangle the linear variation of ϕ is, of course,

$$\phi(x, y) = \alpha_1 + \alpha_2 x + \alpha_3 y \tag{5.8a}$$

which we have used several times. For other elements in the series we refer to Figure 5.8 to write the complete polynomial of the desired order. For the second element of the series, the six-node triangular element, the quadratic variation of ϕ is written as

$$\phi(x, y) = \alpha_1 + \alpha_2 x + \alpha_3 y + \alpha_4 x^2 + \alpha_5 xy + \alpha_6 y^2 \tag{5.8b}$$

and for the 10-node triangle the cubic variation of ϕ is written as

$$\phi(x, y) = \alpha_1 + \alpha_2 x + \alpha_3 y + \alpha_4 x^2 + \alpha_5 xy + \alpha_6 y^2$$
$$+ \alpha_7 x^3 + \alpha_8 x^2 y + \alpha_9 xy^2 + \alpha_{10} y^3 \tag{5.8c}$$

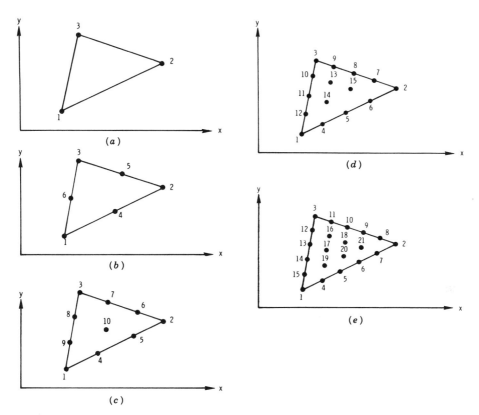

Figure 5.24 Linear and higher-order triangular elements with ϕ specified at the nodes (the nodes along any line are equally spaced): (*a*) Linear (3 nodes); (*b*) Quadratic (6 nodes); (*c*) Cubic (10 nodes); (*d*) Quartic (15 nodes); (*e*) Quintic (21 nodes).

The interpolation functions $N_i(x, y)$ could be derived using the direct approach described in Section 5.4.3. However, this is a tedious, inefficient approach which we seek to avoid. For this family of triangles, Irons et al. [7] have demonstrated that the interpolation functions for the higher-order elements can be derived in terms of the natural coordinates L_i for the linear triangle. We discuss in detail a procedure advanced by Silvester [12]. Seeking an orderly method for designating nodes within higher-order triangles, Silvester introduced a triple-index numbering scheme. After we discuss this numbering scheme we can use it to express the interpolation functions for any order of triangle. We first describe the general approach for any order triangle and then present the interpolation functions for the quadratic and cubic triangles.

The nodes of the elements in Figure 5.24 can be given the three-digit label $\alpha\beta\gamma$, where α, β, and γ are integers satisfying $\alpha + \beta + \gamma = n$, and n is the

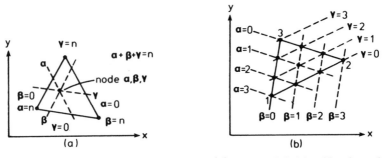

Figure 5.25 Node identification with digits: (*a*) Three-digit identification of a node within a triangle; (*b*) Example of digit node identification for the 10-node cubic triangle.

order of the interpolation polynomial for the triangle. These integers designate constant coordinate lines in the area coordinate system and indicate the number of steps or levels by which a particular node is located from a side of the triangle. Figure 5.25*a* shows this designation for a typical node within a triangle, while Figure 5.25*b* shows the complete specification of the nodes for a 10-node cubic triangle (the interior node, for instance, has the designation 111). We may use the same digit notation for the interpolation functions for the element. Employing a triple subscript, we may write $N_{\alpha\beta\gamma}(L_1, L_2, L_3)$ to denote the interpolation function for node $\alpha\beta\gamma$ as a function of the area coordinates L_1, L_2, and L_3.

Silvester [12] has shown that the interpolation functions for an *n*th-order triangular element may be expressed by the following simple and convenient formula:

$$N_{\alpha\beta\gamma}(L_1, L_2, L_3) = N_{\alpha}(L_1)N_{\beta}(L_2)N_{\gamma}(L_3)$$

where

$$N_{\alpha}(L_1) = \prod_{i=1}^{\alpha}\left(\frac{nL_1 - i + 1}{i}\right), \qquad \alpha \geq 1 \qquad (5.39)$$

$$= 1, \qquad\qquad\qquad\qquad \alpha = 0 \qquad (5.40)$$

For $N_{\beta}(L_2)$ and $N_{\gamma}(L_3)$ the formula has the same form. The symbol $\prod$ signifies as before, the product of all the terms.

Equations 5.39 and 5.40 now provide the means for constructing the interpolation functions for all higher-order elements of this series. Suppose, for example, that we want to develop the interpolation functions for the six node triangle element shown in Figure 5.26. Since the functions are quadratic, $n = 2$, and with the three-digit node notation the interpolation functions we seek are $N_{200}, N_{020}, N_{002}$ and $N_{110}, N_{011}, N_{101}$ for the three corner nodes and

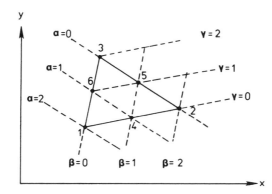

Figure 5.26 Node identification for a six-node quadratic triangle.

the three side nodes, respectively. Consider node 1 in Figure 5.26, which we designate as N_{200}, that is, $\alpha = 2$, $\beta = 0$, $\gamma = 0$. From equations 5.39 and 5.40,

$$N_\alpha = N_2 = \prod_{i=1}^{2} \left(\frac{2L_1 - i + 1}{i} \right)$$

$$= \left(\frac{2L_1 - 1 + 1}{1} \right) \left(\frac{2L_1 - 2 + 1}{2} \right)$$

$$= L_1(2L_1 - 1)$$

$$N_\beta = N_0 = 1$$

$$N_\gamma = N_0 = 1$$

Hence

$$N_{200} = N_2(L_1) N_0(L_2) N_0(L_3)$$

$$= L_1(2L_1 - 1) \tag{5.41}$$

Similarly, it is easy to show that

$$N_{020} = L_2(2L_2 - 1)$$

$$N_{002} = L_3(2L_3 - 1)$$

$$N_{110} = 4L_1 L_2 \tag{5.42}$$

$$N_{011} = 4L_2 L_3$$

$$N_{101} = 4L_1 L_3$$

Thus for the six-node quadratic triangular element shown in Figure 5.24b,

$$N_i = L_i(2L_i - 1), \qquad i = 1, 2, 3$$
$$N_4 = 4L_1L_2$$
$$N_5 = 4L_2L_3$$
$$N_6 = 4L_1L_3$$

$$(5.43)$$

In a similar way we may show for the 10 node cubic triangle in Figure 5.24c that the interpolations functions are

$$N_i = \tfrac{1}{2}L_i(3L_i - 1)(3L_i - 2), \qquad i = 1, 2, 3$$
$$N_4 = \tfrac{9}{2}L_1L_2(3L_1 - 1) \qquad N_7 = \tfrac{9}{2}L_2L_3(3L_3 - 1)$$
$$N_5 = \tfrac{9}{2}L_1L_2(3L_2 - 1) \qquad N_8 = \tfrac{9}{2}L_3L_1(3L_3 - 1) \qquad (5.44)$$
$$N_6 = \tfrac{9}{2}L_2L_3(3L_2 - 1) \qquad N_9 = \tfrac{9}{2}L_3L_1(3L_1 - 1)$$
$$N_{10} = 27L_1L_2L_3$$

Rectangular Elements   Interpolation functions for rectangular elements with sides parallel to the global axes are easily developed using the Lagrangian interpolation concepts discussed in Sections 5.6.1. Although we discussed Lagrange interpolation in only one dimension, we may generalize these concepts to two or more dimensions simply by forming products of the functions that hold for the individual one-dimensional coordinate directions.

Suppose, for instance, that we wish to derive the four interpolation functions for the four-node rectangle in Figure 5.27. We may proceed as in Section 5.4.3, but an easier way is to use Lagrangian interpolation functions and write them directly. After we define the local coordinates as shown in Figure 5.27 we may write

$$\phi(\xi, \eta) = N_1(\xi, \eta)\phi_1 + N_2(\xi, \eta)\phi_2 + N_3(\xi, \eta)\phi_3 + N_4(\xi, \eta)\phi_4$$

where

$$N_1(\xi, \eta) = L_1(\xi)L_1(\eta), \qquad N_2(\xi, \eta) = L_2(\xi)L_2(\eta), \qquad \text{etc.}$$

and the L_i are the Lagrange polynomials defined in equation 5.33.

Interpolation functions formed as products in this way satisfy the requirements of possessing unit value at the node for which they are defined and zero value at the other nodes. For instance, $L_1(\xi_1) = 1$ and $L_1(\xi_2) = L_1(\xi_3) = L_1(\xi_4) = 0$. Also, L_1 varies linearly between nodes 1 and 2 and hence preserves C^0 continuity along edge 1–2. Similar comments hold for the other

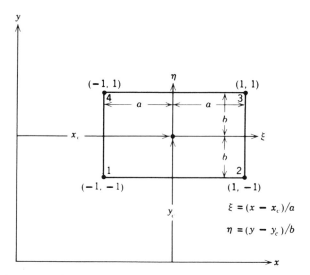

Figure 5.27 A rectangular element showing the relation between local and global coordinates. This local coordinate system is convenient for developing Lagrangian interpolation functions for rectangular elements.

first-order Lagrange polynomials for the other nodes. Referring to equation 5.33, we can write the explicit expressions at once:

$$N_1(\xi, \eta) = L_1(\xi)L_1(\eta) = \frac{\xi - \xi_2}{\xi_1 - \xi_2} \times \frac{\eta - \eta_4}{\eta_1 - \eta_4} = \frac{\xi - 1}{-1 - 1} \times \frac{\eta - 1}{-1 - 1}$$

$$(5.45a)$$

$$N_2(\xi, \eta) = L_2(\xi)L_2(\eta) = \frac{\xi - \xi_1}{\xi_2 - \xi_1} \times \frac{\eta - \eta_3}{\eta_2 - \eta_3} = \frac{\xi + 1}{1 - 1(-1)} \times \frac{\eta - 1}{-1 - 1}$$

$$(5.45b)$$

$$N_3(\xi, \eta) = L_3(\xi)L_3(\eta) = \frac{\xi - \xi_4}{\xi_3 - \xi_4} \times \frac{\eta - \eta_2}{\eta_3 - \eta_2} = \frac{\xi + 1}{1 - (-1)} \times \frac{\eta + 1}{1 - (-1)}$$

$$(5.45c)$$

$$N_4(\xi, \eta) = L_4(\xi)L_4(\eta) = \frac{\xi - \xi_3}{\xi_4 - \xi_3} \times \frac{\eta - \eta_1}{\eta_4 - \eta_1} = \frac{\xi - 1}{-1 - 1} \times \frac{\eta + 1}{1 - (-1)}$$

$$(5.45d)$$

These interpolation functions are called bilinear because each factor is a linear function of a natural coordinate. The interpolation functions given in equation 5.45 are identical to the interpolation functions that we wrote for the quadrilateral elements in natural coordinates in equation 5.23c.

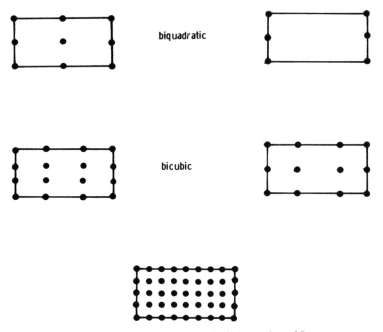

Figure 5.28   A sample of elements from the infinite series of Lagrange rectangles.

Other higher-order rectangular elements can be formulated in precisely the same way. Figure 5.28 shows just a sample of the infinite number of elements that can be constructed. Regardless of the number of nodes in the ξ and η directions, we can write the interpolation function for node k as

$$N_k(\xi, \eta) = L_k(\xi)L_k(\eta) \tag{5.46}$$

and the variation of the field variable within the element as

$$\phi(\xi, \eta) = \sum_{k=1}^{n} L_k(\xi)L_k(\eta)\phi_k$$

where n is the total number of nodes assigned to the element. The functions given by equation 5.46 are expressed in terms of local coordinates, but a simple substitution recovers their form in the global coordinate system. Note that the order of the polynomials given by equation 5.46 is all that is needed to ensure C^0 continuity along element boundaries.

The usefulness of Lagrangian elements is limited, however, because the higher-order elements contain a large number of interior nodes. These may be eliminated by the condensation process of Section 5.7 when desired, but extra manipulation is required. Another important aspect of Lagrangian elements is that the interpolation functions are never complete polynomials,

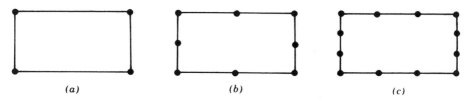

(a)   (b)   (c)

Figure 5.29 Useful elements for the "serendipity" family [13]: (a) Linear; (b) Quadratic; (c) Cubic.

and they possess geometric isotropy only when equal numbers of nodes are used in the x and y directions.

A more useful set of rectangular elements is that known as the "serendipity family,"[3] devised by Ergatoudis et al. [13] and shown in Figure 5.29. These elements contain only exterior nodes and their interpolation functions were derived by inspection. In terms of the natural coordinate system defined in Figure 5.27, the interpolation functions are, from reference 13, follows:

1. Linear element

$$N_i(\xi, \eta) = \tfrac{1}{4}(1 + \xi\xi_i)(1 + \eta\eta_i) \qquad (5.47)$$

2. Quadratic element
   a. For nodes at $\xi = \pm 1$, $\eta = \pm 1$

   $$N_i(\xi, \eta) = \tfrac{1}{4}(1 + \xi\xi_i)(1 + \eta\eta_i)(\xi\xi_i + \eta\eta_i - 1)$$

   b. For nodes at $\xi = 0$, $\eta = \pm 1$

   $$N_i(\xi, \eta) = \tfrac{1}{2}(1 - \xi^2)(1 + \eta\eta_i)$$

   c. For nodes at $\xi = \pm 1$, $\eta = 0$

   $$N_i(\xi, \eta) = \tfrac{1}{2}(1 + \xi\xi_i)(1 - \eta^2) \qquad (5.48)$$

3. Cubic element
   a. For nodes at $\xi = \pm 1$, $\eta = \pm 1$

   $$\dot{N}_i(\xi, \eta) = \tfrac{1}{32}(1 + \xi\xi_i)(1 + \eta\eta_i)\left[9(\xi^2 + \eta^2) - 10\right]$$

   b. For nodes at $\xi = \pm 1$, $\eta = \pm \tfrac{1}{3}$

   $$N_i(\xi, \eta) = \tfrac{9}{32}(1 + \xi\xi_i)(1 - \eta^2)(1 + 9\eta\eta_i) \qquad (5.49)$$

   and similarly for the other side nodes.

[3]This terminology, coined in reference 13, stems from a fairy tale of old Ceylon (once called Serendip), where there were once three princes "who in their travels were always discovering, by chance or by sagacity, [agreeable] things they did not seek."

Each of these elements preserves C^0 continuity along the element boundaries, because the interpolation functions are complete polynomials in the linear coordinates along the boundaries.

Rectangular elements, in general, are of limited use because they are not well suited for representing curved boundaries. However, an assemblage of rectangular and triangular elements, with triangular elements near the boundary can be very effective. The appeal of rectangular elements stems from the ease with which interpolation functions can be found for them.

Arbitrary Quadrilateral Elements   A major use of the Lagrangian and serendipity interpolation functions for rectangular elements is in the development of arbitrary quadrilateral elements by mapping rectangular elements into other shapes following the approach shown in Figure 5.15. The mapping approach is widely used to develop general arbitrary quadrilateral elements with both straight and curved edges. We present a detailed description applicable to arbitrary quadrilaterals in Section 5.10.

5.8.2   Elements for C^1 Problems

Constructing two-dimensional elements that can be used for problems requiring continuity of the field variable ϕ as well as its normal derivative $\partial\phi/\partial n$ along elements boundaries is far more difficult than constructing elements for C^0 continuity alone. To preserve C^1 continuity we must be sure that ϕ and $\partial\phi/\partial n$ are uniquely specified along the element boundaries by the degrees of freedom assigned to the nodes along a particular boundary.

Analysts first began to encounter difficulties in formulating elements for C^1 problems when they attempted to apply finite element techniques to plate-bending problems. For such problems the displacement of the midplane of the plate for Kirchhoff plate-bending theory is the field variable in each element, and interelement continuity of the displacement and its slope is a desirable *physical* requirement. Also, since the functional for plate bending involves second-order derivatives, continuity of slope at element interfaces is a mathematical requirement because it ensures convergence as element size is reduced. For these reasons analysts have labored to find elements giving continuity of slope and value, and we now discuss several that they have found.

Rectangular Elements   Whereas triangles are the simplest element shapes to establish C^0 continuity in two dimensions, rectangles with sides parallel to the global axes are the simplest element shapes of C^1 continuity in two dimensions. But to develop a rectangle with C^1 continuity is not so obvious as we might first surmise. Consider the rectangular element shown in Figure 5.30a. The element has length a and width b, and for convenience we

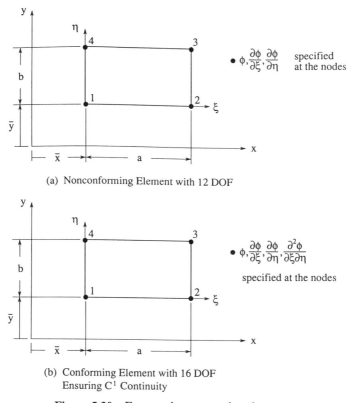

(a) Nonconforming Element with 12 DOF

(b) Conforming Element with 16 DOF
Ensuring C^1 Continuity

Figure 5.30 Four-node rectangular elements.

assume the origin of a local ξ, η coordinate system at node 1. At each node we assume as degrees of freedom ϕ, $\partial\phi/\partial x$, and $\partial\phi/\partial y$. Since we have three unknowns at each node, the element has 12 degrees of freedom. With reference to Pascal's triangle (Figure 5.8), we see that we must select an incomplete polynomial as the interpolation function. A popular choice is

$$\phi(\xi, \eta) = C_1 + C_2\xi + C_3\eta + C_4\xi^2 + C_5\xi\eta + C_6\eta^2 + C_7\xi^3$$
$$+ C_8\xi^2\eta + C_9\xi\eta^2 + C_{10}\eta^3 + C_{11}\xi^3\eta + C_{12}\xi\eta^3 \quad (5.50)$$

which is an incomplete quartic polynomial with three terms missing. With the assumed polynomial having 12 unknown constants C_1–C_{12}, we can, in principle, using the approach of Section 5.4.3 to derive equations for the interpolation function N_i, $i = 1, 2, \ldots, 12$. The question we wish to address is, will the assumed polynomial for the 12 degrees of freedom rectangular element provide C^1 continuity?

If the element provides C^1 continuity, then on a typical element edge the nodal degrees of freedom should determine ϕ, $\partial\phi/\partial\xi$, and $\partial\phi/\partial\eta$ uniquely.

Consider the edge along $\xi = 0$ connecting nodes 1 and 4. On this edge

$$\phi(0, \eta) = C_1 + C_3\eta + C_6\eta^2 + C_{10}\eta^3$$

$$\frac{\partial \phi}{\partial \eta}(0, \eta) = C_3 + 2C_6\eta + 3C_{10}\eta^2 \tag{a}$$

The four constants C_1, C_3, C_6, and C_{10} are uniquely determined by the four known degrees of freedom at nodes 1 and 4. In other words, we can write

$$\phi(0, 0) = \phi_1 = C_1$$

$$\frac{\partial \phi}{\partial \eta}(0, 0) = \frac{\partial \phi_1}{\partial \eta} = C_3$$

$$\phi(0, b) = \phi_4 = C_1 + C_3 b + C_6 b^2 + C_{10} b^3$$

$$\frac{\partial \phi}{\partial \eta}(0, b) = \frac{\partial \phi_4}{\partial \eta} = C_3 + 2C_6 b + 3C_{10} b^2 \tag{b}$$

From these four simultaneous equations we can solve for the four constants. Thus the dependent variable ϕ and the derivative $\partial \phi / \partial \eta$ are uniquely determined by the nodal degrees of freedom. But what about the normal derivative $\partial \phi / \partial \xi$ that is needed for C^1 continuity? From the assumed polynomial,

$$\frac{\partial \phi}{\partial \xi}(0, \eta) = C_2 + C_5\eta + C_9\eta^2 + C_{12}\eta^3 \tag{c}$$

Thus, in equation c for the normal derivative we have four new constants to determine: C_2, C_5, C_9, and C_{12}. But on the edge there are only two other nodal degrees of freedom that we can use:

$$\frac{\partial \phi}{\partial \xi}(0, 0) = \frac{\partial \phi_1}{\partial \xi} = C_2$$

and

$$\frac{\partial \phi}{\partial \xi}(0, b) = \frac{\partial \phi_4}{\partial \xi} = C_2 + C_5 b + C_9 b^2 + C_{12} b^3 \tag{d}$$

Hence we have two more constants than equations, and we cannot solve for C_2, C_5, C_9, and C_{12} uniquely. The result is that on the interface between two adjacent elements the normal slope is not uniquely specified by the common nodal degrees of freedom shared by the elements. Therefore the slope normal to the interface may be discontinuous, and C_1 continuity is *not* attained. An element that provides continuity of dependent variable and first derivatives at the nodes but fails to provide normal derivative continuity along an element edge is called a *nonconforming* or *incompatible* element. We discuss nonconforming elements briefly at the close of this section.

The failure of the assumed polynomial given in equation 5.50 to provide C^1 continuity suggests using additional nodal degrees of freedom and adding more terms to the assumed polynomial. Figure 5.30*b* shows a rectangle with ϕ, $\partial\phi/\partial\xi$, $\partial\phi/\partial\eta$, and $\partial^2\phi/\partial\xi\,\partial\eta$ specified at the corner nodes. This is a 16 degrees of freedom element, and we need to assume an incomplete quintic polynomial (21 terms). In principle, we can then use the approach of Section 5.4.3 to solve for the 16 unknown coefficients in terms of the nodal degrees of freedom and obtain the element interpolation functions. To avoid these difficulties we make use of the Hermite interpolation functions discussed in Section 5.6.2. Just as we have done with Lagrange formulas, we form products of one-dimensional Hermite polynomials and obtain interpolation functions ensuring continuity of ϕ and its normal derivatives $\partial\phi/\partial n$ along element boundaries. Bogner et al. [14] have shown that appropriate cubic interpretation polynomials in the expansion for ϕ are

$$\phi(\xi,\eta) = \sum_{j=1}^{4}\left[N_{1j}\phi_j + N_{2j}\left(\frac{\partial\phi}{\partial\xi}\right)_j + N_{3j}\left(\frac{\partial\phi}{\partial\eta}\right)_j + N_{4j}\left(\frac{\partial^2\phi}{\partial\xi\,\partial\eta}\right)_j \right] \quad (5.51a)$$

where

$$N_{1j}(\xi,\eta) = H^1_{0j}(\xi)H^1_{0j}(\eta)$$

$$N_{2j}(\xi,\eta) = H^1_{0j}(\xi)H^1_{1j}(\eta)$$

$$N_{3j}(\xi,\eta) = H^1_{1j}(\xi)H^1_{0j}(\eta)$$

$$N_{4j}(\xi,\eta) = H^1_{1j}(\xi)H^1_{1j}(\eta) \quad (5.51b)$$

and the Hermite polynomials are defined in equation 5.34. With the additional degrees of freedom, $\partial^2\phi/\partial\xi\,\partial\eta$, at the nodes, and the Hermite polynomials this element provides C^1 continuity and is a fully compatible element. Reference 15 discusses the element in more detail and gives explicit algebraic equations for the interpolation functions.

Triangular Elements   A number of investigators have devised complex triangular elements for C^1 continuity. Two approaches can be taken. In one approach we specify ϕ, as well as all first and second derivatives as nodal degrees of freedom. Note that the use of elements that impose continuity of second derivatives at nodes is limited to situations in which such continuity is indeed physically possible. In some cases, such as in the analysis of plate bending with inhomogeneous materials, plate properties may vary abruptly from one element to another, and the imposition of continuous second derivatives is not permissible. In a second approach we employ only ϕ and first derivatives as nodal degrees of freedom and develop the element interpolation functions by subdividing the element into subtriangles.

To illustrate the first approach, consider the six-node triangle shown in Figure 5.31a. When ϕ as well as first and second derivatives are specified at the corner nodes, there are 18 degrees of freedom. To specify a 21-term complete quintic polynomial, three more degrees of freedom are needed. The three extra degrees of freedom are obtained by specifying normal derivatives $\partial\phi/\partial n$ at the midside nodes. The element guarantees continuity of ϕ along element boundaries because along a boundary where s is the linear coordinate, ϕ varies in s as a quintic function (see Figure 5.8) with six constants that are uniquely determined by six nodal values, namely ϕ, $\partial\phi/\partial s$, and $\partial^2\phi/\partial s^2$ at each node. Slope continuity is also assured because the normal slope along each edge varies as a quartic function that is uniquely determined by five nodal variables, namely, $\partial\phi/\partial n$ and $\partial^2\phi/\partial n^2$ at each node plus $\partial\phi/\partial n$ at the midside node. Thus the element shown in Figure 5.31a provides full C^1 continuity. Further details of the element interpolation functions are given in references 16–18.

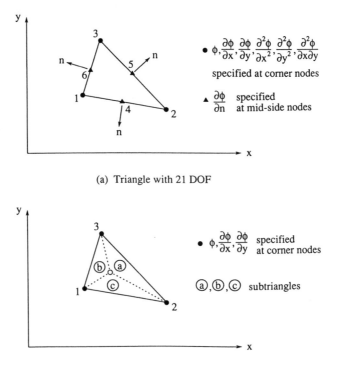

(a) Triangle with 21 DOF

(b) Triangle with 9 DOF (HCT)

Figure 5.31 Triangles with C^1 continuity.

An illustration of the second approach for developing triangles with C^1 continuity is shown in Figure 5.31b. The triangle has nine degrees of freedom, with ϕ, $\partial\phi/\partial x$, and $\partial\phi/\partial y$ specified at corner nodes. It is known as the HCT triangle (Hsieh–Clough–Tocher) and was first developed in reference 19. The approach employs three interior triangles, as shown in Figure 5.31b. Within each triangle an incomplete cubic polynomial is assumed. Internal compatibility requirements between each subtriangle are used to determine unknown coefficients in the assumed polynomials. Further details appear in the original paper [19] and the texts by Gallagher [20] and Yang [15].

Formulation of quadrilateral elements with C^1 continuity is as difficult as forming triangular elements. The most common approach is to form a quadrilateral element by combining two or more conforming triangles (Figure 5.32). For example, Clough and Felippa [21] presented a formulation for a quadrilateral element based on the combination of four conforming triangles. The final quadrilateral element has 12 degrees of freedom and provides C^1 continuity.

Because full slope continuity is difficult to achieve, many investigators have experimented with elements that guarantee slope continuity at the nodes and satisfy the other requirements but violate slope continuity along

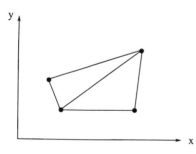

(a) Two Three-Node HCT Triangles

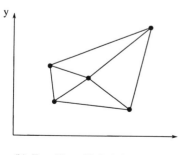

(b) Four Three-Node HCT Triangles

Figure 5.32 Quadrilateral elements giving C^1 continuity formed by assembling triangles.

the element boundaries. These are the so-called *incompatible* or *nonconforming* elements. Incompatible elements have been used with success in plate-bending problems. Although the interpolation functions for plate-bending problems must satisfy the C^1 continuity and completeness requirements of Section 3.4.5 to ensure convergence, it is sometimes possible to achieve convergence without preserving C^1 continuity. Apparently, C^1 continuity is not always a necessary condition for convergence. Experience indicates that convergence is more dependent on completeness than on the compatibility property of the element. Excellent discussions of incompatible elements appear in the texts by Gallagher [20] and Zienkiewicz [22].

The behavior of plate-bending elements has been investigated extensively. A survey [23] lists over 150 references and identifies nearly 90 different plate-bending elements.

5.9 THREE-DIMENSIONAL ELEMENTS

5.9.1 Elements for C^0 Problems

Constructing three-dimensional elements to give C^0 continuity at element interfaces follows immediately from a natural extension of the corresponding elements in two dimensions. Instead of requiring continuity of the field variable along the *edges* of the element, we require continuity on the *faces* of the elements. We must now ensure that the nodes on element interfaces uniquely define the variation of the function on the interfaces.

The three-dimensional counterparts of the triangle and the quadrilateral are the tetrahedron and the hexahedron, respectively. We shall indicate the development of interpolation functions for these elements in a fashion directly analogous to that used for the two-dimensional elements.

Tetrahedral Elements   The simplest element in three dimensions is the four-node tetrahedron (Figure 5.5*a*). By adding more nodes to this basic tetrahedron in the manner indicated in Figure 5.33 we may form an infinite family of higher-order three-dimensional elements. If we take the nodal value of the field variable ϕ as the only degree of freedom at the nodes, each layer of

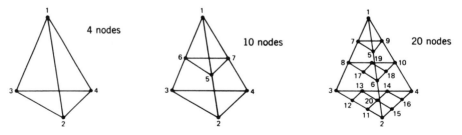

Figure 5.33   First three elements of the tetrahedral family.

nodes contains just enough degrees of freedom to uniquely specify a complete polynomial of the next highest order. To have a complete polynomial of nth order the tetrahedron must have $\frac{1}{6}(n + 1)(n + 2)(n + 3)$ nodes when ϕ is the only nodal variable. Again, this family of tetrahedral elements satisfies the compatibility, completeness, and isotropy requirements.

A linear variation of ϕ within the element would be expressed as

$$\phi = \alpha_1 + \alpha_2 x + \alpha_3 y + \alpha_4 z$$

whereas a quadratic variation is given by

$$\phi = \alpha_1 + \alpha_2 x + \alpha_3 y + \alpha_4 z + \alpha_5 x^2 + \alpha_6 y^2$$
$$+ \alpha_7 z^2 + \alpha_8 xy + \alpha_9 xz + \alpha_{10} yz$$

and so on for the higher-order elements of this series.

Interpolation functions for this family of elements can be obtained by the usual procedure involving the calculation of $[G]^{-1}$, but Silvester [24] has introduced a greatly simplified approach by providing convenient formulas analogous to those for the triangular element family. Using a four-digit subscript notation[4] for the interpolation functions, we may write

$$N_{\alpha\beta\gamma\delta}(L_1, L_2, L_3, L_4) = N_\alpha(L_1) N_\beta(L_2) N_\gamma(L_3) N_\delta(L_4) \quad (5.52a)$$

where

$$N_\alpha(L_1) = \prod_{i=1}^{\alpha} \left(\frac{nL_1 - i + 1}{i} \right), \qquad \text{for } \alpha \geq 1$$

$$= 1, \qquad \text{for } \alpha = 0 \qquad (5.52b)$$

and n is the order of the polynomial. The formulas for N_β, N_γ, and N_δ are similar. Silvester notes that these interpolation functions when evaluated on any face of the tetrahedron degenerate into precisely the interpolation functions defined previously for the corresponding triangular elements.

Hexahedral Elements   The concepts of Lagrange and Hermite interpolation for two-dimensional elements extend also to hexahedral elements in three dimensions. The first three numbers of the Lagrange hexahedral family (right prisms) are shown in Figure 5.34. As before, interpolation functions for this family of elements may be written as the product of the Lagrange polynomials in all of the orthogonal coordinate directions ξ, η, ζ (origin at the

---

[4]Since this notation is an obvious extension of that used for the triangular elements, we shall not elaborate on it further.

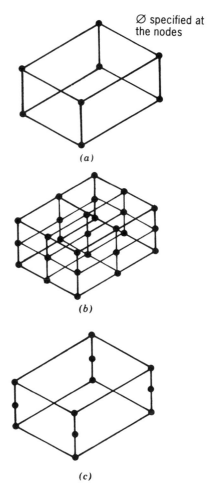

∅ specified at the nodes

(a)

(b)

Figure 5.34 Some hexahedral elements of the Lagrange family: (a) Bilinear (8 degrees of freedom); (b) Biquadratic (27 degrees of freedom); (c) Mixed linear and quadratic (12 degrees of freedom).

(c)

centroid of the element). Hence for node k we have

$$N_k(\xi, \eta, \zeta) = L_k(\xi) L_k(\eta) L_k(\zeta) \qquad (5.53)$$

where it is understood that each function L_k is properly formed to account for the number of nodes in the particular coordinate direction. As in the case of the Lagrange rectangular element, the Lagrange hexahedron contains undesirable interior nodes when higher-order elements are formed (quadratic, cubic, etc.). These interior nodes may be "condensed out" by the standard procedure (Section 5.7), but an alternative is to construct element shape functions directly, using only exterior nodes as shown in Figure 5.35.

The interpolation functions for these so-called serendipity elements are incomplete polynomials and were derived by inspection. We quote the results

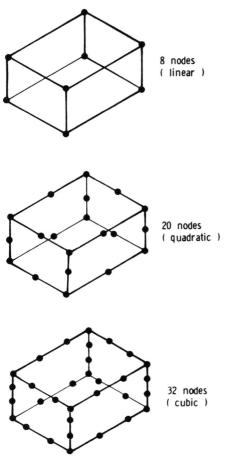

8 nodes
(linear)

20 nodes
(quadratic)

32 nodes
(cubic)

Figure 5.35 Hexahedral elements of the serendipity family containing only exterior nodes.

in the following with notation analogous to that used in Section 5.8.1:

1. Linear element

$$N_i = \tfrac{1}{8}(1 + \xi\xi_i)(1 + \eta\eta_i)(1 + \zeta\zeta_i) \qquad (5.54)$$

2. Quadratic element
   a. Corner nodes

$$N_i = \tfrac{1}{8}(1 + \xi\xi_i)(1 + \eta\eta_i)(1 + \zeta\zeta_i)(\xi\xi_i + \eta\eta_i + \zeta\zeta_i - 2) \quad (5.55a)$$

   b. Typical midside node, $\xi_i = 0$, $\eta_i = \pm1$, $\zeta_i = \pm1$

$$N_i = \tfrac{1}{4}(1 - \xi^2)(1 + \eta\eta_i)(1 + \zeta\zeta_i) \qquad (5.55b)$$

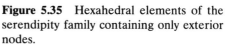

3. Cubic element
   a. Corner nodes

$$N_i = \tfrac{1}{64}(1 + \xi\xi_i)(1 + \eta\eta_i)(1 + \zeta\zeta_i)\left[9(\xi^2 + \eta^2 + \zeta^2) - 19\right] \quad (5.56a)$$

   b. Typical midside node, $\xi_i = \pm \tfrac{1}{3}$, $\eta_i = \pm 1$, $\zeta_i = \pm 1$

$$N_i = \tfrac{9}{64}(1 - \xi^2)(1 + 9\xi\xi_i)(1 + \eta\eta_i)(1 + \zeta\zeta_i) \quad (5.56b)$$

The interpolation functions given in equations 5.54–5.56 are based on a ξ, η, ζ coordinate system with the origin at the centroid of the hexahedral element. The values of ξ_i, η_i, ζ_i are the natural coordinates of the nodes.

Triangular Prisms Modeling complex-shaped, three-dimensional solution domains with hexahedral elements may pose some difficulties because these "brick"-shaped elements may not fit the boundary well. Rather than using a large number of small "bricks," it is advantageous to mix hexahedra and triangular prisms to obtain a good fit.

Interpolation functions for a family of triangular prism elements can be easily generated by forming products of the interpolation functions for the triangular cross sections with Lagrange functions or serendipity functions in the length dimension. The resulting triangular prism elements may then be assembled with Lagrange hexahedra or serendipity hexahedra, respectively. These two types of element families are shown in Figure 5.36.

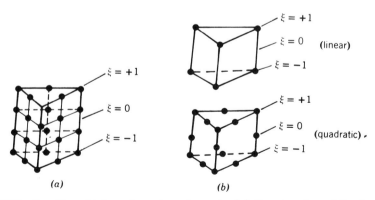

Figure 5.36 Families of triangular prism elements: (*a*) One member of the Lagrange family. (*b*) Two members of the serendipity family.

As an example of the form of the interpolation functions, we quote general results for the quadratic prism of the serendipity type (Figure 5.36b):

1. Corner nodes

$$N_i = \tfrac{1}{2}L_i(2L_i - 1)(1 + \xi) - \tfrac{1}{2}L_i(1 - \xi^2) \qquad (5.57a)$$

2. Midsides of triangles

$$N_i = 2L_iL_j(1 + \xi) \qquad (5.57b)$$

3. Midsides of rectangles

$$N_i = L_i(1 - \xi^2) \qquad (5.57c)$$

The reader may wish to verify, as an exercise, that these functions are indeed suitable.

5.9.2   Elements for C^1 Problems

We shall not discuss the formulation of three-dimensional elements for problems requiring C^1 continuity because no practical elements have yet been developed for this case. The excessive number of degrees of freedom required for such elements precludes their usefulness.

5.10   CURVED ISOPARAMETRIC ELEMENTS FOR C^0 PROBLEMS

Fitting a curved boundary with straight-sided elements (like those dealt with thus far in this chapter) often leads to a satisfactory representation of the boundary, but better fitting would be possible if curve-sided elements could be formulated for the task. If curve sided elements were available, it would be permissible to use a smaller number of larger elements and still achieve a close boundary representation. Also, in practice three-dimensional problems where the great number of degrees of freedom can overburden even the largest computers, it is sometimes essential to have a means for reducing problem size by using fewer elements. The reasoning encouraged the development of curve-sided elements. Taig [25] was the first to introduce curved quadrilateral elements; later his ideas were generalized by Irons [26] and Ergatoudis et al. [27] to other element configurations.

5.10.1   Coordinate Transformation

The essential idea underlying the development of curved-sided elements centers on mapping or transforming simple geometric shapes in some local coordinate system into distorted shapes in the global Cartesian coordinate

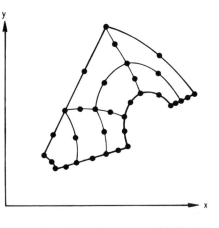

Figure 5.37 A two-dimensional domain represented by curved quadrilateral elements.

system and then evaluating the element equations for the curve-sided elements that result. An example will help to clarify the concepts. For the purpose of discussion we restrict our example to two dimensions, but all the concepts extend immediately to one dimension or three dimensions.

Suppose that we wish to represent a solution domain in x–y Cartesian coordinates by a network of curved-sided quadrilateral elements and furthermore we desire the field variable ϕ to have a quadratic variation within each element. According to our previous discussion, if we choose the nodal values of ϕ as degrees of freedom, three nodes must be associated with each side of the element. The solution domain and the desired finite element model might appear as shown in Figure 5.37. To construct one typical element of this assemblage we focus our attention on the simpler "parent" element in the ξ–η local coordinate system shown in Figure 5.38. We know from Section 5.8.1 that this element is the second member of the serendipity family of rectangular elements, and the quadratic variation of ϕ within the element may be expressed as

$$\phi(\xi, \eta) = \sum_{i=1}^{8} N_i(\xi, \eta) \phi_i \tag{5.58}$$

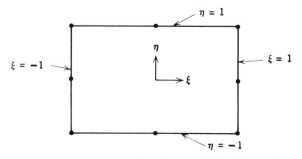

Figure 5.38   Parent rectangular element in local coordinates.

where N_i are the serendipity functions given in equations 5.48. The nodes in the $\xi-\eta$ plane may be mapped into corresponding nodes in the $x-y$ plane by defining the relations

$$x = \sum_{i=1}^{8} F_i(\xi, \eta) x_i, \qquad y = \sum_{i=1}^{8} F_i(\xi, \eta) y_i \qquad (5.59)$$

The extension of this mapping procedure to elements with a different number of nodes and elements in other dimensions is obvious.

For this example the mapping functions F_i must be quadratic, since the curved boundaries of the element in the $x-y$ plane need three points for their unique specification, and the F_i should take on the proper values of unity and zero when evaluated at the nodes in the $\xi-\eta$ plane. Functions meeting these requirements are precisely the quadratic serendipity interpolation functions presented in Section 5.8.1. Hence we can write

$$x = \sum_{i=1}^{8} N_i(\xi, \eta) x_i \qquad y = \sum_{i=1}^{8} N_i(\xi, \eta) y_i \qquad (5.60)$$

where the N_i are given by equations 5.48. The mapping defined in equations 5.60 results in a curved-side quadratic element of the type shown in Figure 5.39. For this particular element the functional representation of the field variable and the functional representation of its curved boundaries are expressed by interpolation functions of the same order. Curve-sided elements formulated in this way are called *isoparametric* elements. Different terminology is used to describe curve-sided elements whose geometry and field variable representations are described by polynomials of different order. The number of nodes used to define a curved element may be different from the number at which the element degrees of freedom are specified. In contrast to isoparametric elements, we define *subparametric* elements as elements whose geometry is described by a polynomial of order lower than that used for the field variable, and *superparametric* elements as those whose geometry is described by a higher-order polynomial. Of the three categories of curve-sided elements, isoparametric elements are the most commonly used, but some

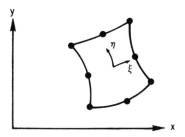

Figure 5.39 A curve-sided quadrilateral element resulting from mapping the rectangular parent element.

forms of subparametric and superparametric elements are employed. In this discussion we confine our attention to isoparametric elements.

When writing equations like equations 5.60, we assume that indeed the transformation between the local ξ–η coordinates and the global x–y coordinates is unique; that is, we assume that each point in one system has a corresponding point in the other system. If the transformation is nonunique, we can expect violent and undesirable distortions in the x–y system that may fold the curved element back upon itself. A method for checking for nonuniqueness and violent distortion will be pointed out later when we discuss the derivation of the curved-element properties.

An important consideration in the construction of curved elements is preservation of the continuity conditions in the global coordinate system. In this regard, Zienkiewicz et al. [28] has advanced three useful guidelines:

1. If two adjacent curved elements are generated from parent elements whose interpolation functions satisfy interelement continuity, these curved elements will be continuous.

2. If the interpolation functions are given in the local coordinate system and they ensure continuity of ϕ in the parent element, ϕ will also be continuous in the curved element.

3. The isoparametric element formulation offers the attractive feature that if the completeness criterion (Section 3.4.5) is satisfied in the parent element, it is automatically satisfied in the curved element.

5.10.2   Evaluation of Element Matrices

Having described the element shape via equations of the form of equation 5.59, we face the task of evaluating the element equations by carrying out the usual integrations appearing in them. In general, the terms in the element equations will contain integrals of the form

$$\int_{A^{(e)}} f\left(\phi, \frac{\partial \phi}{\partial x}, \frac{\partial \phi}{\partial y}\right) dx\, dy \qquad (5.61)$$

where $A^{(e)}$ is the area of the curved element. Since ϕ is expressed as a function of the local coordinates ξ and η as in equation 5.58, it is necessary to express $\partial\phi/\partial x$, $\partial\phi/\partial y$, and $dx\, dy$ in terms of ξ and η also. This can be done as follows. From equation 5.58,

$$\frac{\partial \phi}{\partial x} = \sum_{i=1}^{8} \frac{\partial N_i}{\partial x}\phi_i \qquad \frac{\partial \phi}{\partial y} = \sum_{i=1}^{8} \frac{\partial N_i}{\partial y}\phi_i \qquad (5.62a)$$

Hence we must express $\partial N_i/\partial x$ and $\partial N_i/\partial y$ in terms of ξ and η. Because of the inverse form of equations 5.60, we write, by the chain rule of differentiation,

$$\frac{\partial N_i}{\partial \xi} = \frac{\partial N_i}{\partial x}\frac{\partial x}{\partial \xi} + \frac{\partial N_i}{\partial y}\frac{\partial y}{\partial \xi}$$

$$\frac{\partial N_i}{\partial \eta} = \frac{\partial N_i}{\partial x}\frac{\partial x}{\partial \eta} + \frac{\partial N_i}{\partial y}\frac{\partial y}{\partial \eta}$$

or

$$\left\{\begin{matrix}\frac{\partial N_i}{\partial \xi}\\[2mm]\frac{\partial N_i}{\partial \eta}\end{matrix}\right\} = \begin{bmatrix}\frac{\partial x}{\partial \xi} & \frac{\partial y}{\partial \xi}\\[2mm]\frac{\partial x}{\partial \eta} & \frac{\partial y}{\partial \eta}\end{bmatrix}\left\{\begin{matrix}\frac{\partial N_i}{\partial x}\\[2mm]\frac{\partial N_i}{\partial y}\end{matrix}\right\} = [J]\left\{\begin{matrix}\frac{\partial N_i}{\partial x}\\[2mm]\frac{\partial N_i}{\partial y}\end{matrix}\right\} \qquad (5.62b)$$

where $[J]$ defines a Jacobian matrix,

$$[J] = \begin{bmatrix}\frac{\partial x}{\partial \xi} & \frac{\partial y}{\partial \xi}\\[2mm]\frac{\partial x}{\partial \eta} & \frac{\partial y}{\partial \eta}\end{bmatrix} \qquad (5.63)$$

We evaluate an element Jacobian matrix using the coordinate transformation equation 5.60. For the eight-node quadratic element

$$[J(\xi, \eta)] = \begin{bmatrix}\sum_{i=1}^{8}\frac{\partial N_i}{\partial \xi}(\xi, \eta)x_i & \sum_{i=1}^{8}\frac{\partial N_i}{\partial \xi}(\xi, \eta)y_i\\[2mm]\sum_{i=1}^{8}\frac{\partial N_i}{\partial \eta}(\xi, \eta)x_i & \sum_{i=1}^{8}\frac{\partial N_i}{\partial \eta}(\xi, \eta)y_i\end{bmatrix} \qquad (5.64)$$

To define the desired derivative of ϕ we must invert equation 5.62b, and this involves finding the inverse of the Jacobian matrix as indicated:

$$\left\{\begin{matrix}\frac{\partial N_i}{\partial x}\\[2mm]\frac{\partial N_i}{\partial y}\end{matrix}\right\} = [J]^{-1}\left\{\begin{matrix}\frac{\partial N_i}{\partial \xi}\\[2mm]\frac{\partial N_i}{\partial \eta}\end{matrix}\right\}, \qquad i = 1, 2, \ldots, 8 \qquad (5.65)$$

Now with this equation expressions for $\partial\phi/\partial x$ and $\phi\phi/\partial y$ can be found directly. From equation 5.62a,

$$\left\{\begin{array}{c} \dfrac{\partial \phi}{\partial x} \\[2mm] \dfrac{\partial \phi}{\partial y} \end{array}\right\} = \left[\begin{array}{cccc} \dfrac{\partial N_1}{\partial x} & \dfrac{\partial N_2}{\partial x} & \cdots & \dfrac{\partial N_8}{\partial x} \\[3mm] \dfrac{\partial N_1}{\partial y} & \dfrac{\partial N_2}{\partial y} & \cdots & \dfrac{\partial N_8}{\partial y} \end{array}\right] \left\{\begin{array}{c} \phi_1 \\ \phi_2 \\ \vdots \\ \phi_8 \end{array}\right\}$$

hence, using equation 5.65, we find

$$\left\{\begin{array}{c} \dfrac{\partial \phi}{\partial x} \\[2mm] \dfrac{\partial \phi}{\partial y} \end{array}\right\} = [J]^{-1} \left[\begin{array}{cccc} \dfrac{\partial N_1}{\partial \xi} & \dfrac{\partial N_2}{\partial \xi} & \cdots & \dfrac{\partial N_8}{\partial \xi} \\[3mm] \dfrac{\partial N_1}{\partial \eta} & \dfrac{\partial N_2}{\partial \eta} & \cdots & \dfrac{\partial N_8}{\partial \eta} \end{array}\right] \left\{\begin{array}{c} \phi_1 \\ \phi_2 \\ \vdots \\ \phi_8 \end{array}\right\} \tag{5.66}$$

To complete the evaluation of the integral we need to express the element of area $dx\,dy$ in terms of $d\xi\,d\eta$. Texts on advanced calculus show that

$$dx\,dy = |J|\,d\xi\,d\eta \tag{5.67}$$

where $|J|$ is the determinant of $[J]$. The operations indicated in equations 5.65 and 5.67 depend on the existence of $[J]^{-1}$ for each element of the assemblage, and the coordinate mapping described by equations 5.60 is unique only if $[J]^{-1}$ exists. We can test for uniqueness and acceptable distortion by evaluating $|J|$ and checking its sign. If $|J| \neq 0$ and the sign of $|J|$ does not change in the solution domain, we can be assured of an acceptable mapping.

With these transformations integrals such as expression 5.61 reduce to the form

$$\int_{-1}^{1}\int_{-1}^{1} f'(\xi, \eta)\,d\xi\,d\eta \tag{5.68}$$

where f' is the transformed function f.

Although the integration limits are now those of the simple parent element, the transformed integrand f' is not a simple function that permits closed-form integration. For this reason it is necessary to resort to numerical integration, but this poses no particular difficulty. Always associated with the procedure of numerical integration is the question of how accurately the integration needs to be done to ensure convergence and to guard against making the resulting system matrix singular. Irons [26, 27] provides the following guideline for isoparametric element formulations: Convergence of the finite element process should occur if the numerical integration is accurate enough to evaluate the area or volume of the curved element exactly.

The sample isoparametric element we have considered is just one of many possibilities. Actually, we can start with basic elements in any dimension, elements that may be described by local coordinates ξ, η, ζ or natural coordinates, L_1, L_2, L_3, and so on, and transform these into curved elements. When the parent element such as a triangle or a tetrahedron is expressed in terms of natural coordinates, it is necessary to express one of the L's in terms of the others, since not all the natural coordinates are independent. Also, in this case the limits of integration must be changed to correspond to the boundaries of the element. In any event, numerical integration is still necessary.

5.10.3   Example of Isoparametric Element Matrix Evaluation

To illustrate the evaluation of an isoparametric element matrix we consider the coefficient matrix $[K]^{(e)}$ derived for Poisson's equation in Section 4.2.2. In matrix form a typical element coefficient matrix can be written (with $k_x = k_y = 1$) as

$$[K]^{(e)} = \iint_{A^{(e)}} [B(x,y)]^T [B(x,y)] \, dx \, dy \qquad \text{(a)}$$

where

$$[B(x,y)] = \begin{bmatrix} \dfrac{\partial N_1}{\partial x} & \dfrac{\partial N_2}{\partial x} & \cdots & \dfrac{\partial N_r}{\partial x} \\[2mm] \dfrac{\partial N_1}{\partial y} & \dfrac{\partial N_2}{\partial y} & \cdots & \dfrac{\partial N_r}{\partial y} \end{bmatrix}$$

and r denotes the number of element nodes. Using equation 5.65, we express $[B(x, y)]$ in terms of the ξ, η coordinates as

$$[B(\xi,\eta)] = [J(\xi,\eta)]^{-1} \begin{bmatrix} \dfrac{\partial N_1}{\partial \xi} & \dfrac{\partial N_2}{\partial \xi} & \cdots & \dfrac{\partial N_r}{\partial \xi} \\[2mm] \dfrac{\partial N_1}{\partial \eta} & \dfrac{\partial N_2}{\partial \eta} & \cdots & \dfrac{\partial N_r}{\partial \eta} \end{bmatrix} \qquad \text{(b)}$$

where the Jacobian matrix is defined by equation 5.64 but with the summation over r element nodes. Using the area transformation, equation 5.67, we can write

$$[K]^{(e)} = \int_{-1}^{1} \int_{-1}^{1} [B(\xi,\eta)]^T [B(\xi,\eta)] \, |J(\xi,\eta)| \, d\xi \, d\eta \qquad \text{(c)}$$

where $[B(\xi, \eta)]$ is defined by equation b and the integration occurs over the parent element (Figure 5.38). The terms of the coefficient matrix are computed by numerical integration. The technique most often used is the Legendre–Gauss method, which appears in Section 10.6. Equation c, when evaluated by the Gauss numerical integration formula, is

$$[K]^{(e)} = \sum_{i=1}^{NG} \sum_{j=1}^{NG} W_i W_j [B(\xi_i, \eta_j)]^T [B(\xi_i, \eta_j)] |J(\xi_i, \eta_j)| \qquad \text{(d)}$$

where W_i and W_j denote Gauss weight factors, ξ_i and η_j denote Gauss integration points, and NG is the number of Gauss points in each direction. Gauss integration point coordinates and weight factors appear in Table 10.1. Equation d is a typical form employed in the computation of a coefficient matrix of an isoparametric element. With a digital computer such element matrices are easily and routinely evaluated. Subroutines for a four-node isoparametric quadrilateral element appear in the computer program of Chapter 10. Readers may wish to study these routines to gain a more complete understanding of the practical implementation of isoparametric elements.

5.11 CLOSURE

Many different finite elements suitable for a wide variety of problems were presented in this chapter. Beginning with the basic ideas of polynomial interpolation functions and the selection of the order of the polynomial, we discussed general procedures for constructing finite elements. From linear interpolation concepts we saw how local, natural coordinate systems can be established for line, triangle, and tetrahedron elements. With these natural coordinates we saw how it is possible to write down interpolation functions at once for any members of the higher-order elements in the triangular and tetrahedral families. The interpolation concepts associated with Lagrange and Hermite polynomials also permit direct construction of elements. Though the list of elements we compiled for C^0 and C^1 problems is by no means exhaustive, it is sufficiently extensive to make possible the finite element solution of many problems of continuum mechanics.

Now that the physical and mathematical foundations of the finite element method have been presented, we consider in the following chapters applications of the method in the realms of elasticity, general field problems, heat transfer, and fluid mechanics. In these application chapters we discuss the special techniques employed for each class of problem.

REFERENCES

1. E. D. Rainville, *Elementary Differential Equations*, 6th ed., Macmillan, New York, 1981.
2. B. Carnahan, H. A. Luther, and J. O. Wilkes, *Applied Numerical Methods*, Wiley, New York, 1969.

3. J. Krahula and J. Polhemus, "Use of Fourier Series in the Finite Element Method," *AIAA J.*, Vol. 6, No. 4, April 1968, pp. 726–727.

4. S. Chakrabarti, "Trigonometric Function Representations for Rectangular Plate Bending Elements," *Int. J. Numer. Methods Eng.*, Vol. 3, No. 2, 1971, pp. 261–273.

5. P. Dunne, "Complete Polynomial Displacement Fields for the Finite Element Method," *Aeronaut. J.*, Vol. 72, March 1968, pp. 246–247.

6. R. L. Taylor, "On Completeness of Shape Functions for Finite Element Analysis," *Int. J. Numer. Methods in Eng.*, Vol. 4, No. 1, 1972, pp. 17–22.

7. M. B. Irons, J. G. Ergatoudis, and O. C. Zienkiewicz, Discussion of the Paper by P. Dunne, *J. Royal Aeronaut. Soc.*, Vol. 72, March 1968, pp. 709–711.

8. M. A. Eisenberg and L. E. Malvern, "On Finite Element Integration in Natural Coordinates," *Int. J. Numer. Methods Eng.*, Vol. 7, 1973, pp. 574–575.

9. T. H. H. Pian, "Derivation of Element Stiffness Matrices," *AIAA J.*, Vol. 2, No. 3, 1964, pp. 576–577.

10. E. L. Wilson, "The Static Condensation Algorithm," *Int. J. Numer. Methods Eng.*, Vol. 8, No. 1, 1974, pp. 198–203.

11. S. D. Hansen, G. L. Anderton, N. E. Connacher, and C. S. Dougherty, "Analysis of the 747 Aircraft Wing–Body Intersection," *Proceedings of 2nd Conference on Matrix Methods in Structural Mechanics*, Wright–Patterson Air Force Base, Dayton, OH, October 15–17, 1968.

12. P. Silvester, "Higher-Order Polynomial Triangular Finite Elements for Potential Problems," *Int. J. Eng. Sci.*, Vol. 7, No. 8, pp. 849–861.

13. J. G. Ergatoudis, B. M. Irons, and O. C. Zienkiewicz, "Curved Isoparametric Quadrilateral Elements for Finite Element Analysis," *Int. J. Solids Struct.*, Vol. 4, 1968, pp. 31–42.

14. F. K. Bogner, R. L. Fox, and L. A. Schmit, "The Generation of Inter-element–Compatible Stiffness and Mass Matrices by the Use of Interpolation Formulae," *Proceedings of 1st Conference on Matrix Methods in Structural Mechanics*, Wright–Patterson Air Force Base, Dayton, OH, 1968.

15. T. Y. Yang, *Finite Element Structural Analysis*, Prentice-Hall, Englewood Cliffs, NJ, 1986.

16. J. H. Argyris, I. Fried, and D. W. Scharpf, "The TUBA Family of Plate Elements for the Matrix Displacement Method," Technical Note, *J. Royal Aeronaut. Soc.*, Vol. 72, August 1968, pp. 701–709.

17. K. Bell, "A Refined Triangular Plate Bending Finite Element," *Int. J. Numer. Methods Eng.*, Vol. 1, No. 1, January 1969, pp. 101–122.

18. B. M. Irons, "A Conforming Quartic Triangular Element for Plate Bending," *Int. J. Numer. Methods Eng.*, Vol. 1, 1969, pp. 29–46.

19. R. W. Clough and J. L. Tocher, "Finite Stiffness Matrices for the Analysis of Plate Bending," *Proceedings of 1st Conference on Matrix Methods in Structural Mechanics*, Wright–Patterson Air Force Base, Dayton, OH, October 1965.

20. R. H. Gallagher, *Finite Element Analysis Fundamentals*, Prentice-Hall, Englewood Cliffs, NJ, 1975.

21. R. W. Clough and C. A. Felippa, "A Refined Quadrilateral Element for Analysis of Plate Bending," *Proceedings of 2nd Conference on Matrix Methods in Structural Mechanics*, Wright–Patterson Air Force Base, Dayton, OH, October 1968.

22. O. C. Zienkiewicz, *The Finite Element Method*, 3d, ed., McGraw-Hill, New York, 1977.
23. M. M. Hrabok and T. M. Hrudey, "A Review and Catalog of Plate Bending Finite Elements," *Computers and Structures*, Vol. 19, No. 3, 1984, pp. 479–495.
24. P. Silvester, "Tetrahedral Polynomial Finite Elements for the Helmholtz Equation," *Int. J. Numer. Methods Eng.*, Vol. 4, No. 3, 1972, pp. 405–413.
25. I. C. Taig, "Structural Analysis by the Matrix Displacement Method," *English Electric Aviation Rep.*, SO-17, 1961.
26. B. M. Irons, "Engineering Application of Numerical Integration in Stiffness Method," *AIAA J.*, Vol. 14, 1966, pp. 2035–2037.
27. J. Ergatoudis, B. Irons, and O. C. Zienkiewicz, "Curved Isoparametric Quadrilateral Elements for Finite Element Analysis," *Int. J. Solids Struct.*, Vol. 4, 1968, pp. 31–42.
28. O. C. Zienkiewicz, B. M. Irons, J. Ergatoudis, S. Ahmad, and F. C. Scott, "Iso Parameter and Associated Element Families for Two- and Three-Dimensional Analysis," in *Finite Elements Methods in Stress Analysis*, I. Holand and K. Bell (eds.), Tapir Press, Trondheim, Norway, 1969, pp. 383–432.

PROBLEMS

1. Use the method of Section 5.4.3 to derive the interpolation functions N_i, $i = 1, 2, 3$ for a three-node quadratic line element. Use the node locations shown in Figure 3.7.

2. Repeat Problem 1, but use the Lagrange polynomial approach described in Section 5.6.1.

3. Use the integration formula, equation 5.14, to evaluate the following integrals for the two-node linear one-dimensional element shown in Figure 5.12:

   a. $\displaystyle\int_{x_1}^{x_2} \{L\}\lfloor L \rfloor\, dx$

   b. $\displaystyle\int_{x_1}^{x_2} \{L\}\, dx$

   c. $\displaystyle\int_{x_1}^{x_2} (L_1\phi_1 + L_2\phi_2)\begin{Bmatrix} L_1 \\ L_2 \end{Bmatrix}\lfloor L_1 \quad L_2 \rfloor\, dx$

4. Use the integration formula, equation 5.22, to evaluate the following integrals for the three-node linear triangle shown in Figure 5.13:

   a. $\displaystyle\int_A \lfloor L \rfloor (x_1 L_1 + x_2 L_2 + x_3 L_3)\, dA$

   b. $\displaystyle\int_A \{L\}(x_1 L_1 + x_2 L_2 + x_3 L_3)\, dA$

5. A cantilever beam of length $3L$ with flexural rigidity EI is loaded with a tip force P. A finite element model with three nodes (Figure P5.5) yields the following equilibrium equations:

$$\frac{EI}{L^3}\begin{bmatrix} 24 & 0 & -12 & -6L & 0 & 0 \\ 0 & 8L^2 & 6L & 2L^2 & 0 & 0 \\ -12 & 6L & 24 & 0 & -12 & -6L \\ -6L & 2L^2 & 0 & 8L^2 & 6L & 2L^2 \\ 0 & 0 & -12 & 6L & 12 & 6L \\ 0 & 0 & -6L & 2L^2 & 6L & 4L^2 \end{bmatrix}\begin{Bmatrix} w_1 \\ \theta_1 \\ w_2 \\ \theta_2 \\ w_3 \\ \theta_3 \end{Bmatrix} = \begin{Bmatrix} 0 \\ 0 \\ 0 \\ 0 \\ P \\ 0 \end{Bmatrix}$$

where w_i, $i = 1, 2, 3$ are the beam nodal displacements and θ_i are the beam nodal rotations. Use the condensation procedure described in Section 5.7 to eliminate the rotational degrees of freedom θ_i and compute a condensed coefficient matrix $[\bar{K}]$ and condensed load vector $\{\bar{R}\}$ for computing the unknown displacements w_i. Do not solve for w_i.

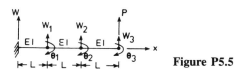

Figure P5.5

6. Derive the interpolation functions $N_i(\xi, \eta)$, $i = 1, 2, 3, 4$, for the four-node triangular element shown in Figure P5.6. Assume that node 4 is located at the element centroid.

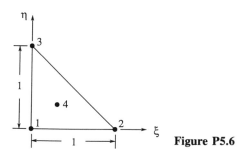

Figure P5.6

7. For the five-node rectangular element shown in Figure P5.7, an interpolation polynomial

$$\phi = C_1 + C_2\xi + C_3\eta + C_4\xi\eta + C_5(\xi^2 + \eta^2)$$

is proposed. Derive the interpolation functions N_i, $i = 1, 2, 3, 4, 5$.

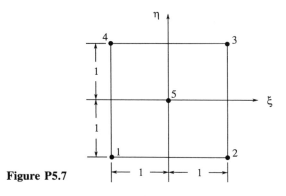

Figure P5.7

8. Verify the interpolation functions, equation 5.43, for the six-node quadratic triangular element.

9. Verify the interpolation functions, equation 5.44, for the 10-node cubic triangular element.

10. Evaluate the integrals

   a. $\int_A N_1 N_2 \, dA$

   b. $\int_A \{N\} \, dA$
   where N_i, $i = 1, 2, \ldots, 6$ are the interpolation functions for the six-node quadratic triangle (Figure 5.26).

11. Use Lagrange interpolation polynomials to derive equations for the interpolation functions N_i, $i = 1, 2, \ldots, 9$ for the nine-node biquadratic Lagrange rectangular element (Figure 5.28). Sketch the shapes of typical interpolation functions for corner and midside nodes.

12. Sketch the shapes of typical interpolation functions of the linear, quadratic, and cubic serendipity family (equations 5.47–5.49).

13. For the 16 degrees of freedom rectangle shown in Figure 5.30b, the general form of the interpolation functions are given in equations 5.51. Derive the explicit equations for the interpolation functions $N_{1j}, N_{2j}, N_{3j}, N_{4j}$, $j = 1, 2, 3, 4$.

14. For the 16 degrees of freedom rectangular element of the preceding problem show that ϕ, $\partial\phi/\partial\xi$, and $\partial\phi/\partial\eta$ on the element edge $\xi = 0$ are determined uniquely by the nodal degrees of freedom at nodes 1 and 4.

15. Derive the interpolation functions for the linear serendipity triangular prism element shown in Figure 5.36b.

16. For the four-node quadrilateral element shown in Figure P5.16 we are given that the values of the dependent variable ϕ at the nodes are $\phi_1 = 3.0$, $\phi_2 = 3.5$, $\phi_3 = 4.0$, and $\phi_4 = 5.0$. Compute the values of the derivatives $\partial\phi/\partial x$ and $\partial\phi/\partial y$ at the element centroid. Use the approach described in Section 5.10.2.

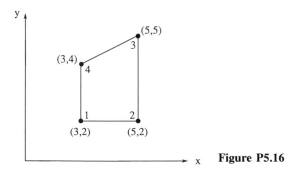

Figure P5.16

17. For the four-node isoparametric quadrilateral element with interpolation functions given by equation 5.47 set up the equations to evaluate the "mass" matrix

$$[M] = \iint_{A^{(e)}} \{N\}\lfloor N\rfloor\, dx\, dy$$

18. Set up the equations for computing the load vector

$$\{R\} = \int_{A^{(e)}} \{N\}\, dA$$

for Poisson's equation for the linear isoparametric quadrilateral shown in Figure 5.15. How would you modify the computation for an eight-node quadratic isoparametric element?

19. Write a computer program to evaluate the coefficient matrix $[K]^{(e)}$, equation d in Section 5.10.3, for a four-node linear isoparametric quadrilateral element. Investigate the number of Gauss points required to compute $[K]^{(e)}$ by comparing the computer solution with the exact coefficient matrix for a linear square element. Use one, four, and nine Gauss points in your evaluation. For a linear square element

$$[K]^{(e)} = \frac{1}{6}\begin{bmatrix} 4 & -1 & -2 & -1 \\ -1 & 4 & -1 & -2 \\ -2 & -1 & 4 & -1 \\ -1 & -2 & -1 & 4 \end{bmatrix}$$

PART II

6

ELASTICITY PROBLEMS

6.1 INTRODUCTION

Most of the past applications of the finite element method to solid mechanics problems have relied on a variational principle for the derivation of the element equations. This is due in large part to the fact that most problems in solid mechanics have several variational principles based on physical interpretations with which engineers are quite comfortable. In deference to this classical approach we derive the element equations for general three-dimensional linear elasticity using the minimum potential energy principle covered in Appendix C. The unknowns for which we solve in this formulation are the displacements, and this formulation is known as the *displacement formulation*. Other variational principles naturally lead to *force formulations*, where the forces are the unknowns, and some to *mixed* or *hybrid formulations* where there is some combination of both forces and displacements as unknowns. Pian and Tong [1] have tabulated several variational formulations for finite element applications in elasticity (see Table 6.1). For particular problems one principle may be more suitable than another, but for a large class of problems the displacement formulation is the simplest to apply. Consequently, it remains the most widely used.

After deriving a displacement formulation with the minimum potential energy principle, we derive the same element equations from the three -dimensional governing differential equations using a more general mathematical approach in the form of the Galerkin method so that the reader is able to see the details of both approaches. The introductory nature of this text precludes the discussion of more advanced approaches beyond the displacement formulation, but readers interested in other methods may consult the references given in Table 6.1.

6.2 GENERAL FORMULATION FOR THREE-DIMENSIONAL PROBLEMS

6.2.1 Problem Statement

We consider a continuous three-dimensional elastic solid with a volume Ω and a surface S (Figure 6.1). Within the volume the solid experiences body forces $\mathbf{F}_b$. On the surface S displacement components are prescribed on an area S_1, and surface tractions $\mathbf{T}$ are prescribed on an area S_2. We wish to determine the unknown displacements and stress components throughout the solid. In the following finite element derivations we make frequent use of the basic equations of elasticity from Appendix C.

6.2.2 The Variational Method

To model three-dimensional linear elasticity we employ Hooke's law with an initial strain term

$$\{\sigma\} = [C]\{\epsilon\} - [C]\{\epsilon_0\} \tag{6.1}$$

TABLE 6.1. Classification of Finite Element Formulations in Linear Elasticity Based on Variational Principles

Model	Variational Principle	Inside Each Element	Along Interelement Boundary	Unknown in Final Equations	References
Displacement	Minimum potential energy	Continuous displacements	Displacement compatibility	Nodal displacements	2
Force	Minimum complementary energy	Continuous and equilibrating stresses	Equilibrium boundary tractions	Stress parameters Generalized nodal displacements	3, 4 5
Hybrid 1	Modified complementary energy	Continuous and equilibrating stresses	Assumed compatible displacements	Nodal displacements	6
Hybrid 2	Modified potential energy	Continuous displacements	Assumed equilibrating boundary tractions	Displacement parameters and boundary forces	7
Hybrid 3	Modified potential energy	Continuous displacements	Assumed boundary tractions for each element and assumed boundary displacements	Nodal displacements	8
Mixed (plate-bending problems)	Reissner's principle	Continuous stresses and displacements	Combinations of boundary displacements and tractions	Combination of boundary displacements and tractions	9

Source. After Pian and Tong [1].

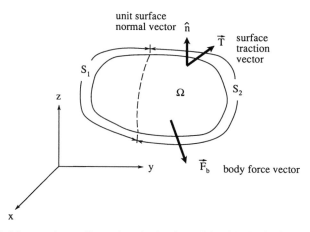

Figure 6.1 Arbitrary three-dimensional elastic solid with body forces and surface tractions.

where the vector of stress components is

$$\{\sigma\} = \begin{Bmatrix} \sigma_x \\ \sigma_y \\ \sigma_z \\ \tau_{xy} \\ \tau_{xz} \\ \tau_{yz} \end{Bmatrix}$$

The vector of infinitesimal strain components is

$$\{\epsilon\} = \begin{Bmatrix} \epsilon_x \\ \epsilon_y \\ \epsilon_z \\ \epsilon_{xy} \\ \epsilon_{xz} \\ \epsilon_{yz} \end{Bmatrix} = \begin{bmatrix} \dfrac{\partial}{\partial x} & 0 & 0 \\ 0 & \dfrac{\partial}{\partial y} & 0 \\ 0 & 0 & \dfrac{\partial}{\partial z} \\ \dfrac{\partial}{\partial y} & \dfrac{\partial}{\partial x} & 0 \\ \dfrac{\partial}{\partial z} & 0 & \dfrac{\partial}{\partial x} \\ 0 & \dfrac{\partial}{\partial z} & \dfrac{\partial}{\partial y} \end{bmatrix} \{\delta\} \tag{6.2}$$

where

$$\{\delta\} = \begin{Bmatrix} u(x, y, z) \\ v(x, y, z) \\ w(x, y, z) \end{Bmatrix}$$

is the displacement vector with components u, v, w, in the x, y, z coordinate directions, respectively. The matrix $[C]$ is the linear elastic modulus matrix, and the vector $\{\epsilon_0\}$ is the initial strain due to nonuniform temperature distributions, shrink fits, and so on.

The strain energy for a three-dimensional linear elastic body may be written using equation C.20a (Appendix C) as

$$U = \frac{1}{2} \int_\Omega (\lfloor \epsilon \rfloor [C]\{\epsilon\} - 2\lfloor \epsilon \rfloor [C]\{\epsilon_0\}) \, d\Omega \tag{6.3}$$

where Ω denotes the volume of the solid. As in Appendix C, we include energy attributable to the body forces and boundary forces to get the total potential energy of the system:

$$\Pi(u, v, w) = \frac{1}{2} \int_\Omega (\lfloor \epsilon \rfloor [C]\{\epsilon\} - 2\lfloor \epsilon \rfloor [C]\{\epsilon_0\}) \, d\Omega$$

$$- \int_\Omega \lfloor \delta \rfloor \{F_b\} \, d\Omega - \int_{S_2} \lfloor \delta \rfloor \{T\} \, d\Gamma \tag{6.4}$$

where

$$\{F_b\} = \lfloor X \quad Y \quad Z \rfloor^T$$

is the vector of body force components per unit volume, and

$$\{T\} = \lfloor T_x \quad T_y \quad T_z \rfloor^T$$

is the vector of surface traction components per unit area on portion S_2 of the boundary. At equilibrium, the displacement field (u, v, w) in the body is such that the total system potential energy assumes a minimum value.

Assume that the volume Ω (Figure 6.1) is divided into M discrete elements. We may write the potential energy of the assemblage of elements as the sum of the potential energies of all elements, provided that the interpolation functions expressing the variation of displacement within each element satisfy the compatability and completeness requirements specified in

Chapter 3. In other words, to write

$$\Pi(u,v,w) \approx \sum_{e=1}^{M} \Pi^{(e)}(u,v,w) \tag{6.5}$$

and to be rigorously assured of convergence as element mesh size decreases, the interpolation functions must satisfy the following requirements.

Compatability: Since only first-order derivatives of displacement appear in the integrand of the functional for the potential energy, the interpolation functions must be such that at element interfaces the displacement is continuous. In other words, we must have C^0 continuity in displacement at element interfaces.

Completeness: The interpolation functions representing the displacement within an element must be such that rigid-body displacements (uniform states of displacement) and constant-strain states (uniform first-derivative states of displacement) are represented in the limit as element size is reduced.

For plane stress and plane strain as well as three-dimensional elasticity problems polynomial interpolation functions satisfy the compatibility and completeness requirements when the polynomials contain at least the constant and linear terms (see Figures 5.8 and 5.9).

If we select interpolation functions according to the above guidelines, we may focus our attention on an isolated element. To express $\Pi^{(e)}(u,v,w)$, the potential energy functional for one element, in terms of discrete values of displacement components, we assume that within each element having r nodes the displacement field is approximately related to its nodal values by r interpolating functions $N_i(x,y,z)$, so that the distributed displacement field is expressed as

$$\{\tilde{\delta}\}^{(e)} = \left\{ \begin{array}{c} \sum\limits_{i=1}^{r} N_i(x,y,z)u_i \\[2mm] \sum\limits_{i=1}^{r} N_i(x,y,z)v_i \\[2mm] \sum\limits_{i=1}^{r} N_i(x,y,z)w_i \end{array} \right\} = [N]\{\delta\}^{(e)} \tag{6.6}$$

where $[N]$ is the interpolation function matrix given by

$$[N] = \begin{bmatrix} N_1 & 0 & 0 & N_2 & 0 & 0 & \cdots & N_r & 0 & 0 \\ 0 & N_1 & 0 & 0 & N_2 & 0 & \cdots & 0 & N_r & 0 \\ 0 & 0 & N_1 & 0 & 0 & N_2 & \cdots & 0 & 0 & N_r \end{bmatrix}$$

and the element displacement vector is

$$\{\delta\}^{(e)} = \begin{Bmatrix} u_1 \\ v_1 \\ w_1 \\ \vdots \\ u_r \\ v_r \\ w_r \end{Bmatrix}$$

If we substitute equation 6.6 into equation 6.2, we may express the element strain vector as

$$\{\epsilon\}^{(e)} = [B]\{\delta\}^{(e)} \qquad (6.7a)$$

where the strain interpolation matrix $[B]$ is evaluated as

$$[B] = \begin{bmatrix} \dfrac{\partial N_1}{\partial x} & 0 & 0 & \dfrac{\partial N_2}{\partial x} & 0 & 0 & \cdots & \dfrac{\partial N_r}{\partial x} & 0 & 0 \\[2mm] 0 & \dfrac{\partial N_1}{\partial y} & 0 & 0 & \dfrac{\partial N_2}{\partial y} & 0 & \cdots & 0 & \dfrac{\partial N_r}{\partial y} & 0 \\[2mm] 0 & 0 & \dfrac{\partial N_1}{\partial z} & 0 & 0 & \dfrac{\partial N_2}{\partial z} & \cdots & 0 & 0 & \dfrac{\partial N_r}{\partial z} \\[2mm] \dfrac{\partial N_1}{\partial y} & \dfrac{\partial N_1}{\partial x} & 0 & \dfrac{\partial N_2}{\partial y} & \dfrac{\partial N_2}{\partial x} & 0 & \cdots & \dfrac{\partial N_r}{\partial y} & \dfrac{\partial N_r}{\partial x} & 0 \\[2mm] \dfrac{\partial N_1}{\partial z} & 0 & \dfrac{\partial N_1}{\partial x} & \dfrac{\partial N_2}{\partial z} & 0 & \dfrac{\partial N_2}{\partial x} & \cdots & \dfrac{\partial N_r}{\partial z} & 0 & \dfrac{\partial N_r}{\partial x} \\[2mm] 0 & \dfrac{\partial N_1}{\partial z} & \dfrac{\partial N_1}{\partial y} & 0 & \dfrac{\partial N_2}{\partial z} & \dfrac{\partial N_2}{\partial y} & \cdots & 0 & \dfrac{\partial N_r}{\partial z} & \dfrac{\partial N_r}{\partial y} \end{bmatrix} \qquad (6.7b)$$

We can see a principal difference between the finite element formulations to three-dimensional elasticity problems and the finite element formulations we considered in earlier chapters. Before, the unknown field variable was a scalar, and at each nodal point there was only one unknown—the nodal value of the field variable. In the present elasticity problem, however, the unknown field variable (the displacement) is a vector with three components. Thus at each node point there are three unknowns, the three nodal values of the components of displacement.

By substituting these quantities into the potential energy functional, equation 6.4, we obtain the potential energy of a single element in terms of the nodal values of the displacement field for that element. Since the displacement field for the element has been expressed in terms of known interpolation functions and unknown nodal displacements, the potential energy functional will be similarly expressed. Thus for an element (e), the discretized functional is

$$\Pi^{(e)} = \Pi^{(e)}(u_1, \ldots, u_r, v_1, \ldots, v_r, w_1, \ldots, w_r) \qquad (6.8)$$

or, more explicitly, equation 6.4 for a single element may be written as

$$\Pi^{(e)} = \frac{1}{2} \int_{\Omega^{(e)}} \left(\lfloor \delta \rfloor^{(e)} [B]^T [C][B]\{\delta\}^{(e)} - 2\lfloor \delta \rfloor^{(e)} [B]^T [C]\{\epsilon_0\} \right) d\Omega$$

$$- \int_{\Omega^{(e)}} \lfloor \delta \rfloor^{(e)} [N]^T \{F_b\}^{(e)} \, d\Omega - \int_{S_2^{(e)}} \lfloor \delta \rfloor^{(e)} [N]^T \{T\}^{(e)} \, d\Gamma \qquad (6.9)$$

where the integrations are performed over the volume $\Omega^{(e)}$ and surface $S^{(e)}$ of a single element. We know that at equilibrium the potential energy of the *system* assumes a minimum value. Because of the summation principle expressed in equation 6.5, we may carry out the minimization process element by element. We note that the potential energy of the discretized *system* assumes a minimum value when the first variation of the functional vanishes, that is,

$$\delta\Pi(u, v, w) = \sum_{e=1}^{M} \delta\Pi^{(e)}(u, v, w) = 0 \qquad (6.10)$$

where

$$\delta\Pi^{(e)}(u, v, w) = \sum_{i=1}^{r} \frac{\partial\Pi^{(e)}}{\partial u_i} \delta u_i + \sum_{i=1}^{r} \frac{\partial\Pi^{(e)}}{\partial v_i} \delta v_i + \sum_{i=1}^{r} \frac{\partial\Pi^{(e)}}{\partial w_i} \delta w_i \qquad (6.11)$$

But δu_i, δv_i, and δw_i are independent (not necessarily zero) variations, hence we must have

$$\frac{\partial\Pi^{(e)}}{\partial u_i} = \frac{\partial\Pi^{(e)}}{\partial v_i} = \frac{\partial\Pi^{(e)}}{\partial w_i} = 0, \qquad i = 1, 2, \ldots, r \qquad (6.12)$$

for every element (e) of the system. Equations 6.12 express the conditions we use to find the element equations. This is equivalent to minimizing the potential energy by differentiating $\Pi^{(e)}$ with respect to the vector $\{\delta\}^{(e)}$ as in Appendix A and setting the result equal to zero:

$$\frac{\partial\Pi^{(e)}}{\partial\{\delta\}^{(e)}} = 0 \qquad (6.13)$$

Section A.9 of Appendix A describes the minimization of a functional of the form of equation 6.9 with respect to several variables as indicated by equation 6.13. Following this procedure and minimizing equation 6.9 gives

$$\int_{\Omega^{(e)}} [B]^T [C][B]\{\delta\}^{(e)} \, d\Omega - \int_{\Omega^{(e)}} [B]^T [C]\{\epsilon_0\} \, d\Omega$$

$$- \int_{\Omega^{(e)}} [N]^T \{F_b\}^{(e)} \, d\Omega - \int_{S_2^{(e)}} [N]^T \{T\}^{(e)} \, d\Gamma = 0 \quad (6.14)$$

or

$$[K]^{(e)} \{\delta\}^{(e)} = \{F\}^{(e)} \quad\quad (6.15)$$

where the element stiffness matrix is

$$[K]^{(e)} = \int_{\Omega^{(e)}} [B]^T [C][B] \, d\Omega \quad\quad (6.16a)$$

The element force vector is

$$\{F\}^{(e)} = \{F_0\}^{(e)} + \{F_B\}^{(e)} + \{F_T\}^{(e)} \quad\quad (6.16b)$$

where the element initial force vector is

$$\{F_0\}^{(e)} = \int_{\Omega^{(e)}} [B]^T [C]\{\epsilon_0\} \, d\Omega \quad\quad (6.16c)$$

the element body force vector is

$$\{F_B\}^{(e)} = \int_{\Omega^{(e)}} [N]^T \{F_b\}^{(e)} \, d\Omega \quad\quad (6.16d)$$

and the element force vector due to surface loading (present only for boundary elements) is

$$\{F_T\}^{(e)} = \int_{S_2^{(e)}} [N]^T \{T\}^{(e)} \, d\Gamma \quad\quad (6.16e)$$

Equation 6.15 expresses the equilibrium equations for a single element.

In Chapter 2 we introduced structural finite elements for simple elements including a truss element and a plane stress element. In Section 2.2.3 we derived the element stiffness matrices for these elements using the principle of minimum potential energy. Equation 6.16a is a generalization of the stiffness matrix for a structural finite element, and equations 6.16c–6.16e are generalized equations for evaluating element load vectors. In other words, the element stiffness matrices we derived in Chapter 2 are special cases of the more general form of the element stiffness matrix presented in equation 6.16.

6.2.3 The Galerkin Method

In the preceding sections we used the minimum potential energy principle to derive the element equations for three-dimensional elasticity, but we may also use the method of weighted residuals. Both of these methods are used extensively in elasticity and result in the same set of matrix equations, so it is worthwhile to see how the method of weighted residuals may be used to derive the same element equations for three-dimensional elasticity. As mentioned in Chapter 4, there are a number of weighted residual methods that we might choose, but we will use the Galerkin (Bubnov–Galerkin) method, because it is the most common. In later chapters of this book we will use the Galerkin method to develop finite element formulations for heat transfer and fluid mechanics problems where variational principles do not exist.

The first step is to begin with the governing differential equation from which we may obtain the weak form. For elasticity the governing differential equations are the equilibrium equations, C.11:

$$\frac{\partial \sigma_x}{\partial x} + \frac{\partial \tau_{xy}}{\partial y} + \frac{\partial \tau_{xz}}{\partial z} + X = 0$$

$$\frac{\partial \tau_{xy}}{\partial x} + \frac{\partial \sigma_y}{\partial y} + \frac{\partial \tau_{yz}}{\partial z} + Y = 0$$

$$\frac{\partial \tau_{xz}}{\partial x} + \frac{\partial \tau_{yz}}{\partial y} + \frac{\partial \sigma_z}{\partial z} + Z = 0$$

which may be written in matrix form as

$$[L]^T\{\sigma\} - \{F_b\} = 0 \tag{6.17}$$

where $[L]$ is a differential operator matrix defined as

$$[L] = \begin{bmatrix} \dfrac{\partial}{\partial x} & 0 & 0 \\[2mm] 0 & \dfrac{\partial}{\partial y} & 0 \\[2mm] 0 & 0 & \dfrac{\partial}{\partial z} \\[2mm] \dfrac{\partial}{\partial y} & \dfrac{\partial}{\partial x} & 0 \\[2mm] \dfrac{\partial}{\partial z} & 0 & \dfrac{\partial}{\partial x} \\[2mm] 0 & \dfrac{\partial}{\partial z} & \dfrac{\partial}{\partial y} \end{bmatrix}$$

The second equation employed is the strain–displacement relationship:

$$\{\epsilon\} = [L]\{\delta\} \tag{6.18}$$

where the strain vector $\{\epsilon\}$ and the displacement vector $\{\delta\}$ were defined in equation 6.2. The third equation we employ is the linear elastic constitutive equation, equation 6.1:

$$\{\sigma\} = [C]\{\epsilon\} - [C]\{\epsilon_0\} \tag{6.1}$$

where $[C]$ is the elastic modulus matrix and $\{\epsilon_0\}$ is the initial strain vector as in the previous section. Now we substitute the strain–displacement equation into the constitutive equation to get

$$\{\sigma\} = [C][L]\{\delta\} - [C]\{\epsilon_0\}$$

which we then substitute into the differential equilibrium equation to yield

$$[L]^T([C][L]\{\delta\} - [C]\{\epsilon_0\}) - \{F_b\} = 0 \tag{6.19}$$

Equation 6.19 is a matrix representation of the three equilibrium equations of linear elasticity. The parentheses are inserted to emphasize that the different operator $[L]^T$ operates on all quantities within the parentheses.

We may now cast this equation in integral form as a weighted residual method to give

$$\int_\Omega [W][L]^T([C][L]\{\delta\} - [C]\{\epsilon_0\}) \, d\Omega - \int_\Omega [W]\{F_b\} \, d\Omega = 0 \tag{6.20}$$

where we have premultiplied by $[W]$, a 3×3 matrix of as yet unspecified weighting functions

$$[W] = \begin{bmatrix} W_1 & 0 & 0 \\ 0 & W_2 & 0 \\ 0 & 0 & W_3 \end{bmatrix}$$

We integrate the first term of equation 6.20 by parts to get the weak form

$$\int_\Omega ([L][W]^T)^T [C][L]\{\delta\} \, d\Omega - \int_\Omega ([L][W]^T)^T [C]\{\epsilon_0\} \, d\Omega$$

$$+ \int_\Omega [W]\{F_b\} \, d\Omega - \int_{S_2} [W]\{T\} \, d\Gamma = 0 \tag{6.21}$$

where the surface tractions are

$$\{T\} \equiv [n][C][L]\{\delta\}$$

and $[n]$ is the matrix of the components of the unit vector normal to the surface given by

$$[n] = \begin{bmatrix} n_x & 0 & 0 \\ 0 & n_y & 0 \\ 0 & 0 & n_z \end{bmatrix}$$

The next step is to formulate the element equations from the above weak form. We approximate the displacement field within an element by $\{\delta\} \approx \{\tilde{\delta}\}^{(e)}$, where the approximate displacement field is $\{\tilde{\delta}\}^{(e)} = [N]\{\delta\}^{(e)}$ as in equation 6.6. Then we make the general weight function matrix $[W]$ specific by setting it equal to the interpolation matrix $[N]^T$, noting that the dimensions of $[W]$ now change from 3×3 to $3r \times 3$ because although there are only three displacement degrees of freedom in equation 6.21, there are $3r$ displacement degrees of freedom in our Galerkin approximation for a single element.

The product of the differential operator and the interpolation matrix is the element strain interpolation matrix defined by equations 6.6 and 6.7:

$$[B] = [L][N]$$

By substituting these quantities into the global weak form we arrive at the weak form for a single element:

$$\int_{\Omega^{(e)}} [B]^T [C][B]\{\delta\}^{(e)} \, d\Omega - \int_{\Omega^{(e)}} [B]^T [C]\{\epsilon_0\}^{(e)} \, d\Omega$$

$$- \int_{\Omega^{(e)}} [N]^T \{F_b\}^{(e)} \, d\Omega - \int_{S_2^{(e)}} [N]^T \{T\}^{(e)} \, d\Gamma = 0 \quad (6.22)$$

This equation is identical to equation 6.14 derived from the variational principle and results in the identical force–displacement equations for the element as in the previous section:

$$\overset{3r \times 3r}{[K]^{(e)}} \overset{3r \times 1}{\{\delta\}^{(e)}} = \overset{3r \times 1}{\{F\}^{(e)}} \quad (6.23)$$

where the element stiffness matrix is given by

$$[K]^{(e)} = \int_{\Omega^{(e)}} [B]^T [C][B] \, d\Omega \quad (6.24a)$$

the element displacement vector is

$$\{\delta\}^{(e)} = \begin{Bmatrix} u_1 \\ v_1 \\ w_1 \\ \vdots \\ u_r \\ v_r \\ w_r \end{Bmatrix} \tag{6.24b}$$

and the element force vector is

$$\{F\}^{(e)} = \int_{\Omega^{(e)}} [B]^T [C] \{\epsilon_0\}^{(e)} \, d\Omega$$

$$+ \int_{\Omega^{(e)}} [N]^T \{F_b\}^{(e)} \, d\Omega + \int_{S_2^{(e)}} [N]^T \{T\}^{(e)} \, d\Gamma \tag{6.24c}$$

We have shown how the element equations may be derived using the Galerkin method with the resulting element equations being identical to those obtained using the minimum potential energy principle. The variational approach is often considered a classical approach by some engineers, but the Galerkin method is considered to be a more straightforward approach by others. We feel that neither method is superior to the other. We presented both to demonstrate that either approach may be used to develop a finite element formulation for elasticity problems.

Although the consideration of nonlinear structural mechanics problems is beyond the scope of this book, we should note that the principle of minimum potential energy is limited to elastic materials. For elastic materials with geometric nonlinearities, either the principle of minimum potential energy or the Galerkin method may be used to develop finite element equations. However, for inelastic structural mechanics problems, finite element equations are developed by the more fundamental virtual work principle or the Galerkin method. Thus the Galerkin method offers a generality that many analysts prefer. We shall use the Galerkin method exclusively in later chapters dealing with nonlinear problems in heat transfer and fluid mechanics.

6.2.4 The System Equations

Equation 6.16a is the general form of the element stiffness matrix for three-dimensional elasticity problems. The complete force–displacement equations for the discretized elastic solid (the system) are assembled from the sets of element equations like equations 6.15. The assembly process follows the procedure described in Section 2.3. Again, the *system* equations have the same form as the *element* equations except that they are expanded in dimension to include all nodes. Hence, when the discretized system has m nodes, the system equations become

$$\overset{3m \times 3m}{[K]} \; \overset{3m \times 1}{\{\delta\}} \; = \; \overset{3m \times 1}{\{F\}} \tag{6.25}$$

where $\{\delta\}$ is a column vector of nodal displacement components for the entire system, and $\{F\}$ is the column vector of resultant nodal forces.

For the displacement formulation we have developed here, either force or displacement is known at *every* node of the system. If body forces and initial strains are absent, the vector $\{F\}$ has zero components except for the components corresponding to nodes where concentrated external forces *or* displacements are specified. To account for prescribed boundary displacements the system equation 6.25 is modified according to one of the procedures described in Section 2.3.4. After the displacement boundary conditions have been inserted, we may solve equation 6.25 by using any of the standard techniques for solving linear simultaneous algebraic equations (see Section 10.7.1).

Determining the explicit element equations for a given problem involves evaluating equations 6.16, and, in general, this calls for numerical integration. When linear interpolation functions are used for triangular elements, evaluating the necessary integrals can be simplified by using the closed-form integration formula given in equation 5.22. For elements affected by body forces $\{F_b\}$ or boundary elements experiencing surface tractions $\{T\}$, we face the additional task of evaluating the integrals in equations 6.16d and 6.16e. A simplification known as the *lumped force* technique can be used to avoid evaluation of the boundary force integrals. The procedure is to find concentrated nodal force components which are the static equivalents of the given distributed loading. Then we simply allocate these to the appropriate boundary nodes and disregard the line integrals. This intuitive procedure leads to the correct values for the nodal forces when the interpolation function N_i are linear; but when higher-order interpolation functions are used, the lumped force technique leads to "inconsistent" nodal loads.

In contrast to the lumped force technique, we can use the *consistent force* technique. This means we introduce no intuitive reasoning about how the distributed loading should be assigned to the nodes. Instead, we evaluate the line integrals as they appear and allow the mathematics to dictate what the nodal loads should be. If the N_i are nonlinear, the nonlinear weighting of the distributed load is then naturally incorporated. The advantage of the consistent force technique is that the analysis is inherently more rational. The assurance of an upper bound on the potential energy is also preserved, but this is of little engineering value.

Treatment of the body force term depends on how the body force originates. If the body forces are due to dynamic action, D'Alembert's principle can be used to express them in terms of acceleration components. In this case we must consider a dynamics problem. Dynamics problems are discussed later in this chapter in Section 6.7.

For the steady-state problem once the system equations are solved for the nodal displacements, we may return to the basic relations between stress and strain, and strain and displacement, to find the stress at any point in any of the elements. Referring to equations C.19 and C.38, we may write a general equation for the stress components, including stresses due to displacement and initial strains:

$$\{\sigma\}^{(e)} = [C]^{(e)}[B]^{(e)}\{\delta\}^{(e)} - [C]^{(e)}\{\epsilon_0\}^{(e)} \tag{6.26}$$

If any initial stresses are present, these must also be added.

6.3 APPLICATION TO PLANE STRESS AND PLANE STRAIN

Often three-dimensional elasticity problems can be reduced to more tractable two-dimensional problems by recognizing that the essential descriptions of the geometry and the loading require only two independent coordinates. Plane stress and plane strain problems [10, 11] are two examples of this simplification. The stress analysis of very long solids, such as concrete dams or walls whose geometry and loading are constant in the longest dimension, falls into the category of *plane strain* problems. For these kinds of problems we may determine stresses and displacements by studying only a unit-thickness slice of the solid in the x–y plane (Figure 6.2). Similarly, if we are investigating very thin, flat plates whose loading occurs *only* in the x–y plane of the plate and not transverse to the plane, we have a problem in *plane stress* (Figure 6.3).

The basic equations for plane stress and plane strain are summarized in Appendix C. In the following we derive the element equations for the linear three-node triangle in plane stress or plane strain. Substitution of the appropriate matrix of elastic constants $[C]$ specializes the equations for one case or the other.

6.3.1 Displacement Model for a Triangular Element

A typical triangular element with forces and displacements defined at its nodes is shown in Figure 2.6. If we use the natural coordinates described in Section 5.5, we may express the linear variation of the displacement field within the element in terms of the nodal values of horizontal and vertical displacement as follows:

$$\{\tilde{\delta}\}^{(e)} = \begin{Bmatrix} u(x,y) \\ v(x,y) \end{Bmatrix}^{(e)} = \begin{Bmatrix} L_1 u_1 + L_2 u_2 + L_3 u_3 \\ L_1 v_1 + L_2 v_2 + L_3 v_3 \end{Bmatrix}^{(e)} \tag{6.27a}$$

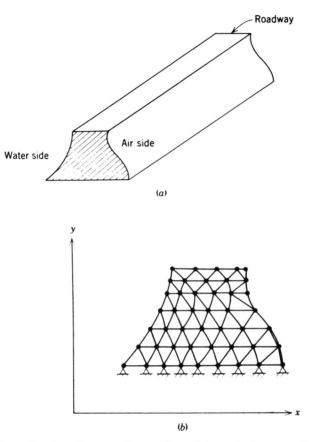

Figure 6.2 Example of a plane strain problem: (*a*) earth dam; (*b*) finite element model of a unit slice of the dam.

or in the notation of equation 6.6

$$\begin{Bmatrix} u \\ v \end{Bmatrix} = \begin{bmatrix} L_1 & 0 & L_2 & 0 & L_3 & 0 \\ 0 & L_1 & 0 & L_2 & 0 & L_3 \end{bmatrix} \begin{Bmatrix} u_1 \\ v_1 \\ u_2 \\ v_2 \\ u_3 \\ v_3 \end{Bmatrix} \qquad (6.27b)$$

and

$$[N] = \begin{bmatrix} L_1 & 0 & L_2 & 0 & L_3 & 0 \\ 0 & L_1 & 0 & L_2 & 0 & L_3 \end{bmatrix} \qquad (6.27c)$$

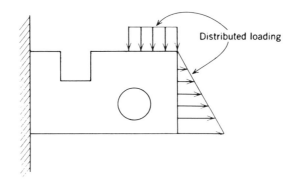

Figure 6.3   Example of a plane stress problem: a thin, flat plate subjected to in-plane loading.

where the interpolation functions are given by equation 5.17:

$$L_i = \frac{1}{2\Delta}(a_i + b_i x + c_i y), \qquad i = 1, 2, 3 \qquad (5.17)$$

$$a_1 = x_2 y_3 - x_3 y_2 \qquad a_2 = x_3 y_1 - x_1 y_3 \qquad a_3 = x_1 y_2 - x_2 y_1$$
$$b_1 = y_2 - y_3 \qquad b_2 = y_3 - y_1 \qquad b_3 = y_1 - y_2$$
$$c_1 = x_3 - x_2 \qquad c_2 = x_1 - x_3 \qquad c_3 = x_2 - x_1$$

and

$$2\Delta = \begin{vmatrix} 1 & x_1 & y_1 \\ 1 & x_2 & y_2 \\ 1 & x_3 & y_3 \end{vmatrix}$$

As noted previously, the linear interpolation model for displacement in equations 6.27 satisfies both the compatibility and the completeness requirements. Clearly this definition for displacement results in an element of constant stress and constant strain.

6.3.2   Element Stiffness Matrix for a Triangle

Having chosen the type of element and the variation of the displacement within the element, we can now determine the element equations from equations 6.16. For plane stress or plane strain there are three strain components: $\epsilon_x, \epsilon_y, \epsilon_{xy}$. Moreover, there are two degrees of freedom per node so that each triangular element has six displacement degrees of freedom. The strain–interpolation matrix $[B]$, equation 6.7, is a 3×6 array defined by

$$[B]^{(e)} = \begin{bmatrix} \dfrac{\partial L_1}{\partial x} & 0 & \dfrac{\partial L_2}{\partial x} & 0 & \dfrac{\partial L_3}{\partial x} & 0 \\[2mm] 0 & \dfrac{\partial L_1}{\partial y} & 0 & \dfrac{\partial L_2}{\partial y} & 0 & \dfrac{\partial L_3}{\partial y} \\[2mm] \dfrac{\partial L_1}{\partial y} & \dfrac{\partial L_1}{\partial x} & \dfrac{\partial L_2}{\partial y} & \dfrac{\partial L_2}{\partial x} & \dfrac{\partial L_3}{\partial y} & \dfrac{\partial L_3}{\partial x} \end{bmatrix}$$

Substituting equation 5.17 and performing the indicated differention yields

$$[B]^{(e)} = \frac{1}{2\Delta} \begin{bmatrix} b_1 & 0 & b_2 & 0 & b_3 & 0 \\ 0 & c_1 & 0 & c_2 & 0 & c_3 \\ c_1 & b_1 & c_2 & b_2 & c_3 & b_3 \end{bmatrix} \tag{6.28}$$

For an isotropic material in plane strain we have, by definition,

$$[C] = \frac{E}{(1+\nu)(1-2\nu)} \begin{bmatrix} 1-\nu & \nu & 0 \\ \nu & 1-\nu & 0 \\ 0 & 0 & \dfrac{1-2\nu}{2} \end{bmatrix} \tag{C.25}$$

while for plane stress

$$[C] = \frac{E}{1-\nu^2} \begin{bmatrix} 1 & \nu & 0 \\ \nu & 1 & 0 \\ 0 & 0 & \dfrac{1-\nu}{2} \end{bmatrix} \tag{C.32}$$

where E is Young's modulus and ν is Poisson's ratio.

To be specific let us consider the case of plane stress. Substituting equations C.32 and 6.28 into equation 6.16a gives

$$[K]^{(e)} = \frac{E}{4\Delta^2(1-\nu^2)} \begin{bmatrix} b_1 & 0 & c_1 \\ 0 & c_1 & b_1 \\ b_2 & 0 & c_2 \\ 0 & c_2 & b_2 \\ b_3 & 0 & c_3 \\ 0 & c_3 & b_3 \end{bmatrix} \begin{bmatrix} 1 & \nu & 0 \\ \nu & 1 & 0 \\ 0 & 0 & \dfrac{1-\nu}{2} \end{bmatrix}$$

$$\times \begin{bmatrix} b_1 & 0 & b_2 & 0 & b_3 & 0 \\ 0 & c_1 & 0 & c_2 & 0 & c_3 \\ c_1 & b_1 & c_2 & b_2 & c_3 & b_3 \end{bmatrix} \iint_{A^{(e)}} t^{(e)} \, dA \tag{6.29}$$

If the thickness of the element $t^{(e)}$ is constant, the integral becomes the area of the triangle Δ, and we have

$$[K]^{(e)} = \frac{Et^{(e)}}{4\Delta(1-v^2)} \begin{bmatrix} b_1 & 0 & c_1 \\ 0 & c_1 & b_1 \\ b_2 & 0 & c_2 \\ 0 & c_2 & b_2 \\ b_3 & 0 & c_3 \\ 0 & c_3 & b_3 \end{bmatrix} \begin{bmatrix} 1 & v & 0 \\ v & 1 & 0 \\ 0 & 0 & \dfrac{1-v}{2} \end{bmatrix}$$

$$\times \begin{bmatrix} b_1 & 0 & b_2 & 0 & b_3 & 0 \\ 0 & c_1 & 0 & c_2 & 0 & c_3 \\ c_1 & b_1 & c_2 & b_2 & c_3 & b_3 \end{bmatrix} \qquad (6.30)$$

If the thickness of the element varies, we must evaluate the integral

$$\iint_{A^{(e)}} t^{(e)} \, dA^{(e)}$$

To do this we may approximately represent the variation of $t^{(e)}$ over the element by writing

$$t^{(e)} \approx t_1^{(e)}L_1 + t_2^{(e)}L_2 + t_3^{(e)}L_3 \qquad (6.31)$$

that is, we can give $t^{(e)}$ linear variation over the element by expressing it in terms of the interpolation function L_i and its nodal values $t_i^{(e)}$. Then

$$\iint_{A^{(e)}} t^{(e)} \, dA \simeq \iint_{A^{(e)}} \left(t_1^{(e)}L_1 + t_2^{(e)}L_2 + t_3^{(e)}L_3 \right) dA^{(e)} \qquad (6.32a)$$

and, using the integration formula from equation 5.22, we have

$$\iint_{A^{(e)}} t^{(e)} \, dA^{(e)} \approx \frac{\Delta}{3} \left(t_1^{(e)} + t_2^{(e)} + t_3^{(e)} \right) \qquad (6.32b)$$

Another approach would be to evaluate the integral numerically, using Gaussian quadrature.

Proceeding on the assumption that our triangular element has a constant thickness, we may carry out the matrix multiplications indicated in equation 6.30 to determine the element stiffness matrix. The result is exactly the same as obtained previously (equation 2.37).

6.3.3   Element Force Vectors for a Triangle

The initial force vector is simply formed by substituting the prescribed initial strain vector $\{\epsilon_0\}$ into equation 6.16c and carrying out the integration over the element. Hence from equation 6.16c

$$\{F_0\}^{(e)} = \frac{E}{2\Delta(1 - \nu^2)} \begin{bmatrix} b_1 & 0 & c_1 \\ 0 & c_1 & b_1 \\ b_2 & 0 & c_2 \\ 0 & c_2 & b_2 \\ b_3 & 0 & c_3 \\ 0 & c_3 & b_3 \end{bmatrix} \begin{bmatrix} 1 & \nu & 0 \\ \nu & 1 & 0 \\ 0 & 0 & \dfrac{1 - \nu}{2} \end{bmatrix} \begin{Bmatrix} \epsilon_{x0} \\ \epsilon_{y0} \\ \epsilon_{xy0} \end{Bmatrix} \iint_{A^{(e)}} t^{(e)} \, dA$$

$$(6.33)$$

The body force vector for constant $t^{(e)}$ is obtained from equation 6.16d

$$\{F_B\} = \int_{\Delta^{(e)}} [N]^T \{F_b\} \, dA$$

and substituting into equation 6.27a:

$$\{F_B\} = \int_{\Delta^{(e)}} \begin{bmatrix} L_1 & 0 \\ 0 & L_1 \\ L_2 & 0 \\ 0 & L_2 \\ L_3 & 0 \\ 0 & L_3 \end{bmatrix} \begin{Bmatrix} X \\ Y \end{Bmatrix} dA \qquad (6.34)$$

The interpolation functions over the triangular element can be evaluated using equation 5.22. For constant thickness $t^{(e)}$ the result is

$$\{F_B\} = \frac{t^{(e)}\Delta}{3} \begin{Bmatrix} X \\ Y \\ X \\ Y \\ X \\ Y \end{Bmatrix}$$

This result shows that the resultant body force in each direction is distributed equally between the three nodes.

Evaluating equation 6.16e for the surface tractions requires a note of explanation. Figure 6.4 shows typical boundary elements where surface

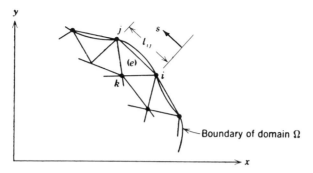

Figure 6.4 Typical boundary elements of domain Ω.

tractions are specified. The actual curved edge of the domain is replaced by the straight edges of the triangles with nodes on the boundary. Consider a typical element e with boundary nodes i and j. The length of the edge ij we denote as l_{ij}. Equation 6.16 provides a means of computing the equivalent nodal forces at nodes i and j due to the surface traction components T_x and T_y acting on this edge of the element. Hence for edge ij we write equation 6.16e as

$$\{F_T\}^{(e)} = \int_0^{l_{ij}} [N]^T \begin{Bmatrix} T_x \\ T_y \end{Bmatrix} t^{(e)} \, ds \tag{6.35}$$

where we denote the edge thickness by $t^{(e)}$. Since $\{F_T\}^{(e)}$ represents the equivalent nodal forces for nodes i and j, it is a vector with four components, two forces at each of the two nodes. Moreover, the interpolation matrix $[N]^T$ is an array with dimensions 4×2. It is not the 6×2 matrix we used in equation 6.34, but rather the smaller 4×2 array, since the integration is performed *only* on the edge of the element. On the edge of the triangular element quantities vary linearly, and we may use as interpolation functions the linear one-dimensional interpolation functions

$$L_i = 1 - \frac{s}{l_{ij}}$$

$$L_j = \frac{s}{l_{ij}}$$

that we customarily use for one-dimensional elements. Thus the integral for the nodal forces becomes

$$\{F_T\}^{(e)} = \int_0^{l_{ij}} \begin{bmatrix} L_i & 0 \\ 0 & L_i \\ L_j & 0 \\ 0 & L_j \end{bmatrix} \begin{Bmatrix} T_x \\ T_y \end{Bmatrix} t^{(e)} \, ds$$

If we assume that the element thickness is constant and that the surface tractions are uniform along the edge, then we may evaluate the integral directly to obtain

$$\{F_T\}^{(e)} = \frac{t^{(e)} l_{ij}}{2} \begin{Bmatrix} T_x \\ T_y \\ T_x \\ T_y \end{Bmatrix} \tag{6.36}$$

Equation 6.36 shows that the force component at each node is one-half the area of the edge times the surface traction (force per unit area) acting on the edge. Thus the nodal force is simply one-half the resultant edge force. The equal division of this resultant between the two nodes occurs because we assumed that the tractions are uniform along the edge. Each element having an edge with surface tractions will contribute a force vector like equation 6.36 to the system equations.

Comments In the foregoing we derived the element equations for a simple linear triangle in plane stress and plane strain. The first successful applications of the finite element method in solid mechanics relied on these equations. Since these earliest applications many other formulations for higher-order triangular elements and quadrilateral elements have been used. Also, special formulations for anisotropic materials, as well as incompressible materials,[1] have appeared. Readers interested in the explicit equations associated with these special topics should see the standard references [10–12] on the subject.

6.4 APPLICATION TO AXISYMMETRIC STRESS ANALYSIS

The axisymmetric solid with axisymmetric loading offers another example of a three-dimensional elasticity problem that can be described by two indepen-

---

[1]Standard potential energy formulations do not hold for incompressible materials, because for these cases Poisson's ratio equals 0.5, and because of the term $(1 - 2\nu)$ in the denominator the usual matrix of elastic constants is undefined. Materials such as rubber, some soils, and solid propellants for rockets are typically considered to be incompressible.

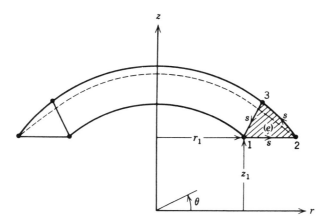

Figure 6.5   Axisymmetric ring element with triangular cross section.

dent variables. Recalling our discussion in Section 5.2, we note that axisymmetry prevails whenever derivatives of the dependent variables with respect to the circumferential coordinate are zero. In other words, when axisymmetric problems are described in terms of cylindrical coordinates r, θ, z, all variables are independent of θ.

The simplest element for general axisymmetric stress problems is a toroidal volume swept by a triangle revolved about the z axis (Figure 6.5). The nodes of this element are then viewed as *circles* in space instead of points. We may derive the element equations by careful application of the formulas in equations 6.16.

6.4.1   Displacement Model for Triangular Toroid

In the axisymmetric problem only the radial and axial displacements are nonzero. For the axial and radial displacements we may use the same linear displacement model we used for the three-node triangle in plane stress and plane strain. The same natural coordinates apply if we recognize that the coordinates r and z correspond to x and y, respectively. Thus we can write the displacement field as

$$\{\tilde{\delta}\}^{(e)} = \left\{ \begin{array}{c} u(r,z) \\ w(r,z) \end{array} \right\}^{(e)} = \left\{ \begin{array}{c} L_1 u_1 + L_2 u_2 + L_3 u_3 \\ L_1 w_1 + L_2 w_2 + L_3 w_3 \end{array} \right\}^{(e)} \qquad (6.37)$$

where

$$L_i(r,z) = \frac{1}{2\Delta}(a_i + b_i r + c_i z), \qquad i = 1, 2, 3$$

$$a_1 = r_2 z_3 - r_3 z_2, \qquad b_1 = z_2 - z_3, \qquad c_1 = r_3 - r_2, \qquad \text{etc.}$$

and

$$2\Delta = \begin{bmatrix} 1 & r_1 & z_1 \\ 1 & r_2 & z_2 \\ 1 & r_3 & z_3 \end{bmatrix}$$

6.4.2   Element Stiffness Matrix for Triangular Toroid

An important difference between the plane stress–plane strain problem and the axisymmetric problem is that in the latter type we must consider four strains, ϵ_r, ϵ_θ, ϵ_z, and ϵ_{rz}, instead of three. Even though the displacement v in the θ direction is identically zero, the strain ϵ_θ is not zero because it is caused by radial displacement. The equations relating strain and displacement in cylindrical coordinates are

$$\epsilon_r = \frac{\partial u}{\partial r} \qquad \epsilon_\theta = \frac{u}{r} \qquad \epsilon_z = \frac{\partial w}{\partial z} \qquad \epsilon_{rz} = \frac{\partial u}{\partial z} + \frac{\partial w}{\partial r}$$

Since we have four strain components, the strain interpolation matrix $[B]$ is now a 4×6 array. Hence we may write the matrix relating strains and displacement as follows:

$$[B]^{(e)} = \begin{bmatrix} \dfrac{\partial L_1}{\partial r} & 0 & \dfrac{\partial L_2}{\partial r} & 0 & \dfrac{\partial L_3}{\partial r} & 0 \\ \dfrac{L_1}{r} & 0 & \dfrac{L_2}{r} & 0 & \dfrac{L_3}{r} & 0 \\ 0 & \dfrac{\partial L_1}{\partial z} & 0 & \dfrac{\partial L_2}{\partial z} & 0 & \dfrac{\partial L_3}{\partial z} \\ \dfrac{\partial L_1}{\partial z} & \dfrac{\partial L_1}{\partial r} & \dfrac{\partial L_2}{\partial z} & \dfrac{\partial L_2}{\partial r} & \dfrac{\partial L_3}{\partial z} & \dfrac{\partial L_3}{\partial r} \end{bmatrix}$$

$$= \frac{1}{2\Delta} \begin{bmatrix} b_1 & 0 & b_2 & 0 & b_3 & 0 \\ \dfrac{2\Delta L_1}{r} & 0 & \dfrac{2\Delta L_2}{r} & 0 & \dfrac{2\Delta L_3}{r} & 0 \\ 0 & c_1 & 0 & c_2 & 0 & c_3 \\ c_1 & b_1 & c_2 & b_2 & c_3 & b_3 \end{bmatrix} \qquad (6.38)$$

Note that the $[B]$ matrix no longer contains only constant terms; it also includes the function $L_i(r, z)/r$. Thus strains and stresses are no longer constant within an element as in plane stress or strain.

The matrix of elastic constants for a linear isotropic solid (see equation C.6) is

$$[C] = \frac{E}{(1 + \nu)(1 - 2\nu)} \begin{bmatrix} 1 - \nu & \nu & \nu & 0 \\ \nu & 1 - \nu & \nu & 0 \\ \nu & \nu & 1 - \nu & 0 \\ 0 & 0 & 0 & \dfrac{1 - 2\nu}{2} \end{bmatrix} \quad (6.39)$$

Before we can apply equations 6.16 to determine the element equations we must specify the infinitesimal volume

$$d\Omega = 2\pi r \, dr \, dz$$

and infinitesimal surface area

$$d\Gamma = 2\pi r \, ds$$

where s is the length coordinate measured along the sides of the triangular cross section (see Figure 6.5). Substitution of the foregoing results into equation 6.16a then yields the stiffness relations:

$$[K]^{(e)} = \frac{E\pi}{2\Delta^2(1 + \nu)(1 - 2\nu)} \iint_{A^{(e)}} \begin{bmatrix} b_1 & \dfrac{2\Delta L_1}{r} & 0 & c_1 \\ 0 & 0 & c_1 & b_1 \\ b_2 & \dfrac{2\Delta L_2}{r} & 0 & c_2 \\ 0 & 0 & c_2 & b_2 \\ b_3 & \dfrac{2\Delta L_3}{r} & 0 & c_3 \\ 0 & 0 & c_3 & b_3 \end{bmatrix}$$

$$\times \begin{bmatrix} 1 - \nu & \nu & \nu & 0 \\ \nu & 1 - \nu & \nu & 0 \\ \nu & \nu & 1 - \nu & 0 \\ 0 & 0 & 0 & \dfrac{1 - 2\nu}{2} \end{bmatrix}$$

$$\times \begin{bmatrix} b_1 & 0 & b_2 & 0 & b_3 & 0 \\ \dfrac{2\Delta L_1}{r} & 0 & \dfrac{2\Delta L_2}{r} & 0 & \dfrac{2\Delta L_3}{r} & 0 \\ 0 & c_1 & 0 & c_2 & 0 & c_3 \\ c_1 & b_1 & c_2 & b_2 & c_3 & b_3 \end{bmatrix} r \, dr \, dz \quad (6.40)$$

It is evident from equation 6.40 that evaluation of the integrals over the element cross section is now complicated by the presence of the radial and axial coordinates in the integrands. This difficulty can be overcome by using numerical integration or by tediously integrating term by term. However, a simpler procedure is to obtain an approximate average value for the matrix $[B]$ by using centroidal values for r and z; that is, for r and z we substitute

$$\bar{r} = \tfrac{1}{3}(r_1 + r_2 + r_3) \tag{6.41}$$

and

$$\bar{z} = \tfrac{1}{3}(z_1 + z_2 + z_3)$$

respectively. The accuracy of this approximation deteriorates for elements close to the axis, because the relative variation of r is greatest there. However, if we use this simplification, we have

$$
[K]^{(e)} = \frac{E\pi}{2\Delta(1+\nu)(1-2\nu)}
\begin{bmatrix}
b_1 & \dfrac{2\Delta \bar{L}_1}{\bar{r}} & 0 & c_1 \\
0 & 0 & c_1 & b_1 \\
b_2 & \dfrac{2\Delta \bar{L}_2}{\bar{r}} & 0 & c_2 \\
0 & 0 & c_2 & b_2 \\
b_3 & \dfrac{2\Delta \bar{L}_3}{\bar{r}} & 0 & c_3 \\
0 & 0 & c_3 & b_3
\end{bmatrix}
$$

$$
\times
\begin{bmatrix}
1-\nu & \nu & \nu & 0 \\
\nu & 1-\nu & \nu & 0 \\
\nu & \nu & 1-\nu & 0 \\
0 & 0 & 0 & \dfrac{1-2\nu}{2}
\end{bmatrix}
$$

$$
\times
\begin{bmatrix}
b_1 & 0 & b_2 & 0 & b_3 & 0 \\
\dfrac{2\Delta \bar{L}_1}{\bar{r}} & 0 & \dfrac{2\Delta \bar{L}_2}{\bar{r}} & 0 & \dfrac{2\Delta \bar{L}_3}{\bar{r}} & 0 \\
0 & c_1 & 0 & c_2 & 0 & c_3 \\
c_1 & b_1 & c_2 & b_2 & c_3 & b_3
\end{bmatrix} \bar{r} \tag{6.42}
$$

where $\bar{L}_i = a_i + b_i \bar{r} + c_i \bar{z}$.

6.4.3   Element Force Vectors for Triangular Toroid

Evaluation of the nodal forces follows immediately from the application of equation 6.16 when the substitutions noted previously are made. The body force usually results from gravity or centrifugal action. The body force vector is obtained by evaluating equation 6.16d with the toroidal element interpolations functions. Integrating over the toroidal volume, we may write

$$\{F_B\}^{(e)} = 2\pi \int_\Delta \begin{bmatrix} N_1 & 0 \\ 0 & N_1 \\ N_2 & 0 \\ 0 & N_2 \\ N_3 & 0 \\ 0 & N_3 \end{bmatrix} \begin{Bmatrix} X \\ Y \end{Bmatrix} r \, dr \, dz \tag{a}$$

where the interpolation functions ($N_i = L_i$, $i = 1, 2, 3$) are given in equation 6.37. The integral may be evaluated by representing r by the linear interpolation equation

$$r = N_1 r_1 + N_2 r_2 + N_3 r_3 \tag{b}$$

Then the integrations of the interpolation functions over the planar element area can be evaluated readily by using equation 5.27. The result may be written as

$$\{F_B\}^{(e)} = \frac{\pi \Delta}{6} \begin{Bmatrix} (2r_1 + r_2 + r_3)X \\ (2r_1 + r_2 + r_3)Y \\ (r_1 + 2r_2 + r_3)X \\ (r_1 + 2r_2 + r_3)Y \\ (r_1 + r_2 + 2r_3)X \\ (r_1 + r_2 + 2r_3)Y \end{Bmatrix} \tag{c}$$

where, as before, Δ is the element area.

For elements with edges on the surface of the axisymmetric solids that experience surface tractions, we need to compute equivalent nodal forces using equation 6.16e. We follow the same procedure that we used in Section 6.3.3 for the triangular element. Thus we write equation 6.16e for an axisymmetric surface as

$$\{F_T\} = 2\pi \int_0^{l_{ij}} [N]^T \begin{Bmatrix} T_x \\ T_y \end{Bmatrix} r \, ds \tag{d}$$

where as before we use linear interpolation functions L_i and L_j expressed in terms of a local surface coordinate s,

$$L_i = 1 - \frac{s}{l_{ij}}$$

$$L_j = \frac{s}{l_{ij}}$$

where l_{ij} is the distance from node i to node j. To evaluate the integral in equation d, we write r as the linear interpolation

$$r = L_i r_i + L_j r_j$$

where r_i and r_j denote the radial coordinates of nodes i and j. The integral for the equivalent nodal forces is then given by

$$\{F_T\} = 2\pi \int_0^{l_{ij}} \begin{bmatrix} L_i & 0 \\ 0 & L_i \\ L_j & 0 \\ 0 & L_j \end{bmatrix} \begin{Bmatrix} T_x \\ T_y \end{Bmatrix} (L_i r_i + L_j r_j)\, ds$$

The integrals of the interpolation functions can be evaluated using equation 5.14. The final result is

$$\{F_T\} = \frac{\pi l_{ij}}{3} \begin{Bmatrix} (2r_i + r_j)T_x \\ (2r_i + r_j)T_y \\ (r_i + 2r_j)T_x \\ (r_i + 2r_j)T_y \end{Bmatrix} \tag{e}$$

Equation e represents the equivalent nodal forces for an axisymmetric element subject to surface tractions T_x and T_y. Each element having an edge with surface tractions will contribute such a force vector to the system equations.

Comments   When establishing the element mesh for axisymmetric problems, care should be taken to avoid positioning elements in such a way that two nodes have the same or nearly the same radial coordinates. If two radial coordinates are close, the calculated difference between them may be grossly in error, and if $r_i = r_j$, some of the integrals become infinite. Another problem can arise if nodes lie on the z axis, where $r = 0$, because then infinite terms also result. This can be avoided by introducing a small core

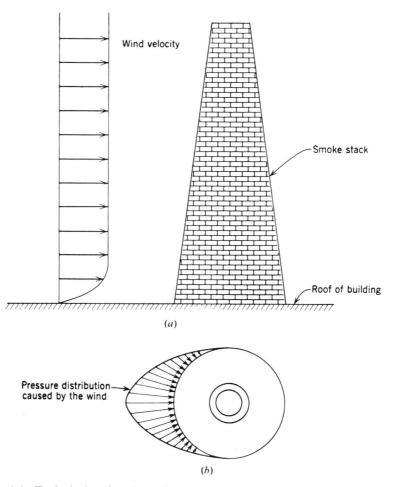

Figure 6.6 Typical situation in which an axisymmetric structure experiences an asymmetric loading: (*a*) smoke stack atop a building; (*b*) plan view showing the wind loading on the stack.

hole along the axis and assigning low values to the radial coordinates that would normally be zero. The radial displacements along the core are then set equal to zero to simulate the actual condition of zero radial displacement at $r = 0$. Details of this procedure and other ones for treating nodes with $r = 0$ are described in references 13–16.

If the geometry of a body is axisymmetric but the loading on the body is not, it is still possible to perform an axisymmetric analysis. When the loads are symmetric about a plane through the axis (see Figure 6.6, for example), the loads as well as the displacement components can be represented by a Fourier series of circular harmonic functions (sine and cosine). Then the corresponding terms of the series are considered individually, and the analysis proceeds in the usual manner. Wilson [13] describes the procedure in detail. For a general nonaxisymmetric loading the displacement components and loads are expressed in terms of a double Fourier series in sine and cosine. The orthogonality property of these trigonometric functions then leaves only $\sin^2 n\theta$ and $\cos^2 n\theta$ terms when integration is performed, and there result $N + 1$ independent *sets* of equations of the form

$$[K]_i\{\delta\}_i = \{F\}_i, \qquad i = 0, 1, 2, \ldots, N$$

where N is the number of terms in the assumed series. After these sets of equations are solved, we return to the series expressions for u and v to find the total displacement field.

Example   The stress distribution that occurs around the thread roots of a bolt–nut system is an important factor in the design of these types of

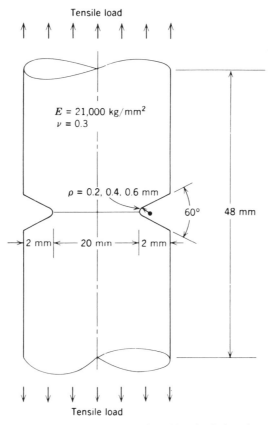

Figure 6.7   Axisymmetric test specimen analyzed by the finite element method [17].

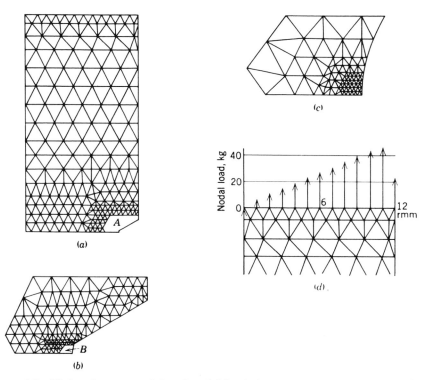

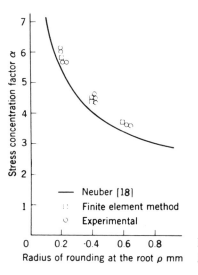

Figure 6.8 Finite element model and nodal loads for three-node ring elements [17]: (*a*) overall mesh; (*b*) magnified figure of *A* (*A* × 4); (*c*) magnified figure of *B* (*B* × 10); (*d*) nodal load assignment.

Figure 6.9 Comparison of results for the V-grooved rod in tension [17].

mechanical fasteners. To study the fundamental nature of these stress distributions Maruyama [17] considered a V-grooved rod under tensile loading. He determined the stress distribution surrounding the annular V groove by first carrying out an axisymmetric finite element analysis and then conducting some special experiments.

The geometry of the test specimen for both the analysis and the experiment is shown in Figure 6.7. For the finite element analysis three-node triangular ring elements were used. Figure 6.8 shows the successive mesh refinement used near the groove tip, where stress gradients are expected to be relatively large. Because of symmetry, only the upper half of the specimen needs to be considered in the analysis.

Calculations were repeated for different mesh sizes near the stress concentration until the numerical values of stress stabilized. The stress assigned to a particular node was computed as the average of the stresses in all of the elements sharing that node. Figure 6.9 shows the good agreement between the experimental results and the results obtained by the finite element method.

6.5   APPLICATION TO PLATE-BENDING PROBLEMS

Many practical structures such as the fuselages of aircraft and spacecraft, the decks, hulls, and bulkheads of ships, and the roofs of buildings contain sections that are thin plates. The shape and loading of these plate sections are often quite complex, but if the thickness is much smaller than the other two dimensions and deflections are small, classical plate theory as described in Appendix C can be applied. Then the problem of determining displacements and stresses in the plates becomes two dimensional, and the unknown field variable is $w(x, y)$, the transverse deflection of the plate.

In this section we show how the finite element method can be used for the analysis of plate-bending problems. Unlike the previous finite element formulations of elasticity problems, the treatment of plate-bending problems involves two levels of approximation: first, classical plate theory itself is an approximation; second, we have the finite element model, which contains its own built-in approximations. Nevertheless, as we shall see, good prediction of displacement and stress is possible.

As we discuss in Appendix C, it is convenient to express plate-bending problems in terms of plate curvatures and moments, that is, generalized strains and stresses, rather than the usual strains and stresses. Thus for plate bending we have the following definitions (equations C.43 and C.44) for generalized strains and stresses:

$$\{\epsilon\} \equiv \left\{ \begin{array}{c} -\dfrac{\partial^2 w}{\partial x^2} \\[2mm] -\dfrac{\partial^2 w}{\partial y^2} \\[2mm] -2\dfrac{\partial^2 w}{\partial x\,\partial y} \end{array} \right\} \tag{6.43}$$

$$\{\sigma\} \equiv \left\{ \begin{array}{c} M_{xx} \\ M_{yy} \\ M_{xy} \end{array} \right\} \tag{6.44}$$

Then a special form of the constitutive relation, equation 6.1, for a plate is used. Thus we take

$$[C] = D \begin{bmatrix} 1 & \nu & 0 \\ \nu & 1 & 0 \\ 0 & 0 & \dfrac{1-\nu}{2} \end{bmatrix} \tag{6.45a}$$

where the flexural rigidity D of the plate is defined by

$$D = \frac{Eh^3}{12(1-\nu^2)} \tag{6.45b}$$

where h is the plate thickness.
We shall use a potential energy formulation and hence develop the displacement method of analysis as we previously did for the three-dimensional elasticity problem. From equation C.45 we have after inserting equations 6.43, 6.44, and C.41, the following potential energy for a classical isotropic plate,

$$\Pi_p = \frac{1}{2} D \int_A \left[\left(\frac{\partial^2 w}{\partial x^2} \right)^2 + 2\nu \frac{\partial^2 w}{\partial x^2} \frac{\partial^2 w}{\partial y^2} + \left(\frac{\partial^2 w}{\partial y^2} \right)^2 + 2(1-\nu)\left(\frac{\partial^2 w}{\partial x\,\partial y} \right)^2 \right] dA$$

$$- \int_{S_q} qw \, dA \tag{6.46}$$

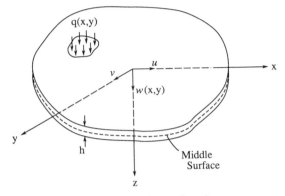

Figure 6.10 Arbitrary flat plate
with transverse loading.

where q is a specified pressure on a portion s_q on the surface of the plate
(Figure 6.10).

The explicit form of the plate potential energy reveals the required order
of the finite element interpolation functions, which we discuss in the next
section.

6.5.1 Requirements for the Displacement Interpolation Functions

Following the standard procedure in finite element analysis, we divide the
region of the plate into elements and then interpolate the displacement (the
transverse deflection $w(x, y)$) in each element. Since the potential energy
functional, equation 6.46, contains second-order derivatives of $w(x, y)$, the
basic compatibility and completeness requirements to be satisfied by the
interpolation functions in the displacement model

$$w^{(e)}(x, y) = [N]\{w\}^{(e)} \tag{6.47}$$

are now more stringent than before. The displacement formulation of plate
bending is a C^1 problem; hence we have the following requirements.

Compatibility: The interpolation functions should be such that $w^{(e)}(x, y)$
is C^1 continuous at element interfaces. This means that along element
interfaces we must have continuity of w as well as continuity of $\partial w/\partial n$,
the derivative of w normal to the interface. In Chapter 5 we saw that a
number of different polygonal elements give C^1 continuity. All of these
elements require specification of w and some or all of its second partial
derivatives at corner nodes (see Section 5.8.2).

Completeness: The interpolations functions should be such that $w^{(e)}(x, y)$
is C^2 continuous within the element. Hence rigid-body, constant-slope,

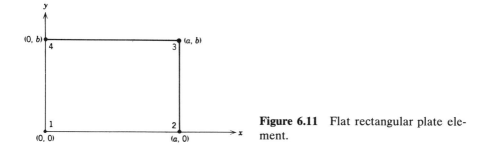

Figure 6.11   Flat rectangular plate element.

and constant-curvature states of deformation should be represented in the displacement model, equation 6.47. This is possible only if the terms 2, x, y, x^2, xy, and y^2 appear in the interpolation polynomials.

If element formulations satisfying these compatibility and completeness requirements are used, we can be sure that successive solutions obtained from element mesh refinement will converge monotonically to the correct displacements and stresses. Convergence of the second derivatives of w (curvatures), however, occurs only in the mean [18]. Though compatibility and completeness guarantee convergence, it is still possible to obtain convergent solutions using incompatible element formulations, and in some cases the solutions with incompatible elements are better because convergence is faster.

6.5.2   Rectangular Plate-Bending Elements

To illustrate the general procedure for deriving element equations for plate-bending problems we consider the simplest case of a rectangular element for an isotropic plate.

Compatible Elements   A rectangular element (see Figure 6.11) capable of preserving C^1 continuity must have the following degrees of freedom defined at its four corner nodes:

$$w(x, y) \qquad \frac{\partial w}{\partial x} \qquad \frac{\partial w}{\partial y} \qquad \frac{\partial^2 w}{\partial x \, \partial y}$$

As we saw in Section 5.8.2, the assumed displacement model, which satisfies the compatibility and completeness requirements, then takes the form

$$w^{(e)}(x, y) = \sum_{i=1}^{4} \left[{}_1N_i w_i + {}_2N_i \left(\frac{\partial w}{\partial x} \right)_i + {}_3N_i \left(\frac{\partial w}{\partial y} \right)_i + {}_4N_i \left(\frac{\partial^2 w}{\partial x \, \partial y} \right)_i \right] \quad (6.48)$$

where the $_jN_i$ are the cubic Hermite interpolation functions given by equation 5.51b. When we substitute the displacement model of equation 6.48 into equation 6.46, we obtain the potential energy of the element in terms of its 16 nodal degrees of freedom. Minimization of the resulting potential energy expression $\Pi_p^{(e)}$ then requires that the first partial derivatives of $\Pi_p^{(e)}$ with respect to each of the degrees of freedom be zero:

$$
\left\{
\begin{array}{c}
\dfrac{\partial \Pi_p^{(e)}}{\partial w_i} \\[2mm]
\dfrac{\partial \Pi^{(e)}}{\partial (\partial w/\partial x)_i} \\[2mm]
\dfrac{\partial \Pi^{(e)}}{\partial (\partial w/\partial y)_i} \\[2mm]
\dfrac{\partial \Pi^{(e)}}{\partial (\partial^2 w/\partial x\, \partial y)_i}
\end{array}
\right\} = \{0\}, \qquad i = 1, 2, 3, 4 \tag{6.49}
$$

The element equations result from carrying out the manipulations indicated in equations 6.49. We shall omit all details of these tedious manipulations given by Bogner et al. [19]. In matrix form equations 6.49 become

$$
\overset{16 \times 16}{[K]}{}^{(e)} \overset{16 \times 1}{\{\delta\}} - \overset{16 \times 1}{\{P\}} = \{0\} \tag{6.50}
$$

The foregoing for the conforming rectangular element in plate bending indicates the complexities involved in formulating the element equations. The more complex conforming elements for C^1 problems discussed in Chapter 5 lead to even greater algebraic complexities. In many cases explicit expressions for the element equations are never fully written out; instead, the terms are obtained by computer as needed. Whether or not the element equations obtained in this way are correct is determined by solving test problems whose exact solutions are known. Figure 6.12 shows the performance of the 16-degrees-of-freedom element when it is used to analyze a centrally loaded plate with simply supported edges. The percentage of error in the calculation of the center displacement is plotted against the parameter NB^2, where N is the number of equations in the final assemblage, and B is the half-bandwidth of the assembled stiffness matrix. Square elements were used, and because of symmetry only one-quarter of the plate was considered.

This displacement element satisfies all the convergence requirements; hence we expect the convergence that is indeed achieved. The solution also

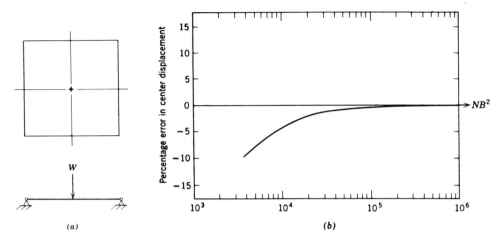

Figure 6.12 Performance of the 16-degree-of-freedom rectangular element [20]: (*a*) centrally loaded, simply supported square plate; (*b*) convergence of center displacement.

gives an upper bound on the potential energy of the system and, for this special example, a lower bound on the displacements.

Incompatible Elements As we previously indicated, some incompatible (sometimes called *nonconforming*) elements may show superior convergence compared to compatible elements. There are also additional practical reasons for considering incompatible plate-bending elements. General-purpose structural analysis programs usually permit six nodal degrees of freedom—three displacements and three rotations. Compatible elements such as the preceding example possess additional nodal degrees of freedom and thus do not fit well into a typical general-purpose program. On the other hand, several incompatible plate elements exist with only nodal displacements and rotations as degrees of freedom. These elements fit well into general-purpose programs and permit the analysis of a variety of general structures. Such elements are particularly appealing because they can be used with beam elements to model stiffened plates and with plane stress elements to model shells.

One of the basic incompatible elements is the rectangular element with the displacement interpolation:

$$w(x, y) = \alpha_1 + \alpha_2 x + \alpha_3 y + \alpha_4 x^2 + \alpha_5 xy + \alpha_6 y^2 + \alpha_7 x^3$$

$$+ \alpha_8 x^2 y + \alpha_9 xy^2 + \alpha_{10} y^3 + \alpha_{11} x^3 y + \alpha_{12} xy^3$$

This element preserves only C^0 continuity and has the following degrees of freedom at its corner nodes

$$w(x, y) \qquad \theta_x = \frac{\partial w}{\partial y} \qquad \theta_y = -\frac{\partial w}{\partial y}$$

where θ_x and θ_y denote plate rotations (positive by the right-hand rule) about the x and y axes, respectively. The coefficients α_i, $i = 1, 2, \ldots, 12$ can be evaluated by the procedure of Section 5.4.3 to yield explicit expressions for N_i. The element stiffness matrix and consistent force vectors can then be derived by minimizing the potential energy, equation 6.46, as previously discussed for the compatible plate element. These derivations have been carried out and explicit equations are available for the 12×12 stiffness matrix for isotropic [21], orthotropic [12], and general anisotropic laminated plates [22].

The element does not preserve normal slope continuity between adjacent elements boundaries and therefore violates one of the conditions for convergence to exact answers with a refined mesh. Convergence, however, has been proved [23], and numerical results [12] convincingly demonstrate the convergence rates and accuracy for both displacements and bending moments.

Comments Here we have considered the simplest plate-bending element and only one plate flexure formulation, namely, the potential energy formulation. Rectangular elements are the most simple example of plate-bending elements, but triangular and quadrilateral elements are more versatile for general structural analysis. Section 5.8.2 presents a further discussion of elements for problems with C^1 continuity. Zienkiewicz and Taylor [12] present an excellent discussion of several other plate-bending element formulations and element types, but such a detailed discussion is beyond the scope of this text. Reference 24 evaluates three-node triangular plate-bending elements and discusses elements that are superior performers. It also contains an excellent reference list on research in plate-bending elements.

6.6 THREE-DIMENSIONAL PROBLEMS

6.6.1 Introduction

In the early sections of this chapter we developed the element equations for three-dimensional elasticity but did not discuss the details of the elements employed in finite element analysis of three-dimensional elasticity problems.

The additional degree of freedom in three-dimensional problems considerably enlarges the magnitude of such problems. To illustrate this point we need consider only the common example of the discretization process for one-, two-, and three-dimensional analyses. Suppose that we pick 10 nodes to define one-dimensional elements for the stress analysis of a rod. Since there is only one degree of freedom per node, we have roughly 10 unknowns in the problem. To perform a stress analysis of a two-dimensional problem to roughly the same degree of accuracy we would need a 10×10 mesh of nodes to define triangular elements for the region. With two degrees of freedom per node the number of unknowns for this problem climbs to about 200. Now for a three-dimensional problem the discretization of the region into tetrahedral elements involves a nodal mesh of $10 \times 10 \times 10$, and with three degrees of freedom per node we have almost 3000 unknowns to find. It is obvious from this exercise that the treatment of realistic three-dimensional problems can readily tax the storage capacity of most computers.

Because three-dimensional problems present such difficulties, much effort has been expended to develop three-dimensional isoparametric elements that give accurate results with fewer degrees of freedom [28–30]. Also, three-dimensional problems helped to motivate the development of efficient data-handling schemes and solution techniques for large-order systems of equations.

6.6.2 Formulation for the Linear Tetrahedral Element

The simplest solid element is the four-node tetrahedron with the displacement field interpolated linearly between nodes. In terms of the natural coordinates described in Section 5.5 we may write the displacement field within a typical element as

$$\begin{Bmatrix} u(x, y, z) \\ v(x, y, z) \\ w(x, y, z) \end{Bmatrix} = \begin{Bmatrix} L_1 u_1 + L_2 u_2 + L_3 u_3 + L_4 u_4 \\ L_1 v_1 + L_2 v_2 + L_3 v_3 + L_4 v_4 \\ L_1 w_1 + L_2 w_2 + L_3 w_3 + L_4 w_4 \end{Bmatrix}$$

where the interpolation functions are

$$L_i = \frac{1}{6V}(a_i + b_i x + c_i y + d_i z), \qquad i = 1, 2, 3, 4$$

V and the coefficients a_i, b_i, c_i, d_i being given by equations 5.26 and 5.27. As mentioned, if the tetrahedron is defined in a right-handed Cartesian coordinate system, the nodes must be numbered so that nodes 1, 2, and 3 are ordered counterclockwise when viewed from node 4. The matrix $[B]^{(e)}$

relating strains and displacements takes the form

$$[B]^{(e)} = \begin{bmatrix} \dfrac{\partial N_1}{\partial x} & 0 & 0 & \dfrac{\partial N_2}{\partial x} & 0 & 0 & \dfrac{\partial N_3}{\partial x} & 0 & 0 & \dfrac{\partial N_4}{\partial x} & 0 & 0 \\[2mm] 0 & \dfrac{\partial N_1}{\partial y} & 0 & 0 & \dfrac{\partial N_2}{\partial y} & 0 & 0 & \dfrac{\partial N_3}{\partial y} & 0 & 0 & \dfrac{\partial N_4}{\partial y} & 0 \\[2mm] 0 & 0 & \dfrac{\partial N_1}{\partial z} & 0 & 0 & \dfrac{\partial N_2}{\partial z} & 0 & 0 & \dfrac{\partial N_3}{\partial z} & 0 & 0 & \dfrac{\partial N_4}{\partial z} \\[2mm] \dfrac{\partial N_1}{\partial y} & \dfrac{\partial N_1}{\partial x} & 0 & \dfrac{\partial N_2}{\partial y} & \dfrac{\partial N_2}{\partial x} & 0 & \dfrac{\partial N_3}{\partial y} & \dfrac{\partial N_3}{\partial x} & 0 & \dfrac{\partial N_4}{\partial y} & \dfrac{\partial N_4}{\partial x} & 0 \\[2mm] \dfrac{\partial N_1}{\partial z} & 0 & \dfrac{\partial N_1}{\partial x} & \dfrac{\partial N_2}{\partial z} & 0 & \dfrac{\partial N_2}{\partial x} & \dfrac{\partial N_3}{\partial z} & 0 & \dfrac{\partial N_3}{\partial x} & \dfrac{\partial N_4}{\partial z} & 0 & \dfrac{\partial N_4}{\partial x} \\[2mm] 0 & \dfrac{\partial N_1}{\partial z} & \dfrac{\partial N_1}{\partial y} & 0 & \dfrac{\partial N_2}{\partial z} & \dfrac{\partial N_2}{\partial y} & 0 & \dfrac{\partial N_3}{\partial z} & \dfrac{\partial N_3}{\partial y} & 0 & \dfrac{\partial N_4}{\partial z} & \dfrac{\partial N_4}{\partial y} \end{bmatrix}$$

$$= \frac{1}{6V} \begin{bmatrix} b_1 & 0 & 0 & b_2 & 0 & 0 & b_3 & 0 & 0 & b_4 & 0 & 0 \\ 0 & c_1 & 0 & 0 & c_2 & 0 & 0 & c_3 & 0 & 0 & c_4 & 0 \\ 0 & 0 & d_1 & 0 & 0 & d_2 & 0 & 0 & d_3 & 0 & 0 & d_4 \\ c_1 & b_1 & 0 & c_2 & b_2 & 0 & c_3 & b_3 & 0 & c_4 & b_4 & 0 \\ d_1 & 0 & b_1 & d_2 & 0 & b_2 & d_3 & 0 & b_3 & d_4 & 0 & b_4 \\ 0 & d_1 & c_1 & 0 & d_2 & c_2 & 0 & d_3 & c_3 & 0 & d_4 & c_4 \end{bmatrix}$$

With the matrix of constitutive properties $[C]$ given by equation C.6, we may write the general submatrix of the element stiffness matrix as

$$[k] = [B]^{T(e)}[C][B]^{(e)}V$$

since the integration over the element volume involves only constant terms. Evaluation of the other element matrices, the initial strain matrix, and the load matrices is also straightforward, because the strain and stress components are constant within each element.

6.6.3   Higher-Order Elements

As indicated in Chapter 5, any higher-order tetrahedral elements can be easily formed using natural coordinates and the recursion relationship between interpolation functions of consecutive order. Lagrangian interpolation also leads to a family of higher-order hexahedral elements. In actual three-dimensional stress analyses, though, only the lower-order elements (linear, quadratic, and cubic) have been used. The natural question as to which elements give the best accuracy per unit of computation time has been asked and partially answered by three investigators [26, 28, 29]. Clough [29] studied the performance of several three-dimensional elements and suggested that the trace[2] of the element stiffness matrix is a reliable indication of the relative quality of different elements. He found that the best elements are characterized by the lowest trace. When the linear displacement hexahedron was compared on this basis with a hexahedron formed from five tetrahedra, it was found to have only about half the trace of the assembled tetrahedra. In most three-dimensional stress analyses hexahedral elements of various orders are preferred over tetrahedral elements.

6.7 INTRODUCTION TO STRUCTURAL DYNAMICS

In many practical design situations it is not sufficient to know just the force–deflection properties of a structure, because the loading of the structure may be a function of time. In this case the designer may need to know the natural frequencies of the structure or perhaps the way in which stresses and deflections are propagated through the structure. Just as the finite element method has been found to be a useful tool for the analysis of static structural problems, it is also most helpful in analyzing the dynamic behavior of structures.

6.7.1 Formulation of Equations

When an elastic structure is subjected to a dynamic load, the displacement field within the structure varies with time and two types of distributed body forces must be taken into account. One body force stems from the inertia of the structure, while the other originates because of internal friction or external damping of some sort. Both forces act in the direction opposite to the direction of motion at a given point in the body. In other words, both types of forces oppose the motion.

The inertia force is brought about by an acceleration of the structure. If $\{\bar{\delta}\}$ is the displacement field of the elastic body, the distributed inertia force acting throughout the body is given by $\rho\{\ddot{\bar{\delta}}\}$, where ρ is the mass density of the material, and $\{\ddot{\bar{\delta}}\} \equiv (\partial^2/\partial t^2)\{\tilde{\delta}\}$ is the column vector of distributed accelerations. According to the well-known D'Alembert principle, this inertia force is statically equivalent to the force $\{F_a\} = -\rho\{\ddot{\bar{\delta}}\}$, that is, $\{F_a\}$ may be treated as a statically imposed body force.

Compared with the inertia force, the damping force is far more difficult to characterize in general. Damping forces may result from friction within a deforming material, from motion through a viscous fluid, or from rubbing with some other bodies in dry contact. Lazan [30] discusses many of the details of these kinds of damping. Usually the damping force is not linearly related to the rate of change of displacement in a body but, rather, has a

---

[2]The *trace* of a square matrix is defined as the sum of the diagonal terms and is equal to the sum of the eigenvalues of the matrix.

more complicated relationship. However, to simplify the analysis of a dynamically loaded structure the conventional procedure is to replace the actual damping force, which is usually unknown, by an approximate viscous damping force proportional to velocity. By this means we write the damping force acting on the body as $\{F_d\} = c\{\dot{\delta}\}$, where c is some known damping coefficient and $\dot{\tilde{\delta}} \equiv (\partial/\partial t)(\tilde{\delta})$. Analytical approaches for formulating and solving structural dynamics problems with damping are discussed by Meirovitch [31]. The development of practical methods for considering damping in realistic problems poses a difficult problem, and much research remains to be done in this area. Some of the basic developments employed in many finite element structural dynamics analyses appeared in papers by Clough et al., for example, references 32 and 33, and are described in the text by Bathe [34].

Assuming that we may describe the inertia and damping forces as indicated, we can derive the equations of motions governing the dynamic behavior of an elastic structure by referring to the discretized equilibrium equations obtained earlier in this chapter. At any instant of time the potential energy of the body is given by equation 6.4. Within a typical three-dimensional element we assume that the displacement field is expressed as

$$\{\tilde{\delta}\}^{(e)} = \begin{Bmatrix} u(x,y,z,t) \\ v(x,y,z,t) \\ w(x,y,z,t) \end{Bmatrix}^{(e)} = \begin{Bmatrix} \sum N_i(x,y,z)u_i(t) \\ \sum N_i(x,y,z)v_i(t) \\ \sum N_i(x,y,z)w_i(t) \end{Bmatrix}^{(e)} \qquad (6.51)$$

where now u, v, and w are the time-dependent components of displacement in the three coordinate directions. Minimization of the potential energy with respect to the nodal values of displacement then leads to force–displacement relations of the form of equation 6.15, except that all the parameters may now be time dependent. The body force vector, however, now includes the inertia and damping forces. Hence we have for the body force term of equation 6.16b the usual term plus two extra terms:

$$\{F_B\} = \int_{\Omega^{(e)}} \left[[N]^T\{F_b\} - c[N]^T\{\dot{\tilde{\delta}}\} - \rho[N]^T\{\ddot{\tilde{\delta}}\} \right] d\Omega \qquad (6.52)$$

The other terms of equation 6.16b remain the same. This leads to element equations of the form

$$[M]^{(e)}\{\ddot{\delta}\}^{(e)} + [C]^{(e)}\{\dot{\delta}\}^{(e)} + [K]^{(e)}\{\delta\}^{(e)} = \{F(t)\}^{(e)} \qquad (6.53)$$

where for an element with r nodes we have

$$[M]^{(e)} = \int_{\Omega^{(e)}} \rho[N]^T[N] \, d\Omega \qquad (6.54)$$

and

$$[C]^{(e)} = \int_{\Omega^{(e)}} c[N]^{T}[N] \, d\Omega^{(e)} \tag{6.55}$$

The displacement interpolation matrix $[N]$ is the same as used previously, and $[M]^{(e)}$ and $[C]^{(e)}$ are the element mass and damping matrices, respectively. Further assembly gives the system equation of motion:

$$[M]\{\ddot{\delta}\} + [C]\{\dot{\delta}\} + [K]\{\delta\} = \{F(t)\} \tag{6.56}$$

The system matrices $[M]$ and $[C]$ are usually sparsely populated (often banded) matrices that, when formed from equations 6.54 and 6.55, are referred to as *consistent* mass and damping matrices. The term *consistent* is used to distinguish these matrices from *lumped* mass or damping matrices, which are obtained by assuming that element mass and damping characteristics are concentrated at the nodes. To form lumped matrices we simply assign to each node of the system an amount of mass or a damping coefficient that can be physically attributed to that location in the body. Usually nonoverlapping yet contiguous regions surrounding the nodes are chosen, and the mass and damping associated with a particular region are assigned to the node in that region. Lumped matrices obtained in this way are diagonal.

Lumped matrices offer computational advantages because they are easy to compute, store, and invert, and they considerably simplify the resulting equations of motion. However, when we use consistent matrices, we can expect a more accurate calculation of eigenvalues and eigenvectors. In general, the errors incurred by lumping increase as the complexity of the element we use increases. Further details of the relative merits of consistent versus lumped matrices can be found in the discussions by Clough [32] and Tong et al. [35].

Once the matrices have been formed, the solution of structural dynamics problems involves the solution of equations of the form of equation 6.56. Prior to a solution the system matrices are modified to account for any boundary conditions not already included as natural boundary conditions. Then the equations are solved subject to specified initial displacements $\{\delta(0)\}$ and velocity distributions $\{\dot{\delta}(0)\}$. A problem may be damped $[C] \neq 0$, or undamped, $[C] = 0$. Also, the discretized forcing function may be zero, harmonic, periodic, aperiodic, or random. Problems with random or nondeterministic forcing functions are not discussed here; the interested reader can find an introductory treatment of this subject in Meirovitch [31].

We shall consider fundamental approaches for solving the discretized equations. More specialized numerical techniques and further details can be found in other finite element texts, for example, reference 34, texts on structural dynamics such as reference 36, and texts on numerical methods [37]. We first review the conventional means for finding natural frequencies and natural modes for a vibrating system; then we consider a popular general procedure for solving the time-dependent equations known as mode superposition; and finally we illustrate an alternative procedure for solving the time-dependent problem that relies on recurrence relations, which permits a time-stepping or time-marching solution. Clearly these solution techniques are not related to the finite element method; however, the techniques are widely used by finite element practitioners in solving structural dynamics problems.

6.7.2 Free Undamped Vibrations

If we imagine the ideal case in which the system has no damping and no external forcing functions, the equations of motion reduce to

$$[M]\{\ddot{\delta}\} + [K]\{\delta\} = \{0\} \tag{6.57}$$

This equation expresses the condition of natural vibration (simple harmonic motion), where at any instant the restoration influences in the system balance the inertia influences. The states of natural vibration are called *natural modes* or principal modes, and the frequencies of vibration are the natural frequencies. The system will have as many natural modes and natural frequencies as it has unconstrained degrees of freedom.

To find these natural modes and frequencies we assume that the displacement vector may be expressed as

$$\{\delta\} = \{\Phi\} e^{i\omega t} = \{\Phi\}(\cos \omega t + i \sin \omega t)$$

where $\{\Phi\}$ is the vector of unknown amplitude at the nodes (modal vector) and ω is one natural frequency. Noting that

$$\{\ddot{\delta}\} = -\omega^2 \{\Phi\} e^{i\omega t}$$

we find upon substitution that equation 6.57 reduces to

$$[[K] - \omega^2 [M]]\{\Phi\} = \{0\} \tag{6.58}$$

Equation 6.58 is recognized as an eigenvalue problem; the equation has a nontrivial solution only when the determinant $|[K] - \omega^2 [M]| = 0$. This is equivalent to the polynomial

$$(\omega^2)^n + (\quad)(\omega^2)^{n-1} + \cdots + (\quad)(\omega^2) + (\quad) = 0 \tag{6.59}$$

For matrices of dimension $n \times n$ there will be n values of ω_i^2 satisfying equation 6.59, and hence n vectors $\{\Phi\}$ that satisfy equation 6.58.

When the ideal system is impulsively excited, it may vibrate in any one of its natural modes, depending on just what initial conditions were imposed. As we shall see, once the natural modes of a system are known, it is possible to use them to uncouple the equations of motion, equation 6.56, and obtain a solution for the dynamic response to many kinds of forcing functions $\{F(t)\}$.

The weighted orthogonality of the natural modes or eigenvectors is an important property that we can use to advantage in solving real dynamics problems. It is easy to show [31] that for two different eigenvectors corresponding to two different frequencies ω_i and ω_j we have

$$\lfloor \Phi \rfloor_i [K]\{\Phi\}_j = 0, \qquad i \neq j$$

or

$$\lfloor \Phi \rfloor_i [M]\{\Phi\}_j = 0, \qquad i \neq j$$

(6.60)

where $[K]$ and $[M]$ are weighting matrices. At a given frequency, say ω_i, we have

$$\lfloor \Phi \rfloor_i [K]\{\Phi\}_i = C_{K_i}$$

$$\lfloor \Phi \rfloor_i [M]\{\Phi\}_i = C_{M_i}$$

(6.61)

where C_{K_i} and C_{M_i} are constants different from zero. Often it is convenient to normalize the eigenvectors so that $C_{K_i} = 1$ or $C_{M_i} = 1$. The constants C_{K_i} and C_{M_i} are called the generalized stiffness and generalized mass, respectively.

The solution of the matrix eigenvalue problem, equation 6.58, for the n natural frequencies and modal vectors is a basic problem in numerical analysis. Popular eigenvalue solution techniques include the Jacobi, Householder, and Givens methods for computing all of the system modes, and the power, inverse power, and subspace iteration methods for computing a selected part of the system modes. Some of these methods are discussed in references 31, 34, 36, and 37.

Example   To illustrate an application of finite elements in free vibration problems we describe a study of the free vibrations of panels fabricated of advanced composite materials. Filamentary composite materials are currently replacing conventional materials where high stiffness and minimum weight are desirable. For example, fishing rods, tennis rackets, and golf clubs are now available that are manufactured from high-modulus graphite composite materials. The aerospace industry has been among the leaders in the utilization of these materials. In aerospace applications, panels are made from laminae with different filament orientations, similar to the way plywood is made from wood laminae. The properties of the composite panel depend on the laminae filament orientations and the "stacking" sequence. In reference 38 a finite element and experimental study was made of the vibration

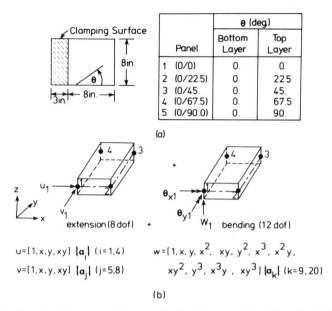

(a)

(b)

Figure 6.13 Laminated composite panel vibrations analyzed by the finite element method: (*a*) panel characteristics; (*b*) rectangular plate-bending element.

behavior of a set of unsymmetrically laminated boron–epoxy panels. Unsymmetrically laminated panels are characterized by a coupling between the bending and extension displacements so that they behave quite differently from comparable panels of isotropic materials.

The geometry of typical test panels are shown in Figure 6.13*a*; unsymmetrical laminations occur because the top lamina of each panel had a filament orientation of $\theta = 0$ but the bottom lamina θ varied from panel to panel. The rectangular plate-bending element used had 20 degrees of freedom, as shown in Figure 6.13*b* and employed the incompatible displacement function $w(x, y)$ given in Section 6.5.2. To determine the panel frequencies and mode shapes, an eigenvalue problem, equation 6.56, of size $n = 360$ was solved for each panel.

Sample experimental and computed vibration frequencies and panel nodal patterns appear in Figure 6.14. Note the significant effect that the bottom lamina orientation has upon the panel frequencies and nodal patterns. Figure 6.14 shows the good agreement between the experimental results and the results obtained by the finite element method.

6.7.3 Finding Transient Motion via Mode Superposition

Once the eigenvalue problem is solved for the natural modes and frequencies, we can use the method of mode superposition to determine the solution

Panel No.	Mode:	Frequency, Hz						Avg. % Difference
		1	2	3	4	5	6	
1	Exp.	41.2	48.7	117	–	262	273	–
(0/0)	F.E.	41.5	53.0	103	211	256	267	5.0
2	Exp.	34.3	59.4	143	216	264	309	–
(0/22.5)	F.E.	34.4	56.1	119	214	235	253	8.8
3	Exp.	24.4	55.1	141	157	208	281	–
(0/45)	F.E.	24.3	48.5	118	151	193	242	9.0
4	Exp.	21.8	43.3	132	149	184	177	–
(0/67.5)	F.E.	21.8	43.4	130	138	177	254	1.4
5	Exp.	21.4	34.3	134	157	169	283	–
(0/90)	F.E.	20.4	38.4	127	153	140	244	9.2

(a)

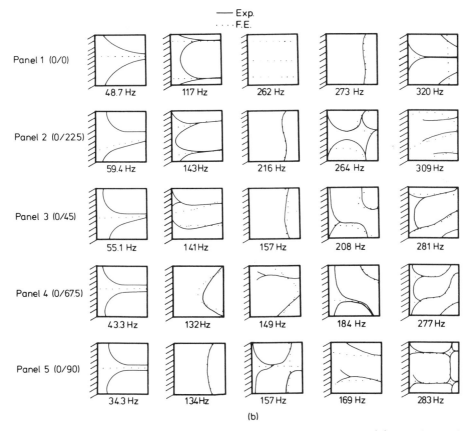

(b)

Figure 6.14 Comparison of results for composite panel vibrations: (*a*) experimental and finite element computed frequencies; (*b*) experimental and finite element computed nodal patterns.

to the complete set of equations with $[C] \neq [0]$ and $\{F(t)\} \neq \{0\}$. The method relies on the basic premise that the solution vector $\{\delta\}$ may be expressed as a linear combination of all n eigenvectors of the system. Hence we set

$$\{\delta(t)\} = [\{\Phi\}_1, \{\Phi\}_2, \{\Phi\}_3, \ldots, \{\Phi\}_n]\{\Lambda(t)\}$$
$$= [A]\{\Lambda(t)\} \tag{6.62}$$

where $[A]$ is a square matrix whose columns are the eigenvectors and $\{\Lambda(t)\}$ is the vector of unknown modal amplitudes. If we substitute equation 6.62 into equation 6.56 and then premultiply the resulting equation by the transpose of the eigenvector matrix $[A]$, we obtain

$$[A]^T[M][A]\{\ddot{\Lambda}\} + [A]^T[C][A]\{\dot{\Lambda}\} + [A]^T[K][A]\{\Lambda\} = [A]^T\{F\}$$

or

$$[M^*]\{\ddot{\Lambda}\} + [C^*]\{\dot{\Lambda}\} + [K^*]\{\Lambda\} = \{F^*\} \tag{6.63}$$

where

$$\begin{aligned}
[M^*] &= [A]^T[M][A] \\
[C^*] &= [A]^T[C][A] \\
[K^*] &= [A]^T[K][A] \\
\{F^*\} &= [A]^T\{F\}
\end{aligned} \tag{6.64}$$

Now, by definition,

$$[K]\{\Phi\}_j = \omega_j^2[M]\{\Phi\}_j$$

Hence

$$K_{ij}^* = \omega_j^2 M_{ij}^*$$

In view of the orthogonality relations expressed by equations 6.60 we see that the matrices $[K^*]$ and $[M^*]$ are diagonal because $K_{ij}^* = 0$, $i \neq j$, and so forth. Furthermore, the equations are *uncoupled* if we assume that $[C]$ is proportional to either $[K]$ or $[M]$. This assumption is usually justified for structural systems, because the actual coupling produced by damping is often negligible. When we assume that $[C]$ is proportional to $[M]$, a typical uncoupled equation for the ith mode has the form

$$M_{ii}^*\ddot{\Lambda}_i + 2\zeta_i\omega_i M_{ii}^*\dot{\Lambda}_i + \omega^2 M_{ii}^*\Lambda_i = F_i^* \tag{6.65}$$

where ζ_i is the damping ratio for mode $\{\Phi\}_i$. A discussion of how the ζ_i can be determined is given in Clough [32]. Dividing through by M_{ii}^* yields the more classical form of the equation of motion to be solved for $\Lambda_i(t)$:

$$\ddot{\Lambda}_i + 2\zeta_i\omega_i\dot{\Lambda}_i + \omega_i^2\Lambda_i = \frac{F_i^*}{M_{ii}^*}, \qquad i = 1, 2, \ldots, n \qquad (6.66)$$

The solution of the complete dynamics problem now involves solving independently n linear, ordinary, uncoupled differential equations like equation 6.66 for $\Lambda_i(t)$ and then combining the results according to equation 6.62. Equation 6.66 may be solved by several different direct numerical integration schemes [37]; however, the Duhamel integral [31] provides a particularly convenient way to obtain the solution. for any forcing function (F_i^*/M_{ii}^*) the time history of Λ_i can be found by evaluating

$$\Lambda_i(t) = \frac{1}{M_{ii}^*\omega_n(1-\zeta_i^2)^{1/2}} \int_0^t \left[\frac{F_i^*(\tau)}{M_{ii}^*}\right] e^{-\zeta_i\omega_i(t-\tau)}$$

$$\times \sin\left[(t-\tau)\omega_n(1-\zeta_i^2)^{1/2}\right] d\tau, \qquad i = 1, 2, 3, \ldots, n \quad (6.67)$$

where t is the particular time at which the response is sought.[3] By evaluating equation 6.67 at different time increments—$0 < t_1 < t_2 < t_3 < t_4 \cdots$ —we can obtain the complete time history for each modal amplitude $\Lambda_i(t)$.

Solving dynamics problems by the method of mode superposition involves considerable computational effort because the eigenvalue problem must first be solved for the natural modes and frequencies. However, once these are found, they may be stored and used later to find the dynamic responses of the system to other forcing functions that may be of interest. It should be noted that mode superposition holds only for linear problems, that is, problems in which $[K]$ and/or $[M]$ do not depend on $\{\delta\}$.

6.7.4 Finding Transient Motion via Recurrence Relations

Instead of solving equation 6.56 by first using the eigenvectors of the system to derive decoupled equations, an alternative is to apply direct numerical integration in time and to solve the full set of coupled equations. The procedure relies on deriving recursion formulas that relate the values of $\{\delta\}$, $\{\dot{\delta}\}$, and $\{\ddot{\delta}\}$ at one instant of time t to the values of these quantities at a later

---

[3]Equation 6.67 results from solving equation 6.66 via Laplace transforms. It was assumed that

$$\Lambda_i(0) = \dot{\Lambda}_i(0) = 0, \qquad i = 1, 2, \ldots, n$$

A slightly different expression would be obtained for nonhomogeneous initial conditions.

time $t + \Delta t$, where Δt is a small time increment. The recursion formulas make it possible for the solution to be "marched out" in time, starting with the initial conditions at time $t = 0$ and continuing step by step until the desired length of time is achieved. At each time $t + \Delta t$ we compute the displacement vector $\{\delta\}_{t+\Delta t}$ and its derivatives, the velocity $\{\dot{\delta}\}_{t+\Delta t}$, and the acceleration $\{\ddot{\delta}\}_{t+\Delta t}$ in terms of the following: (1) the values of these quantities at previous times; (2) the mass, damping; and stiffness matrices of the system, $[M]$, $[C]$, and $[K]$, respectively; and (3) the time-dependent vector of nodal forces $\{F(t)\}$. The recursion formulas (computational algorithms) fall into two general classes, depending on the types of equations that must be solved at each time step. In an *explicit* recursion formula a set of uncoupled algebraic equations is solved at each time step or $\{\delta\}_{t+\Delta t}$; in an *implicit* recursion formula a set of coupled (simultaneous) algebraic equations is solved at each time step for $\{\delta\}_{t+\Delta t}$. Explicit and implicit algorithms have important differences in their performances, since computational efficiency, accuracy, and stability may vary significantly between the two algorithm types. It is beyond the scope of this book to describe these differences in depth; instead, we present an example of each that is popular in finite element work and cite references where further details may be found. We illustrate explicit computational algorithms by the central difference method and implicit algorithms by the Newmark method.

Central Difference Method   By writing series expansions for the vector $\{\delta\}$ at times $t - \Delta t$, t, and $t + \Delta t$, finite difference formulas for $\{\dot{\delta}\}$ and $\{\ddot{\delta}\}$ can be derived, including backward, central, and forward difference approximations for these derivatives [37]. For the second-order equations of structural dynamics the central difference method is effective, but for first-order equations the other difference approximations are also employed. See Chapter 7, Section 7.5, for a closely related discussion on first-order equations. Using the central difference method we can write

$$\{\dot{\delta}\}_t = \frac{\{\delta\}_{t+\Delta t} - \{\delta\}_{t-\Delta t}}{2\,\Delta t} \tag{6.68a}$$

$$\{\ddot{\delta}\}_t = \frac{\{\delta\}_{t+\Delta t} - 2\{\delta\}_t + \{\delta\}_{t-\Delta t}}{\Delta t^2} \tag{6.68b}$$

where the terms neglected in the series used to derive these approximations are of order $(\Delta t)^2$. We obtain the recurrence formula for $\{\delta\}_{t+\Delta t}$ by writing equation 6.56 at time t and then substituting equations 6.68 into it. Thus we write

$$[M]\{\ddot{\delta}\}_t + [C]\{\dot{\delta}\}_t + [K]\{\delta\}_t = \{F(t)\}$$

introduce equations 6.68, and solve for $\{\delta\}_{t+\Delta t}$ to obtain

$$\left[\frac{1}{\Delta t^2}[M] + \frac{1}{2\,\Delta t}[C]\right]\{\delta\}_{t+\Delta t} = \{F(t)\} - \left[[K] - \frac{2}{\Delta t^2}[M]\right]\{\delta\}_t$$

$$- \left[\frac{1}{\Delta t^2}[M] - \frac{1}{2\,\Delta t}[C]\right]\{\delta\}_{t-\Delta t} \quad (6.69)$$

where $\{\delta\}_{t+\Delta t}$ on the left-hand side is unknown and all of the terms of the right-hand side are known. For a given Δt equation 6.69 is a recurrence formula for calculating the vector of nodal unknowns $\{\delta\}_{t+\Delta t}$ at the end of the time step from known displacement values at previous time steps, that is, $\{\delta\}_t$ and $\{\delta\}_{t-\Delta t}$. In general, equation 6.69 is a set of linear algebraic equations of the form

$$[\bar{K}]\{\delta\}_{t+\Delta t} = \{\bar{F}\}, \qquad \text{(implicit)} \qquad (6.70)$$

with

$$[\bar{K}] = \frac{1}{\Delta t^2}[M] + \frac{1}{2\,\Delta t}[C] \qquad (6.71a)$$

and

$$\{\bar{F}\}_t = \{F(t)\} - \left[[K] - \frac{2}{\Delta t^2}[M]\right]\{\delta\}_t - \left[\frac{1}{\Delta t^2}[M] - \frac{1}{2\,\Delta t}[C]\right]\{\delta\}_{t-\Delta t}$$

$$(6.71b)$$

where $[\bar{K}]$ is an effective stiffness matrix, and $\{\bar{F}\}$ is an effective load vector. If we use consistent mass and damping matrices, then equation 6.70 typifies an implicit algorithm, since a set of simultaneous algebraic equations are solved at each time step. If, however, we use lumped mass and damping matrices, $[\bar{K}]$ becomes a diagonal matrix and equation 6.70 uncouples, yielding an explicit algorithm. The explicit algorithm form of equation 6.70 is

$$\left[\diagdown\!\bar{K}\diagdown\right]\{\delta\}_{t+\Delta t} = \{\bar{F}\}, \qquad \text{(explicit)} \qquad (6.72)$$

where

$$\left[\diagdown\!\bar{K}\diagdown\right] = \frac{1}{\Delta t^2}\left[\diagdown\!M\diagdown\right] + \frac{1}{2\,\Delta t}\left[\diagdown\!C\diagdown\right] \qquad (6.73a)$$

and

$$\{\bar{F}\}_t = \{F(t)\} - \left[[K] - \frac{2}{\Delta t^2}\left[\diagdown M\diagdown\right]\right]\{\delta\}_t$$

$$- \left[\frac{1}{\Delta t^2}\left[\diagdown M\diagdown\right] - \frac{1}{2\Delta t}\left[\diagdown C\diagdown\right]\right]\{\delta\}_{t-\Delta t} \qquad (6.73b)$$

Since the explicit recurrence formula, equation 6.72, avoids the solution of simultaneous algebraic equations, the computational effort to compute the transient response via equation 6.72 may be significantly smaller than via the implicit form, equation 6.70. Thus the computational advantage of the explicit central difference recurrence formula is improved efficiency. There is, however, a computational disadvantage of the algorithm that is not immediately obvious. This disadvantage is that we cannot select Δt larger than a critical time step Δt_{cr}; for if we select the time step Δt larger than Δt_{cr}, the computed response may become unstable and computed displacement values will grow without bound as time increases. Thus the disadvantage of the explicit central recurrence formula is that it is only *conditionally stable*. References 12 and 34 show that to obtain a stable solution we must choose

$$\Delta t \leq \Delta t_{cr} = \frac{T_n}{\pi} \qquad (6.74)$$

where T_n is the smallest natural vibration period of the system of size n. The period T_n corresponds to the largest natural frequency, ω_n, that is, $T_n = 2\pi/\omega_n$.

In contrast to the conditionally stable explicit central difference recurrence formula, some implicit recurrence relations are *unconditionally stable*, which means that we can choose Δt without concern for computational instability. The Newmark algorithm to be discussed next is an example of such an unconditionally stable recurrence formula.

Before proceeding to the Newmark algorithm we summarize the steps required to compute the dynamic response by the explicit central difference method. Note that the calculation of $\{\delta\}_{t+\Delta t}$ involves $\{\delta\}_t$ and $\{\delta\}_{t-\Delta t}$ (see equation 6.73b), and therefore a special starting procedure is required. Since $\{\delta\}_{t=0}$ and $\{\dot{\delta}\}_{t=0}$ are known initial conditions, we can compute $\{\ddot{\delta}\}_{t=0}$ from the equation of motion, equation 6.56; then equation 6.68 can be used to compute $\{\delta\}_{-\Delta t}$:

$$\{\delta\}_{-\Delta t} = \{\delta\}_{t=0} - \Delta t\{\dot{\delta}\}_{t=0} + \tfrac{1}{2}\Delta t^2\{\ddot{\delta}\}_{t=0} \qquad (6.75)$$

The solution sequence for the explicit central difference algorithm is summarized in Table 6.2.

TABLE 6.2. Solution Sequence for Explicit Algorithm (lumped mass and damping matrices)

INITIAL CALCULATIONS

1. Form element matrices $[\tilde{}M\tilde{}]^{(e)}$, $[\tilde{}C\tilde{}]^{(e)}$, and $[K]^{(e)}$.
2. Initialize $\{\delta\}_t$ and $\{\dot{\delta}\}_t$ at $t = 0$.
3. Select $\Delta t < \Delta t_{\mathrm{cr}}$.
4. Assemble the effective stiffness matrix $[\bar{K}]$.
5. Modify $[\bar{K}]$ for boundary conditions.
6. Compute $\{\delta\}_{-\Delta t}$.

AT EACH TIME STEP

1. Form the effective element load vector $\{\bar{F}\}_t^{(e)}$
2. Assemble the effective load vector $\{\bar{F}\}_t$.
3. Modify the effective load vector for boundary conditions.
4. Solve for displacements $\{\delta\}_{t+\Delta t}$ at time $t + \Delta t$.
5. Solve for velocities and accelerations $\{\dot{\delta}\}_t$ and $\{\ddot{\delta}\}_t$ at time t.

Newmark Beta Method   The original Newmark method [39] and its generalized modification commonly called the Newmark beta method are implicit time integration methods. The Newmark beta equations for the displacement and velocity at time $t + \Delta t$ are

$$\{\delta\}_{t+\Delta t} = \{\delta\}_t + \{\dot{\delta}\}_t \Delta t + \left[(1 - \beta)\{\ddot{\delta}\}_t + \beta\{\ddot{\delta}\}_{t+\Delta t} \right] \Delta t^2 \quad (6.76a)$$

$$\{\dot{\delta}\}_{t+\Delta t} = \{\dot{\delta}\}_t + \left[(1 - \alpha)\{\ddot{\delta}\}_t + \alpha\{\ddot{\delta}\}_{t+\Delta t} \right] \Delta t \quad (6.76b)$$

where α and β are parameters that control the stability and accuracy of the method. In the original method $\alpha = \frac{1}{2}$ and $\beta = \frac{1}{4}$. We obtain the recurrence formula by writing equation 6.56 at time $t + \Delta t$,

$$[M]\{\ddot{\delta}\}_{t+\Delta t} + [C]\{\dot{\delta}\}_{t+\Delta t} + [K]\{\delta\}_{t+\Delta t} = \{F(t + \Delta t)\}$$

and substituting for $\{\ddot{\delta}\}_{t+\Delta t}$ and $\{\dot{\delta}\}_{t+\Delta t}$. We first solve for $\{\ddot{\delta}\}_{t+\Delta t}$ in terms of $\{\delta\}_{t+\Delta t}$ and other vectors by using 6.76a. Next we insert this result into equations 6.76b to express $\{\dot{\delta}\}_{t+\Delta t}$ in terms of $\{\delta\}_{t+\Delta t}$ and other vectors. Finally, substituting for $\{\dot{\delta}\}_{t+\Delta t}$ and $\{\ddot{\delta}\}_{t+\Delta t}$ in the equation of motion permits us to write the recurrence formula in terms of an effective stiffness and load vector,

$$[\bar{K}]\{\delta\}_{t+\Delta t} = \{\bar{F}\}_{t+\Delta t} \quad (6.77)$$

where the effective stiffness matrix is

$$[\bar{K}] = [K] + \frac{\alpha}{\beta \, \Delta t}[C] + \frac{1}{\beta \, \Delta t^2}[M] \tag{6.78a}$$

and the effective load vector is

$$\{\bar{F}\} = \{F\}_{t+\Delta t}$$
$$+ [C]\left(\frac{1}{\beta \, \Delta t}\{\delta\}_t + \left(\frac{\alpha}{\beta} - 1\right)\{\dot{\delta}\}_t + \frac{\Delta t}{2}\left(\frac{\alpha}{\beta} - 2\right)\{\ddot{\delta}\}_t\right) \tag{6.78b}$$
$$+ [M]\left(\frac{1}{\beta \, \Delta t^2}\{\delta\}_t + \frac{1}{\beta \, \Delta t}\{\dot{\delta}\}_t + \left(\frac{1}{2\beta} - 1\right)\{\ddot{\delta}\}_t\right)$$

In equation 6.77, $\{\delta\}_{t+\Delta t}$ on the left-hand side is unknown and all the terms on the right-hand side are known, hence equation 6.77 represents the recurrence formula for computing $\{\delta\}_{t+\Delta t}$ at the end of a time step. Note that the effective stiffness matrix $[\bar{K}]$ given in equation 6.78a depends on the system stiffness matrix, hence we cannot render $[\bar{K}]$ diagonal by lumping the damping and mass matrices, as we were able to do with the central difference recurrence formula. Consequently, the Newmark algorithm is implicit, and we must solve a set of simultaneous equations at each time step.

The performance of the Newmark beta algorithm has been studied extensively. We know the algorithm to be unconditionally stable for $\alpha = \frac{1}{2}$ and $\beta = \frac{1}{4}$ in linear problems. This choice of α and β is known as the constant acceleration method. Another common choice, $\alpha = \frac{1}{2}$ and $\beta = \frac{1}{6}$ is known as the linear acceleration method. A discussion of the stability of this algorithm is beyond the scope of this text, but details appear in the texts by Zienkiewicz and Taylor [12] and Bathe [34]. Thus we may select the time step Δt in the Newmark algorithm without concern for solution stability, but we must consider the effects of time step size on solution accuracy. The basic guideline is that the time step should be small enough so that the response in all modes that contribute significantly to the total response is calculated accurately. Generally, the time step may be larger than the time step Δt_{cr} required in the central difference method, but selection of a time step in practical problems often represents a compromise between solution accuracy and computational expense. Usually, computational experience is required to make the compromise between these alternatives.

The solution sequence for the implicit Newmark algorithm is summarized in Table 6.3.

Zienkiewicz and Taylor [12] approach the problem of deriving recurrence relations for equation 6.55 by applying the Galerkin form of the method of weighted residuals to the vector $\{\delta\}$ and selecting interpolation functions that are functions only of time. The interpolation functions may span one or more

TABLE 6.3. Solution Sequence for Implicit Algorithms

INITIAL CALCULATIONS

1. Form element matrices $[M]^{(e)}$, $[C]^{(e)}$, and $[K]^{(e)}$.
2. Initialize $\{\delta\}_t$, $\{\dot{\delta}\}_t$, and $\{\ddot{\delta}\}_t$ at $t = 0$.
3. Select Δt.
4. Assemble the effective stiffness matrix $[\bar{K}]$.
5. Modify $[\bar{K}]$ for boundary conditions.
6. Factor $[\bar{K}]$.

AT EACH TIME STEP

1. Form the effective element load vector $\{\bar{F}\}^{(e)}_{t+\Delta t}$.
2. Assemble the effective load vector $\{\bar{F}\}_{t+\Delta t}$.
3. Modify the effective vector for boundary conditions.
4. Solve for displacements $\{\delta\}_{t+\Delta t}$ at time $t + \Delta t$.
5. Solve for velocities and accelerations $\{\dot{\delta}\}_{t+\Delta t}$ and $\{\ddot{\delta}\}_{t+\Delta t}$.

time increments, depending on the order of interpolation chosen. General families of recurrence formulas may be derived by this approach. The central difference and Newmark methods are special cases of this more general approach.

6.8 CLOSURE

In this chapter we examined a general finite element formulation for a variety of elasticity problems, namely, plane stress, plane strain, axisymmetric solids, plate bending, and general three-dimensional solids. Though the element equations derived for these problems were explicitly evaluated only for the simplest types in each case, the equations are general and apply for many element shapes and displacement models. The introductory treatment of finite elements in structural dynamics derived the basic equations for dynamic problems, discussed free-vibration problems, and presented popular methods for finding transient solutions.

The discussion in this chapter does not include several advanced topics in structural mechanics such as bifurcation buckling problems, nonlinear problems due to large displacements and/or material nonlinearity, incompressible materials, and analyses of shells of arbitrary curvature. Such problems are of fundamental importance in structural mechanics and have been the subject of extensive finite element research in recent years. Solutions of these problems are beyond the scope of this introductory text, but detailed information on these appear in the comprehensive text by Zienkiewicz and Taylor [12] and the finite element structural mechanics texts by Cook et al. [16] and Bathe [34]. Our treatment, however, provides the fundamental background for the study of these more advanced topics. Lastly, we have not covered the validation of element performance in modeling particular loading configurations. These validation procedures, called *patch tests*, are used to determine whether an element has the capability of converging when a discretized mesh is sufficiently refined. These tests also are discussed in the three references mentioned above.

REFERENCES

1. T. H. H. Pian and P. Tong, "Basis of Finite Element Methods for Solid Continua." *Int. J. Numer. Methods Eng.*, Vol. 1, No. 1, 1969, p. 26.
2. R. J. Melosh, "Basis for Derivation of Matrices for the Direct Stiffness Method," *J. Am. Inst. Aeronaut. Astron.*, Vol. 1, No. 7, 1963, pp. 1631–1637.
3. Z. M. Elias, "Duality in Finite Element Methods," *Proc. ASCE J. Eng. Mech. Div.*, Vol. 94, No. EM 4, 1968, pp. 931–946.
4. L. S. D. Morley, "The Triangular Equilibrium Element in the Solution of Plate Bending Problems," *Aeronaut. Q.*, 1968, pp. 149–169.
5. B. M. Fraeijs de Veubeke, "Upper and Lower Bounds in Matrix Structural Analysis," *Matrix Methods Struct. Anal.*, AGARD, Vol. 72, 1964, pp. 165–201.
6. T. H. H. Pian, "Derivation of Element Stiffness Matrices by Assumed Stress Distribution," *J. Am. Inst. Aeronaut. Astron.*, Vol. 2, No. 7, 1964, pp. 1333–1336.
7. R. Yamamoto and H. Isshiki, "Variational Principles and Dualistic Scheme for Intersection Problems in Elasticity," *J. Fac. Eng. Univ. Tokyo*, Vol. 30, No. 1, 1969.
8. P. Tong, "New Displacement Hybrid Finite Element Model for Solid Continua," *Int. J. Numer. Methods Eng.*, Vol. 2, No. 1, 1970, pp. 73–84.
9. L. R. Herrmann, "A Bending Analysis for Plates," *Proceedings of 1st Conference on Matrix Methods in Structural Mechanics* (AFFDL-TR-66-80), Wright-Patterson Air Force Base, Dayton, OH, 1965.
10. I. Holand, "The Finite Element Method in Plane Stress Analysis," Chapter 2, in *The Finite Element Method in Stress Analysis*, I. Holand and K. Bell (eds.), Tapir Press, Trondheim, Norway, 1969.
11. R. W. Clough, "The Finite Element in Plane Stress Analysis," *Proceedings of 2d ASCE Conference on Electronic Computation*, Pittsburgh, PA, September 1960.
12. O. C. Zienkiewicz and R. L. Taylor, *The Finite Element Method*, Vol. 1, *Basic Formulation and Linear Problems*, McGraw-Hill, London, 1989, and *The Finite Element Method*, Vol. 2, *Solid and Fluid Mechanics, Dynamics and Non-Linearity*, McGraw-Hill, London, 1991.
13. E. L. Wilson, "Structural Analysis of Axisymmetric Solids," *AIAA J.*, Vol. 3, No. 12, December 1965, pp. 2267–2274.
14. R. Clough and Y. Rashid, "Finite Element Analysis of Axisymmetric Solids," *Proc. ASCE J. Eng. Mech. Div.*, Vol. 91, No. EMI, February 1965, pp. 71–85.
15. S. Utku, "Explicit Expressions for Triangular Torus Element Stiffness Matrix," *AIAA J.*, Vol. 6, No. 6, June 1968, pp. 1174–1175.

16. R. D. Cook, D. S. Malkus, and M. E. Plesha, *Concepts and Applications of Finite Element Analysis*, 3d ed., Wiley, New York, 1989.

17. K. Maruyama, "Stress Analysis of a Bolt–Nut Joint by the Finite Element Method and the Copper-Electroplating Method," *Bull. Jpn. Soc. Mech. Eng.*, Vol. 16, No. 94, April 1973, pp. 671–678.

18. O. C. Zienkiewicz, "The Finite Element Method from Intuition to Generality," *Appl. Mech. Rev.*, Vol. 23, No. 3, March 1970, pp. 249–256.

19. F. K. Bogner, R. L. Fox, and L. A. Schmit, Jr., "The Generation of Interelement-Compatible Stiffness and Mass Matrices by the Use of Interpolation Formulas," *Proceedings of 1st Conference on Matrix Methods in Structural Mechanics* (AFFDL-TR-66-80), Wright-Patterson Air Force Base, Dayton, OH, 1965.

20. R. W. Clough and J. L. Tocher, "Finite Element Stiffness Matrices for Analysis of Plate Bending," *Proceedings of 1st Conference on Matrix Methods in Structural Mechanics*, Wright Patterson Air Force Base, Dayton, OH, November 1965.

21. J. S. Przemieniecki, *Theory of Matrix Structural Analysis*, McGraw-Hill, New York, 1968.

22. E. A. Thornton and S. Tseng, "A Finite Element Analysis of General Laminated Plates," *Old Dominion University School of Engineering Technical Report 74-M4*, June 1974.

23. J. E. Walz, R. E. Fulton, and N. J. Cyrus, "Accuracy and Convergence of Finite Element Approximation," *Proceedings of 2d Conference of Matrix Methods in Structural Mechanics*, Wright Patterson Air Force Base, Dayton, OH, 1968.

24. J. L. Batoz, K. J. Bathe, and L. W. Ho, "A Study of Three-Node Triangular Plate Bending Elements," *Int. J. Numer. Methods Eng.*, Vol. 15, 1980, pp. 1771–1812.

25. J. Ergatoudis, B. M. Irons, and O. C. Zienkiewicz, "Three Dimensional Analysis of Arch Dams and Their Foundations," *Proceedings of Symposium on Architecture of Dams*, Institute of Civil Engineering, UK, 1968.

26. S. Fjeld, "Three Dimensional Theory of Elasticity," in *Finite Element Methods in Stress Analysis*, I. Holand and K. Bell (eds.), Tapir Press, Trondheim, Norway, 1969.

27. Y. Rashid, "Three-Dimensional Analysis of Elastic Solids," *Int. J. Solids Struct.*, Part I, Vol. 5, 1969, pp. 1311–1332; Part II, Vol. 6, 1970, pp. 195–207.

28. R. J. Melosh, "Structural Analysis of Solids," *Proc. ASCE J. Struct. Div.*, Vol. 89, No. ST-4, August 1963, pp. 205–223.

29. R. W. Clough, "Comparison of Three-Dimensional Finite Elements," *Proceedings of Symposium on Application of Finite Element Methods in Civil Engineering*, ASCE–Vanderbilt University, Nashville, TN, November 1969.

30. B. J. Lazan, *Damping of Materials and Members in Structural Mechanics*, Pergamon, New York, 1968.

31. L. Meirovitch, *Analytical Method in Vibrations*, Macmillan, New York, 1967.

32. R. W. Clough, "Analysis of Structural Vibrations and Dynamic Response," in *Recent Advances in Matrix Methods of Structural Analysis and Design*, R. H. Gallagher, Y. Yamada, and J. T. Oden (eds.), University of Alabama Press, Huntsville, AL, 1971.

33. R. W. Clough and K. J. Bathe, "Finite Element Analysis of Dynamic Response," in *Advances in Computational Methods in Structural Mechanics and Design*, J. T.

Oden, R. W. Clough and Y. Yamamato (eds.), University of Alabama Press, Huntsville, AL, 1972.

34. K. J. Bathe, *Finite Element Procedures in Engineering Analysis*, Prentice-Hall, Englewood Cliffs, NJ, 1982.

35. P. Tong, T. H. H. Pian, and L. L. Bucciarelli, "Mode Shapes and Frequencies by Finite Element Method Using Consistent and Lumped Masses," *Comput. Struct.*, Vol. 1, 1971, pp. 623–638.

36. W. C. Hurty and M. F. Rubinstein, *Dynamics of Structures*, Prentice-Hall, Englewood Cliffs, NJ, 1964.

37. B. Carnahan, H. A. Luther, and J. O. Wilkes, *Applied Numerical Methods*. Wiley, New York, 1969.

38. E. A. Thornton, "Free Vibrations of Unsymmetrical Laminated Cantilevered Composite Panels," *Shock Vib. Bull*, Part 2, September 1977, pp. 79–88.

39. N. M. Newmark, "A Method of Computation for Structural Dynamics," *J. Eng. Mech. Div. ASCE*, Vol. 85, No. EM3, 1959, pp. 67–94.

PROBLEMS

1. Verify that the derivation of the element stiffness matrix for a triangle presented in Section 6.3 leads to the result presented in equation 2.37.

2. The single constant-strain triangle of Figure P6.2 is in a state of plane

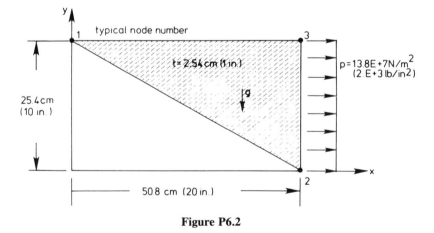

Figure P6.2

stress. Compute the following:

a. The element stiffness matrix.

b. The element equivalent nodal force vector due to the applied load. Assume that $E = 20.7E + 10$ N/m^2 ($30E + 6$ lb/in.^{2}) and $\nu = 0.3$.

3. The plane stress triangle shown in Figure P6.2 experiences the following nodal displacements: $u_1 = v_1 = 0$; $u_2 = 9.30E - 3$ cm $(3.66E - 3$ in.$)$, $v_2 = -0.064E - 3$ cm $(-0.025E - 3$ in.$)$; $u_3 = 10.1E - 3$ cm $(3.98E - 3$ in.$)$, $v_3 = -1.59E - 3$ cm $(-0.625E - 3$ in.$)$. Compute the element stresses σ_x, σ_y, and τ_{xy} due to these displacements, assuming that E and ν are as given in the previous problem.

4. The element shown in Figure P6.2 is subject to a gravity loading in the negative y direction. Assuming that the material specific weight is $2E - 5$ N/m^3 $(0.284$ lb/in^3$)$, compute the equivalent nodal forces due to the gravity body force.

5. A constant-strain triangular element in a state of plane stress is subject to a uniform temperature change of ΔT. Assuming that the coefficient of thermal expansion is α, derive an equation for the equivalent nodal forces due to the temperature change. See Appendix C, Section C.10, for a discussion of thermal effects in elasticity.

6. An additional element is added to Figure P6.2 to give the two-element model of the flat plate shown in Figure P6.6. Assuming a state of *plane strain*, compute the following:

   a. The nodal displacements at nodes 3 and 4.

   b. The reactive forces at nodes 1 and 2.

   c. The stresses σ_x, σ_y, and τ_{xy} in the two elements due to the applied load.

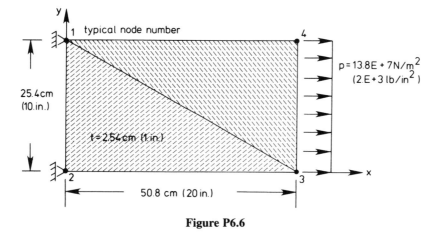

Figure P6.6

7. Compute the nodal displacements and element stresses for the plate shown in Figure P6.7. Assume plane stress, thickness $t = 1.5$ cm $(0.59$ in.$)$, $E = 6.9E + 10$ N/m^2 $(10E + 6$ lb/in^2$)$, and $\nu = 0.25$.

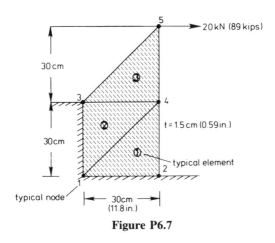

Figure P6.7

8. Consider the rectangular plane stress element shown in Figure P6.8. Element nodal displacements are u_i, v_i, $i = 1, 4$. The element displacement field is assumed to be

$$u = \lfloor N \rfloor \{u\} \qquad v = \lfloor N \rfloor \{v\}$$

with

$$\lfloor N \rfloor = \lfloor (1 - \xi)(1 - \eta) \quad \xi(1 - \eta) \quad \xi\eta \quad (1 - \xi)\eta \rfloor$$

where $\xi = x/a$ and $\eta = y/b$. Evaluate $[B]^{(e)}$, equation 6.7a, for the element. Is the element a constant-strain element? Discuss how element strains and stresses vary within the element and illustrate their variations with sketches.

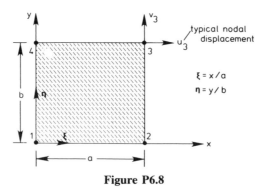

Figure P6.8

9. Using the interpolation functions given in Problem 8, verify the following submatrices of an element stiffness matrix (assume plane stress and

element thickness t; the element aspect ratio is $\beta = b/a$):

$$[k_{11}] = \frac{Et}{12(1-\nu^2)} \begin{bmatrix} 4\beta + \dfrac{2(1-\nu)}{\beta} & \dfrac{3}{2}(1+\nu) \\[3mm] \dfrac{3}{2}(1+\nu) & \dfrac{4}{\beta} + 2(1-\nu)\beta \end{bmatrix} \begin{matrix} u_1 \\[3mm] v_1 \end{matrix}$$

$$\begin{matrix} \quad\;\; u_1 \qquad\qquad\qquad\quad v_1 \end{matrix}$$

$$[k_{12}] = \frac{Et}{12(1-\nu^2)} \begin{bmatrix} -4\beta + \dfrac{1-\nu}{\beta} & -\dfrac{3}{2}(1-3\nu) \\[3mm] \dfrac{3}{2}(1-3\nu) & \dfrac{2}{\beta} - 2(1-\nu) \end{bmatrix} \begin{matrix} u_1 \\[3mm] v_1 \end{matrix}$$

$$\begin{matrix} \quad\;\; u_2 \qquad\qquad\qquad\quad v_2 \end{matrix}$$

10. A frequently used structural element is the beam element shown in Figure P6.10. The element generalized displacements are the transverse displacement $w(x)$ and slope θ, where $\theta = -dw/dx$. The potential energy of the beam subject to a transverse loading $p(x)$ per unit length is

$$\Pi = \frac{1}{2}\int_0^L EI\left(\frac{d^2w}{dx^2}\right)^2 dx - \int_0^L p(x)w(x)\,dx$$

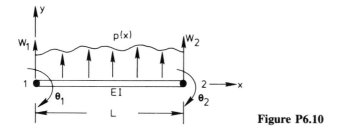

Figure P6.10

where I is the moment of inertia of the cross-sectional area. Assume that

$$w(x) = N_1(x)w_1 + N_2(x)\theta_1 + N_3(x)w_2 + N_4(x)\theta_2$$

where w_1, w_2, θ_1, and θ_2 are the nodal displacements and slopes, respectively. Use the principle of minimum potential energy to derive general equations for a beam element stiffness matrix and an equivalent nodal force vector.

11. For the beam element given in Problem 10 the interpolation functions are the Hermite polynomials, equation 5.35:

$$N_1 = 1 - 3s^2 + 2s^3 \qquad N_2 = -Ls(s - 1)^2$$
$$N_3 = s^2(3 - 2s) \qquad N_4 = -Ls^2(s - 1)$$

where $s = x/L$ and minus signs are used in N_2 and N_4 to account for the θ sign convention. Show that the element stiffness matrix is

$$
[k] = \frac{2EI}{L^3}
\begin{array}{cccc}
w_1 & \theta_1 & w_2 & \theta_2 \\
\end{array}
\begin{bmatrix}
6 & -3L & 6 & -3L \\
-3L & 2L^2 & 3L & L^2 \\
6 & 3L & 6 & 3L \\
-3L & L^2 & 3L & 2L^2
\end{bmatrix}
\begin{array}{c}
w_1 \\
\theta_1 \\
w_2 \\
\theta_2
\end{array}
$$

12. The beam element shown in Figure P6.10 is subjected to a uniform load per unit length $p(x) = p_0$. Show that the element equivalent nodal force vector is

$$
\{F\} = \frac{p_0 L}{2}
\begin{Bmatrix}
1 \\
-\dfrac{L}{6} \\
1 \\
\dfrac{L}{6}
\end{Bmatrix}
$$

13. Use calculus of variations (see Appendix B) to minimize the beam potential energy that appears in Problem 10. Show that the Euler–Lagrange equation is

$$\frac{d^2}{dx^2}\left(EI\frac{d^2w}{dx^2}\right) - p(x) = 0$$

which is the governing differential equation of the beam.

14. Apply the method of weighted residuals to the governing differential equation for the beam given in Problem 13 and derive general equations for the beam element stiffness matrix and equivalent nodal force vector.

15. Use the beam element stiffness matrix and equivalent load vector given in Problems 11 and 12, respectively, to solve for the slope θ_2 at the center support of the beam shown in Figure P6.15.

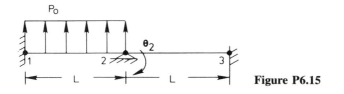

Figure P6.15

16. The differential equation governing a beam subjected to an axial compressive load P is

$$\frac{d^2}{dx^2}\left(EI\frac{d^2w}{dx^2}\right) + P\frac{d^2w}{dx^2} = 0$$

Use the method of weighted residuals to derive general equations for the element matrices that govern beam buckling. Assume the shape for $w(x)$ given in Problem 10.

17. Explain why the displacement function $w(x, y)$ for the incompatible rectangular plate-bending element in Section 6.5 can lead to discontinuous normal slopes at element interfaces.

18. Show that the consistent mass matrix for the two-node rod element in Figure P6.18 is

$$[M]^{(e)} = \rho\frac{AL}{6}\begin{matrix}u_1 & u_2 \\ \begin{bmatrix} 2 & 1 \\ 1 & 2 \end{bmatrix} & \begin{matrix}u_1 \\ u_2\end{matrix}\end{matrix}$$

where $u(x) = (1 - x/L)u_1 + (x/L)u_2$.

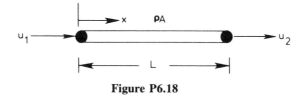

Figure P6.18

19. Read the paper "Consistent Mass Matrix for Distributed Mass Systems" by J. S. Archer, *Journal of the Structural Division*, *ASCE*, Vol. 89, No. ST4, August 1963, pp. 161–178. Write a brief summary of the paper and discuss the conclusions. Verify equations (16) and (17) in the paper.

20. Consider the longitudinal vibrations of the rod shown in Figure P6.20. Compute the vibration frequencies of the rod in terms of $(k/m)^{1/2}$

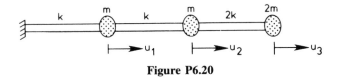

Figure P6.20

assuming that the rod mass is small in comparison to the three lumped masses shown. Compute and sketch the corresponding mode shapes.

21. A scale model of an early design of the Space Shuttle orbiter had the properties shown in Figure P6.21. Write a computer program to compute the natural frequencies and mode shapes of the model. Assume the model to have "free-free" boundary conditions. Use a consistent mass matrix (see the paper cited in Problem 19 for the consistent mass matrix) and the beam stiffness matrix given in Problem 11. Assume the properties of aluminum. Sketch the vibration mode shapes and explain the physical significance of the lowest two eigenvalues and eigenvectors.

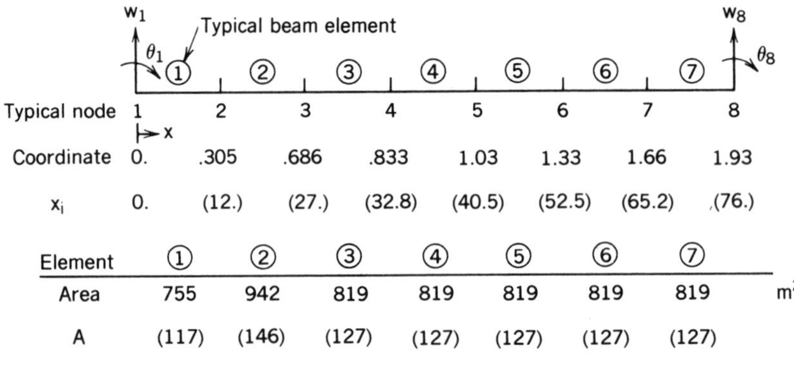

Typical node	1	2	3	4	5	6	7	8	
Coordinate	0.	.305	.686	.833	1.03	1.33	1.66	1.93	m
x_i	0.	(12.)	(27.)	(32.8)	(40.5)	(52.5)	(65.2)	(76.)	(in.)

Element	①	②	③	④	⑤	⑥	⑦	
Area	755	942	819	819	819	819	819	$m^2 \times 10^{-4}$
A	(117)	(146)	(127)	(127)	(127)	(127)	(127)	(in^2)
Inertia	131	256	169	169	169	169	169	$m^4 \times 10^{-6}$
I	(316)	(614)	(407)	(407)	(407)	(407)	(407)	(in^4)

Figure P6.21

22. Read the paper "Direct Integration Methods in Structural Dynamics" by R. E. Nickell, *Journal of Engineering Mechanics Division*, *ASCE*, Vol. 99, April 1973, pp. 303–317.

    a. What is the basic dilemma involved in integrating the structural equations of motion?

    b. What is meant by *instability* in numerical integration?

c. What are some differences between implicit and explicit methods of numerical integration?

d. Of the integration methods presented, which numerical method is found to be the most attractive? Discuss the justification of this conclusion.

23. Consider a system with two degrees of freedom for which the equations of motion are

$$\begin{bmatrix} 0.5 & 0 \\ 0 & 0.25 \end{bmatrix} \begin{Bmatrix} \ddot{u}_1 \\ \ddot{u}_2 \end{Bmatrix} + \begin{bmatrix} 4 & -2 \\ -2 & 2 \end{bmatrix} \begin{Bmatrix} u_1 \\ u_2 \end{Bmatrix} = \begin{Bmatrix} 0 \\ 0 \end{Bmatrix}$$

Verify that the free-vibration periods for the system are $T_1 = 4.1$ and $T_2 = 1.7$. Use the central difference method with time steps (a) $\Delta t = T_2/5$ and (b) $\Delta t = 5T_2$ to calculate the response of the system for 10 steps. Assume that $u_1(0) = 1$, $u_2(0) = 0$, and $\dot{u}_1(0) = \dot{u}_2(0) = 0$. Sketch the displacement histories and discuss the stability of the computed displacements.

24. Use the Newmark method to solve Problem 6.23. Discuss the relative computational effort required to solve the problem by the central difference and Newmark methods. Sketch the displacement histories predicted by the Newmark method and discuss the comparative performance of the two algorithms.

7

GENERAL FIELD PROBLEMS

7.1 INTRODUCTION

In this chapter we present finite element formulations for a significant class of physical problems known as field problems. Heat conduction problems, electrostatic fields, and irrotational flows are just a few of the many important field problems that engineers encounter. Field problems share the common characteristic of being governed by similar partial differential equations for the field variable ϕ. Hence we can discuss the solution of these problems by focusing attention on the equations of ϕ without identifying ϕ as a particular physical quantity for a particular problem.

As in Chapter 6, our goal here is to derive element equations in terms of general interpolation functions N_i. We shall do this by applying either a classical variational principle or the method of weighted residuals with Galerkin's criterion. The formulations are for three-dimensional problems, but equations for problems of other dimensions follow immediately. After the element equations are known we can evaluate them for the element type of our choice. We shall consider some examples of this evaluation process for simple linear elements.

We begin the chapter with a discussion of general equilibrium problems characterized by quasiharmonic equations and eigenvalue problems governed by the Helmholtz equation. Next we consider practical time-dependent propagation field problems whose solutions ultimately involve the solution of simultaneous ordinary differential equations with time as the independent variable. Finally, we describe two procedures for obtaining the solution of such equations.

7.2 EQUILIBRIUM PROBLEMS

Equilibrium problems require the determination of the steady-state (time-independent) spatial distribution of a field variable. The equilibrium problems considered here include problems of steady-state temperature distributions, steady flows, and torsion of an elastic structural member.

7.2.1 Quasiharmonic Equations

Suppose that the field variable ϕ is to be found in a three-dimensional solution domain Ω bounded by a surface Γ (Figure 7.1). For steady-state (time-independent) problems the field equation to be solved is the quasiharmonic equation, expressed in general terms as

$$\frac{\partial}{\partial x}\left(k_x \frac{\partial \phi}{\partial x}\right) + \frac{\partial}{\partial y}\left(k_y \frac{\partial \phi}{\partial y}\right) + \frac{\partial}{\partial z}\left(k_z \frac{\partial \phi}{\partial z}\right) = f(x, y, z) \qquad (7.1)$$

where k_x, k_y, k_z, and f are given functions independent of ϕ, and the coefficients k_x, k_y, and k_z are bounded away from zero in Ω. The physical interpretation of the parameters in equation 7.1 depends on the particular physical problem. Table 7.1 lists a number of typical field problems and indicates the meaning of ϕ as well as the other parameters for each problem.

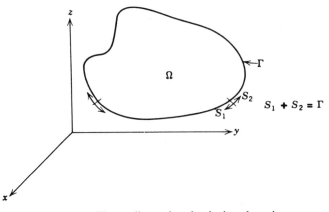

Figure 7.1   Three-dimensional solution domain.

Note that for an inhomogeneous and/or nonisotropic medium, equation 7.1 implies that the coordinates x, y, and z coincide with the principal coordinates. For a homogeneous medium k_x, k_y, and k_z are constant, while for an isotropic medium $k_x = k_y = k_z = k$.

7.2.2   Boundary Conditions

The description of the field problem is not complete until boundary conditions are specified; that is, equation 7.1 must be solved subject to additional constraints imposed on the bounding surface. Usually on some part of the boundary the value of ϕ is a specified function such as

$$\phi = \Phi(x, y, z) \qquad \text{on } S_1 \tag{7.2}$$

while on the remaining part of the boundary we have the condition

$$k_x \frac{\partial \phi}{\partial x} n_x + k_y \frac{\partial \phi}{\partial y} n_y + k_z \frac{\partial \phi}{\partial z} n_z + g(x, y, z) + h(x, y, z)\phi = 0 \qquad \text{on } S_2$$

$$\tag{7.3}$$

where g and h are known a priori and n_x, n_y, and n_z are the direction cosines of the outward normal to the surface. The union of S_1 and S_2 forms the complete boundary Γ, $S_1 \cup S_2 = \Gamma$.

The boundary condition expressed by equation 7.2 is known as the *Dirichlet* condition, and the specified function Φ is sometimes called Dirichlet data. Equation 7.3 is the *Cauchy* boundary condition. If the functions $g = h = 0$, the Cauchy condition reduces to the *Neumann* boundary condition, sometimes known as the "essential" boundary condition. A field prob-

TABLE 7.1. Identification of Physical Parameters

Problem	ϕ	k_x, k_y, k_z	f	g	h
Diffusion flow in porous media	Hydraulic head	Hydraulic conductivity	Internal source flow	Boundary flow	—
Electric conduction	Voltage	Electric conductivity	Internal current source	Externally applied boundary current	—
Electrostatic field	Electric force field intensity	Permittivity	Internal current source	—	—
Fluid-film lubrication	Pressure	k_x, k_y are functions of film thickness and viscosity, $k_z = 0$	Net flow due to various actions	Boundary flow	—
Membrane deflection	Transverse deflection	k_x, k_y are membrane tensions $k_z = 0$	Transverse distributed load	—	—
Gravitation	Component of gravitational force vector per unit mass	—	—	—	—
Heat conduction	Temperature	Thermal conductivity	Internal heat generation	Boundary heat flux	Convective heat transfer coefficient
Irrotational flow	Velocity potential or stream function	—	0	Boundary velocity	0
Torsion	Stress function	Reciprocal of shear modulus	Angle of twist per unit length	—	—
Seepage	Pressure	Permeability	Internal flow source	—	—
Magnetostatics	Magnetomotive force	Magnet permeability	Internal magnetic field source	Externally applied magnetic field intensity	—

lem is said to have mixed boundary conditions when some portions of Γ have Dirichlet boundary conditions while other portions have Cauchy boundary conditions.

Equations 7.1–7.3 comprise a well-posed elliptic boundary value problem whose solution by finite element methods can be based on a classical variational principle. The elliptic classification of partial differential equations is briefly discussed in Section 3.2.3. Elliptic boundary value problems have a closed solution domain and their solution must simultaneously satisfy the governing partial differential equation and all boundary conditions. We have formulated this field problem as a general case that can easily be specialized to particular cases. For example, if $k_x = k_y = k_z = k = $ constant, equation 7.1 reduces to

$$\nabla^2 \phi = \frac{f}{k}(x, y, z) \tag{7.4}$$

which is known as *Poisson's* equation. Furthermore, if $f = 0$, we have the well-known *Laplace* equation:

$$\nabla^2 \phi = 0 \tag{7.5}$$

We shall encounter this equation again in Chapter 9 when we investigate potential flow problems.

As we pointed out in Chapter 3, the finite element method is only one of many different ways to obtain approximate numerical solutions to these equations. For many years analysts used finite difference techniques, but when nonuniform meshes are needed for problems with irregular geometry, the finite element method is better suited. The finite element equations for this general field problem can be easily derived from a variational principle by following the procedures of Chapter 3. Identical element equations are obtained if instead of the variational approach we use the method of weighted residuals with Galerkin's criterion (Chapter 4).

7.2.3 Variational Principle

It can be shown [1, 2] that the function $\phi(x, y, z)$ that satisfies equations 7.1–7.3 also minimizes the functional

$$I(\phi) = \frac{1}{2}\int_\Omega \left[k_x \left(\frac{\partial \phi}{\partial x} \right)^2 + k_y \left(\frac{\partial \phi}{\partial y} \right)^2 + k_z \left(\frac{\partial \phi}{\partial z} \right)^2 + 2f\phi \right] dx\,dy\,dz$$

$$+ \int_{S_2} \left(g\phi + \frac{1}{2}h\phi^2 \right) dS_2 \tag{7.6}$$

Following the procedures of Appendix B, it is easy to show that equation 7.1 is the Euler–Lagrange condition for the functional of equation 7.6. The variational principle on which we can base the derivation of the element equation is

$$\delta I(\phi) = 0 \tag{7.7}$$

7.2.4 Element Equations

Suppose that the solution domain Ω is divided into M elements of r nodes each. By the usual procedure we may express the behavior of the unknown function ϕ within each element as

$$\phi^{(e)} = \sum_{i=1}^{r} N_i \phi_i = \lfloor N \rfloor \{\phi\}^{(e)} \tag{7.8}$$

where ϕ_i is the nodal value of ϕ at node i. Equation 7.8 implies that only nodal values of ϕ are taken as nodal degrees of freedom, but derivatives of ϕ may also be used as nodal parameters without changing the procedure to be followed. The quantity ϕ_i may be thought of as a general nodal parameter.

Since the functional $I(\phi)$ contains only first-order derivatives, we have a C^0 problem, and the N_i must be chosen to preserve at least continuity of ϕ at element interfaces. If the interpolation functions guarantee C^0 continuity, we may focus attention on one element, because the integral $I(\phi)$ can be represented as the sum of integrals over all the elements:

$$I(\phi) = \sum_{e=1}^{M} I(\phi^{(e)}) \tag{7.9}$$

The discretized form of the functional for one element is obtained by substituting equation 7.8 into equation 7.6. Then the minimum condition $\delta I(\phi) = 0$ for one element becomes

$$\frac{\partial I(\phi^{(e)})}{\partial \phi_i} = 0, \qquad i = 1, 2, \ldots, r \tag{7.10}$$

Consider first elements with edges on S_2. For a node i on boundary S_2, from equation 7.6 we have

$$\frac{\partial I(\phi^{(e)})}{\partial \phi_i} = 0$$

$$= \int_{\Omega^{(e)}} \left[k_x \frac{\partial \phi^{(e)}}{\partial x} \frac{\partial}{\partial \phi_i} \left(\frac{\partial \phi^{(e)}}{\partial x} \right) + k_y \frac{\partial \phi^{(e)}}{\partial y} \frac{\partial}{\partial \phi_i} \left(\frac{\partial \phi^{(e)}}{\partial y} \right) \right.$$

$$\left. + k_z \frac{\partial \phi^{(e)}}{\partial z} \frac{\partial}{\partial \phi_i} \left(\frac{\partial \phi^{(e)}}{\partial z} \right) + f \frac{\partial \phi^{(e)}}{\partial \phi_i} \right] d\Omega$$

$$+ \int_{S_2^{(e)}} \left(g \frac{\partial \phi^{(e)}}{\partial \phi_i} + h \phi^{(e)} \frac{\partial \phi^{(e)}}{\partial \phi_i} \right) dS_2 \qquad \text{on surface } S_2^{(e)} \quad (7.11)$$

If an element does not have an edge on S_2, the second integral does not appear.

Referring to equation 7.8, we may evaluate each of the derivatives in 7.11. These typically become

$$\frac{\partial \phi^{(e)}}{\partial x} = \sum_{i=1}^{r} \frac{\partial N_i}{\partial x} \phi_i = \left\lfloor \frac{\partial N}{\partial x} \right\rfloor \{\phi\}^{(e)}$$

$$\frac{\partial}{\partial \phi_i} \left(\frac{\partial \phi^{(e)}}{\partial x} \right) = \frac{\partial N_i}{\partial x}$$

$$\frac{\partial \phi^{(e)}}{\partial \phi_i} = N_i$$

Thus we have

$$\frac{\partial I(\phi^{(e)})}{\partial \phi_i} = 0$$

$$= \int_{\Omega^{(e)}} \left[k_x \left\lfloor \frac{\partial N}{\partial x} \right\rfloor \{\phi\}^{(e)} \frac{\partial N_i}{\partial x} + k_y \left\lfloor \frac{\partial N}{\partial y} \right\rfloor \{\phi\}^{(e)} \frac{\partial N_i}{\partial y} \right.$$

$$\left. + k_z \left\lfloor \frac{\partial N}{\partial z} \right\rfloor \{\phi\}^{(e)} \frac{\partial N_i}{\partial z} + f N_i \right] d\Omega$$

$$+ \int_{S_2^{(e)}} \left[g N_i + h \lfloor N \rfloor N_i \{\phi\}^{(e)} \right] dS_2 \qquad \text{on surface } S_2^{(e)} \quad (7.12)$$

Combining all of the equations like equation 7.12 for all of the nodes of the element gives the following set of element equations:

$$\left\{ \frac{\partial I}{\partial \phi^{(e)}} \right\}^{(e)} = \left\{ \begin{array}{c} \dfrac{\partial I(\phi^{(e)})}{\partial \phi_1} \\[2mm] \dfrac{\partial I(\phi^{(e)})}{\partial \phi_2} \\[2mm] \vdots \\[2mm] \dfrac{\partial I(\phi^{(e)})}{\partial \phi_r} \end{array} \right\} = \overset{r \times r}{[K]}^{(e)} \overset{r \times 1}{\{\phi\}}^{(e)} + \overset{r \times r}{[K_{S_2}]}^{(e)} \overset{r \times 1}{\{\phi\}}^{(e)} + \overset{r \times 1}{\{R\}}^{(e)} = 0$$

$$(7.13)$$

where the coefficients of the matrices $[K]^{(e)}$, $\{R\}^{(e)}$, and $[K_{S_2}]^{(e)}$ are given by

$$k_{ij} = \int_{\Omega^{(e)}} \left(k_x \frac{\partial N_i}{\partial x} \frac{\partial N_j}{\partial x} + k_y \frac{\partial N_i}{\partial y} \frac{\partial N_j}{\partial y} + k_z \frac{\partial N_i}{\partial z} \frac{\partial N_j}{\partial z} \right) d\Omega \quad (7.14)$$

$$k_{S_{2ij}} = \int_{S_2^{(e)}} hN_i N_j \, dS_2 \quad (7.15)$$

$$R_i = \int_{\Omega^{(e)}} fN_i \, d\Omega + \int_{S_2^{(e)}} gN_i \, dS_2 \quad (7.16)$$

We emphasize again that the matrix $[K_{S_2}]^{(e)}$ and the last term in equation 7.16 appear only if element (e) contributes to the definition of the boundary portion S_2. We see that $[K_{S_2}]^{(e)}$ is actually a square matrix that can be added to $[K]^{(e)}$ to form the following element equations:

$$\left[[K]^{(e)} + [K_{S_2}]^{(e)} \right] \{\phi\}^{(e)} + \{R\} = 0 \quad (7.17)$$

Assembly of these element equations to obtain the system equations follows the standard procedure. It is important to note that since the nodal unknowns ϕ_i are scalars, no transformation of matrices computed in local coordinates is necessary before assembly of the global matrices. The matrix $[K_{S_2}]^{(e)}$ is associated with $h(x, y, z)$ given in the boundary condition, equation 7.3. Referring to Table 7.1, we note that $h(x, y, z)$ is identified as the convective heat transfer coefficient in heat transfer. The matrix $[K_{S_2}]^{(e)}$ is characteristic of convective boundary conditions in heat transfer and will be discussed further in Chapter 8. We note that for the other interpretations of the quasiharmonic equation shown in Table 7.1, $h(x, y, z)$ does not appear, and therefore the computation of $[K_{S_2}]^{(e)}$ is not required.

7.2.5   Element Equations in Two Dimensions

Plane Quasiharmonic Equations   To illustrate the application of the foregoing equations we shall use linear triangular elements and develop the matrix equations for two-dimensional problems with general boundary conditions. The boundary condition corresponding to convective heat transfer is omitted, since it is considered in detail in Chapter 8. The governing equation is

$$\frac{\partial}{\partial x} \left(k_x \frac{\partial \phi}{\partial x} \right) + \frac{\partial}{\partial y} \left(k_y \frac{\partial \phi}{\partial y} \right) = f(x, y) \quad (7.18)$$

and the boundary conditions segregate as follows:

$$\phi = \phi(x, y) \qquad \text{on } S_1 \quad (7.19)$$

$$k_x \frac{\partial \phi}{\partial x} n_x + k_y \frac{\partial \phi}{\partial y} n_y + g(x, y) = 0 \qquad \text{on } S_2 \quad (7.20)$$

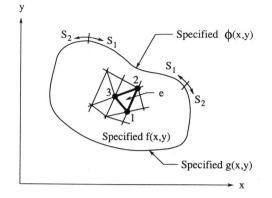

Figure 7.2   Solution domain for general two-dimensional field problems.

where ϕ, k_x, k_y, f, and g have the possible physical identifications indicated in Table 7.1. Figure 7.2 shows the two-dimensional solution domain, the boundary segments, and a typical finite element.

If we divide the region into triangular elements and assume that ϕ varies linearly within each element, we may then write immediately

$$\phi^{(e)}(x, y) = \sum_{i=1}^{3} L_i \phi_i$$

where the L_i are the natural coordinates for the triangle given by equations 5.17–5.19:

$$L_i = \frac{a_i + b_i x + c_i y}{2\Delta}, \qquad i = 1, 2, 3$$

After identifying the interpolation functions N_i with L_i we may refer to equations 7.14 and 7.15 to write the element matrices. Since

$$\frac{\partial N_i}{\partial x} = \frac{b_i}{2\Delta} \qquad \frac{\partial N_j}{\partial x} = \frac{b_j}{2\Delta}$$

$$\frac{\partial N_i}{\partial y} = \frac{c_i}{2\Delta} \qquad \frac{\partial N_j}{\partial y} = \frac{c_j}{2\Delta} \tag{7.21}$$

we have from equation 7.14

$$k_{ij} = \int_{\Delta} \left(k_x \frac{b_i}{2\Delta} \frac{b_j}{2\Delta} + k_y \frac{c_i}{2\Delta} \frac{c_j}{2\Delta} \right) t \, dx \, dy$$

where t is the thickness of an element (usually taken as unity), and Δ is the area of the triangle.

Hence

$$k_{ij} = \frac{t}{4\Delta^2} \left(b_i b_j \int_\Delta k_x \, dx \, dy + c_i c_j \int_\Delta k_y \, dx \, dy \right) \tag{7.22}$$

If k_x and k_y can be assigned constant average values within the element, then evaluation of the integrals is trivial and we have

$$k_{ij} = \frac{t}{4\Delta} (k_x b_i b_j + k_y c_i c_j) \tag{7.23}$$

Since h is identically zero, the matrix $[K_{S_2}]^{(e)}$ is not required, and equation 7.23 is the complete coefficient matrix.

The vector R_i given by equation 7.15 has contributions from all elements within the domain from $f(x, y)$ and contributions from the edges of elements on the boundary due to $g(x, y)$. For an interior element such as the one shown in Figure 7.2 we need to evaluate

$$R_i = \int_\Delta f L_i \, dx \, dy$$

We do this for two cases: (1) constant f within an element, and (2) a linear variation of f within an element.

Constant f. If f is constant within an element, we have from the integration formula of equation 5.22

$$R_i = f \int_\Delta L_i \, dx \, dy = \frac{f\Delta}{3} \tag{7.24}$$

and a similar result is found for nodes j and k.

Linear f. If we linearly interpolate f in terms of its nodal values as $f = f_1 L_1 + f_2 L_2 + f_3 L_3$, we have

$$R_i = \int_\Delta (f_1 L_1 L_i + f_2 L_2 L_i + f_3 L_3 L_i) \, dx \, dy$$

and again employing the formula of equation 5.22

$$\{R\}^{(e)} = \frac{\Delta}{12} \begin{bmatrix} 2 & 1 & 1 \\ 1 & 2 & 1 \\ 1 & 1 & 2 \end{bmatrix} \begin{Bmatrix} f_1 \\ f_2 \\ f_3 \end{Bmatrix} \tag{7.25}$$

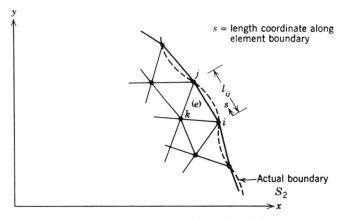

Figure 7.3   Boundary elements where $g(x, y)$ is specified.

Now suppose that element (e) forms part of the boundary curve S_2 as shown in Figure 7.3. In this case we must evaluate the boundary integral

$$R_i = \int_{C_2} gL_i \, dS_2$$

Consider the typical boundary element (e) with nodes i and j that lie on a boundary segment. We evaluate the integral for two cases: (1) constant g on the boundary segment and (2) a linear variation of g on the boundary. In Figure 7.3 we define the length coordinate s and the length of the boundary segment as l_{ij}. Along the boundary segment the interpolation functions may be taken as the natural coordinates in one dimension given by equation 5.11:

$$L_i(s) = 1 - \frac{s}{l_{ij}} \qquad L_j(s) = \frac{s}{l_{ij}}$$

Constant g. If g is constant along the boundary, then at node i

$$R_i = g \int_{S_2} L_i \, dS_2 = g \int_0^{l_{ij}} \left(1 - \frac{s}{l_{ij}}\right) ds = \frac{1}{2} g l_{ij} \qquad (7.26)$$

and similarly for node j.

Linear g. If we linearly interpolate g in terms of its nodal values as

$$g = \left(1 - \frac{s}{l_{ij}}\right) g_i + \frac{s}{l_{ij}} g_j$$

then at node i

$$R_i = \int_0^{l_{ij}} \left[\left(1 - \frac{s}{l_{ij}}\right)^2 g_i + \frac{s}{l_{ij}}\left(1 - \frac{s}{l_{ij}}\right) g_j \right] ds$$

and at node j

$$R_j = \int_0^{l_{ij}} \left[\left(1 - \frac{s}{l_{ij}}\right)\frac{s}{l_{ij}} g_i + \left(\frac{s}{l_{ij}}\right)^2 g_j \right] ds$$

Typical terms are evaluated as follows:

$$\int_0^{l_{ij}} \left(1 - \frac{s}{l_{ij}}\right)^2 ds = \frac{l_{ij}}{3}$$

$$\int_0^{l_{ij}} \left(1 - \frac{s}{l_{ij}}\right)\frac{s}{l_{ij}} ds = \frac{l_{ij}}{6}$$

hence the complete vector is

$$\{R_1\}^{(e)} = \frac{l_{ij}}{6}\begin{bmatrix} 2 & 1 \\ 1 & 2 \end{bmatrix}\begin{Bmatrix} g_i \\ g_j \end{Bmatrix} \tag{7.27}$$

This completes the evaluation of the element matrices for plane linear triangular elements. Note that the element is assumed to lie in the x–y plane and that the coefficient matrix k_{ij} given in equation 7.23 is computed in global coordinates. In an alternative approach not presented here an equivalent element coefficient matrix may be derived in element local coordinate axes. Since the nodal unknowns ϕ_i are scalars, transformation to global coordinates is not required. The contributions to the vector R_i given in equations 7.24–7.27 are likewise scalar quantities and require no transformation to global coordinates. Assembly and solution of the system equations are the remaining two steps.

Axisymmetric Quasiharmonic Equations   For a three-dimensional domain with geometric symmetry about the z axis a field problem is conveniently formulated in cylindrical coordinates r, θ, z. If, in addition, all field functions and coefficients are independent of θ, the dependent variable ϕ is a function of only r and z. Then the domain can be represented by axisymmetric ring elements and analyzed as a two-dimensional problem. Examples of axisymmetric ring elements are shown in Figure 5.4; note that the nodes of the finite element model lie in the r–z plane.

The governing quasiharmonic equation is

$$\frac{\partial}{\partial r}\left(k_r r \frac{\partial \phi}{\partial r}\right) + \frac{\partial}{\partial z}\left(k_z r \frac{\partial \phi}{\partial z}\right) = rf(r, z) \qquad (7.28)$$

with boundary conditions

$$\phi = \phi(r, z) \qquad \text{on } S_1 \qquad (7.29)$$

$$k_r r \frac{\partial \phi}{\partial r} n_r + k_z r \frac{\partial \phi}{\partial z} n_z + rg(r, z) = 0 \qquad \text{on } S_2 \qquad (7.30)$$

Now if typical coefficients such as k_x and k_y are taken as $k_r r$ and $k_z r$, respectively, the previous formulation for a two-dimensional problem in Cartesian coordinates can be used and the element matrices will have the same form as equations 7.14 and 7.15. The coordinates x and y are identified with r and z, respectively, and the integrations are carried out in the r–z plane. The coefficients of the element equations are given by

$$k_{ij} = \int_{A^{(e)}}\left(k_r \frac{\partial N_i}{\partial r}\frac{\partial N_j}{\partial r} + k_z \frac{\partial N_i}{\partial z}\frac{\partial N_j}{\partial z}\right) r \, dr \, dz \qquad (7.31)$$

$$R_i = \int_{A^{(e)}} fN_i r \, dr \, dz + \int_{S_2^{(e)}} gN_i r \, dS_2 \qquad (7.32)$$

To evaluate the integrals appearing in the element equations when triangular ring elements are used we express r as a linear function of the three vertex nodal values of r and write

$$r = \sum_{i=1}^{3} L_i r_i \qquad (7.33)$$

Using this substitution and the integration formula of equation 5.22 yields

$$k_{ij} = \frac{(r_1 + r_2 + r_3)}{12\Delta}(k_r b_i b_j + k_z c_i c_j) \qquad (7.34)$$

where the b and c coefficients and the area Δ of the triangle are given by equations 5.18 and 5.19 expressed in r–z coordinates.

The vector R_i has contributions from all elements within the domain from $f(r, z)$ and contributions from the edges of elements on the boundary due to $g(r, z)$. For an interior element we evaluate

$$R_i = \int_{\Delta} fL_i r \, dr \, dz$$

For convenience we assume that f is constant within an element. Substituting for r using equation 7.33 and employing the integration formula of equation 5.22 yields

$$\{R\}^{(e)} = \frac{f\Delta}{12}\begin{bmatrix} 2 & 1 & 1 \\ 1 & 2 & 1 \\ 1 & 1 & 2 \end{bmatrix}\begin{Bmatrix} r_1 \\ r_2 \\ r_3 \end{Bmatrix} \tag{7.35}$$

For an element (e) that forms part of the boundary as shown in Figure 7.3 we evaluate the contribution from $g(r, z)$ as

$$R_i = \int_{S_2} rgL_i \, dS_2$$

We consider a typical boundary segment with nodes i and j. Assuming that g is constant, we evaluate the integral by employing the length coordinate s and integrating over the surface of the boundary segment. Then at node i

$$R_i = g\int_0^{l_{ij}} rL_i \, ds$$

which we evaluate using the procedures employed previously to yield

$$\{R\}^{(e)} = \frac{gl_{ij}}{6}\begin{bmatrix} 2 & 1 \\ 1 & 2 \end{bmatrix}\begin{Bmatrix} r_i \\ r_j \end{Bmatrix} \tag{7.36}$$

This completes the evaluation of the element matrices for the triangular ring element. Note that the element matrices are derived in global coordinates. The evaluation of the element integrals was facilitated by expressing r as a linear function of the three vertex nodal values of r, equation 7.33. This useful substitution makes it possible to apply the integration formula of equation 5.22 and thus avoid the need for numerical integration. An explicit method for deriving matrix equations for triangular ring elements of arbitrarily high order is given by Silvester and Konard [3].

Readers wishing to solve various forms of the two-dimensional quasiharmonic equation can benefit from the experience reported by Emery and Carson [4]. They studied the accuracy and efficiency of the finite element method in comparison to the finite difference method for the computation of temperature. Although they were primarily interested in transient problems, they also investigated methods for computing steady-state temperature distributions. Working with linear, quadratic, and cubic triangular elements, they chose three test problems and concluded that the quadratic element is the most accurate and desirable. However, linear finite elements are popular with many analysts and are effective for a variety of practical problems.

7.3 EIGENVALUE PROBLEMS

Eigenvalue problems require the determination of critical values of certain parameters and the corresponding spatial distributions of a field variable. Classical examples of eigenvalue problems are mechanical and structural vibrations, buckling and stability of structures, and resonance problems in acoustics and electrical circuits. The eigenvalue problems considered here include standing waves in shallow water, electromagnetic waves, and acoustic vibrations.

7.3.1 Helmholtz Equations

Another class of field equations closely related to the quasiharmonic equation consists of Helmholtz equations, given generally by

$$\frac{\partial}{\partial x}\left(k_x\frac{\partial \phi}{\partial x}\right) + \frac{\partial}{\partial y}\left(k_y\frac{\partial \phi}{\partial y}\right) + \frac{\partial}{\partial z}\left(k_z\frac{\partial \phi}{\partial z}\right) + \lambda\phi = 0 \qquad (7.37)$$

with Dirichlet- and Neumann-type boundary conditions. For inhomogeneous and anisotropic media x, y, and z are the principal coordinates. The parameter λ is unknown. Equations of the form of equation 7.37 arise in propagation problems involving wave motion of one type or another. This will become evident when we consider transient field problems in the following sections. Before developing the element equations for the Helmholtz problem we consider several special forms of equation 7.37. The following are some practical examples of particular forms of Helmholtz equations.

Seiche Motion   Standing waves on a bounded shallow body of water are governed by the equation [5]

$$\frac{\partial}{\partial x}\left(h\frac{\partial w}{\partial x}\right) + \frac{\partial}{\partial y}\left(h\frac{\partial w}{\partial y}\right) + \frac{4\pi^2}{gT^2}w = 0 \qquad (7.38)$$

where h = water depth at the quiescent state,

w = elevation of the free surface above the quiescent level,

g = acceleration of gravity,

T = period of oscillation.

Equation 7.38 holds under the following assumptions: (1) the flow is frictionless (no damping), (2) the fluid inertia is small, and (3) fluid velocities are constant through the depth h. For equation 7.38 we have the Neumann boundary condition

$$\frac{\partial w}{\partial x}n_x + \frac{\partial w}{\partial y}n_y = 0 \qquad (7.39)$$

to be satisfied at solid boundaries.

Electromagnetic Waves   The propagation of electromagnetic waves in a waveguide filled with a dielectric material obeys the equation [6]

$$\frac{\partial}{\partial x}\left(\frac{1}{\epsilon_d}\frac{\partial\phi}{\partial x}\right) + \frac{\partial}{\partial y}\left(\frac{1}{\epsilon_d}\frac{\partial\phi}{\partial y}\right) + \frac{\partial}{\partial z}\left(\frac{1}{\epsilon_d}\frac{\partial\phi}{\partial z}\right) + \omega^2\mu_0\epsilon_0\phi = 0 \quad (7.40)$$

where ϕ = a component of the magnetic field strength vector **H** or a component of the electric field vector **E**,

ω = wave frequency,

μ_0 = permeability of free space,

ϵ_0 = permittivity of free space,

ϵ_d = permittivity of the dielectric.

Equation 7.40 can be derived from Faraday's law and Maxwell's equations, which are explained in the introductory text by Haytt [7]. If ϕ represents an **H** wave component, say $\phi = H_x$, then ϕ must satisfy the Neumann boundary condition at solid boundaries. But if ϕ represents an **E** wave component, then ϕ satisfies the Dirichlet boundary condition. Equation 7.40 is normally not solved for both the **E** and **H** waves, because one vector field can be obtained from the other via the equation

$$\nabla \times \mathbf{E} = -\sqrt{-1}\,\omega\mu_0\mathbf{H}$$

Acoustic Vibrations   A fluid vibrating in a closed volume represents a sonic field of spherical waves governed by the equation [8]

$$\frac{\partial^2 P}{\partial x^2} + \frac{\partial^2 P}{\partial y^2} + \frac{\partial^2 P}{\partial z^2} + \frac{\omega^2}{c^2}P = 0 \quad (7.41)$$

where P = pressure excess above ambient pressure,

ω = wave frequency,

c = wave velocity in the medium.

The derivation of equation 7.41 is based on a combination of three basic equations for a fluid: (1) the continuity equation, (2) the equation expressing the elastic properties of the fluid, and (3) an elemental force balance equation. The derivation also relies on the following assumptions: (1) the process is adiabatic, (2) the local density changes are small, and (3) the displacement and velocity of the fluid particles are small.

We note that in each of these special cases the coefficient corresponding to λ in equation 7.37 is nonnegative.

7.3.2 Variational Principle

Returning now to the general Helmholtz equation, we can apply the techniques of Appendix B to show that equation 7.37 is the Euler–Lagrange equation for the functional

$$I(\phi) = \frac{1}{2} \int_\Omega \left[k_x \left(\frac{\partial \phi}{\partial x} \right)^2 + k_y \left(\frac{\partial \phi}{\partial y} \right)^2 + k_z \left(\frac{\partial \phi}{\partial z} \right)^2 - \lambda \phi^2 \right] d\Omega \quad (7.42)$$

The function $\phi(x, y, z)$ that minimizes $I(\phi)$ in equation 7.42 and satisfies the given Dirichlet boundary condition also satisfies equation 7.37. No special consideration is needed for the Neumann boundary condition, because this is naturally taken into account in the functional, the additional term being identically zero.

7.3.3 Element Equations

The procedure discussed in Section 7.2.4 for deriving element equations from a variational principle applies again in this case. By expressing ϕ in terms of its nodal values within each element (equation 7.8) and then minimizing the corresponding discretized functional, we obtain element equations of the form

$$[K]^{(e)}\{\phi\}^{(e)} - \lambda[M]^{(e)}\{\phi\}^{(e)} = \{0\} \quad (7.43)$$

where the terms of the matrices $[K]^{(e)}$ and $[M]^{(e)}$ are given by

$$k_{ij} = \int_{\Omega^{(e)}} \left(k_x \frac{\partial N_i}{\partial x} \frac{\partial N_j}{\partial x} + k_y \frac{\partial N_i}{\partial y} \frac{\partial N_j}{\partial y} + k_z \frac{\partial N_i}{\partial z} \frac{\partial N_j}{\partial z} \right) d\Omega \quad (7.44a)$$

$$m_{ij} = \int_{\Omega^{(e)}} N_i N_j \, d\Omega \quad (7.44b)$$

The system equations have the form

$$[[K] - \lambda[M]]\{\phi\} = 0 \quad (7.45)$$

where the global $[K]$ and $[M]$ matrices are assembled from element contributions in the usual manner. Also, $\{\phi\}$ is the column vector of nodal values of ϕ. Before solving equations 7.45 we must modify them according to the procedures of Section 2.3.4 to account for the Dirichlet boundary conditions.

Equations 7.45 are a set of, say n,[1] linear homogeneous algebraic equations in the nodal values of ϕ; but they are different from any that we have

---

[1]Here n is the number of unconstrained degrees of freedom of the problem.

encountered thus far, because λ is, in general, unknown. The problem we have here is called a general *eigenvalue* or *characteristic value* problem, and the λ values are termed *eigenvalues* or *characteristic values*. For each different value of λ_i there is a different column vector $\{\phi\}_i$ that satisfies equation 7.45. The vector $\{\phi\}_i$ that corresponds to a particular value λ_i is called an *eigenvector, characteristic* vector, or *modal* vector.

Determining the eigenvalues constitutes part of the problem. From the fundamentals of linear algebra we know that there will be a nontrivial solution to equation 7.45; in other words, $\{\phi\} \neq \{0\}$ if and only if the determinant (called the *characteristic* determinant) is zero:

$$|[K] - \lambda[M]| = 0 \qquad (7.46)$$

In principle, this is the equation used to find the eigenvalues. If we were to expand equation 7.46, we would obtain an nth-order polynomial in λ such as

$$a_n\lambda^n + a_{n-1}\lambda^{n-1} + \cdots + a_1\lambda + a_0 = 0$$

The fundamental theorem of algebra assures us that this polynomial has n roots λ_i. When the λ_i are substituted into equation 7.45, we have n *sets* of equations to be solved for the n eigenvectors $\{\phi\}_i$. Essentially the complete solution of an eigenvalue problem requires about n times as much computational effort as is needed to find just one vector $\{\phi\}$.

The finite element discretization of the Helmholtz equation leads to matrices $[K]$ and $[M]$, which are both symmetric and positive definite. In this case all the eigenvalues λ_i are distinct, real, positive numbers, and the corresponding eigenvectors $\{\phi\}_i$ are all independent. For each λ_i it is impossible to determine uniquely the n components of $\{\phi\}_i$, because the set of equations is homogeneous. The usual procedure is to assign an arbitrary value to one component of the vector $\{\phi\}_i$ and to then solve the remaining $n - 1$ equations for the other components. The consequence of this fact is that the natural modal vectors $\{\phi\}_i$ of the wave motion are known only within an arbitrary multiplicative constant. The constant is selected by normalizing the modes in some convenient manner. For example, the constant can be chosen such that the largest modal component is unity.

Either direct or iterative numerical methods are available for solving eigenvalue problems such as equation 7.45. When we seek only the first few eigenvalues and eigenvectors rather than the complete set, iterative methods are best. The interested reader can find discussions of direct and iterative methods in standard textbooks [9].

7.3.4 Examples

Triangular Elements   The equations for the coefficients k_{ij} given in equation 7.44a are the same as the coefficients derived for the equilibrium problem given in equation 7.14. Hence for a triangular element the coefficients k_{ij} have been derived previously and are given in equation 7.23. The coefficients m_{ij}, equation 7.44b, are computed by identifying the interpolation function N_i with L_i, where

$$L_i = \frac{a_i + b_i x + c_i y}{2\Delta}$$

and Δ is the area of the triangle, as previously discussed in the plane equilibrium problem. Then

$$m_{ij} = t \int_\Delta L_i L_j \, dx \, dy$$

which is evaluated using the formula of equation 5.22. Hence

$$m_{ij} = \frac{t\Delta}{12} \begin{cases} 2, & i = j \\ 1, & i \neq j \end{cases} \tag{7.47}$$

Tetrahedral Elements   The simplest finite element discretization for the Helmholtz equation in three dimensions is obtained by using four-node linear tetrahedral elements. From Section 5.5.3 we know that the linear variation of ϕ within a tetrahedron may be expressed in terms of the four nodal values of ϕ as

$$\phi^{(e)}(x, y, z) = \sum_{i=1}^{4} L_i(x, y, z) \phi_i$$

where the interpolation functions L_i are the natural coordinates for the tetrahedron. The element matrices may be found from equations 7.44 after we note that $N_i = L_i$, and from equations 5.29

$$\frac{\partial N_i}{\partial x} = \frac{\partial L_i}{\partial x} = \frac{b_i}{6V}$$

$$\frac{\partial N_i}{\partial y} = \frac{\partial L_i}{\partial y} = \frac{c_i}{6V}$$

$$\frac{\partial N_i}{\partial z} = \frac{\partial L_i}{\partial z} = \frac{d_i}{6V}$$

where b_i, c_i, d_i, and V are defined by equations 5.27. Hence we have from equation 7.44a

$$k_{ij} = \frac{1}{36V^2} \int_V (k_x b_i b_j + k_y c_i c_j + k_z d_i d_j)\, d\Omega$$

and if k_x, k_y, and k_z are constant within the element, the integration over the volume becomes trivial and we can write

$$k_{ij} = \frac{1}{36V} (k_x b_i b_j + k_y c_i c_j + k_z d_i d_j) \qquad (7.48)$$

Furthermore

$$m_{ij} = \int_V L_i L_j\, d\Omega$$

which can be conveniently evaluated by the formula of equation 5.30 and found to be

$$m_{ij} = \frac{V}{20} \begin{cases} 2, & i = j \\ 1, & i \neq j \end{cases} \qquad (7.49)$$

7.3.5 Sample Problem

The analysis and control of passenger compartment noise in an automobile constitutes a problem [10] of considerable interest to automotive engineers. The acoustic field in the compartment is governed by the form of Helmholtz equation expressed in equation 7.41, but the solution domain has an exceedingly irregular shape. Shuku and Ishihara [10] recognized that the finite element method is ideally suited to solve the Helmholtz equation over an irregular domain.

Before directly approaching the problem of noise inside cars they conducted some accuracy and convergence studies for different calculation techniques. As a test problem the normal frequencies of a closed rectangular room of dimensions $l_x \times l_y$ were calculated and compared with the exact results given by

$$f_{ij} = \frac{c}{2} \left[\left(\frac{i}{l_x} \right)^2 + \left(\frac{j}{l_y} \right)^2 \right]^{1/2}$$

where f_{ij} is the natural frequency of the (i, j) mode, and c is the speed of sound in air.

Solution domain (closed room) Mesh configuration (size: 2.0 × 1.1 m)	No. of nodes	No. of elements	Normal frequencies (Hz)		
			(1, 0) mode	(0, 1) mode	(2, 0) mode
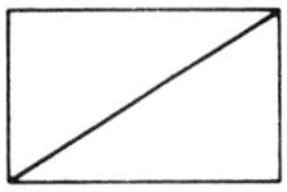	4	2	88.3	165.6	189.9
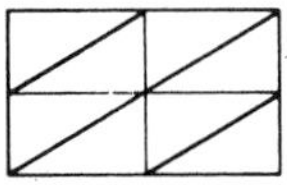	9	8	85.2	154.5	176.2
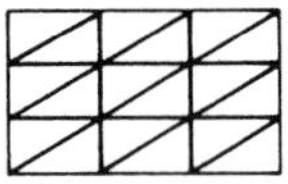	16	18	85.0	154.5	171.8
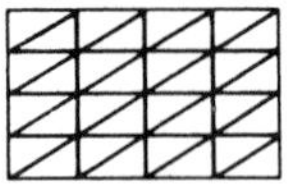	25	32	85.0	154.5	170.7
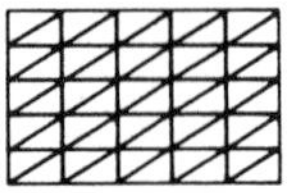	36	50	85.0	154.5	170.3
Exact			85.0	154.5	170.0

Figure 7.4   Effect of mesh size on the computation of normal frequencies of a closed room when modified cubic interpolation is used [10].

Three-node triangular elements with modified cubic interpolation functions were used for the calculations. The pressure within each element was expressed as

$$P(x, y) = a_1 + a_2 x + a_3 y + a_4 x^2 + a_5 xy + a_6 y^2$$
$$+ a_7 x^3 + a_8(x^2 y + xy^2) + a_9 y^3$$

with P, $\partial P/\partial x$, and $\partial P/\partial y$ taken as nodal degrees of freedom at the vertices of the triangle. We recall from Chapter 5 that this element model provides C^0 continuity plus continuity of pressure derivatives at the nodes.

Figure 7.4 shows a comparison of exact results with computed results for various mesh sizes, and Figure 7.5 compares a finite difference solution [11] with linear and modified cubic finite element solutions. It can be seen that the cubic element gives decidedly more accurate results.

The cubic finite element model was used to compute the frequencies and modes of an acoustic field in the interior of a car modeled as shown in

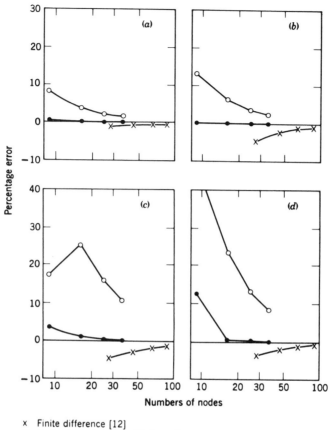

x   Finite difference [12]
o   Finite element model with linear interpolation over triangles
•   Finite element model with modified cubic interpolation over triangles

Figure 7.5 Accuracy comparison for different calculation methods [10]: (*a*) (1, 0) mode; (*b*) (0, 1) mode; (*c*) (2, 0) mode; (*d*) (1, 1) mode.

Figure 7.6. In reality a car interior has soft boundaries, but in the model the boundaries were assumed to be rigid. Solid and dotted lines indicate the loci of points where the sound-pressure level is a minimum for the first three modes. Good agreement between calculated and experimental results is shown.

Applications of three-dimensional acoustic finite elements to automobile passenger compartments and engine combustion chambers appear in reference 12 along with comparisons of finite element solutions to experimental data. Reference 12 also contains a list of references of other finite element acoustic applications.

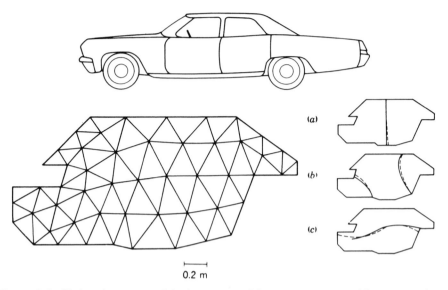

Figure 7.6 Finite element model of an automobile compartment, with a comparison of calculated and experimental results [10]: (*a*) first mode: 86.8 Hz (exp. 87.5 Hz); (*b*) second mode: 138.0 Hz (exp. 138.5 Hz); (*c*) third mode: 154.6 Hz (exp. 157 Hz).

7.4 PROPAGATION PROBLEMS

Propagation problems require the determination of the time-dependent spatial distribution of a field variable. Propagation problems are initial value problems and include transient and unsteady phenomena. Classical examples include diffusion and wave propagation. The propagation problems considered here include a broad class of diffusion and wave propagation problems encountered by engineers.

7.4.1 General Time-Dependent Field Problems

A large class of transient field problems is governed by equations in domain Ω of the form

$$\frac{\partial}{\partial x}\left(k_x\frac{\partial \phi}{\partial x}\right) + \frac{\partial}{\partial y}\left(k_y\frac{\partial \phi}{\partial y}\right) + \frac{\partial}{\partial z}\left(k_z\frac{\partial \phi}{\partial z}\right) = f(x,y,z,t) + c\dot{\phi} + m\ddot{\phi}$$

$$(7.50)$$

with boundary conditions

$$\phi = \Phi(x, y, z, t) \qquad \text{on } S_1, \quad t > 0 \tag{7.51}$$

$$k_x \frac{\partial \phi}{\partial x} n_x + k_y \frac{\partial \phi}{\partial y} n_y + k_z \frac{\partial \phi}{\partial z} n_z + q(x, y, z, t) + h(x, y, z, t)\phi = 0$$

$$\text{on } S_2, \quad t > 0 \tag{7.52}$$

and initial conditions

$$\phi = \phi_0(x, y, z) \qquad \text{in } \Omega, \quad t = 0 \tag{7.53}$$

$$\dot{\phi} = \dot{\phi}_0(x, y, z) \qquad \text{in } \Omega, \quad t = 0 \tag{7.54}$$

The parameters c and m determine the character of equation 7.50. If $m = 0$ and $c \neq 0$, equation 7.50 is parabolic; whereas if $m \neq 0$, the equation is hyperbolic. The parabolic and hyperbolic classification of partial differential equations is briefly discussed in Section 3.2.3. For both classes of equations the solution marches out into the time domain guided in transit by the spatial boundary conditions. Parabolic equations are characterized by the fact that disturbances propagate into the solution domain with infinite speed. For parabolic equations the nature of the disturbance may be drastically altered as it propagates into the solution domain. For example, according to the parabolic heat conduction equation, heat applied suddenly to the boundary of a region will cause temperatures at interior points to rise immediately. In contrast, hyperbolic equations are characterized by the fact that disturbances propagate into the solution domain with finite speed. For hyperbolic equations the nature of the disturbance may remain unaltered as it propagates into the solution domain. For example, according to the hyperbolic wave equation, a small deformation introduced initially on an elastic string propagates unchanged with a finite wave speed. Further solution properties and classical solution procedures for parabolic and hyperbolic equations are discussed by Crandall [13]. Practical examples of the diffusion equation and wave equation will be presented to illustrate particular examples of propagation problems described by equation 7.50.

Diffusion Equation   Time-dependent diffusion processes occur in a wide variety of phenomena in physics, chemistry, and engineering. Some examples cited by Sneddon [14] include the slowing down of neutrons in matter, the diffusion of substances in physical chemistry, the diffusion of vorticity in viscous fluid flow, and heat conduction in solids. For a homogeneous isotropic medium diffusion processes are governed by the equation

$$D\left(\frac{\partial^2 \phi}{\partial x^2} + \frac{\partial^2 \phi}{\partial y^2} + \frac{\partial^2 \phi}{\partial z^2}\right) = \frac{\partial \phi}{\partial t} \qquad (7.55)$$

where D is a diffusion coefficient. The diffusion equation, equation 7.55, is parabolic.

One example of diffusion is the movement of a chemical species from a region of high concentration to a region of low concentration [15]. The process is described by Fick's first law of diffusion

$$\mathbf{J} = -D_{AB}\left(\frac{\partial C_A}{\partial x}\hat{i} + \frac{\partial C_A}{\partial y}\hat{j} + \frac{\partial C_A}{\partial z}\hat{k}\right) = -D_{AB}\nabla C_A$$

and conservation of mass for the substance

$$\frac{\partial C_A}{\partial t} + \nabla \cdot \mathbf{J} = 0$$

which, when combined, yield

$$D_{AB}\left(\frac{\partial^2 C_A}{\partial x^2} + \frac{\partial^2 C_A}{\partial y^2} + \frac{\partial^2 C_A}{\partial z^2}\right) = \frac{\partial C}{\partial t} \qquad (7.56)$$

where C_A = molar concentration of A in a binary system of substances A
            and B,
        $\mathbf{J}$ = molar diffusion flux vector,
    D_{AB} = mass diffusivity in binary system AB.

Fick's law of diffusion states that species A diffuses in the direction of decreasing concentration of A just as heat flows by conduction in the direction of decreasing temperature. The diffusivity D_{AB} may depend on the concentration and may be dependent on the spatial coordinates.

Another example of diffusion is flow in porous media [16]. The flow is described by Darcy's law, expressed as

$$\mathbf{V} = -\frac{k}{\mu\rho}\nabla\phi$$

and conservation of mass. The governing equations may take several forms [16], but for saturated flow in a rigid isotropic medium a governing diffusion equation often employed [17] has the form

$$k\left(\frac{\partial^2\phi}{\partial x^2} + \frac{\partial^2\phi}{\partial y^2} + \frac{\partial^2\phi}{\partial z^2}\right) = n\frac{\partial\phi}{\partial t} \tag{7.57}$$

where $\phi = p/\rho g + z$, the piezometric head,
   p = fluid pressure,
   μ = fluid viscosity,
   ρ = fluid density,
   g = acceleration of gravity,
   z = elevation coordinate,
   k = permeability of the porous media,
   n = effective porosity of the porous media,
   $\mathbf{V}$ = flow velocity vector.

A third example of a diffusion process is heat conduction, which is based on Fourier's law and conservation of energy. The governing diffusion equation, the heat conduction equation, appears in Appendix E.

Wave Equation   Wave propagation, whether for electromagnetic waves, acoustic waves, surface waves, or any of the many other types, is a time-dependent phenomenon that for a homogeneous isotropic medium is governed by the equation

$$\frac{\partial^2\phi}{\partial x^2} + \frac{\partial^2\phi}{\partial y^2} + \frac{\partial^2\phi}{\partial z^2} = c\dot{\phi} + m\ddot{\phi} \tag{7.58}$$

where

$$\dot{\phi} = \frac{\partial\phi}{\partial t} \qquad \ddot{\phi} = \frac{\partial^2\phi}{\partial t^2}$$

and the coefficients c and m can, in general, be functions of time. Both Dirichlet and Neumann boundary conditions may apply to equation 7.58, depending on the type of boundary considered; and for the problem statement to be complete, initial conditions on ϕ and $\dot{\phi}$ must be given. The first term on the right-hand side represents the energy dissipation or damping associated with the propagation process, and the second term represents the inertia of the time-dependent motion.

As a particular example of the time-dependent wave equation we consider the Schrödinger equation [18], the heart of quantum mechanics. The wave nature of a single particle is described by the equation

$$\frac{\partial^2 U}{\partial x^2} + \frac{\partial^2 U}{\partial y^2} + \frac{\partial^2 U}{\partial z^2} = \frac{2m(h\nu - V)}{h^2\nu^2}\frac{\partial^2 U}{\partial t^2} \qquad (7.59)$$

where U = wave function whose amplitude squared is proportional to the
          probability of finding the particle at any given point in space,
     m = particle mass,
     V = potential energy of the particle,
     h = Planck's constant,
     ν = wave frequency.

The central problem in wave mechanics is to assign an appropriate character to the potential energy function V and then solve equation 7.59 for U. Since, in general, only wave functions U that are harmonic functions of time are physically important [18], we seek solutions in terms of combinations of $\sin 2\pi\nu t$ and $\cos 2\pi\nu t$. Noting that

$$e^{i2\pi\nu t} = \cos 2\pi\nu t + i \sin 2\pi\nu t$$

we assume solutions of the form

$$U(x, y, z, t) = u(x, y, z)e^{i2\pi\nu t}$$

where $i = \sqrt{-1}$. Substituting into equation 7.59 and noting that

$$\frac{\partial^2 U}{\partial t^2} = -4\pi^2\nu^2 u e^{i2\pi\nu t} = -4\pi^2\nu^2 U$$

gives

$$\frac{\partial^2 U}{\partial x^2} + \frac{\partial^2 U}{\partial y^2} + \frac{\partial^2 U}{\partial z^2} = -\frac{8m\pi^2(h\nu - V)}{h^2}U \qquad (7.60a)$$

or for the wave amplitude only we have

$$\frac{\partial^2 u}{\partial x^2} + \frac{\partial^2 u}{\partial y^2} + \frac{\partial^2 u}{\partial z^2} = -\frac{8m\pi^2(h\nu - V)}{h^2}u \qquad (7.60b)$$

Again, this is the familiar Helmholtz equation, and we see how it arises in wave propagation problems.

7.4.2 Finite Element Equations

Several approaches can be used to construct finite element equations for time-dependent problems. The most common approach is to consider the problem at one instant of time and to assume that the time derivatives such as $\dot{\phi}$ and $\ddot{\phi}$ at that instant are functions of the spatial coordinates only. Then we construct a finite element model by expressing the field variable for a typical element with r degrees of freedom as

$$\phi^{(e)}(x, y, z, t) = \sum_{i=1}^{r} N_i(x, y, z)\phi_i(t) \qquad (7.61)$$

where the N_i are the usual interpolation functions and $\phi_i(t)$ are the time-dependent nodal parameters. At this point we can derive the element equations by using an appropriate variational principle if one exists for our contrived steady-state problem, or we can resort to the method of weighted residuals with Galerkin's criterion. In either case what results after assembly of the element equations is a set of ordinary differential equations in $\{\phi\}$, which we can solve by a variety of techniques to be discussed in Section 7.5.

Another approach (suggested by Oden [19]) is to view the general time-dependent problem in a four-dimensional space–time domain and represent the field variable within a typical element as

$$\phi^{(e)}(x, y, z, t) = \sum_{i=1}^{r} N_i(x, y, z, t)\phi_i$$

where the interpolation functions N_i incorporate both space and time. The idea is simply a natural extension of the interpolation concepts normally used in one-, two-, or three-dimensional problems. Of course, with each added dimension the cost of computation greatly increases, hence a full four-dimensional problem could easily exceed the realm of the practical.

No classical variational principle exists for the problem described by equations 7.50–7.54, so we employ the Galerkin approach to derive finite element equations. Within a typical element we assume

$$\phi^{(e)} = \sum_{i=1}^{r} N_i(x, y, z)\phi_i(t) = \lfloor N \rfloor\{\phi\}^{(e)}$$

where the interpolation functions N_i need only preserve C^0 continuity. The Galerkin criterion then requires that at any instant of time

$$\int_{\Omega^{(e)}} N_i \left[\frac{\partial}{\partial x}\left(k_x \frac{\partial \phi^{(e)}}{\partial x} \right) + \frac{\partial}{\partial y}\left(k_y \frac{\partial \phi^{(e)}}{\partial y} \right) + \frac{\partial}{\partial z}\left(k_z \frac{\partial \phi^{(e)}}{\partial z} \right) \right.$$
$$\left. -f - c\dot{\phi}^{(e)} - m\ddot{\phi}^{(e)} \right] d\Omega = 0, \qquad i = 1, 2, \ldots, r \quad (7.62)$$

where $\Omega^{(e)}$ is the domain for element (e). Now, following the standard procedures fully discussed in Chapter 4, we integrate the terms such as

$$\int_{\Omega^{(e)}} N_i \frac{\partial}{\partial x} \left(k_x \frac{\partial \phi^{(e)}}{\partial x} \right) d\Omega$$

by parts to introduce the boundary conditions on S_2, and, finally, after the manipulations, the resulting element equations become

$$[M]^{(e)}\{\ddot{\phi}\}^{(e)} + [C]^{(e)}\{\dot{\phi}\}^{(e)}$$

$$+ [K]^{(e)}\{\phi\}^{(e)} + [K_{S_2}]^{(e)}\{\phi\}^{(e)} + \{R(t)\}^{(e)} = \{0\} \quad (7.63)$$

where

$$m_{ij} = \int_{\Omega^{(e)}} m N_i N_j \, d\Omega \qquad (7.64a)$$

$$c_{ij} = \int_{\Omega^{(e)}} c N_i N_j \, d\Omega \qquad (7.64b)$$

$$k_{ij} = \int_{\Omega^{(e)}} \left(k_x \frac{\partial N_i}{\partial x} \frac{\partial N_j}{\partial x} + k_y \frac{\partial N_i}{\partial y} \frac{\partial N_j}{\partial y} + k_z \frac{\partial N_i}{\partial z} \frac{\partial N_j}{\partial z} \right) d\Omega \qquad (7.64c)$$

$$k_{S_{2ij}} = \int_{S_2^{(e)}} h N_i N_j \, dS \qquad (7.64d)$$

$$R_i = \int_{\Omega^{(e)}} f N_i \, d\Omega + \int_{S_2^{(e)}} q_i \, dS \qquad (7.64e)$$

Assembly of the element equations then leads to a system of ordinary differential equations of the same form as equation 7.63. The problem solution is complete when these equations are solved for the nodal parameters $\{\phi\}$ subject to the discretized initial conditions. Solution techniques are discussed in Section 7.5.

7.4.3   Element Equations in One Space Dimension

To illustrate the use of the general element equations developed for time-dependent field problems we consider here a problem governed by the following differential equation:

$$\frac{\partial}{\partial x} \left(k_x \frac{\partial \phi}{\partial x} \right) = f(x, t) + c_x \dot{\phi} + m_x \ddot{\phi}$$

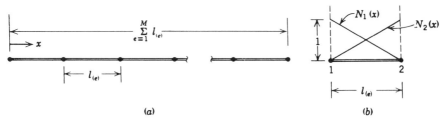

Figure 7.7 Finite element model for a one-dimensional field problem: (*a*) discretized solution domain containing *M* elements; (*b*) linear interpolation functions over a typical element.

with general boundary conditions of the form of equations 7.51 and 7.52, except that in equation 7.52 we assume $h \equiv 0$. General initial conditions of the form of equations 7.53 and 7.54 are possible.

The first step toward finding $\phi(x, t)$ by the finite element method is to divide the spatial solution domain into elements. For simplicity we select *M* elements with linear interpolation functions (Figure 7.7). For a typical element the interpolation functions are, from equations 5.11,

$$N_1(x) = L_1(x) = \frac{x - x_2}{x_1 - x_2}$$

$$N_2(x) = L_2(x) = \frac{x - x_1}{x_2 - x_1}$$

so that the interpolation model is

$$\phi^{(e)}(x, t) = L_1 \phi_1(t) + L_2 \phi_2(t)$$

Equation 7.64 allows us to write the terms of the element equations at once, as follows:

$$m_{ij} = \int_{x_1}^{x_2} m_x L_i L_j \, dx$$

$$c_{ij} = \int_{x_1}^{x_2} c_x L_i L_j \, dx$$

$$k_{ij} = \int_{x_1}^{x_2} k_x \frac{\partial L_i}{\partial x} \frac{\partial L_j}{\partial x} \, dx = \frac{(-1)^{i+j}}{(x_2 - x_1)^2} \int_{x_1}^{x_2} k_x \, dx \qquad (7.65)$$

$$k_{S_{2ij}} = 0 \quad \text{(since } h \equiv 0)$$

$$R_i = \int_{x_1}^{x_2} f(x, t) L_i \, dx$$

If we assume that m_x, c_x, and k_x are constant within an element and $f(x, t)$ varies linearly, that is,

$$f(x, t) = f(x_1, t)L_1 + f(x_2, t)L_2 = f_1L_1 + f_2L_2$$

then the above equations reduce to

$$[M]^{(e)} = \frac{m_x^{(e)}(x_2 - x_1)}{6} \begin{bmatrix} 2 & 1 \\ 1 & 2 \end{bmatrix}$$

$$[C]^{(e)} = \frac{c_x^{(e)}(x_2 - x_1)}{6} \begin{bmatrix} 2 & 1 \\ 1 & 2 \end{bmatrix}$$

$$[K]^{(e)} = \frac{k_x^{(e)}}{x_2 - x_1} \begin{bmatrix} 1 & -1 \\ -1 & 1 \end{bmatrix} \qquad (7.66)$$

$$\left[K_{S_2}\right]^{(e)} = [0]$$

$$\{R\}^{(e)} = \frac{x_2 - x_1}{6} \begin{bmatrix} 2 & 1 \\ 1 & 2 \end{bmatrix} \begin{Bmatrix} f_1 \\ f_2 \end{Bmatrix}$$

where the superscript (e) designates the value for element (e). From equation 7.63 and the above equations we can write the equations for element (e) as

$$[M]^{(e)} \begin{Bmatrix} \ddot{\phi}_1 \\ \ddot{\phi}_2 \end{Bmatrix}^{(e)} + [C]^{(e)} \begin{Bmatrix} \dot{\phi}_1 \\ \dot{\phi}_2 \end{Bmatrix}^{(e)} + [K]^{(e)} \begin{Bmatrix} \phi_1 \\ \phi_2 \end{Bmatrix}^{(e)} + \{R\}^{(e)} = \{0\} \quad (7.67)$$

The problem solution is complete when the number of elements is selected, the element equations are evaluated and assembled, and the resulting system equations are solved.

7.5   SOLVING THE DISCRETIZED TIME-DEPENDENT EQUATIONS

We found in the preceding section that solving propagation problems by the finite element method reduces ultimately to the solution of simultaneous ordinary differential equations of the form

$$[M]\{\ddot{\phi}\} + [C]\{\dot{\phi}\} + [K]\{\phi\} = \{R(t)\} \qquad (7.68)$$

The physical interpretation of the various matrices in this equation depends on the particular type of problem, that is, the nature of the field variable ϕ.

Generally, $[M]$ represents system inertia, $[C]$ is a capacitance or damping matrix, and $[K]$ is a type of stiffness matrix. In some problems either $[C] = [0]$ or $[M] = [0]$. For example, in diffusion problems $[M] = [0]$. Of course, if both are zero, the dynamic character of the problem is absent even though the right-hand side may remain a function of time. Also, depending on the type of problem, the discretized forcing function $\{R(t)\}$ may be zero, harmonic, periodic, aperiodic, or random. Problems with deterministic forcing functions are discussed here; readers interested in random or nondeterministic forcing functions can find an introductory treatment of this subject in Meirovitch [20].

Propagation problems arising from hyperbolic partial differential equations lead to second-order matrix differential equations of the form of equation 7.68, with $[M] \neq [0]$. The wave equation is an example of a typical hyperbolic equation. In addition to hyperbolic field equations, structural dynamics problems formulated with finite elements also lead to discretized second-order equations of form identical to equation 7.68. Solution methods for second-order simultaneous ordinary differential equations are presented in Chapter 6. Readers interested in solution methods for these equations should refer to Section 6.7. The solution methods presented here will focus on the first-order form of equation 7.68 that occurs when $[M] = [0]$.

7.5.1 Solution Methods for First-Order Equations

Solution methods for the first-order simultaneous ordinary differential equations are the subjects of this section:

$$[C]\{\dot{\phi}\} + [K]\{\phi\} = \{R(t)\} \tag{7.69}$$

After formation of equation 7.69 the matrices must be modified to account for any boundary conditions not already included as natural boundary conditions. In addition to the boundary conditions, we must also know the initial conditions $\{\phi(0)\}$.

There are many general methods and several special techniques for solving first-order matrix differential equations—so many, in fact, that it is impractical to survey all methods here. Instead, we consider only two fundamental approaches that have proven popular in finite element analysis. More general surveys and further details of specialized techniques appear in books on numerical computation such as references 21–24. We consider first the mode superposition technique and then the solution methods that rely on recurrence relations which permit time-stepping (time-marching) procedures. Clearly, these solution techniques are not restricted to the finite element method. They apply to equations 7.68 regardless of their origin.

7.5.2 Finding Transient Response via Mode Superposition

Mode superposition is more widely used for second-order matrix equations than for first-order equations. For example, mode superposition is widely used for the second-order equations encountered in structural dynamics. One reason is that in structural dynamics the eigenvalues and eigenvectors of the modes have a clear physical identity as natural frequencies and vibration shapes. For the first-order equations described here the modes have no clear physical interpretation and exist only as mathematical quantities, not as physical entities. However, mode superposition for first-order systems is an effective solution procedure; in addition, the approach also gives valuable insight into the behavior of the recurrence schemes to be presented in the next section. The mode superposition procedure for first-order equations is almost identical to the procedures used for second-order equations presented in Section 6.7. Only the significant steps and basic equations of the technique are presented for first-order systems; for further details readers should refer to Section 6.7, where the technique is discussed in greater detail.

Eigenvalue Problem We consider the case in which the system has no external forcing function and assume that the field variable for a free response is

$$\{\phi\} = \{\Phi\}e^{-\lambda t} \tag{7.70}$$

where $\{\Phi\}$ is a modal vector of unknown amplitude, and λ is a modal decay constant analogous to the natural frequency for a second-order system. The system of equations will have as many modal vectors and decay constants as it has unconstrained degrees of freedom. We find on substitution into equation 7.69, with $R \equiv 0$, that the matrix equation reduces to

$$[[K] - \lambda[C]]\{\Phi\} = 0 \tag{7.71}$$

Equation 7.71 is an eigenvalue problem similar to the one that arose from the solution to the Helmholtz equation (Section 7.3). As before, for matrices of dimension $n \times n$ there will be n values of λ_i, which are the eigenvalues, and n values of $\{\Phi\}_i$, which are the eigenvectors. A pair made up of λ_i and a corresponding $\{\Phi\}_i$ is a mode.

The weighted orthogonality of the modes is an important property we employ in using the system modes to solve equation 7.64. It is easy to show [20] that for two different eigenvectors $\{\Phi\}_i$ and $\{\Phi\}_j$ corresponding to two different eigenvalues λ_i and λ_j we have

$$\lfloor \Phi \rfloor_i [K]\{\Phi\}_j = 0, \quad i \neq j$$

$$\lfloor \Phi \rfloor_i [C]\{\Phi\}_j = 0, \quad i \neq j \tag{7.72}$$

where $[K]$ and $[C]$ have the role of weighting matrices. For $i = j$ we have

$$\lfloor \Phi \rfloor_i [K]\{\Phi\}_i = K_{ii}^*$$
$$\lfloor \Phi \rfloor_i [C]\{\Phi\}_i = C_{ii}^* \tag{7.73}$$

where K_{ii}^* and C_{ii}^* are constants different from zero. In structural dynamics K_{ii}^* and C_{ii}^* are the structural generalized stiffness and generalized mass, respectively. Often the mode shapes are normalized so that the generalized mass is unity, that is, $C_{ii}^* = 1$. For first-order systems these constants have no particular labels.

Dynamic Response   Once the eigenvalue problem is solved for the eigenvectors and eigenvalues, the method of mode superposition gives the complete solution of the set of equations 7.69, with $\{R(t)\} \neq 0$. The method relies on the basic premise that the solution vector $\{\phi\}$ may be expressed as a linear combination of all n eigenvectors of the system. Hence we set

$$\{\phi(t)\} = [\{\Phi\}_1, \{\Phi\}_2, \{\Phi\}_3, \ldots, \{\Phi\}_n]\{a(t)\}$$
$$= [A]\{a(t)\} \tag{7.74}$$

where $[A]$ is a square matrix whose columns are the eigenvectors, and $\{a(t)\}$ is a vector of generalized modal unknowns. The modal expansion, equation 7.74, is substituted into equation 7.69 and the resulting equation is premultiplied by the transpose of the eigenvector matrix. Then the orthogonality relations, equations 7.72, are used to uncouple the resulting equations. The details of these operations appear in Section 6.7. A typical uncoupled equation is

$$\dot{a}_i + \lambda_i a_i = \frac{R_i^*}{C_{ii}^*}, \qquad i = 1, 2, \ldots, n \tag{7.75}$$

where

$$\lambda_i = \frac{K_{ii}^*}{C_{ii}^*}$$

and

$$R_i^* = \lfloor \Phi \rfloor_i \{R\}$$

The term R_i^* on the right-hand side of equation 7.75 is a generalized forcing function.

The initial conditions for a_i are determined from the initial conditions for ϕ_i. From equation 7.74

$$\{\phi(0)\} = [A]\{a(0)\}$$

Although $\{a(0)\}$ may be computed using the inverse of $[A]$, a more efficient computational approach may be formulated by using orthogonality of the mode shapes. Premultiplying by $[A]^T[C]$ yields

$$[A]^T[C][A]\{a(0)\} = [A]^T[C]\{\phi(0)\}$$

and we recognize from the orthogonality equations 7.72–7.73 that

$$[A]^T[C][A] = [\ulcorner C^* \lrcorner]$$

where $[\ulcorner C^* \lrcorner]$ is a diagonal matrix of generalized masses. This last result permits us to write an uncoupled set of equations:

$$\{C^* a(0)\} = [A]^T[C]\{\phi(0)\} \tag{7.76}$$

Thus the initial conditions for a_i are determined directly by equation 7.76 using matrix multiplication.

The solution for the dynamic response now involves solving independently n linear ordinary uncoupled differential equations like equation 7.75 for $a_i(t)$ and then combining the results according to equation 7.74. The final solution is a linear superposition of the response in each mode. Equation 7.75 is solved either explicitly or numerically, depending on the complexity of the forcing function.

The method of mode superposition involves considerable computational effort because the eigenvalue problem must first be solved for the eigenvectors and eigenvalues. However, once these are found, they may be saved and later used to find the response to other forcing functions of interest. This derivation of mode superposition holds only for linear problems, that is, problems in which $[C]$ and $[K]$ do not depend on $\{\phi\}$. The extension of modal superposition for nonlinear problems is discussed in Morris [25].

7.5.3   Finding Transient Response via Recurrence Relations

Instead of solving equations 7.69 by using the eigenvectors of the system to derive uncoupled equations for the generalized modal unknowns, an alternative approach is to apply direct numerical integration in time to solve the original set of coupled equations. The procedure relies on deriving recursion formulas that relate the values of $\{\phi\}$ at one instant of time t to the values of $\{\phi\}$ at a later time $t + \Delta t$, where Δt is the time step. The recursion formulas make it possible for the solution to be "marched out" in time, starting from the initial conditions at time $t = 0$ and continuing step by step until reaching the desired duration.

Consider a popular family of time-marching algorithms [26–30] based on finite difference methods. Let t_n denote a typical time in the response so that $t_{n+1} = t_n + \Delta t$, where $n = 0, 1, 2, \ldots, N$. The algorithms represent the first-order matrix equations by finite difference approximations at an intermediate time t_θ within each time step. A general family of algorithms results by introducing a parameter θ such that $t_\theta = t_n + \theta \Delta t$, where $0 \le \theta \le 1$. We write equation 7.69 at time t_θ as

$$[C]\{\dot{\phi}\}_\theta + [K]\{\phi\}_\theta = \{R(t_\theta)\} \tag{7.77a}$$

and introduce the approximations

$$\{\dot{\phi}\}_\theta = \frac{\{\phi\}_{n+1} - \{\phi\}_n}{\Delta t} \tag{7.77b}$$

$$\{\phi\}_\theta = (1 - \theta)\{\phi\}_n + \theta\{\phi\}_{n+1} \tag{7.77c}$$

$$\{R(t_\theta)\} = (1 - \theta)\{R\}_n + \theta\{R\}_{n+1} \tag{7.77d}$$

Substituting equations 7.77b–7.77d into equation 7.77a gives

$$\left[\theta[K] + \frac{1}{\Delta t}[C]\right]\{\phi\}_{n+1} = \left[-(1 - \theta)[K] + \frac{1}{\Delta t}[C]\right]\{\phi\}_n$$
$$+ (1 - \theta)\{R\}_n + \theta\{R\}_{n+1} \tag{7.78}$$

where the $\{\phi\}_{n+1}$ on the left-hand side of the equation are unknowns, and all of the terms on the right-hand side are known. Equation 7.78 represents a general family of recurrence relations; a particular algorithm depends on the value of θ selected. If $\theta = 0$, the algorithm is the forward difference method (Euler method); if $\theta = \frac{1}{2}$, the algorithm is the Crank–Nicolson method; if $\theta = \frac{2}{3}$, the algorithm is the Galerkin method; and if $\theta = 1$, the algorithm is the backward difference method.

For a given θ, equation 7.78 is a recurrence relation for calculating the vector of nodal values $\{\phi\}_{n+1}$ at the end of the time step from known values of $\{\phi\}_n$ at the beginning of the time step. In general, equation 7.78 is a set of linear algebraic equations of the form

$$[\bar{K}]\{\phi\}_{n+1} = \{\bar{R}\}_{n+1} \tag{7.79}$$

where

$$[\bar{K}] = \theta[K] + \frac{1}{\Delta t}[C] \tag{7.80a}$$

$$\{\bar{R}\}_{n+1} = \left[-(1 - \theta)[K] + \frac{1}{\Delta t}[C]\right]\{\phi\}_n + (1 - \theta)\{R\}_n + \theta\{R\}_{n+1} \tag{7.80b}$$

TABLE 7.2. Solution Sequence for Implicit Algorithms

INITIAL CALCULATIONS

1. Form element matrices $[C]^{(e)}$ and $[K]^{(e)}$.
2. Select θ.
3. Initialize $\{\phi\}_n$ at $t = 0$, $\{\phi\}_0$.
4. Select the time step Δt.
5. Assemble the effective coefficient matrix $[\bar{K}]$.
6. Modify $[\bar{K}]$ for boundary conditions.
7. Factor $[\bar{K}]$.

AT EACH TIME STEP

1. Form element vectors $\{R\}_n^{(e)}$ and $\{R\}_{n+1}^{(e)}$.
2. Assemble the effective vector $\{\bar{R}\}_{n+1}$.
3. Modify $\{\bar{R}\}_{n+1}$ for boundary conditions.
4. Solve for $\{\phi\}_{n+1}$ at the end of the time step.

Consequently, the family of algorithms requires the solution of a set of simultaneous equations at each time step. A time-marching algorithm that requires the solution of simultaneous algebraic equations at each time step is an *implicit* algorithm. The solution of the algebraic equations is simplified by keeping the time step constant within the duration of the response. Then the coefficient matrix $[\bar{K}]$ needs to be formed and factored (see Chapter 10 for a brief discussion of solution techniques for algebraic equations) only once at the beginning of the solution sequence. The solution sequence for constant Δt is summarized in Table 7.2.

Since the implicit family of algorithms requires the repetitive solution of simultaneous algebraic equations, the computational effort for transient solutions is considerably larger than for the corresponding equilibrium problem. One method of reducing the computational effort for the transient solution is to employ an explicit algorithm. In an *explicit* algorithm the nodal unknowns $\{\phi\}$ at each time step are computed from uncoupled algebraic equations. Since an explicit algorithm avoids the solution of simultaneous algebraic equations, the computational effort for the transient response is reduced significantly. An explicit form of the recurrence formula 7.78 can be derived for $\theta = 0$ (forward difference method), provided that we approximate the coefficient matrix $[C]$. For $\theta = 0$ we note from equation 7.80a that the effective coefficient matrix depends only on $[C]$. This matrix, sometimes called the mass or capacitance matrix, is approximated by using a diagonal or "lumped" form. Such lumped mass matrices are employed in structural dynamics and are discussed in Section 6.7. If the lumped mass approach is used, then the effective coefficient matrix $[C]$ is diagonal and each nodal unknown can be computed individually from the uncoupled equations. The explicit algorithm form of equation 7.79 is

$$[^\backsim C_\backsim]\{\phi\}_{n+1} = \{\bar{R}\}_n \tag{7.81}$$

TABLE 7.3. Solution Sequence for Explicit Algorithm

INITIAL CALCULATIONS

1. Form element matrices $[^\frown C_\smallsmile]^{(e)}$ and $[K]^{(e)}$.
2. Initialize $\{\phi\}_n$ at $t = 0$, $\{\phi\}_0$.
3. Select the time step Δt.
4. Assemble the coefficient matrix $[^\frown C_\smallsmile]$.

AT EACH TIME STEP

1. Form element vector $\{R\}_n^{(e)}$.
2. Assemble the effective vector $\{\overline{R}\}_n$.
3. Solve for $\{\phi\}_{n+1}$ at the end of the time step.
4. Correct $\{\phi\}_{n+1}$ for boundary conditions.

where

$$\{\overline{R}\}_n = [-\Delta t[K] + [^\frown C_\smallsmile]]\{\phi\}_n + \Delta t\{R\}_n \tag{7.82}$$

The solution sequence for the explicit algorithm is summarized in Table 7.3.

The explicit algorithm clearly avoids the solution of simultaneous algebraic equations and requires substantially less computational effort than the implicit algorithm. We shall see in the next section that this computational advantage is offset by the disadvantage that the time step Δt must be selected to be less than a critical value for the response to remain stable. If the time step for the explicit algorithm is selected arbitrarily, the computed response may become unstable, and the computed values will grow without bound as time increases.

7.5.4 Oscillation and Stability of Transient Response

The direct integration of the first-order matrix equations by the recurrence formula, equation 7.78 introduces numerical errors in the computed transient response because of the finite difference approximations employed in equations 7.77b–7.77d. The transient response computed by the recurrence formula will therefore approximate the exact solution to the original matrix differential equations, but it will approach the exact solution arbitrarily closely as $\Delta t \to 0$. In contrast, for large Δt the transient response computed by the recurrence formula may exhibit entirely unrealistic behavior, including nonphysical oscillations, and in some instances the transient response may become unstable. An understanding of when the computed response can oscillate or become unstable is consequently of great practical importance. The study of the stability of numerical solution algorithms for ordinary differential equations is an important topic in numerical analysis, and the interested reader should consult numerical analysis texts, for example, reference 22. Herein the oscillation tendencies and stability of recurrence formula equation 7.78 are evaluated by an elementary but lucid technique based on the mode superposition approach described in Section 7.5.2.

Consider free response in a typical mode. From equation 7.75, with $R_i^* = 0$, a typical generalized modal amplitude satisfies the uncoupled differential equation

$$C_{ii}^* \dot{a}_i + K_{ii}^* a_i = 0, \qquad i = 1, 2, \ldots, n \qquad (7.75)$$

where the eigenvalue of the ith mode is

$$\lambda_i = \frac{K_{ii}^*}{C_{ii}^*}$$

with generalized constants K_{ii}^* and C_{ii}^* computed from the system coefficients and eigenvectors by equation 7.73. Using the recurrence formula, equation 7.78, to solve the uncoupled modal equation 7.75, we can write

$$\left(\theta K_{ii}^* + \frac{1}{\Delta t} C_{ii}^* \right) (a_i)_{n+1} = \left(-(1 - \theta) K_{ii}^* + \frac{1}{\Delta t} C_{ii}^* \right) (a_i)_n$$

Next, since the coefficients are no longer matrices but scalars, we can solve for $(a_i)_{n+1}$. Using the definition of λ_i above, we can therefore write the simple recurrence formula

$$(a_i)_{n+1} = r_i (a_i)_n \qquad (7.83a)$$

where

$$r_i = \frac{1 - (1 - \theta)\lambda_i \Delta t}{1 + \theta \lambda_i \Delta t} \qquad (7.83b)$$

Note that r_i represents the ratio of the modal amplitude at successive instants of time in the response. For real systems the response in the absence of a forcing function decays smoothly with increasing time. This simple but important observation places a limitation on the range of r_i, and as a result the permissible values of Δt. If $|r_i| > 1$, the response grows with time and eventually becomes unbounded. In addition, if $-1 < r_i < 0$, the modal amplitudes oscillate between plus and minus values. For this latter case the response may remain bounded, but the oscillatory response will not represent the true response realistically. Thus for the solution to remain bounded and to ensure a stable response,

$$|r_i| < 1$$

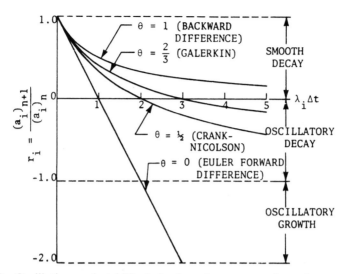

Figure 7.8 Oscillation and stability behavior of recurrence formula, equation 7.83, for typical modal response.

If this constraint is imposed on equation 7.83b, we find that

$$\lambda_i \Delta t (-1 + 2\theta) > -2 \qquad (7.84)$$

The stability requirement expressed above is always satisfied if $\theta \geq \frac{1}{2}$, and algorithms corresponding to this range of θ are *unconditionally stable*. For $\theta < \frac{1}{2}$, the algorithms are *conditionally stable*, because the stability requirement, equation 7.84, limits the maximum size of the time step that can be employed. For conditionally stable algorithms the critical time step Δt_{cr} is determined from equation 7.84 as

$$\Delta t_{cr} = \frac{2}{1 - 2\theta} \frac{1}{\lambda_i} \qquad \left(0 \leq \theta \leq \frac{1}{2}\right) \qquad (7.85)$$

For time steps $\Delta t \leq \Delta t_{cr}$ the computed response will remain bounded, but for time steps $\Delta t > \Delta t_{cr}$ the computed response will grow without bound. Further insight into the oscillation tendencies and stability of the transient response emerges from Figure 7.8. Plotting the ratio r_i against $\lambda_i \Delta t$ for four values of θ shows that the Euler forward difference algorithm is stable only if the time step is less than a critical value, but the Crank–Nicolson, Galerkin, and backward difference algorithms are unconditionally stable. Figure 7.8 also shows that the conditionally stable Euler, the unconditionally stable Crank–Nicolson, and the Galerkin algorithms will exhibit oscillations if the time step is too large. In particular, the Euler algorithm predicts an oscillatory response if $\lambda_i \Delta t > 1$; the Crank–Nicolson algorithm predicts an oscillatory response if $\lambda_i \Delta t > 2$; and the Galerkin algorithm predicts an oscillatory response if $\lambda_i \Delta t > 3$. The unconditionally stable backward difference method always predicts a smooth decay.

In Section 7.5.3 implicit and explicit algorithms were discussed, and we noted that an explicit Euler forward difference algorithm based on $\theta = 0$ offers significant savings in computational effort, since the nodal unknowns at each time step are computed from uncoupled algebraic equations. The explicit algorithm has the disadvantage that it is conditionally stable, and the time step must be selected to be less than the critical value specified by equation 7.85, with $\theta = 0$:

$$\Delta t_{cr} = \frac{2}{\lambda_i} \quad (\theta = 0, \text{ Euler forward difference})$$

The limitation on the time step for the explicit Euler forward difference algorithm means that the explicit algorithm requires smaller time steps than an implicit algorithm. Nevertheless, the conditionally stable explicit algorithm may exhibit significant computational savings in practical problems.

The discussion above has shown the importance of the time-step size in the step-by-step computation of the transient response of a typical mode. The time-step size is limited by oscillation and stability considerations. To solve the complete set of equations by step-by-step integration we must consider the transient response in all modes. For the complete set of equations the use of the eigenvalue of the highest mode in the selection of the time step will insure desirable computational behavior in all modes. In particular, for the conditionally stable Euler forward difference algorithm the critical time step is determined from equation 7.85 as

$$\Delta t_{cr} = \frac{2}{\lambda_{max}} \quad (\theta = 0, \text{ Euler forward difference}) \qquad (7.86)$$

where λ_{max} is the maximum system eigenvalue. The selection of the time step for the implicit unconditionally stable algorithms is less critical but is still an important practical consideration. For these algorithms the time-step selection represents a compromise between choosing the largest possible step to reduce computational expense and eliminating spurious oscillations from the response. If the time step is based on the largest eigenvalue of the significant lower modes, a large time step is permitted, but at the possible expense of objectionable numerical oscillations in the higher modes. If the time step is based on the largest system eigenvalue, the response is free from oscillations, but excessive computational effort results due to a small time step in integration of the lower modes. Usually a compromise resulting from trial and error between these alternatives is best.

Stiff systems of equations are sets of equations with widely separated eigenvalues. Stiff systems arise in many finite element applications, and the time integration of stiff equations may be expensive because of the requirement of very small time steps. Time integration algorithms that are effective for stiff equations have been developed by Gear and others and are described in references 31–33.

7.5.5   Algorithm Order

In the preceding discussion we learned that the time-marching algorithm is unconditionally stable for $\theta \geq \frac{1}{2}$ and conditionally stable for $\theta < \frac{1}{2}$. For $\theta = 0$ with a diagonalized coefficient matrix $[C]$, an explicit algorithm results. For other values of θ, a family of implicit algorithms is available. We discussed the compromises between computational effort and stability that arise in selecting either an explicit or an implicit algorithm.

An additional factor we must consider in selecting a time-marching scheme is the algorithm order. When we derived the time-marching scheme beginning in equations 7.77, temporal approximations were introduced. These approximations introduce errors that depend on the time step Δt, but as Δt approaches zero, the temporal errors approach zero. Practically, however, the roundoff error of finite arithmetic increases as Δt approaches zero. So the total numerical error never goes to zero but reaches a minimum value past which it actually begins to increase as Δt approaches zero due to the prominence of the roundoff error. Nevertheless, the algorithm order signifies the rate at which the temporal errors decrease with Δt. If the temporal errors decrease linearly with Δt, the algorithm is first-order accurate; if the temporal errors decrease quadratically with Δt, the algorithm is second-order accurate, and so on. In selecting a time-marching scheme, we clearly wish to have a higher-order scheme, for then as we reduce the time-step size, temporal errors reduce rapidly.

To understand the order of the θ family of algorithms we return to the beginning of the derivation of equations 7.77 and consider the approximations more carefully. As before, we evaluate the matrix set of equations at an intermediate time t_θ between times t_n and t_{n+1} to produce equation 7.77a. We wish to evaluate the order of the errors inherent in equations 7.77b–7.77c. To do this, we write approximations for $\{\phi\}_n$ and $\{\phi\}_{n+1}$, in terms of Taylor series expansions about time t_θ. These expansions are

$$\{\phi\}_n = \{\phi\}_\theta - \theta \, \Delta t \{\dot\phi\}_\theta + \tfrac{1}{2}\theta^2 \, \Delta t^2 \{\ddot\phi\}_\theta - \tfrac{1}{6}\theta^3 \, \Delta t \{\dddot\phi\}_\theta + 0(\Delta t^4) \tag{7.87a}$$

$$\{\phi\}_{n+1} = \{\phi\}_\theta + (1 - \theta) \, \Delta t \{\dot\phi\}_\theta + \tfrac{1}{2}(1 - \theta)^2 \, \Delta t^2 \{\ddot\phi\}_\theta$$
$$+ \tfrac{1}{6}(1 - \theta)^3 \, \Delta t^3 \{\dddot\phi\}_\theta + O(\Delta t^4) \tag{7.87b}$$

where $O(\Delta t^4)$ indicates that the first term in the remainder of the series is of order Δt^4. We evaluate the error in the approximation given in equation 7.77b by subtracting 7.87a from 7.87b and solving for $\{\phi\}_\theta$. After dividing by Δt, we find

$$\{\phi\}_\theta = \frac{\{\phi\}_{n+1} - \{\phi\}_n}{\Delta t} - \frac{1}{2}(1 - 2\theta)\,\Delta t\{\phi\}_\theta$$
$$- \frac{1}{6}(1 - 3\theta + 3\theta^2)\,\Delta t^2\{\phi\}_\theta + O(\Delta t^3) \qquad (7.88)$$

From equation 7.88, we observe that the error in the approximation for $\{\phi\}_\theta$ from retaining only the first term in the expansion depends on θ and Δt. If we choose $\theta = \frac{1}{2}$ the second term on the right-hand side of equation 7.88 vanishes, and the error in the approximation given in equation 7.77b is $O(\Delta t^2)$; otherwise, the error is $O(\Delta t)$.

We may identify the error in the approximation of equation 7.77c by using the Taylor expansions, equations 7.87, and solving for $\{\phi\}_\theta$. We multiply equation 7.87a by $1 - \theta$, multiply equation 7.87b by θ, and add to obtain

$$\{\phi\}_\theta = (1 - \theta)\{\phi\}_n + \theta\{\phi\}_{n+1} - \tfrac{1}{2}\theta(1 - \theta)\,\Delta t^2\{\phi\}_\theta + O(\Delta t^3) \quad (7.89)$$

Here we see that the error introduced by interpolating $\{\phi\}_\theta$ according to equation 7.77c is $O(\Delta t^2)$. In a similar way Taylor series expansions for $\{R\}$ show that the interpolation given in equation 7.77d is $O(\Delta t^2)$.

From consideration of the errors present in the approximations introduced in equations 7.77b–7.77c, we conclude that only the implicit Crank–Nicolson algorithm for $\theta = \frac{1}{2}$ is second-order accurate. All other algorithms of the θ family are first-order accurate. For this reason, many analysts regard the selection of an algorithm from the θ family as the choice between the explicit, conditionally stable, first-order accurate Euler ($\theta = 0$) scheme, or the implicit, unconditionally stable, second-order accurate Crank–Nicolson ($\theta = \frac{1}{2}$) scheme.

The computation of the transient response via recurrence relations is important and is the subject of current research. We presented a family of relatively simple but popular algorithms used in integrating first-order equations arising in finite element applications. More sophisticated methods are the subject of current research, particularly for systems of nonlinear equations and for "stiff" systems of equations [31, 32]. We discuss the time integration of nonlinear first-order equations in Chapter 8 on heat transfer.

7.5.6 Sample Problem

To illustrate the preceding methods of computing the transient response of first-order systems arising in field problems, we describe solutions of a special case of the one-dimensional example presented in Section 7.4.3. Consider a problem governed by the one-dimensional diffusion equation

$$\frac{\partial^2 \phi}{\partial x^2} = \frac{\partial \phi}{\partial t}$$

with boundary conditions

$$\frac{\partial \phi(0, t)}{\partial x} = 0, \qquad t < 0$$

$$-\frac{\partial \phi(0, t)}{\partial x} = 1, \qquad t > 0$$

$$\frac{\partial \phi(L, t)}{\partial x} = 0$$

and initial condition

$$\phi(x, 0) = 0$$

A physical interpretation is the one-dimensional heat conduction problem of a constant unit heat flux applied suddenly to one face of a solid bounded by two parallel planes a distance L apart. The problem was solved [28] with a spatial discretization of 20 elements with linear interpolation functions. Typical element matrices appear in Section 7.4.3. For the one-dimensional diffusion equation above, a typical element (e) equation is

$$\frac{l}{6} \begin{bmatrix} 2 & 1 \\ 1 & 2 \end{bmatrix} \begin{Bmatrix} \dot{\phi}_1 \\ \dot{\phi}_2 \end{Bmatrix}^{(e)} + \frac{1}{l} \begin{bmatrix} 1 & -1 \\ -1 & 1 \end{bmatrix} \begin{Bmatrix} \phi_1 \\ \phi_2 \end{Bmatrix}^{(e)} = \{0\}$$

where l is the element length. The problem was solved by mode superposition and by three forms ($\theta = \frac{1}{2}$, $\theta = \frac{2}{3}$, $\theta = 1$) of the recurrence formula, equation 7.78.

For solution via mode superposition the matrix eigenvalue problem, equation 7.71, was first solved for the assembled equations. The first 10 eigenvalues computed from the finite element model are compared with the exact eigenvalues of the boundary value problem in Table 7.4.

The eigenvalues computed from the 20-element model show excellent agreement for the lower modes, but the error increases with modal number. We note that the finite element-computed eigenvalues are upper bounds on the exact values, which is characteristic of finite element eigenvalue solutions with consistent mass (capacitance) matrices. The transient response for $\phi(0, t)$ computed by the mode superposition method and the three forms of the recurrence formula are compared with the analytical solution for two different time steps in Tables 7.5 and 7.6, where the differences in the finite element solutions and the analytical solutions appear in parentheses.

TABLE 7.4. **Eigenvalues of the One-Dimensional Diffusion Problem**

Mode	Eigenvalues λ_i	
i	Finite Element	Exact[a]
1	0.618	0.617
2	2.488	2.467
3	5.655	5.552
4	10.198	9.870
5	16.229	15.422
6	23.894	22.207
7	33.375	30.226
8	44.888	39.478
9	58.678	49.965
10	75.000	61.685

[a] Exact values computed from $i^2\pi^2/L^2$, with $L = 4$.

Since the forcing function arising from the suddenly applied boundary condition remains constant throughout the solution, the mode superposition method gives a solution with only errors introduced by the spatial discretization. The recurrence formula solutions have different behavior. For the larger time step (Table 7.5) the Crank–Nicolson scheme has rather severe oscillations at the beginning that tend to die out with time. The Galerkin scheme also has oscillations, but less severe than those of the Crank–Nicolson scheme. The backward difference scheme does not oscillate (see Figure 7.8), but underestimates the response. With a smaller time step (Table 7.6) the Crank–Nicolson scheme shows significant improvement, but the Galerkin scheme still exhibits better short-time accuracy. The backward difference scheme is only slightly improved.

TABLE 7.5. **Transient Response $\phi(0, t)$ for $\Delta t = 0.10$**

Time t	Analytical Solution	Mode Superposition	Crank–Nicolson $\theta = \frac{1}{2}$	Galerkin $\theta = \frac{2}{3}$	Backward Difference $\theta = 1$
0.1	0.357	0.354(−0.003)	0.433(+0.076)	0.378(+0.021)	0.311(−0.046)
0.2	0.505	0.503(−0.002)	0.460(−0.045)	0.486(−0.019)	0.472(−0.033)
0.3	0.618	0.616(−0.002)	0.652(+0.034)	0.613(−0.005)	0.591(−0.027)
0.4	0.714	0.712(−0.002)	0.688(−0.026)	0.705(−0.009)	0.690(−0.024)
0.5	0.798	0.797(−0.001)	0.818(+0.020)	0.791(−0.007)	0.777(−0.021)
0.6	0.874	0.873(−0.001)	0.857(−0.017)	0.867(−0.007)	0.855(−0.019)
0.7	0.944	0.943(−0.001)	0.957(+0.013)	0.938(−0.006)	0.926(−0.018)
0.8	1.009	1.008(−0.001)	0.998(−0.011)	1.003(−0.006)	0.993(−0.016)
0.9	1.070	1.069(−0.001)	1.079(+0.009)	1.067(−0.003)	1.055(−0.015)
1.0	1.128	1.127(−0.001)	1.120(−0.008)	1.123(−0.005)	1.114(−0.014)

Note. Numbers in parentheses denote differences between the finite element solutions and the analytical solution.

TABLE 7.6. Transient Response $\phi(0, t)$ **for** $\Delta t = 0.05$

Time t	Crank–Nicolson $\theta = \frac{1}{2}$	Galerkin $\theta = \frac{2}{3}$	Backward Difference $\theta = 1$
0.1	0.332(−0.025)	0.344(−0.013)	0.332(−0.025)
0.2	0.493(−0.012)	0.498(−0.007)	0.487(−0.018)
0.3	0.612(−0.006)	0.612(−0.006)	0.603(−0.015)
0.4	0.710(−0.004)	0.709(−0.005)	0.701(−0.013)
0.5	0.795(−0.003)	0.793(−0.005)	0.787(−0.011)
0.6	0.872(−0.002)	0.870(−0.004)	0.864(−0.010)
0.7	0.943(−0.001)	0.940(−0.004)	0.935(−0.009)
0.8	1.008(−0.001)	1.006(−0.003)	1.000(−0.009)
0.9	1.069(−0.001)	1.067(−0.003)	1.062(−0.008)
1.0	1.127(−0.001)	1.125(−0.003)	1.120(−0.008)

Note. Numbers in parentheses denote differences between the finite element solutions and the analytical solution.

Reference 28 concludes that mode superposition is convenient for situations where a small number of modes is required, such as for long-time solutions, but the Crank–Nicolson and Galerkin schemes are attractive for other cases, with the Galerkin superior for short-time accuracy. The superiority of the Galerkin method for short-time accuracy for problems with quickly varying boundary conditions is also noted in reference 34.

7.6 CLOSURE

The finite element formulation for a wide variety of continuum problems commonly called field problems was explored in this chapter. Rather than considering individual physical problems, we examined classes of problems governed by similar differential equations and their associated boundary conditions. General expressions were derived for the element equations. Clearly, these can be specialized for any particular element type having at least C^0 continuity (see Chapter 5). Several helpful numerical techniques for solving discretized time-dependent equations were also presented. Although the treatment is far from exhaustive, it provides a convenient starting point.

Our discussion here of steady and unsteady field problems considered only the most commonly encountered situations. Also, we restricted our attention to *linear* second-order partial differential equations. A number of practical situations arise where the governing equations are of the same form as those discussed in this chapter but are inherently nonlinear. For example, problems governed by the quasiharmonic equation become nonlinear when the coefficients k_x, k_y, and k_z are functions of ϕ or its first derivatives. Heat conduction problems with temperature-dependent thermal conductivities and electromagnetic problems with field-dependent magnetic permeability are just two typical cases. Element equations for problems of this type can be derived by using the method of weighted residuals with Galerkin's criterion. The matrix equations obtained this way have the same forms as those given in this chapter; however, the coefficient matrices $[K]$, $[C]$, and $[M]$ are then functions of ϕ and/or its gradients. Assembly of the element equations to form the system equations is the same, but the resulting system equations are nonlinear and must be solved iteratively.

REFERENCES

1. S. G. Mikhlin, *Variational Methods in Mathematical Physics*, Macmillan, New York, 1964 (English translation of the 1957 edition).

2. R. S. Schechter, *The Variational Method in Engineering*, McGraw-Hill, New York, 1967.

3. P. Silvester and A. Konard, "Axisymmetric Triangular Finite Elements for the Scalar Helmholtz Equation," *Int. J. Numer. Methods Eng.*, Vol. 5, No. 3, 1973, pp. 481–497.

4. A. F. Emery and W. W. Carson, "An Evaluation of the Use of the Finite Element Method in the Computation of Temperature," *J. Heat Transfer*, May 1971, pp. 136–145.

5. C. Taylor, B. S. Patil, and O. C. Zienkiewicz, "Harbor Oscillation: A Numerical Treatment for Undamped Natural Modes," *Proc. Inst. Civil Eng.*, Vol. 43, 1969, pp. 141–155.

6. P. L. Arlett, A. K. Bahrani, and O. C. Zienkiewicz, "Application of Finite Element Method to the Solution of Helmholtz's Equation," *Proc. IEEE*, Vol. 115, No. 12, 1968, pp. 1762–1766.

7. W. H. Haytt, Jr., *Engineering Electromagnetics*, McGraw-Hill, New York, 1958.

8. L. E. Kinsler, A. R. Frey, A. B. Coppens and J. V. Sanders, *Fundamentals of Acoustics*, Third Edition, Wiley, New York, 1982.

9. J. H. Wilkinson, *The Algebraic Eigenvalue Problem*, Oxford University Press, London, 1965.

10. T. Shuku and K. Ishihara, "The Analysis of the Acoustic Field in Irregularly Shaped Rooms by the Finite Element Method," *J. Sound Vib.*, Vol. 29, 1973, pp. 67–76.

11. T. Shuku, "Finite Difference Analysis of the Acoustic Field in Irregular Rooms," *J. Acoust. Soc. Jpn.*, Vol. 28, 1972, pp. 5–12.

12. S. H. Sung, "Automotive Applications of Three-Dimensional Acoustics Finite Elements," *Society of Automotive Engineers, Paper No. 810397*, February 1981.

13. S. H. Crandall, *Engineering Analysis*, McGraw-Hill, New York, 1956.

14. I. N. Sneddon, *Elements of Partial Differential Equations*, McGraw-Hill, New York, 1957.

15. R. B. Bird, W. E. Stewart, and E. N. Lightfoot, *Transport Phenomena*, Wiley, New York, 1960.

16. J. Bear, *Dynamics of Fluids in Porous Media*, American Elsevier, New York, 1972.

17. C. S. Desai, "Finite Element Methods for Flow in Porous Media," in *Finite Elements in Fluids*, Vol. 1, R. H. Gallagher, J. T. Oden, C. Taylor, and O. C. Zienkiewicz (eds.), Wiley, London, 1975.

18. H. Semat, *Introduction to Atomic and Nuclear Physics*, Holt, Rinehart & Winston, New York, 1962, p. 218.

19. J. T. Oden, "A General Theory of Finite Elements. II: Applications," *Int. J. Numer. Methods Eng.*, Vol 1, No. 3, 1969, pp. 247–259.

20. L. Meirovitch, *Analytical Methods in Vibrations*, Macmillan, New York, 1967.

21. L. Fox and D. F. Mayers, *Computing Methods for Scientists and Engineers*, Oxford University Press, London, 1968.

22. R. Richtmyer and K. Morton, *Difference Methods for Initial Value Problems*, Wiley–Interscience, New York, 1967.

23. G. Forsythe and W. Wasow, *Finite Difference Methods for Partial Differential Equations*, Wiley, New York, 1960.

24. R. L. Ketter and S. Prawel, *Modern Methods of Engineering Computation*, McGraw-Hill, New York, 1969.

25. N. F. Morris, "The Use of Modal Superposition in Nonlinear Dynamics," *Comput. Struct.*, Vol. 7, 1977, pp. 65–72.

26. T. Belytschko and T. J. R. Hughes (eds.), *Computational Methods for Transient Analysis*, Elsevier, New York, 1983.

27. W. L. Wood and R. W. Lewis, "A Comparison of Time Marching Schemes for the Transient Heat Conduction Equation," *Int. J. Numer. Methods Eng.*, Vol. 9, 1975, pp. 679–689.

28. M. A. Hogge, "Integration Operators for First Order Linear Matrix Differential Equations," *Comput. Methods Appl. Mech. Eng.*, Vol. 11, 1977, pp. 281–294.

29. O. C. Zienkiewicz, *The Finite Element Method*, 3d ed., McGraw-Hill, New York, 1977, pp. 569–580.

30. A. J. Baker and M. O. Soliman, "Utility of a Finite Element Solution Algorithm for Initial-Value Problems," *J. Comput. Phys.*, Vol. 32, 1979, pp. 289–324.

31. L. F. Shampine and C. W. Gear, "A User's View of Solving Stiff Ordinary Differential Equations," *SIAM Rev.*, Vol. 21, No. 1, January 1979, pp. 1–11.

32. D. C. Krinke and R. L. Huston, "An Analysis of Algorithms for Solving Differential Equations," *Comput. Struct.*, Vol. 11, 1980, pp. 69–74.

33. C. W. Gear, "The Automatic Integration of Ordinary Differential Equations," *Commun. ACM*, Vol. 14, No. 3, 1971, pp. 176–190.

34. J. Donea, "On the Accuracy of Finite Element Solutions to the Transient Heat Conduction Equation," *Int. J. Numer. Methods Eng.*, Vol. 8, 1974, pp. 103–110.

PROBLEMS

1. Solve Laplace's equation

$$\frac{\partial^2 \phi}{\partial x^2} + \frac{\partial^2 \phi}{\partial y^2} = 0$$

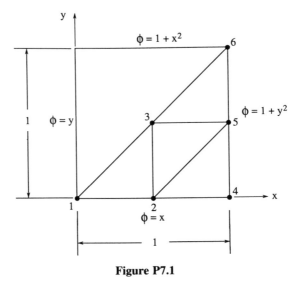

Figure P7.1

for the region shown in Figure P7.1 with the indicated boundary conditions. Exploit symmetry and use triangular elements to compute the nodal value ϕ_3 for the mesh shown.

2. Obtain an approximate solution for the quasiharmonic equation

$$k\left(\frac{\partial^2 \phi}{\partial x^2} + \frac{\partial^2 \phi}{\partial y^2}\right) + Q = 0$$

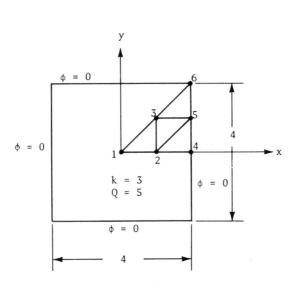

Figure P7.2

subject to the boundary conditions $\phi = 0$ on the boundary of the square region shown in Figure P7.2. Exploit symmetry and use triangular elements to compute values for ϕ_i at the nodes for the mesh shown.

3. Solve the quasiharmonic equation with the boundary conditions of Problem 2 using the mesh of square elements shown in Figure P7.3.

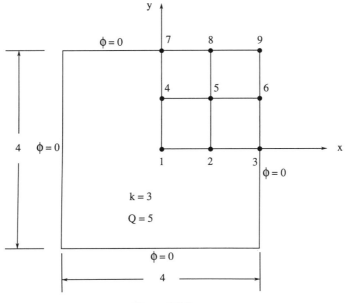

Figure P7.3

4. In the theory of elasticity the torsion of a noncircular shaft by a torque M_t is formulated as an elliptic boundary value problem with a stress function $\phi(x, y)$ as the basic unknown. (See Appendix C for a brief description of the basic equations of the theory of elasticity.) The equilibrium equations are satisfied if the stress function satisfies

$$\frac{\partial^2 \phi}{\partial x^2} + \frac{\partial^2 \phi}{\partial y^2} + 2G\theta = 0 \qquad \text{in } \Omega$$

where G is the shear modulus of elasticity and θ is the angle of twist per unit length. For a solid cross section the requirement of a stress-free boundary yields the boundary condition

$$\phi = 0 \qquad \text{on } \Gamma$$

Within the cross section the stresses are computed from the stress function by differentiation,

$$\tau_{zx} = \frac{\partial \phi}{\partial y} \qquad \tau_{zy} = -\frac{\partial \phi}{\partial x}$$

At the ends of the shaft the first moment of stresses integrated over the cross-sectional area A must equal the twisting moment. This requirement gives

$$M_t = 2 \int_A \phi \, dA$$

and the twisting moment is related to the angle of twist by

$$M_t = GJ\theta$$

where J is the torsional constant. Use the variational approach described in Section 7.2 to develop a finite element formulation of the torsion problem, with the stress function ϕ as the nodal unknown. Derive general equations for the element matrices similar to equations 7.14–7.16.

5. Evaluate the element equations derived in Problem 4 for a plane triangular element. Develop equations for computing element stresses and the torsional constant J for the cross section.

6. Use the formulation and equations developed in Problems 4 and 5 to obtain an approximate solution for the torsion of a shaft with an equilateral triangular cross section (Figure P7.6). Use the three-element mesh shown to compute shaft stresses and the torsional constant J. Discuss the accuracy of the solution.

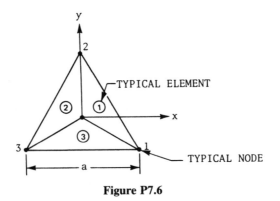

Figure P7.6

7. Read the paper "Torsion of Nonhomogeneous Anisotropic Bars" by S. Valliappan and V. A. Pulmano, *Journal of Structural Division, ASCE,*

Vol. 100, No. ST1, January 1974, pp. 286–295. Describe the finite element approach that is developed and discuss a typical example to illustrate the results presented. Summarize the conclusions of the paper.

8. Finite element methods for flow in a porous medium are described in reference 17. A governing equation for saturated flow can be expressed as

$$\frac{\partial}{\partial x}\left(k_x\frac{\partial \phi}{\partial x}\right) + \frac{\partial}{\partial y}\left(k_y\frac{\partial \phi}{\partial y}\right) + Q = 0 \qquad \text{in } \Omega$$

where k_x and k_y are coefficients of permeability, ϕ is the piezometric head, and Q is the internal fluid source. Boundary conditions are (Figure 7.1)

$$\phi = \phi_1 \qquad \text{on } S_1$$

$$k_x\frac{\partial \phi}{\partial x}n_x + k_y\frac{\partial \phi}{\partial y}n_y + q = 0 \qquad \text{on } S_2$$

where q represents boundary flow out of the region. Use the variational approach described in Section 7.2 to develop a finite element formulation for the porous medium flow. Derive general equations for the element matrices similar to equations 7.14–7.16.

9. Repeat Problem 8, but use the method of weighted residuals (see Chapter 4) to derive the equations for the element matrices.

10. Derive the matrix equations for the axisymmetric triangular element presented in Section 7.2.5. Verify the coefficient matrix given in equation 7.34 and the vectors given in equations 7.35 and 7.36.

11. Use the techniques of Appendix C to show that the Helmholtz equation, equation 7.37, is the Euler–Lagrange equation for the functional given in equation 7.42.

12. The vibrating string is a classical example of an eigenvalue problem. A string (Figure P7.12) of mass per unit length m is stretched between two fixed supports by a large initial tension T. For small motions the equation of motion of the string is the wave equation

$$\frac{\partial^2 w}{\partial x^2} = \frac{m}{T}\frac{\partial^2 w}{\partial t^2}$$

where $w(x,t)$ is the transverse displacement of the string. For free oscillations the transverse displacement may be expressed as

$$w(x,t) = \phi(x)\sin(\omega t + \theta)$$

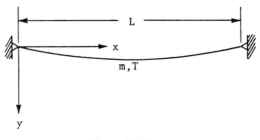

Figure P7.12

where $\phi(x)$ is a vibration shape (eigenfunction), ω is the natural frequency, and θ is a phase angle. Substituting into the equation of motion yields the governing equation for free oscillations,

$$\frac{d^2\phi}{dx^2} + \frac{m\omega^2}{T}\phi = 0$$

The boundary conditions require the amplitude of the vibration be zero at the supports:

$$\phi(0) = 0 \qquad \phi(L) = 0$$

Use the variational approach described in Section 7.3 to develop a finite formulation for the freely vibrating string. Derive general equations for the element matrices similar to equations 7.44. Evaluate the element matrices for a string element with two nodes and a linear interpolation of the amplitude $\phi(x)$.

13. Use the element equations derived in Problem 12 to compute the first three natural frequencies and mode shapes for a vibrating string. Use four elements of length $L/4$. Compare the finite element predicted results with the exact analytical solution. Sketch the mode shapes.

14. Use the element equations derived in Problem 12 to compute approximate natural frequencies for a vibrating string. Use two elements to represent one-half of the string $(0 < x < L/2)$ and employ symmetry. Calculate the natural frequencies and mode shapes for the first two symmetrical modes. Sketch the mode shapes. What would the boundary condition be at $x = L/2$ if we wished to compute the frequencies and mode shapes of antisymmetric modes?

15. The natural torsional vibrations of an elastic shaft supporting a rigid end disk (Figure P7.15) is described by the partial differential equation

$$JG\frac{d^2\phi}{dx^2} + \rho\omega^2\phi = 0$$

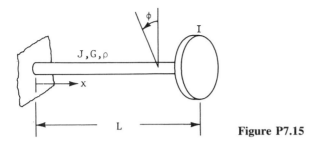

Figure P7.15

where G is the shear modulus, J is the shaft torsional constant, ρ is the material density, $\phi(x)$ is the modal amplitude of the angle of twist, and ω is the natural frequency. The mass moment inertia of the disk is I. The boundary conditions are

$$\phi(0) = 0 \qquad GJ\frac{d\phi(L)}{dx} = I\omega^2\phi(L)$$

Use the variational approach to develop a finite element formulation for free oscillations. Derive general element equations similar to equations 7.44. How does the boundary condition at $x = L$ change the element matrices?

16. Repeat Problem 15, but use the method of weighted residuals to derive the element equations.

17. A vibration problem of an elastic shaft with a rigid end disk is described by

$$\frac{d^2\phi}{dx^2} + \lambda^2\phi = 0$$

$$\phi(0) = 0 \qquad \frac{d\phi}{dx}(L) = \alpha\phi(L)$$

where λ is the eigenvalue, and α is known. Using two linear finite elements with length $L/2$, set up the matrix equations for determining the eigenvalues (natural frequencies) and eigenvectors (mode shapes). For $L = 6$ and $\alpha = 3$ determine the eigenvalues and eigenvectors. Sketch the possible vibration shapes.

18. A uniform elastic membrane has mass per unit area m and is stretched with uniform tension T over a rigid circular frame of radius $r = a$. In polar coordinates the hyperbolic equation of motion of the membrane is

given by

$$T\left(\frac{\partial^2 w}{\partial r^2} + \frac{1}{r}\frac{\partial w}{\partial r} + \frac{1}{r^2}\frac{\partial^2 w}{\partial \theta^2}\right) + p = m\frac{\partial^2 w}{\partial t^2}$$

where $w(r, \theta, t)$ is the transverse deflection and $p(r, \theta, t)$ is the external surface pressure. For axisymmetric motions the deflection and load are independent of θ, that is, $w = w(r, t)$ and $p = p(r, t)$. Formulate the eigenvalue problem for axisymmetric free vibrations and develop a finite element formulation for the problem. How would the finite element equations be changed by the addition of a point mass M located at the center of the membrane?

19. Consider the axisymmetric motions of an elastic membrane as described in Problem 14. Develop a finite element formulation for the time-dependent response of the membrane, given $p(r, t)$ to be a prescribed function of r and t.

20. The longitudinal motion of a thin rod of variable cross section is governed by the hyperbolic equation of motion

$$\frac{\partial}{\partial x}\left[EA(x)\frac{\partial u}{\partial x}\right] = \rho A(x)\frac{\partial^2 u}{\partial t^2}$$

where $u(x, t)$ is the longitudinal displacement, E is the modulus of elasticity, $A(x)$ is the cross-sectional area, and ρ is the material density. Develop the finite element equations for computing the time-dependent transient response of the rod. Derive general element equations similar to equations 7.64.

21. Evaluate the element matrices for the motion of the rod described in Problem 20 for an element with two nodes. Develop element matrices for two cases:

   a. A constant cross-sectional area.

   b. A linear variation of cross-sectional area $A(x)$, where

$$A(x) = N_1(x)A_1 + N_2(x)A_2$$

   with N_i ($i = 1, 2$) as the element interpolation functions and A_i as the element cross-sectional areas at the nodes.

22. Develop the finite element equations of motion for computing the transient response of a vibrating string. The differential equation describing the string motion is given in Problem 12, and the string is shown in Figure P7.12.

23. A string of mass per unit length m is suspended from a ceiling and has attached to its bottom end a mass M (Figure P7.23). The string tension T is assumed to be uniform along the string length and constant with time.

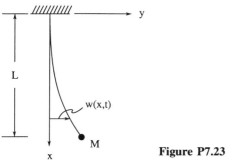

Figure P7.23

The boundary/initial value problem is described by

$$\frac{\partial^2 w}{\partial x^2} = \frac{m}{T}\frac{\partial^2 w}{\partial t^2}$$

where $w(x, t)$ is the transverse displacement of the string. The initial conditions are

$$w(x, 0) = f(x)$$

$$\frac{\partial w}{\partial t}(x, 0) = g(x)$$

where $f(x)$ and $g(x)$ are known functions. The boundary conditions are

$$w(0, t) = 0$$

$$T\frac{\partial w}{\partial x}(L, t) + M\frac{\partial^2 w}{\partial t^2}(L, t) = 0$$

Use the method of weighted residuals to develop a finite element formulation for the problem that will lead to matrix equations that can be solved for the nodal displacement histories. How does the tip mass M affect the matrix equations of motion?

24. An infinitely long hollow cylinder of inner radius a and outer radius b (Figure P7.24) that is initially at uniform temperature is suddenly subjected to convective heating on its inner surface due to the internal flow of a hot fluid at a uniform temperature T_1. The temperature within the wall of the cylinder is given by $T(r, t)$, where r is the radial coordinate. The governing equation is the parabolic heat conduction equation (see Appendix E):

$$k\left(\frac{\partial^2 T}{\partial r^2} + \frac{1}{r}\frac{\partial T}{\partial r}\right) = \rho c\frac{\partial T}{\partial t}$$

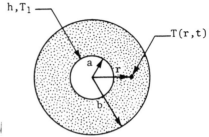

Figure P7.24

where k is the thermal conductivity, ρ is the density, and c is the specific heat. The initial condition is

$$T(r,0) = T_0, \qquad a \le r \le b$$

and the boundary conditions are

$$-k\frac{\partial T}{\partial r} = h(T - T_1), \qquad r = a$$

$$\frac{\partial T}{\partial r} = 0, \qquad r = b$$

Develop a finite element formulation for determining the transient temperature distribution within the cylinder.

25. A substance diffuses from a solution into a dry porous slab of thickness L immersed in this solution (Figure P7.25). The concentration of the solution is assumed to remain constant at all times and we wish to determine the concentration $c(x, t)$ within the slab. The boundary value

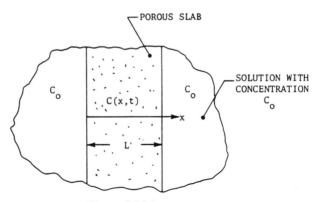

Figure P7.25

problem that must be solved is the parabolic diffusion equation

$$D\frac{\partial^2 C}{\partial x^2} = \frac{\partial C}{\partial t}$$

where D is the diffusivity. The initial condition is

$$C(x,0) = 0, \qquad 0 \le x \le L$$

and the boundary conditions are

$$C(0,t) = C_0, \qquad C(L,t) = C_0$$

Develop a finite element formulation for determining the transient concentration within the slab. Derive the element matrices for an element with two nodes.

26. Consider the one-dimensional diffusion problem given by

$$k_x\frac{\partial^2 \phi}{\partial x^2} = f_x + c_x\frac{\partial \phi}{\partial t}$$

with $k_x = 1$ and $c_x = \frac{3}{2}$. The initial condition is

$$\phi(x,0) = 0, \qquad 0 \le x \le L$$

with $L = 8$. The boundary conditions are

$$\phi(0,t) = 0 \qquad \phi(L,t) = 0$$

The forcing function

$$f_x(x,0) = \begin{Bmatrix} 0 \\ f_0 \end{Bmatrix}\begin{Bmatrix} t < 0 \\ t \ge 0, \end{Bmatrix} \quad 0 \le x \le L$$

with $f_0 = -\frac{5}{4}$. Use a two-element model and symmetry to derive the matrix equations of motion

$$\frac{1}{2}\begin{bmatrix} 4 & 1 \\ 1 & 2 \end{bmatrix}\begin{Bmatrix} \dot{\phi}_2 \\ \dot{\phi}_3 \end{Bmatrix} + \frac{1}{2}\begin{bmatrix} 2 & -1 \\ -1 & 1 \end{bmatrix}\begin{Bmatrix} \phi_2 \\ \phi_3 \end{Bmatrix} = \frac{5}{4}\begin{Bmatrix} 2 \\ 1 \end{Bmatrix}$$

27. Compute the transient response of the equations given in Problem 26 by the method of mode superposition. First solve the matrix eigenvalue problem, equation 7.71, for the system eigenvalues and eigenvectors; then derive and solve the uncoupled modal equations, equations 7.75;

finally, obtain equations for the nodal response by using mode superposition via equation 7.74. Compute and plot the time history of the nodal variables for $0 < t < 50$. Use a time step $\Delta t = 1$.

28. Solve for the transient response of the equations given in Problem 26 by using the explicit Euler method of time integration. First form the "lumped mass" matrix for the system by (1) adding elements of each row of the coefficient matrix of $\{\dot{\phi}\}$, (2) placing these values on the diagonal, and (3) zeroing the off-diagonal terms. What is the critical time step for the Euler method? Compute and plot the time history of the nodal variables for $0 < t < 50$.

29. Solve for the transient response of the equations given in Problem 26 by using the implicit Crank–Nicolson method ($\theta = \frac{1}{2}$). Compute and plot the time history of the nodal variables for $0 < t < 50$. Use a time step $\Delta t = 1$.

8

HEAT TRANSFER PROBLEMS

8.1 INTRODUCTION

In heat transfer analysts have relied on traditional finite difference methods to obtain computer solutions to difficult problems. Finite difference methods coupled with lumped parameter concepts resulted in effective general-purpose computer programs widely used in a variety of industry and government offices.

Heat transfer analysts calculate temperature distributions as a first step in the selection of materials for structures that may experience abnormal temperature levels during their service life. Analysts must model realistically environmental boundary conditions, represent complicated geometry, and analyze a variety of materials (solids and fluids) that deviate from simple constant-property isotropic materials. Textbook problems often assume away these realistic challenges. Because finite difference approaches do not always meet these challenges adequately, there is an increasing interest in finite element methods. In some cases finite elements can help, but we should not expect finite elements to achieve success in every instance. Yet there are several good reasons to consider finite elements for thermal analysis. In addition to the ability to treat irregular geometry, finite elements can offer improved accuracy and in some cases improved efficiency for the same accuracy when compared to finite difference methods. Also, thermal–structural analysts argue that finite element methods permit a common discretization for the thermal and mechanical analysis of structures.

Finite element researchers in the late 1960s sometimes regarded heat transfer analysis as simply a special example of a field problem. Often this view was understandable, because many of the early finite element researchers had structural backgrounds and failed to appreciate the subtleties of thermal-fluid analysis. However, in the last decade, as more individuals with a thermal science background entered the finite element field, there has been a growing recognition of the effectiveness of finite elements in the solution of heat transfer problems.

This chapter describes finite element methods for realistic heat transfer problems. For conduction, convection, and radiation we present the theoretical background, develop element equations, describe solution methods, and provide illustrative examples. We begin with general steady-state and transient conduction problems. Then we consider conduction with radiation boundary conditions. Next we describe problems characterized by the convective–diffusion equation. The chapter concludes with a brief description of free- and forced-convection problems. Along the way the reader will find many practical suggestions for applying these techniques to real problems.

8.2   CONDUCTION

We present a finite element formulation for computation of the steady-state temperature distribution $T(x, y, z)$ and/or transient temperature distribution $T(x, y, z, t)$ for solids with general surface heat transfer. Chapter 7 presents the fundamentals of finite element heat conduction, where we discuss general field problems. Linear steady-state and transient heat conduction with convection boundary conditions appear as special cases of the general field problems discussed there. Some of the early developments in finite element heat transfer analysis appear in references 1–5 and references contained therein. We devote this section to a more detailed discussion of the finite element solution of practical problems in conduction heat transfer.

8.2.1 Problem Statement

Consider steady-state and/or transient heat transfer in a three-dimensional anisotropic solid Ω bounded by a surface Γ (Figure 8.1). The problem is governed by the energy equation (see Appendix E):

$$-\left(\frac{\partial q_x}{\partial x} + \frac{\partial q_y}{\partial y} + \frac{\partial q_z}{\partial z}\right) + Q = \rho c \frac{\partial T}{\partial t} \tag{8.1}$$

where q_x, q_y, and q_z are components of the heat flow rate vector per unit area in Cartesian coordinates (x, y, z), $Q(x, y, z, t)$ is the internal heat generation rate per unit volume, ρ is the density, and c is the specific heat. For an anisotropic medium Fourier's law is

$$q_x = -\left(k_{11}\frac{\partial T}{\partial x} + k_{12}\frac{\partial T}{\partial y} + k_{13}\frac{\partial T}{\partial z}\right)$$

$$q_y = -\left(k_{21}\frac{\partial T}{\partial x} + k_{22}\frac{\partial T}{\partial y} + k_{23}\frac{\partial T}{\partial z}\right) \tag{8.2}$$

$$q_z = -\left(k_{31}\frac{\partial T}{\partial x} + k_{32}\frac{\partial T}{\partial y} + k_{33}\frac{\partial T}{\partial z}\right)$$

where k_{ij} is the symmetric conductivity tensor. The material properties ρ, c, and k_{ij} may be temperature dependent. If we substitute Fourier's law, equation 8.2, into the energy equation, equation 8.1, we obtain the parabolic heat conduction equation. The heat conduction equation is solved subject to an initial condition and boundary conditions on all portions of the surface Γ. The initial condition specifies the temperature distribution at time zero,

$$T(x, y, z, 0) = T_0(x, y, z) \tag{8.3}$$

Heat conduction boundary conditions take several forms (see Appendix E); consider the frequently encountered conditions of specified surface temperature, specified surface heat flow, convective heat exchange, and radiation

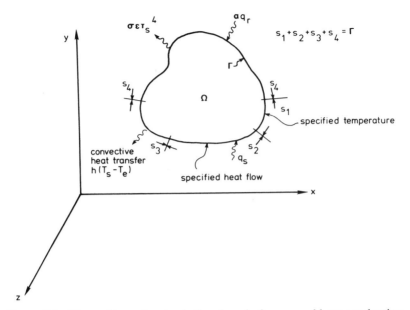

Figure 8.1 Three-dimensional solution domain for general heat conduction.

heat exchange. The boundary conditions (Figure 8.1) are

$$T_s = T_1(x, y, z, t) \qquad \text{on } S_1 \qquad (8.4a)$$

$$q_x n_x + q_y n_y + q_z n_z = -q_s \qquad \text{on } S_2 \qquad (8.4b)$$

$$q_x n_x + q_y n_y + q_z n_z = h(T_s - T_e) \qquad \text{on } S_3 \qquad (8.4c)$$

$$q_x n_x + q_y n_y + q_z n_z = \sigma \epsilon T_s^4 - \alpha q_r \qquad \text{on } S_4 \qquad (8.4d)$$

where T_1 is the specified surface temperature, which may vary with position and time, n_x, n_y, and n_z are the direction cosines of the outward normal to the surface, q_s is the specified heat flow rate per unit area (positive into the surface), h is a convective heat transfer coefficient that may be a function of the convective exchange temperature T_e, T_s is the unknown surface temperature, σ is the Stefan–Boltzmann constant, ϵ is the surface emissivity, which may be a function of the surface temperature, α is the surface absorptivity, and q_r is the incident radiant heat flow rate per unit area.

The governing equation and boundary conditions are similar to the time-dependent field problem formulated in Chapter 7. The principal differences are that this formulation permits temperature-dependent anisotropic material properties, and the boundary conditions include temperature-dependent convection coefficients and nonlinear radiation boundary conditions. Temperature-dependent material properties and nonlinear boundary conditions signify that the problem is inherently nonlinear. Element equations are derived by the method of weighted residuals with Galerkin's criteria. Several of the element matrices have the same form as those given in Chapter 7; however, nonlinear terms arise from the radiation boundary conditions. Assembly of element matrices is the same as in previous chapters, but the nonlinear equations must be solved iteratively. In the following sections we present a derivation of heat conduction finite elements in terms of general interpolation functions N_i and element matrices frequently used in heat conduction elements. Solution methods for steady-state and transient heat conduction and illustrative examples also appear.

8.2.2 Finite Element Formulation

The solution domain Ω is divided into M elements of r nodes each. By the usual procedure we express the temperature and temperature gradients within each element as

$$T^{(e)}(x, y, z, t) = \sum_{i=1}^{r} N_i(x, y, z) T_i(t) \tag{8.5a}$$

$$\frac{\partial T^{(e)}}{\partial x}(x, y, z, t) = \sum_{i=1}^{r} \frac{\partial N_i}{\partial x}(x, y, z) T_i(t) \tag{8.5b}$$

$$\frac{\partial T^{(e)}}{\partial y}(x, y, z, t) = \sum_{i=1}^{r} \frac{\partial N_i}{\partial y}(x, y, z) T_i(t) \tag{8.5c}$$

$$\frac{\partial T^{(e)}}{\partial z}(x, y, z, t) = \sum_{i=1}^{r} \frac{\partial N_i}{\partial z}(x, y, z) T_i(t) \tag{8.5d}$$

or in matrix notation

$$T^{(e)}(x, y, z, t) = \lfloor N(x, y, z) \rfloor \{T(t)\} \tag{8.6a}$$

$$\begin{Bmatrix} \dfrac{\partial T}{\partial x}(x, y, z, t) \\[2mm] \dfrac{\partial T}{\partial y}(x, y, z, t) \\[2mm] \dfrac{\partial T}{\partial z}(x, y, z, t) \end{Bmatrix} = [B(x, y, z)]\{T(t)\} \tag{8.6b}$$

where $\lfloor N \rfloor$ is the temperature interpolation matrix, $[B]$ is the temperature-gradient interpolation matrix

$$\lfloor N(x, y, z) \rfloor = \lfloor N_1 \quad N_2 \quad \cdots \quad N_r \rfloor \tag{8.7a}$$

$$[B(x, y, z)] = \begin{bmatrix} \dfrac{\partial N_1}{\partial x} & \dfrac{\partial N_2}{\partial x} & \cdots & \dfrac{\partial N_r}{\partial x} \\[2mm] \dfrac{\partial N_1}{\partial y} & \dfrac{\partial N_2}{\partial y} & \cdots & \dfrac{\partial N_r}{\partial y} \\[2mm] \dfrac{\partial N_1}{\partial z} & \dfrac{\partial N_2}{\partial z} & \cdots & \dfrac{\partial N_r}{\partial z} \end{bmatrix} \tag{8.7b}$$

$T_i(t)$ is the value of the temperature at each node, and $\{T(t)\}$ is the vector of element nodal temperatures. The second-order heat conduction equation requires only C^0 continuity, and we may use temperature as the only nodal unknown. We focus on a single element and for simplicity omit the superscript (e). The method of weighted residuals is used to derive the element equations starting with the energy equation, equation 8.1. For a review of the method of weighted residuals readers should consult Chapter 4. Section 4.2 applies the method to one- and two-dimensional heat conduction problems. The method of weighted residuals requires

$$\int_{\Omega^{(e)}} \left(\frac{\partial q_x}{\partial x} + \frac{\partial q_y}{\partial y} + \frac{\partial q_z}{\partial z} - Q + \rho c \frac{\partial T}{\partial t} \right) N_i \, d\Omega = 0 \tag{8.8}$$

where $\Omega^{(e)}$ is the domain for element (e). Following the procedures discussed in Chapter 4, we integrate the term

$$\int_{\Omega^{(e)}} \left(\frac{\partial q_x}{\partial x} + \frac{\partial q_y}{\partial y} + \frac{\partial q_z}{\partial z} \right) N_i \, d\Omega$$

by Gauss's theorem, which introduces surface integrals of the heat flow across the element boundary $\Gamma^{(e)}$. We write the result in the rearranged form

$$\int_{\Omega^{(e)}} \rho c \frac{\partial T}{\partial t} N_i \, d\Omega - \int_{\Omega^{(e)}} \left\lfloor \frac{\partial N_i}{\partial x} \quad \frac{\partial N_i}{\partial y} \quad \frac{\partial N_i}{\partial z} \right\rfloor \begin{Bmatrix} q_x \\ q_y \\ q_z \end{Bmatrix} d\Omega$$

$$= \int_{\Omega^{(e)}} Q N_i \, d\Omega - \int_{\Gamma^{(e)}} (\mathbf{q} \cdot \hat{n}) N_i \, d\Gamma, \qquad i = 1, 2, \ldots, r \tag{8.9}$$

Next we express the surface integral as the sum of integrals over S_1, S_2, S_3, and S_4 and introduce the boundary conditions, equations 8.4. Thus

$$\int_{\Omega^{(e)}} \rho c \frac{\partial T}{\partial t} N_i \, d\Omega - \int_{\Omega^{(e)}} \left| \frac{\partial N_i}{\partial x} \quad \frac{\partial N_i}{\partial y} \quad \frac{\partial N_i}{\partial z} \right| \begin{Bmatrix} q_x \\ q_y \\ q_z \end{Bmatrix} d\Omega$$

$$= \int_{\Omega^{(e)}} Q N_i \, d\Omega - \int_{S_1} (\mathbf{q} \cdot \hat{n}) N_i \, d\Gamma + \int_{S_2} q_s N_i \, d\Gamma - \int_{S_3} h(T_s - T_e) N_i \, d\Gamma$$

$$- \int_{S_4} (\sigma \epsilon T_s^4 - \alpha q_r) N_i \, d\Gamma, \qquad i = 1, 2, \ldots, r \tag{8.10}$$

As the last step we introduce the element temperatures from equation 8.6a and heat flow components from Fourier's law, equation 8.2. For convenience we first write equation 8.2 in matrix form,

$$\begin{Bmatrix} q_x \\ q_y \\ q_z \end{Bmatrix} = - \begin{bmatrix} k_{11} & k_{12} & k_{13} \\ k_{21} & k_{22} & k_{23} \\ k_{31} & k_{32} & k_{33} \end{bmatrix} \begin{Bmatrix} \dfrac{\partial T}{\partial x} \\ \dfrac{\partial T}{\partial y} \\ \dfrac{\partial T}{\partial z} \end{Bmatrix} \tag{8.11}$$

where $[k]$ is the thermal conductivity matrix, and then express the temperature gradients in terms of the nodal temperatures through equation 8.6b:

$$\begin{Bmatrix} q_x \\ q_y \\ q_z \end{Bmatrix} = -[k][B]\{T\} \tag{8.12}$$

After some manipulation the resulting element equations become

$$[C]\left\{ \frac{dT}{dt} \right\} + [[K_c] + [K_h]]\{T\}$$

$$= \{R_T\} + \{R_Q\} + \{R_q\} + \{R_h\} + \{R_\sigma\} + \{R_r\} \tag{8.13}$$

where

$$[C] = \int_{\Omega^{(e)}} \rho c \{N\} \lfloor N \rfloor \, d\Omega \qquad (8.14a)$$

$$[K_c] = \int_{\Omega^{(e)}} [B]^T [k][B] \, d\Omega \qquad (8.14b)$$

$$[K_h] = \int_{S_3} h\{N\} \lfloor N \rfloor \, d\Gamma \qquad (8.14c)$$

$$\{R_T\} = -\int_{S_1} (\mathbf{q} \cdot \hat{n})\{N\} \, d\Gamma \qquad (8.15a)$$

$$\{R_Q\} = \int_{\Omega} Q\{N\} \, d\Omega \qquad (8.15b)$$

$$\{R_q\} = \int_{S_2} q_s\{N\} \, d\Gamma \qquad (8.15c)$$

$$\{R_h\} = \int_{S_3} hT_e\{N\} \, d\Gamma \qquad (8.15d)$$

$$\{R_\sigma\} = -\int_{S_4} \sigma \epsilon T_s^4\{N\} \, d\Gamma \qquad (8.15e)$$

$$\{R_r\} = \int_{S_4} \alpha q_r\{N\} \, d\Gamma \qquad (8.15f)$$

The coefficient matrix $[C]$ of the time derivative of the nodal temperatures is the element capacitance matrix. The coefficient matrices $[K_c]$ and $[K_h]$ are element conductance matrices and relate to conduction and convection, respectively. The convection matrix is computed only for elements with surface convection. The vectors $\{R_T\}$, $\{R_Q\}$, $\{R_q\}$, and $\{R_h\}$, are heat load vectors arising from specified nodal temperatures, internal heat generation, specified surface heating and surface convection, respectively. The vectors $\{R_\sigma\}$ and $\{R_r\}$ arise from surface radiation. The vectors $\{R_T\}$ represents unknown nodal heat loads applied to maintain the nodes on the surface S_1 at specified temperatures. These heat loads may be computed, if desired, after the assembly of the element equations by the procedure described in Section 2.3.4. The integral definition of $\{R_T\}$, equation 8.15a, is not evaluated and is not considered in subsequent discussion. The convection and radiation heat load vectors are computed only for elements with surface convection and/or radiation.

Equations 8.13 are a general nonlinear formulation of element equations for transient heat conduction in an anisotropic medium. Assembly of the element equations to obtain the system equations follows the standard procedure. Note that since the nodal unknowns T_i are scalars, no transformations of matrices computed in local coordinates are necessary before assembly of the global matrices.

For analysis of practical heat conduction problems special cases of the general equations are usually considered because solution algorithms depend on whether a problem is steady state or transient, linear or nonlinear. For subsequent discussion the following cases are identified:

Linear steady-state analysis:

$$[[K_c] + [K_h]]\{T\} = \{R_Q\} + \{R_q\} + \{R_h\} \qquad (8.16)$$

Linear transient analysis:

$$[C]\{\dot{T}(t)\} + [[K_c] + [K_h(t)]]\{T(t)\}$$
$$= \{R_Q(t)\} + \{R_q(t)\} + \{R_h(t)\} \qquad (8.17)$$

Nonlinear steady-state analysis:

$$[[K_c(T)] + [K_h(T)]]\{T\}$$
$$= \{R_Q(T)\} + \{R_q(T)\} + \{R_h(T)\} + \{R_\sigma(T)\} + \{R_r(T)\} \quad (8.18)$$

Nonlinear transient analysis:

$$[C(T)]\{\dot{T}\} + [[K_c(T)] + [K_h(T,t)]]\{T(t)\}$$
$$= \{R_Q(T,t)\} + \{R_q(T,t)\} + \{R_h(T,t)\} + \{R_\sigma(T,t)\}$$
$$+ \{R_r(T,t)\} \qquad (8.19)$$

For linear steady-state analysis equation 8.16 shows that the element conductance matrix has contributions from conduction and convection, and the heat load vector has contributions from internal heat generation, surface heating, and surface convection. For a linear steady-state analysis, element matrices and heat load vectors are constant and a linear solution of a set of simultaneous equations is required. For a linear transient analysis, equation 8.17 shows that element capacitance matrices are also required, element convection matrices and heat load vectors are time dependent, and a solution of the equations by a time-marching scheme is required. For a nonlinear steady-state analysis, equation 8.18 shows that element matrices and heat load vectors have contributions from radiation, and the matrices and vectors are temperature dependent, thus the equations are nonlinear and require solution by an iterative scheme. For the general nonlinear transient case,

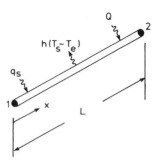

Figure 8.2 Rod heat transfer element.

equation 8.19 shows that element matrices and heat load vectors are both temperature and time dependent, and solution by a time-marching scheme is required. The details of each of these analyses appear in subsequent sections. However, before we consider solution methods, the next section presents typical element matrices employed in many of these analyses.

8.2.3 Element Equations

To illustrate application of the foregoing element equations we consider one- and two-dimensional conduction elements with specified heating and surface convection and develop element capacitance and conductance matrices and heat load vectors. Element thermal properties, internal heat generation, surface heating, and surface convection are assumed constant for an element. The contributions of radiation heat transfer are presented in Section 8.3, which discusses conduction with surface radiation.

One-Dimensional Rod Element A one-dimensional two-node rod element with conduction, internal heat generation, specified surface heating, and surface convection is shown in Figure 8.2. The rod has cross-sectional area A and perimeter p. In local coordinates the element interpolation functions are given by equations 5.11:

$$N_1(x) = L_1(x) = 1 - \frac{x}{L}$$

$$N_2(x) = L_2(x) = \frac{x}{L} \qquad (8.20a)$$

and by equation 8.7b

$$B_1 = \frac{\partial N_1}{\partial x} = -\frac{1}{L}$$

$$B_2 = \frac{\partial N_2}{\partial x} = \frac{1}{L} \qquad (8.20b)$$

Element matrices, equations 8.14a–8.14c, and heat load vectors, equations 8.15b–8.15d, are readily evaluated using $[k] = k$, $d\Omega = A\,dx$ and $d\Gamma = p\,dx$:

$$[C] = \int_0^L \rho c\{N\}\lfloor N\rfloor A\,dx = \frac{\rho cAL}{6}\begin{bmatrix} 2 & 1 \\ 1 & 2 \end{bmatrix} \quad \text{(consistent)} \quad (8.21a)$$

$$[K_c] = \int_0^L k\{B\}\lfloor B\rfloor A\,dx = \frac{kA}{L}\begin{bmatrix} 1 & -1 \\ -1 & 1 \end{bmatrix} \quad (8.21b)$$

$$[K_h] = \int_0^L h\{N\}\lfloor N\rfloor p\,dx = \frac{hpL}{6}\begin{bmatrix} 2 & 1 \\ 1 & 2 \end{bmatrix} \quad (8.21c)$$

$$\{R_Q\} = \int_0^L Q\{N\}A\,dx = \frac{QAL}{2}\begin{Bmatrix} 1 \\ 1 \end{Bmatrix} \quad (8.21d)$$

$$\{R_q\} = \int_0^L q_s\{N\}p\,dx = \frac{q_s pL}{2}\begin{Bmatrix} 1 \\ 1 \end{Bmatrix} \quad (8.21e)$$

$$\{R_h\} = \int_0^L hT_e\{N\}p\,dx = \frac{hT_e pL}{2}\begin{Bmatrix} 1 \\ 1 \end{Bmatrix} \quad (8.21f)$$

The rod element capacitance matrix, equation 8.21a, is known as a consistent capacitance matrix, because it is derived using the general finite element matrix equation, equation 8.14a, which is consistent with the other element matrix definitions given in equation 8.14. An alternative approach [5, 6] is to "lump" the capacitance at each node, thereby producing a diagonal capacitance matrix. A diagonal capacitance matrix is convenient in transient analysis because it permits the use of an explicit time integration algorithm. See Section 8.2.4 for a discussion of time-integration algorithms. The lumped capacitance matrix for a rod element is

$$[C] = \frac{\rho cAL}{2}\begin{bmatrix} 1 & 0 \\ 0 & 1 \end{bmatrix} \quad \text{(lumped)} \quad (8.21g)$$

which physically means that the element capacitance ρcAL is divided equally between the two nodes.

Triangular Element A three-node triangular element of thickness t with conduction, internal heat generation, specified surface heating, and surface convection is shown in Figure 8.3. The element may experience surface heating and convective heat exchanges on each edge; for simplicity these heat exchanges are shown only for a typical edge. In global coordinates the

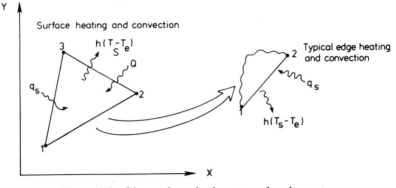

Figure 8.3 Plane triangular heat transfer element.

element interpolation functions for the triangle are given by equations 5.17–5.19:

$$N_i = L_i = \frac{a_i + b_i x + c_i y}{2\Delta}, \qquad i = 1, 2, 3 \qquad (8.22a)$$

Since

$$\frac{\partial N_i}{\partial x} = \frac{b_i}{2\Delta} \qquad \frac{\partial N_i}{\partial y} = \frac{c_i}{2\Delta}$$

equation 8.7b yields

$$[B] = \frac{1}{2\Delta} \begin{bmatrix} b_1 & b_2 & b_3 \\ c_1 & c_2 & c_3 \end{bmatrix} \qquad (8.22b)$$

where Δ is the area of the triangle. Note that $[B]$ is a matrix of constants. Element matrices, equations 8.14a–8.14c, and heat load vectors, equations 8.15b–8.15d, are readily evaluated either by using the integration formula for area coordinates, equation 5.22, or by direct integration. For example, the elements of the consistent capacitance matrix are computed from equation 8.14a as

$$C_{ij} = \int_\Delta \rho c N_i N_j t \, dA$$

which we evaluate by using equation 5.22 to yield

$$[C] = \frac{\rho c t \Delta}{12} \begin{bmatrix} 2 & 1 & 1 \\ 1 & 2 & 1 \\ 1 & 1 & 2 \end{bmatrix} \qquad \text{(consistent)} \qquad (8.23a)$$

Note that t is used in this context to denote the element thickness. The conduction matrix is computed from equation 8.14b as

$$[K_c] = \int_\Delta [B]^T [k][B]\, t\, dA$$

but since the elements of the temperature-gradient interpolation matrix $[B]$ given in equation 8.22b are constant, the integrand is constant, and we obtain directly

$$[K_c] = t\Delta [B]^T [k][B] \qquad (8.23b)$$

The surface convection matrix and the heat load vectors for internal heat generation, surface heating, and convection over the surface area Δ are derived similarly:

$$[K_h] = \int_\Delta h\{N\}\lfloor N\rfloor\, dA = \frac{h\Delta}{12}\begin{bmatrix} 2 & 1 & 1 \\ 1 & 2 & 1 \\ 1 & 1 & 2 \end{bmatrix} \qquad (8.23c)$$

$$\{R_Q\} = \int_\Delta Q\{N\}t\, dA = \frac{Qt\Delta}{3}\begin{Bmatrix} 1 \\ 1 \\ 1 \end{Bmatrix} \qquad (8.23d)$$

$$\{R_q\} = \int_\Delta q_s\{N\}\, dA = \frac{q_s\Delta}{3}\begin{Bmatrix} 1 \\ 1 \\ 1 \end{Bmatrix} \qquad (8.23e)$$

$$\{R_h\} = \int_\Delta hT_e\{N\}\, dA = \frac{hT_e\Delta}{3}\begin{Bmatrix} 1 \\ 1 \\ 1 \end{Bmatrix} \qquad (8.23f)$$

The lumped capacitance matrix for a triangular element is

$$[C] = \frac{\rho c t\Delta}{3}\begin{bmatrix} 1 & 0 & 0 \\ 0 & 1 & 0 \\ 0 & 0 & 1 \end{bmatrix} \quad \text{(lumped)} \qquad (8.23g)$$

where the capacitance is equally divided between the three nodes. Note that this result can be obtained from equation 8.23a by adding the elements of a row, placing the result on the diagonal, and zeroing the off-diagonal elements.

If the edge of a triangle coincides with the boundary of a region with surface heat transfer, an additional conductance matrix and heat load vector are required. The straight edge of a typical triangle resembles the foregoing rod element, hence we can use the corresponding matrices for the rod element, provided that we employ the proper surface area. The surface area of the rod element is pL, and the surface of a typical triangle edge is tl_{12}, where l_{12} is the length of the edge. Using this substitution, we obtain from equations 8.21c, 8.21e, and 8.21f

$$[K_h] = \frac{htl_{12}}{6} \begin{bmatrix} 2 & 1 \\ 1 & 2 \end{bmatrix} \tag{8.24a}$$

$$\{R_q\} = \frac{q_s tl_{12}}{2} \begin{Bmatrix} 1 \\ 1 \end{Bmatrix} \tag{8.24b}$$

$$\{R_h\} = \frac{hT_e tl_{12}}{2} \begin{Bmatrix} 1 \\ 1 \end{Bmatrix} \tag{8.24c}$$

An alternative, more general procedure for developing these equations evaluates the general element integrals for a typical edge by using a local coordinate s measured along the element edge. This procedure is presented in Section 7.2.5, which interested readers should consult for further details.

Axisymmetric Triangular Element   Heat transfer in a three-dimensional domain with geometric symmetry about the z axis is formulated conveniently in cylindrical coordinates r, θ, z. If, in addition, all heat transfer and thermal properties are independent of θ, then the temperature is a function of only r and z, and the domain can be represented by axisymmetric ring elements and analyzed as a two-dimensional problem. A three-node axisymmetric triangular element with conduction, internal heat generation, specified edge heating, and edge convection is shown in Figure 8.4. Note that the element lies in the $r-z$ plane, and edge heating and convection are shown on a typical edge. The derivation of element matrices is similar to the plane triangular element; the principal difference is that integrations are carried out in the $r-z$ plane and in cylindrical coordinates $d\Omega = 2\pi r \, dr \, dz$. The derivation of some element

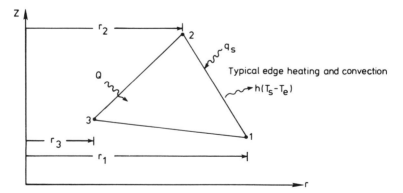

Figure 8.4   Axisymmetric triangular heat transfer element.

matrices for an axisymmetric element for field problems appears in Section 7.2.5; hence, for brevity, details are omitted here. The element matrices are

$$[C] = \frac{\rho c \Delta}{60} \begin{bmatrix} 6r_1 + 2r_2 + 2r_3 & 2r_1 + 2r_2 + r_3 & 2r_1 + r_2 + 2r_3 \\ & 2r_1 + 6r_2 + 2r_3 & r_1 + 2r_2 + 2r_3 \\ \text{Symmetric} & & 2r_1 + 2r_2 + 6r_3 \end{bmatrix}$$

(8.25a)

$$[K_c] = \frac{(r_1 + r_2 + r_3)\Delta}{3}[B]^T[k][B]$$

(8.25b)

$$\{R_Q\} = \frac{Q\Delta}{12} \begin{bmatrix} 2 & 1 & 1 \\ 1 & 2 & 1 \\ 1 & 1 & 2 \end{bmatrix} \begin{Bmatrix} r_1 \\ r_2 \\ r_3 \end{Bmatrix}$$

(8.25c)

where r_i, $i = 1, 2, 3$, are the radial nodal coordinates, and Δ is the area of the triangle; Δ and the elements of $[B]$ are computed from nodal coordinates using equations 5.18 and 5.19, respectively. The contributions from convection heat transfer and surface heating along a typical element edge 1–2 are

$$[K_h] = \frac{hl_{12}}{12} \begin{bmatrix} 3r_1 + r_2 & r_1 + r_2 \\ r_1 + r_2 & r_1 + 3r_2 \end{bmatrix}$$

(8.25d)

$$\{R_q\} = \frac{q_s l_{12}}{6} \begin{bmatrix} 2 & 1 \\ 1 & 2 \end{bmatrix} \begin{Bmatrix} r_1 \\ r_2 \end{Bmatrix}$$

(8.25e)

$$\{R_h\} = \frac{hT_e l_{12}}{6} \begin{bmatrix} 2 & 1 \\ 1 & 2 \end{bmatrix} \begin{Bmatrix} r_1 \\ r_2 \end{Bmatrix}$$

(8.25f)

where l_{12} is the length of the edge. A factor of 2π that appears in each of equations 8.25 has been canceled, since it appears on both sides of equation 8.13 in the final solution.

Isoparametric Elements   The foregoing rod, plane, and axisymmetric triangular elements with two and three nodes, respectively, are based on simple temperature interpolation functions that permit explicit definitions of the element matrices for linear analysis, equations 8.23–8.25. The elements were among the first used for finite element thermal analysis because of their relative simplicity and ease of programming. However, the elements have relatively low accuracy because of the lower-order interpolation functions, and in modern programs they are supplemented and/or replaced by higher-order isoparametric elements (see Sections 5.9–5.10) that have higher accuracy and the capability to approximate curved boundaries closely. The

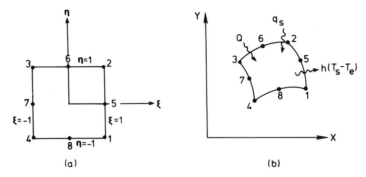

Figure 8.5   Isoparametric heat transfer element.

disadvantage of isoparametric elements is that element matrices can no longer be derived explicitly; instead, elements matrices must be computed by numerical integration. However, these integration procedures have become standard and require only a small penalty in computer time compared to element matrices computed from explicit element equations.

To illustrate the isoparametric element formulation for heat conduction we consider the serendipity element family (Section 5.9) in a two-dimensional solution domain. Serendipity elements may have either 4, 8, or 12 nodes; the popular eight-node quadratic element is shown in Figure 8.5. Figure 8.5a shows the "parent" element in the $\xi-\eta$ plane and a typical quadratic element in the $x-y$ plane. Although only the quadratic element is shown, the element formulation is presented for an element with r nodes. The formulation is illustrated for two-dimensional elements, but the extension to three-dimensional elements follows directly.

Using the isoparametric approach, the nodes in the $\xi-\eta$ plane are mapped onto corresponding nodes in the $x-y$ plane by using the same interpolation functions that are used to interpolate the temperature. From equation 5.60,

$$x(\xi, \eta) = \sum_{i=1}^{r} N_i(\xi, \eta) x_i \tag{8.26a}$$

$$y(\xi, \eta) = \sum_{i=1}^{r} N_i(\xi, \eta) y_i \tag{8.26b}$$

$$T(\xi, \eta) = \sum_{i=1}^{r} N_i(\xi, \eta) T_i \tag{8.26c}$$

where the interpolation functions $N_i(\xi, \eta)$ appear in equations 5.47–5.49.
Element matrices involving volume integrals are evaluated by writing

$$d\Omega = t \, dx \, dy = t |J(\xi, \eta)| d\xi \, d\eta$$

where we use equation 5.67 to express the element of area in the ξ–η plane; $|J|$ is the determinant of the Jacobian, equation 5.64:

$$[J(\xi, \eta)] = \begin{bmatrix} \dfrac{\partial x}{\partial \xi} & \dfrac{\partial y}{\partial \xi} \\[2mm] \dfrac{\partial x}{\partial \eta} & \dfrac{\partial y}{\partial \eta} \end{bmatrix}$$

which can be evaluated using equations 8.26a and 8.26b. This transformation permits the element integrals to be evaluated by integration over the unit square of the parent element (Figure 8.5a). For example, the element capacitance matrix is

$$[C] = \int_{-1}^{1}\int_{-1}^{1} \rho ct\{N(\xi, \eta)\}\lfloor N(\xi, \eta)\rfloor |J(\xi, \eta)|\, d\xi\, d\eta$$

which is normally evaluated by numerical integration. A popular method of numerical integration is Gauss–Legendre quadrature, which appears in Chapter 10. Using this approach, the capacitance matrix is

$$[C] = \sum_{i=1}^{NG}\sum_{j=1}^{NG} W_i W_j \rho ct\{N(\xi_i, \eta_j)\}\lfloor N(\xi_i, \eta_j)\rfloor |J(\xi_i, \eta_j)| \qquad (8.27)$$

where W_i and W_j are Gauss weights, ξ_i and η_j are the coordinates of the Gauss points, and NG is the number of Gauss points in each integration direction. Gauss weights and coordinates appear in Table 10.1.

The element conduction matrix is computed in a similar manner, except the temperature-gradient interpolation matrix $[B]$, which appears in the integrand of equation 8.14b, is evaluated in terms of ξ and η. From equation 8.7b

$$[B(x, y)] = \begin{bmatrix} \dfrac{\partial N_1}{\partial x} & \dfrac{\partial N_2}{\partial x} & \cdots & \dfrac{\partial N_r}{\partial x} \\[2mm] \dfrac{\partial N_1}{\partial y} & \dfrac{\partial N_2}{\partial y} & \cdots & \dfrac{\partial N_r}{\partial y} \end{bmatrix}$$

but from equation 5.67

$$\left\{ \begin{array}{c} \dfrac{\partial N_i}{\partial x} \\[2mm] \dfrac{\partial N_i}{\partial y} \end{array} \right\} = [J(\xi, \eta)]^{-1} \left\{ \begin{array}{c} \dfrac{\partial N_i}{\partial \xi} \\[2mm] \dfrac{\partial N_i}{\partial \eta} \end{array} \right\}$$

which yields

$$[B(\xi,\eta)] = [J(\xi,\eta)]^{-1} \begin{bmatrix} \dfrac{\partial N_1}{\partial \xi} & \dfrac{\partial N_2}{\partial \xi} & \cdots & \dfrac{\partial N_r}{\partial \xi} \\ \dfrac{\partial N_1}{\partial \eta} & \dfrac{\partial N_2}{\partial \eta} & \cdots & \dfrac{\partial N_r}{\partial \eta} \end{bmatrix} \qquad (8.28)$$

The element conduction matrix, equation 8.14b, is

$$[K_c] = \int_{-1}^{1} \int_{-1}^{1} t [B(\xi,\eta)]^T [k][B(\xi,\eta)] |J(\xi,\eta)| d\xi \, d\eta$$

which is evaluated by Gauss–Legendre quadrature

$$[K_c] = \sum_{i=1}^{NG} \sum_{j=1}^{NG} W_i W_j t [B(\xi_i,\eta_j)]^T [k][B(\xi_i,\eta_j)] |J(\xi_i,\eta_j)| \quad (8.29)$$

In the capacitance and conduction matrices the element thickness may be constant or vary over the element area, that is, $t(\xi,\eta)$; however, in typical elements t is constant and is brought outside of the summations in equations 8.27 and 8.29.

The remaining element matrices are evaluated similarly by Gauss–Legendre quadrature in the $\xi-\eta$ plane. Integrals along an element edge are evaluated using a local coordinate s; the distance $d\Gamma$ along an element edge is expressed in terms of ds, where $ds = (dx^2 + dy^2)^{1/2}$. Using equations 8.26a and 8.26b, the integration along an element edge is carried out along the corresponding edge of the parent element by Gauss–Legendre quadrature. The details are left as an exercise and appear as a problem at the end of the chapter.

8.2.4 Linear Steady-State and Transient Solutions

Linear heat transfer problems consist of solving equation 8.16 for steady-state temperatures and solving equation 8.17 for transient temperatures, where each equation is subject to appropriate initial and boundary conditions. Solutions are similar to the equilibrium and propagation field problems discussed in Chapter 7. The principal differences are that surface convection introduces conductance matrices and heat load vectors for boundary elements, and general anisotropic thermal conductivity properties are permitted in the element formulations in this chapter. The numerical solution of linear steady-state heat conduction problems poses no special difficulties, since the symmetric system of algebraic equations is solved conveniently using standard methods such as Gauss elimination (see Chapter 10 for further details of solution methods for linear algebraic equations). The numerical solution of transient heat conduction problems requires solving a set of first-order simultaneous ordinary differential equations. In Section 7.5.1 we describe two fundamental approaches popular in finite element analysis: modal superposition and numerical integration (time-marching) procedures. The numerical integration procedures are the most popular approach and rely on recursion formulas that permit the solution to be "marched out" in time, starting from an initial temperature distribution. A general family of algorithms is discussed in detail in Chapter 7; herein we summarize the two common approaches for heat transfer analysis: the explicit forward difference scheme and implicit one-parameter "θ" schemes. The explicit forward difference (Euler) scheme requires a lumped capacitance matrix and computes temperatures at time t_{n+1} from a set of *uncoupled* algebraic equations (see equations 7.79–7.80):

$$[^{\frown}C_{\lrcorner}]\{T\}_{n+1} = \{\overline{R}\}_{n+1} \qquad (8.30a)$$

where

$$\{\overline{R}\}_{n+1} = [[^{\frown}C_{\lrcorner}] - \Delta t[K]]\{T\}_n + \Delta t\{R\}_n \qquad (8.30b)$$

The explicit scheme offers significant computational savings, since nodal temperatures are computed directly from the uncoupled equations, equation 8.30a, without the expense of solving simultaneous algebraic equations. The disadvantage of the explicit scheme is that the algorithm is conditionally stable, and the time step Δt must be selected so as to be smaller than a critical time step Δt_{cr} given by

$$\Delta t_{cr} = \frac{2}{\lambda_m} \qquad (7.86)$$

where λ_m is the maximum system eigenvalue. See Chapter 7 for further discussion of the system matrix eigenvalue problem.

The implicit θ numerical integration algorithm permits either lumped or consistent capacitance matrices and computes temperatures at time t_{n+1} from a set of *coupled* algebraic equations:

$$[\overline{K}]\{T\}_{n+1} = \{\overline{R}\}_{n+1} \qquad (8.31a)$$

where

$$[\overline{K}] = \theta[K] + \frac{1}{\Delta t}[C] \qquad (8.31b)$$

$$\{\overline{R}\}_{n+1} = \left[-(1-\theta)[K] + \frac{1}{\Delta t}[C]\right]\{T\}_n + (1-\theta)\{R\}_n + \theta\{R\}_{n+1}$$

$$(8.31c)$$

The parameter θ may be chosen to give different algorithms. If $\theta = \frac{1}{2}$, the algorithm is the Crank–Nicolson method; if $\theta = \frac{2}{3}$, the algorithm is the Galerkin method; and if $\theta = 1$, the algorithm is the backward difference method. These algorithms are unconditionally stable, but too large a time step may introduce spurious oscillations in the computed response, hence time-step selection is an important practical consideration for both implicit and explicit algorithms.

To gain insight into the parameters that influence time-step size, let us consider a conduction problem modeled with a uniform mesh of one-dimensional elements. A theorem due to Irons and Ahmad [7] states that the maximum system eigenvalue is always less than the maximum element eigenvalue. Thus we can obtain a conservative estimate of the maximum system eigenvalue by considering a single element. For an element, we write equation 7.71 as

$$\left[[K]^{(e)} - \lambda [C]^{(e)} \right] \{\Phi\}^{(e)} = \{0\} \qquad (7.71)$$

Next we substitute the element matrices and set the determinant of the square matrix to zero and obtain the characteristic equation. The roots of this equation are the element eigenvalues. For a rod element with a lumped capacitance matrix, we use equations 8.21a and 8.21b to write the determinant:

$$\left\| \begin{bmatrix} 1 & -1 \\ -1 & 1 \end{bmatrix} - \lambda \frac{\rho c L^2}{k} \begin{bmatrix} \frac{1}{2} & 0 \\ 0 & \frac{1}{2} \end{bmatrix} \right\| = 0$$

Expanding the determinant produces a quadratic equation for λ which has the roots

$$\lambda_1 = 0 \qquad \lambda_2 = \frac{4k}{\rho c L^2} \qquad \text{(lumped capacitance)}$$

Proceeding in a similar way, but using a consistent capacitance matrix we find

$$\lambda_1 = 0 \qquad \lambda_2 = \frac{12k}{\rho c L^2} \qquad \text{(consistent capacitance)}$$

We may use these estimates of the maximum system eigenvalue to obtain the critical time step for the explicit Euler algorithm and to estimate desirable time steps for the implicit Crank–Nicolson algorithm.

Euler   The Euler algorithm uses a lumped capacitance matrix and

$$\Delta t_{cr} = \frac{2}{\lambda_m}$$

Substituting λ_m from above we then obtain

$$\Delta t_{cr} = \frac{1}{2}\frac{\rho c}{k}L^2$$

which can be written as

$$\Delta t_{cr} = \frac{1}{2}\frac{L^2}{\alpha} \tag{8.32}$$

where α is the thermal diffusivity, $\alpha = k/\rho c$. This result shows the respective roles of the thermal diffusivity and the mesh size in determining the critical time step. The mesh size is particularly important since the element length appears squared. Thus if the element length is halved, the critical time step is reduced by a factor of four.

Crank–Nicolson   With reference to Figure 7.8 we see that although the algorithm is unconditionally stable, oscillations will occur for $\lambda \Delta t > 2$. Thus if we use the previously determined eigenvalues we can estimate the time-step size for which we may expect oscillations to commence. Substituting the eigenvalues yields,

$$\Delta t_{osc} = \frac{1}{2}\frac{L^2}{\alpha} \qquad \text{(lumped)} \tag{8.33a}$$

$$\Delta t_{osc} = \frac{1}{6}\frac{L^2}{\alpha} \qquad \text{(consistent)} \tag{8.33b}$$

To avoid oscillations in the computed response, we should select our time step smaller than these values. *A significant conclusion is that the consistent capacitance formulation requires a time step that is smaller than the time step required for a lumped capacitance approach.*

The application of Irons' theorem to the estimation of time steps for one-dimensional conduction elements shows that the important parameters are the thermal diffusivity and the element length. Equations 8.32 and 8.33 may be used to estimate practical time steps for computations with the Euler and Crank–Nicolson algorithms. These equations may also be used as approximations for two-dimensional elements if the length L is replaced by a characteristic element dimension. For instance, for a triangular element, a characteristic length could be the distance from the element centroid to the most distant element node. We also note that computational experience on graded meshes indicates that Irons' theorem may be overly conservative. Reference 8 presents an alternative method for estimating time-step sizes for a heat conduction problem and discusses the problem in further detail.

The choice of an explicit algorithm in heat conduction requires the use of a lumped capacitance matrix, whereas the choice of an implicit algorithm allows either a lumped or consistent capacitance matrix. The question of whether to use a lumped or consistent matrix is intriguing, and there are opposing views on the subject. The relative gain in computational speed and computer storage offered by the lumped capacitance method has motivated some analysts, for example, Wilson et al. [5], to use the lumped approach exclusively. This view is supported by computational experience that indicates an insignificant loss in accuracy by the lumped approach. Furthermore, additional justification for the lumped approach is that virtually all finite difference methods use it. Another argument for use of the lumped approach is that the consistent capacitance matrix approach may predict temperature distributions with unrealistic oscillations [9]. This anomalous behavior is typically encountered for problems with sharp thermal transients; for example, temperatures near suddenly cooled boundaries show an initial increase. The same problem when solved with a lumped capacitance matrix shows temperatures near the cooled boundary decreasing, as intuitively expected.

Yet there is merit to arguments in support of the consistent capacitance approach [10, 11]. First, there is an arbitrariness in the methods for "lumping" the capacitance matrix. Two common methods are to sum the coefficients of the rows of the consistent capacitance matrix or to compute the diagonal terms of the consistent capacitance matrix and scale these terms to preserve the total element capacitance. For simple elements such as the rod and triangle we presented in Section 8.2.3, these approaches lead to "intuitively obvious" lumped matrices, which often predict transient temperatures with negligible error. However, for higher-order isoparametric elements the choice of a method to lump the capacitance is not clear, and the method of lumping can have a significant effect on the accuracy of computed temperatures [12]. Moreover, the consistent capacitance matrix predicts more accurate temperatures than the lumped capacitance matrix approach, *provided* that an adequate mesh is employed to represent the spatial temperature distribution. The importance of mesh refinement is emphasized in reference 11, where it is argued that the unrealistic temperature oscillations (wiggles) are a positive feature of the consistent approach, since the wiggles are an indication that there is a problem (often an inadequate mesh) that is causing these wiggles. It is argued that suppression of the wiggles by lumping the capacitance matrix is inappropriate because it can lead to an intuitively realistic but actually inaccurate result.

Some of these points are illustrated by comparing lumped and consistent finite element solutions with an exact solution for one-dimensional transient conduction [11]. The problem is stated by

$$\frac{\partial^2 T}{\partial x^2} = \frac{\partial T}{\partial t}$$

$$T(x,0) = 0, \qquad 0 \le x \le 1$$

$$T(0,t) = 1, \qquad t > 0$$

$$T(1,t) = 0, \qquad t \ge 0$$

Comparative temperatures at two times in the response are shown in Figure 8.6; temperatures were computed with equally spaced linear elements. At the early time, $t = 0.004$, the consistent capacitance solution displays spurious wiggles and a negative temperature, while the lumped capacitance results are smooth and "reasonable." At the later time, $t = 0.04$, all temperatures from the consistent capacitance are positive and remain positive thereafter. In reference 11 the authors argue that the wiggles mean that "for this problem

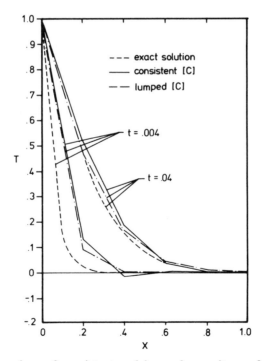

Figure 8.6 Comparison of consistent and lumped capacitance finite element solutions for transient heat conduction.

and this mesh (or any mesh) there is a minimum 'time of believability'. For times less than this, the discretized solution is too inaccurate to be useful."

Clearly, there are pros and cons of consistent versus lumped capacitance approaches, and each has its merits and demerits. The lumped capacitance approach has clear advantages in computational efficiency, but with some loss of accuracy. For highly complex realistic problems the loss of accuracy is likely to be insignificant relative to other uncertainties in the thermal model (e.g., material properties), and the gain in computation efficiency and attendent reduction in computer costs will justify the use of this approach. For more regular solution domains or in problems where high accuracy of the temperature distribution is required (e.g., in some thermal stress problems), the use of the more accurate though more expensive consistent approach is justified.

8.2.5   Nonlinear Steady-State Solutions

Equation 8.18 describes nonlinear steady-state heat transfer problems. The three commonly encountered sources of nonlinearities are temperature-dependent properties, radiation heat transfer, and temperature-dependent heating. The thermal conductivity and the convection coefficient may be temperature dependent and affect the solution through the conduction and convection matrices, equations 8.14b and 8.14c, respectively, and the convection heat load vector, equation 8.15d. Radiation heat transfer is inherently nonlinear and affects the solution through the radiation vector, equation 8.15e, and the incident heat load vector, equation 8.15f. The internal heat generation and surface heating rates may also be temperature dependent and affect the solution through the heat load vectors, equations 8.15b and 8.15c.

Newton–Raphson Formulation   The most popular method for solving the nonlinear equations is the Newton–Raphson iteration method, which we describe in Chapter 10. The algorithm is

$$[J]^m \{\Delta T\}^{m+1} = -\{F\}^m \tag{8.34a}$$

$$\{T\}^{m+1} = \{T\}^m + \{\Delta T\}^{m+1} \tag{8.34b}$$

where the superscript m denotes the mth iteration, $[J]^m$ is the Jacobian matrix, and $\{\Delta T\}^{m+1}$ is a vector of nodal temperature increments due to an unbalanced or residual load vector $\{F\}^m$ in the matrix equations at the mth iteration. In its most general form the Jacobian and unbalanced load vector are recomputed at each iteration, but a modified Newton–Raphson algorithm is also used in which the Jacobian is computed only once and thereafter held constant. Convergence of the solution is usually specified by requiring the temperature increments (or a convenient norm of these increments) to meet a specified convergence criterion. The Newton–Raphson iteration algorithm usually converges faster with fewer iterations, but it may be computationally slower since the Jacobian matrix must be recomputed at each iteration. Also, there are occasions when the Newton–Raphson iteration method will converge but the modified Newton–Raphson algorithm will fail to converge.

To apply the Newton–Raphson method to nonlinear heat transfer problems we derive general forms for the Jacobian and unbalanced load vector. Equation 8.18 now becomes

$$\{F\} = [K(T)]\{T\} - \{R(T)\} \tag{8.35a}$$

where

$$[K(T)] = [K_c] + [K_h] \tag{8.35b}$$

and

$$\{R(T)\} = \{R_Q(T)\} + \{R_q(T)\} + \{R_h(T)\} + \{R_\sigma(T)\} + \{R_r(T)\} \tag{8.35c}$$

For an arbitrary set of nodal temperatures the vector $\{F\}$ in equation 8.35a represents an unbalance in nodal heat loads, and we write a typical unbalanced equation as

$$F_i = \sum_{j=1}^{r} K_{ij}T_j - R_i$$

which permits computation of the Jacobian by the definition $J_{ij} = \partial F_i / \partial T_j$. Recalling that K_{ij} is a function of T_j, we differentiate the term within the summation as a product and then write the result as

$$[J] = [K] + [\Delta K] + [\Delta R] \tag{8.36a}$$

$$\Delta K_{ij} = \sum_{l=1}^{r} \frac{\partial K_{il}}{\partial T_j} T_l \tag{8.36b}$$

$$\Delta R_{ij} = -\frac{\partial R_i}{\partial T_j} \tag{8.36c}$$

Equation 8.36 shows that the Jacobian matrix has three parts: the element conductance matrix $[K]$, equation 8.35b; an increment in this matrix $[\Delta K]$, equation 8.36b; and a contribution from temperature-dependent heat vectors, equation 8.36c. Observe that the Jacobian matrix is asymmetric due to $[\Delta K]$ and $[\Delta R]$, and consequently the iterative solution requires an asymmetric equation solver. If we neglect $[\Delta K]$ and $[\Delta R]$ and approximate the

Jacobian by the conductance matrix, we retain matrix symmetry, but convergence of the iterative solution is slower.

Further evaluation of the Jacobian requires additional knowledge about the nonlinearities and identification of element types. In the following example we illustrate a nonlinear solution resulting from temperature -dependent properties. In Section 8.3 we consider conduction with radiation boundary conditions. An example of temperature-dependent internal heat generation appears in reference 13.

Example: Temperature-Dependent Properties   In practical heat transfer problems [14] material thermal properties are typically available as experimental curves or tabular data; therefore in finite element analysis a piecewise linear variation of properties is normally used. There is, however, some latitude in modeling the spatial variation of thermal properties within an element. In simple lower-order elements, such as the rod and triangular elements, which appear in Section 8.2.3, we may assume that thermal properties are constant within an element. In higher-order isoparametric elements thermal properties should be permitted to vary within an element. Thornton and Wieting [14] use the former approach to develop a solution procedure for several temperature-dependent parameters, and Lyness et al. [15] use the latter approach for nonlinear solutions to the quasiharmonic equation with 12-node isoparametric elements.

Following reference 14, we assume the following: (1) the heat load vectors are constant; (2) the thermal properties are constant within an element; (3) element thermal properties depend only on the average element temperature T_a, where

$$T_a = \frac{1}{r} \sum_{i=1}^{r} T_i$$

and (4) element conductance matrices can be expressed as

$$[K] = \text{TP}[\bar{K}]$$

where $[\bar{K}]$ is the element conductance matrix for a unit thermal parameter and TP is the thermal parameter.

An element Jacobian matrix is then given by equation 8.36a, with $[\Delta R]$ equal to zero because of assumption (1). Assumptions (2) and (3) permit a simple computation of $[\Delta K]$ by expressing the differentiation in equation 8.36b by the chain rule of differentiation. Thus using the preceding definition of T_a:

$$\Delta K_{ij} = \sum_{l=1}^{r} \frac{\partial K_{il}}{\partial T_j} T_l = \sum_{l=1}^{r} \frac{dK_{il}}{dT_a} \frac{dT_a}{dT_j} T_l = \frac{1}{r} \sum_{l=1}^{r} \frac{dK_{il}}{dT_a} T_l$$

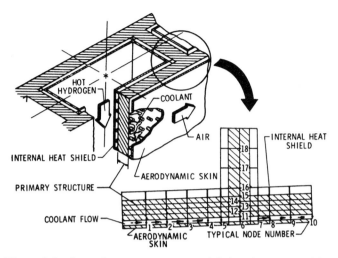

Figure 8.7   Scramjet strut segment and finite element model.

Next, using the definition of a unit conductance matrix, we can write this last result as

$$\Delta K_{ij} = \frac{1}{r} \frac{d(\text{TP})}{dT_a} \sum_{l=1}^{r} \overline{K}_{il} T_l$$

The derivative $d(\text{TP})/dT_a$ represents the slope of the thermal parameter curve evaluated at the average element temperature. Note that ΔK_{ij} is a matrix with equal columns, since the right-hand side is independent of j. Combining the equations for K_{ij} and ΔK_{ij} gives an explicit definition for the Jacobian:

$$J_{ij} = \text{TP}\overline{K}_{ij} + \frac{1}{r} \frac{d(\text{TP})}{dT_a} \sum_{l=1}^{r} \overline{K}_{il} T_l \tag{8.37}$$

This form of the Jacobian is computationally convenient, since it is valid for all element types, and element unit conductance matrices are computed only once.

In reference 14 the computational procedure based on the Jacobian given in equation 8.37 is applied to a segment of a strut for a hydrogen-cooled supersonic combustion ramjet (scramjet) engine (Figure 8.7). Scramjet engines are under development for the National Aero-Space Plane (X-30). The strut is heated by hot hydrogen flowing in internal manifolds and external air flow, and the structure is cooled by cold hydrogen flowing between the structure and aerodynamic skin. The finite element model utilizes rod and quadrilateral conduction elements to represent the heat shield, primary

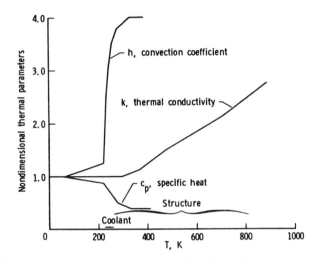

Figure 8.8   Temperature variation of hypothetical thermal parameters.

structure, and aerodynamic skin, and special heat-exchanger elements represent the coolant passage. The temperatures on the boundary of the segment are specified. The temperature variation of three hypothetical thermal parameters are shown in Figure 8.8; hypothetical parameters are used to give a significant variation of the thermal parameters and a severe test of the algorithm. Element Jacobian matrices and heat load vectors are computed at each iteration and assembled to yield the corresponding system matrices.

The calculated temperatures at selected nodes (Figure 8.7) appear in Table 8.1. The temperatures converged in four iterations, even though there was more than a 300% variation in the convection coefficient. In other applications with more realistic thermal parameters convergence was even more rapid. This example illustrates the typically good convergence of the Newton–Raphson method for temperature-dependent properties. The modified Newton–Raphson method with a constant Jacobian also works well in these applications but requires a few more iterations.

8.2.6 Nonlinear Transient Solutions

Solution Algorithms   Equation 8.19 describes nonlinear transient heat transfer problems. As in steady-state solutions, the commonly encountered sources of nonlinearities are temperature-dependent properties (including the specific heat for transient solutions), radiation heat transfer, and temperature-dependent heating. Integration techniques for transient nonlinear solutions [16–20] are typically a combination of the methods for linear transient solutions and steady-state nonlinear solutions. The transient solution of the

TABLE 8.1. Finite Element-Predicted Temperatures

Nodes	Iteration			
	0	1	2	3, 4
Coolant Temperatures (K)				
1	223.7	225.2	225.2	225.6
2	225.3	228.6	228.9	229.0
3	226.9	232.3	233.3	233.4
4	228.7	236.3	237.6	237.8
5	229.9	239.3	241.2	241.6
6	231.4	244.1	244.8	247.2
7	233.4	249.2	252.7	253.0
8	234.4	251.7	255.0	255.4
9	236.1	256.0	259.7	259.9
10	237.9	260.0	262.7	263.0
Primary Structure Temperatures (K)				
11	264.7	262.9	264.2	264.3
12	293.2	310.9	312.1	312.2
13	324.4	360.6	360.8	360.9
14	360.7	414.5	412.2	412.2
15	405.9	469.8	466.3	466.5
16	480.7	546.9	544.4	544.4
17	639.4	683.9	688.3	683.3
18	721.1	741.7	741.1	741.1

nonlinear ordinary differential equations is computed by a numerical integration method with iterations at each time step to correct for nonlinearities. Explicit or implicit one-parameter θ schemes are often used as the time integration method, and Newton–Raphson or modified Newton–Raphson methods are used for the iterations. We shall describe the extension of the θ scheme for nonlinear transient problems and illustrate typical algorithm performance with a one-degree-of-freedom example.

First we write the system of governing equations as

$$[C(T)]\{\dot{T}(t)\} + [K(T,t)]\{T(t)\} = \{R(T,t)\}$$

where the components of the conductance matrix and heat load vector appear in equation 8.19. As in Chapter 7, let t_n denote a typical time in the response so that $t_{n+1} = t_n + \Delta t$, where $n = 0, 1, 2, \ldots, N$. A general family of algorithms results by introducing a parameter θ such that $t_\theta = t_n + \theta \Delta t$, where $0 \leq \theta \leq 1$. We evaluate the system of equations at time t_θ,

$$[C(T_\theta)]\{\dot{T}\}_\theta + [K(T_\theta, t_\theta)]\{T\}_\theta = \{R(T_\theta, t_\theta)\}$$

where the subscript θ indicates the temperature vector $\{T(t_\theta)\}$ at time t_θ and introduce the approximations

$$\{\dot{T}\}_\theta = \frac{\{T\}_{n+1} - \{T\}_n}{\Delta t}$$

$$\{T\}_\theta = (1 - \theta)\{T\}_n + \theta\{T\}_{n+1}$$

Substituting the expressions for $\{\dot{T}\}_\theta$ and $\{T\}_\theta$ gives

$$\left[\theta[K(T_\theta, t_\theta)] + \frac{1}{\Delta t}[C(T_\theta)]\right]\{T\}_{n+1}$$

$$= \left[-(1 - \theta)[K(T_\theta, t_\theta)] + \frac{1}{\Delta t}[C(T_\theta)]\right]\{T_n\}$$

$$+ \{R(T_\theta, t_\theta)\} \tag{8.38}$$

where $\{T\}_{n+1}$ and $\{T\}_\theta$ are unknowns and $\{T\}_n$ is known from the previous time step. If $\theta = 0$ and we use lumped capacitance matrices, the algorithm is explicit and reduces to a set of uncoupled algebraic equations similar to the explicit linear solution algorithm, equation 8.30. For $\theta > 0$, the algorithm is implicit and requires solution of a set of coupled algebraic equations similar to the implicit linear solution algorithm, equation 8.31. For $\theta > 0$, we must solve equation 8.38 iteratively, since the coefficient matrices $[K(T_\theta, t_\theta)]$, $[C(T_\theta)]$, and the heat load vector $\{R(T_\theta, t_\theta)\}$ are functions of $\{T\}_\theta$. The Newton–Raphson iteration method, Section 8.2.5, is often used to solve the nonlinear equations at each time step.

Hughes [16] shows the algorithm to be unconditionally stable for $\theta \geq \frac{1}{2}$ as in the corresponding linear algorithm. For $\theta < \frac{1}{2}$ the algorithm is only conditionally stable, and the time step must be chosen smaller than a critical time step

$$\Delta t_{\text{cr}} = \frac{2}{1 - 2\theta} \frac{1}{\lambda_m} \tag{8.39}$$

where λ_m is the largest eigenvalue of the current eigenvalue problem. The explicit and implicit algorithms have the same trade-offs as occur for linear transient solutions. The explicit algorithm requires less computational effort, but it is conditionally stable; the implicit algorithm requires much greater computational effort, but it is unconditionally stable. The nonlinear implicit algorithm requires even greater computational effort than in linear implicit solutions because of the need for iterations at each time step. Thus the selection of a transient solution algorithm for a nonlinear thermal problem is even more difficult than in linear solutions. For further discussion see references 16–20 and references 31–34 in Chapter 7.

Example: A Model Equation   Equation 8.38 represents a family of algorithms for $0 \leq \theta \leq 1$, and as in the corresponding linear algorithm, the performance of the algorithm depends upon the value of θ chosen. For example, all of the schemes are first-order accurate, except for $\theta = \frac{1}{2}$ in which the method is second-order accurate (see Section 7.5.5). The terms *first-order* and *second-order accurate* refer to the truncation errors in the finite difference approximations introduced by replacing the original equations 8.36 by the approximate equations 8.38. First-order accuracy means that the truncation error is proportional to the first power of Δt, and second-order accuracy means that the error is proportional to the second power of Δt.

To evaluate the relative performance of the algorithms for different values of θ, Hogge [17] considers a one-degree-of-freedom model equation for which there exists an exact solution. The problem takes the form

$$\dot{T}(t) + [\lambda_0 + \lambda_1 T(t)]T(t) = 0 \qquad (8.40a)$$

$$T(0) = T_0 \qquad (8.40b)$$

corresponding to equation 8.36 for a one-degree-of-freedom homogeneous situation. A heat transfer problem represented by the model equation is the cooling of a small metal casting or billet in a quenching bath after its removal from a hot furnace. The coefficient $(\lambda_0 + \lambda_1 T)$ of $T(t)$ may be regarded as a temperature-dependent convection coefficient. The exact solution $T_E(t)$ to this problem is

$$T_E(t) = \frac{T_0 \lambda_0 e^{-\lambda_0 t}}{\lambda_0 + \lambda_1 T_0 (1 - e^{-\lambda_0 t})} \qquad (8.41)$$

Equation 8.38 yields an approximate solution for the model equation if we take $[K] = \lambda_0 + \lambda_1 T_\theta$, $[C] = 1$, and $\{R\} = 0$. Thus

$$(1 + \theta \lambda_\theta \, \Delta t) T_{n+1} = [1 - (1 - \theta)\lambda_\theta \, \Delta t] T_n \qquad (8.42a)$$

where

$$\lambda_\theta = \lambda_0 + \lambda_1 T_\theta \qquad (8.42b)$$

$$T_\theta = (1 - \theta) T_n + \theta T_{n+1} \qquad (8.42c)$$

Although the nonlinear equation 8.42 can be solved in closed form for T_{n+1} in terms of T_n, for this analysis we obtain the solution at each time step by Newton–Raphson iteration. For the $(n + 1)$th time step, equation 8.32 gives

$$J_{n+1}^m \, \Delta T_{n+1}^{m+1} = -F_{n+1}^m \qquad (8.43a)$$

$$T_{n+1}^{m+1} = T_{n+1}^m + \Delta T_{n+1}^m \qquad (8.43b)$$

where the superscript m denotes the mth iteration. For this problem we can derive simple algebraic equations for the Jacobian J_{n+1}^m and unbalanced load F_{n+1}^m. For example, from equation 8.34 the Jacobian is

$$J_{n+1}^m = K_{n+1}^m + \frac{\partial K_{n+1}^m}{\partial T_{n+1}} T_{n+1} - \frac{\partial R_{n+1}^m}{\partial T_{n+1}}$$

which by combining equations 8.42a–8.42c and performing the indicated operations yields

$$J_{n+1}^m = 1 + \lambda_0 \theta\, \Delta t + 2\lambda_1 \theta (1 - \theta)\, \Delta t T_n + 2\lambda_1 \theta^2\, \Delta t T_{n+1}^m$$

Further details are left as an exercise (see the problems at the end of the chapter). At each time step we use the temperature from the preceding time as an initial guess and iterate until ΔT_{n+1}^{m+1} meets a convergence tolerance of $10^{-2}\%$. Transient solutions to the model equation were computed by this algorithm for three values of θ ($\theta = \frac{1}{2}$, $\frac{2}{3}$, and 1), $\lambda_0 = 1.0$, and $\lambda_1 = 0.03$ for different time steps and compared to the exact solution, equation 8.41, for the initial condition $T_0 = 100$.

The percentage of error in the calculated temperature is plotted against time in Figure 8.9. Error growth is the least for $\theta = \frac{1}{2}$, validating this choice of θ as the most accurate scheme of the θ family. If the time-step size is decreased, the accuracy improves, as expected, demonstrating second-order convergence for $\theta = \frac{1}{2}$ and first-order convergence for $\theta = \frac{2}{3}$ and $\theta = 1$. In a more detailed study with comparisons to other integration algorithms, Hogge [17] recommends the Crank–Nicolson scheme ($\theta = \frac{1}{2}$), presented here, as being the most accurate *one-step* (t_n, t_{n+1}) integration scheme for nonlinear heat transfer. He also notes that more sophisticated schemes spanning several steps (e.g., t_n, t_{n+1}, t_{n+2}) give better short-time and long-time accuracy.

8.3 CONDUCTION WITH SURFACE RADIATION

In the preceding section we presented a formulation for heat conduction in solids with general surface heat exchanges. Surface radiation heat transfer appeared in the initial formulation, but we deferred discussion of this effect. We devote this section to a description of the special considerations necessary to formulate and solve conduction problems with radiation boundary conditions. In Appendix E we summarize basic concepts and terminology of radiation heat transfer for the reader interested in further details. Early applications of finite elements to conduction problems with surface radiation appear in references 21 and 22.

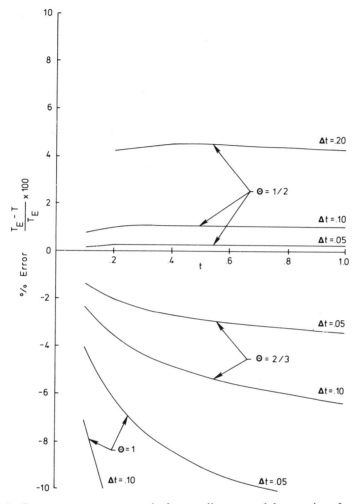

Figure 8.9 Percentage error growth for nonlinear model equation for different algorithms.

8.3.1 Problem Statement

Consider the solid shown in Figure 8.1 experiencing radiation heat transfer on surface S_4. The radiation boundary condition, equation 8.4d, states that the surface heat transfer consists of heat loss by emitted radiation, which is proportional to the fourth power of the surface temperature, $\sigma \epsilon T_s^4$, and heat gain by incident radiation from external sources, αq_r. This statement assumes the radiation surface to be diffuse, gray, and opaque. A diffuse surface emits and reflects incident radiation uniformly in all directions; a gray surface emits and absorbs energy independent of wavelength. A medium with such a surface is called a gray body, and Kirchhoff's law for a gray body states that $\alpha = \epsilon$, where ϵ may be temperature dependent. An opaque medium does not transmit or scatter radiation, and for an opaque medium $\rho + \alpha = 1$, where ρ is the surface reflectivity.

In the absence of radiation heat exchanges between surfaces a simplified approach for representing radiation heat transfer is an equivalent convection boundary condition where the nonlinearity is considered through a temperature-dependent convection coefficient. In this approach [5, 23] the radiation boundary condition, equation 8.4d, is replaced by

$$q_x n_x + q_y n_y + q_z n_z = \sigma \epsilon \left(T_s^4 - T_r^4 \right) \tag{8.44}$$

where T_r is a radiation exchange temperature. For the equivalent convection boundary condition we write

$$q_x n_x + q_y n_y + q_z n_z = h_r(T_s)(T_s - T_r) \tag{8.45a}$$

where

$$h_r(T_s) = \sigma \epsilon \left(T_s^2 + T_r^2 \right)(T_s + T_r) \tag{8.45b}$$

As a result, the radiation vector $\{R_\sigma\}$, equation 8.15e, and the radiation heat load vector $\{R_r\}$, equation 8.15f, are not required. Instead, we use the convection matrix, equation 8.14c, and the convection heat load vector, equation 8.15d, with the temperature-dependent convection coefficient given in equation 8.45b. The advantage of this approach is that no additional element matrices are required, since standard surface convection matrices, for example, equation 8.23c for the plane triangle, may be used. The approach is limited, however, to the special situations (see Appendix E and the discussion preceding equation E.22) for which the boundary condition, equation 8.44, applies. For more general radiation heat exchanges that include radiation heat exchanges between surfaces, a finite element formulation based on the boundary condition of equation 8.4d must be employed.

The heat load vector depends on the incident radiation heat flow rate q_r, which arises from several sources, including (1) directional radiant fluxes from distant sources, (2) radiation exchanges between surfaces of prescribed temperatures, and (3) radiation exchanges between surfaces whose temperatures are unknown a priori. The directional radiant flux poses no special difficulties, because the heat load vector, equation 8.15f, can be evaluated in the usual way. However, the latter two possibilities require special consideration, because radiation exchanges between surfaces must account for the geometrical relationship between surfaces (how one surface "sees" another) and reflected radiation between surfaces. We present a method [21] for considering these additional complications in Section 8.3.3.

8.3.2   Element Equations With Radiation

To illustrate element equations with radiation we consider a steady-state analysis governed by equation 8.18. Element equations for transient problems follow directly when one uses the methods for nonlinear transient solutions discussed in Section 8.2.6.

Steady-State Equations   Equation 8.18 represents the element equations including radiation, but Section 8.2.5 shows that the Newton–Raphson nonlinear solution algorithm, equation 8.34 requires the computation of the unbalanced heat load vector and the Jacobian at each iteration. Including radiation, these are

$$\{F\} = [K_c + K_h]\{T\} - \{R_Q\} - \{R_q\} - \{R_h\} - \{R_\sigma\} - \{R_r\} \quad (8.46a)$$

$$[J] = [K_c + K_h] + [\Delta K] + [\Delta R] \quad (8.46b)$$

where the conductance matrices and heat load vectors are defined in equations 8.14 and 8.15, respectively. As discussed in Section 8.5, the matrix $[\Delta K]$ is the contribution to the Jacobian from temperature-dependent properties in the conductance matrices, and the matrix $[\Delta R]$ is the contribution from temperature-dependent heat load vectors. With radiation, the additional terms to be evaluated are the vectors $\{R_\sigma\}$ and $\{R_r\}$ as well as the contributions of these vectors to the Jacobian. We may expect contributions to the Jacobian from $\{R_\sigma\}$ because this vector depends on the temperature-dependent emissivity and the fourth power of the surface temperature. We note that $\{R_r\}$ may contribute to the Jacobian if the flux depends on the surface temperature because of surface heat exchanges. In practical problems we often assume that q_r is a known heat flux from a previous iteration and neglect this contribution to ΔR.

Radiation Jacobian Matrix   From equation 8.15e the radiation vector is

$$\{R_\sigma\} = -\int_{S_4} \sigma\epsilon(T_s)T_s^4\{N\} \, d\Gamma$$

and the contribution to the unbalanced heat load vector is

$$F_i = \int_{S_4} \sigma\epsilon(T)T^4N_i \, d\Gamma$$

where, for convenience, we omit the subscript s indicating surface temperatures. Then, by definition, the contribution to the Jacobian is

$$J_{ij} = \frac{\partial F_i}{\partial T_j}$$

Performing the differentiation gives

$$J_{ij} = 4 \int_{S_4} \sigma \epsilon T^3 \frac{\partial T}{\partial T_j} N_i \, d\Gamma + \int_{S_4} \sigma \frac{d\epsilon}{dT_j} T^4 N_i \, d\Gamma$$

From equation 8.5a, $\partial T / \partial T_j = N_j$, thus

$$J_{ij} = 4 \int_{S_4} \sigma \epsilon T^3 N_i N_j \, d\Gamma + \int_{S_4} \sigma \frac{d\epsilon}{dT_j} T^4 N_i \, d\Gamma$$

This result is the contribution of the radiation vector to the Jacobian for Newton–Raphson iteration that we note is the $[\Delta R]$ term in equation 8.46b. If the portion due to the emissivity is considered separately, then we may define the *radiation Jacobian* as

$$[J_r] = 4 \int_{S_4} \sigma \epsilon T^3 \{N\} \lfloor N \rfloor \, d\Gamma \qquad (8.47)$$

Because of the presence of the T^3 term in the integrand of the preceding equation, the radiation Jacobian is not constant but depends on an element's nodal temperatures. The T^3 term causes the evaluation of the element integrals to be cumbersome. In one production-type computer program [24] a lumped formulation for the radiation Jacobian is used. The diagonal elements of $[J_r]$ are defined by

$$J_{r_{ii}} = 4 T_i^3 \int_{S_4} \sigma \epsilon N_i \, d\Gamma \qquad \text{(lumped)} \qquad (8.48)$$

and the off-diagonal elements are zero. This approximation has significant practical benefits and is similar to the approximation made by using a lumped capacitance matrix; see Section 8.2.4 for a discussion of consistent and lumped capacitance matrices.

Rod Element Radiation Matrices   To illustrate an application of the foregoing element radiation matrices, consider the one-dimensional two-node rod element shown in Figure 8.2. Element temperature interpolation functions appear in equation 8.20a, and the surface area $d\Gamma = p \, dx$, where p is the rod perimeter. The radiation vector, equation 8.15e, is

$$\{R_\sigma\} = -\sigma \epsilon p \int_0^L (N_1 T_1 + N_2 T_2)^4 \begin{Bmatrix} N_1 \\ N_2 \end{Bmatrix} dx$$

which can be evaluated using equation 5.14 to yield

$$\{R_\sigma\} = -\frac{\sigma \epsilon p L}{30} \begin{Bmatrix} 5T_1^4 + 4T_1^3 T_2 + 3T_1^2 T_2^2 + 2T_1 T_2^3 + T_2^4 \\ T_1^4 + 2T_1^3 T_2 + 3T_1^2 T_2^2 + 4T_1 T_2^3 + 5T_2^4 \end{Bmatrix} \quad (8.49)$$

The Jacobian matrix, equation 8.47, is

$$[J_r] = 4\sigma \epsilon p \int_0^L (N_1 T_1 + N_2 T_2)^3 \begin{Bmatrix} N_1 \\ N_2 \end{Bmatrix} \lfloor N_1 N_2 \rfloor \, dx$$

which produces

$$[J_r] = \frac{\sigma \epsilon p L}{15} \begin{bmatrix} 10T_1^3 + 6T_1^2 T_2 + 3T_1 T_2^2 + T_2^3 & 2T_1^3 + 3T_1^2 T_2 + 3T_1 T_2^2 + 2T_2^3 \\ 2T_1^3 + 3T_1^2 T_2 + 3T_1 T_2^2 + 2T_2^3 & T_1^3 + 3T_1^2 T_2 + 6T_1 T_2^2 + 10T_2^3 \end{bmatrix} \quad (8.50a)$$

The lumped Jacobian matrix, equation 8.48, yields the much simpler approximation

$$[J_r] = 2\sigma \epsilon p L \begin{bmatrix} T_1^3 & 0 \\ 0 & T_2^3 \end{bmatrix} \quad (8.50b)$$

Each of the above matrices clearly shows the strong nonlinearity of a typical radiation term.

8.3.3 Steady-State Solutions

The preceding sections establish the basic equations for the steady-state solution of a conduction problem with radiation boundary conditions. We use the Newton–Raphson iteration method, equation 8.34, with the unbalanced heat load vector and Jacobian defined in equations 8.46. Element radiation Jacobian matrices appear in equations 8.47 and 8.48. The solution procedure depends on the nature of the incident radiation; we consider two cases: (1) incident radiation from distance sources and (2) incident radiation between surfaces.

Incident Radiation From Distance Sources   Incident radiation from distant sources occurs in several applications; solar radiation on the surface of an orbiting space vehicle is one example. We may consider this type of incident radiation in the same way as specified surface heating, except we must take into account the direction of the distant source. Let S denote the intensity of a distant radiant source and a unit vector $\hat{e}$ denote the direction of the distant source. Then the radiation incident on a typical element of surface S_4 is

$$q_i = (\hat{n} \cdot \hat{e})S \qquad (8.51)$$

where $\hat{n}$ is a unit vector normal to S_4. In practical computations q_r is assumed constant over an element surface; hence the radiation heat load vector is readily evaluated. The solution of the problem then proceeds directly by the Newton–Raphson iteration method as previously described.

Incident Radiation Between Surfaces   If there is incident radiation between surfaces, then we must consider the geometrical relationship and reflection of radiant energy between surfaces. Appendix E describes a procedure for the computation of incident radiation under these circumstances. The procedure is based on a radiosity concept that assumes that each surface is isothermal and that the incident heat flow is uniformly distributed over each surface. For N surfaces Appendix E shows that the incident heat flows and temperatures of each surface are related by

$$[[I] - [VF][\diagdown\rho\diagdown]]\{q_r\} = [VF]\{\sigma\epsilon\overline{T}^4\} \qquad (8.52)$$

where $[I]$ is the identity matrix, $[VF]$ is a matrix of view factors, $[\diagdown\rho\diagdown]$ is a diagonal matrix of surface reflectivities, and $\overline{T}_i$ denotes the average surface element temperature. A typical element of the view factor matrix VF_{ij} is the fraction of energy leaving a surface i that arrives at a surface j. Thus the view factors represent the geometrical relationship between surfaces; see Appendix E for further details and references on view factor computation.

The solution procedure consists of solving for the incident heat flows on each of the N surfaces at each iteration. The solution sequence is summarized in Table 8.2.

TABLE 8.2. Solution Sequence for Conduction With Surface Radiation

Initial calculations

    1. Compute the view factor matrix $[VF]$.
    2. Initialize the nodal temperature vector $\{T\}$.

At each iteration

    1. Compute the average surface temperatures $\{\overline{T}\}$.
    2. Solve for incident heat flows $\{q_r\}$.
    3. Form the unbalanced heat load vector $\{F\}$.
    4. Form the Jacobian matrix $[J]$.
    5. Modify $\{F\}$ and $[J]$ for boundary conditions.
    6. Solve for $\{\Delta T\}^{m+1}$ and $\{T\}^{m+1}$.

The solution involves two sets of simultaneous equations: the $N \times N$ set of equations for the surface element incident heat flows, and a standard set of equations for nodal temperatures. The $N \times N$ set of equations 8.52 has a full asymmetric coefficient matrix, but the standard equations for nodal temperatures are banded according to the nodal numbering in the usual way. For small problems equation 8.52 can be solved efficiently by direct elimination, but for large problems an iterative solution method is advantageous [24]. If surface reflections are negligible, that is, if $[\neg\rho\neg] = 0$, then equations 8.52 uncouple and the incident heat flows on each surface are computed directly. If the surface temperatures are known a priori, we may also solve directly for $\{q_r\}$ and omit steps 1 and 2 at each iteration.

The preceding solution method is similar to the approach used in the SPAR thermal analyzer [24]. A closely related solution method is used by the NASTRAN thermal analyzer [22].

8.3.4   Transient Solutions

In Section 8.2.6 we discussed nonlinear transient solutions and presented an application of the θ time-marching algorithm to a scalar model equation. In this section we present two transient applications with radiation. The first application illustrates conduction with emitted radiation, and the second application illustrates radiation with surface reflections.

Example 1: Transient Thermal Response of a Slit Tube   A thin-walled, slit tube (Figure 8.10) initially at uniform temperature is suddenly exposed to intense localized heating. Conduction along the tube is negligible so that the temper-

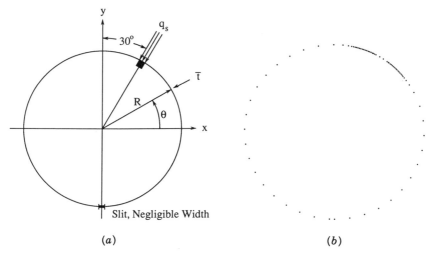

(a) (b)

Figure 8.10   Transient response of slit tube subjected to localized heating: (a) slit tube; (b) finite element model with 73 nodes.

ature varies only around the tube perimeter, that is, $T(\theta, t)$. The problem is motivated by an aerospace application of finite elements to heat transfer in a solar array boom for an orbiting satellite. The problem consists of solving the one-dimensional energy equation

$$\rho c \bar{t} \frac{\partial T}{\partial t} - \frac{k \bar{t}}{R^2} \frac{\partial^2 T}{\partial \theta^2} + \sigma \epsilon T^4 - \alpha q_s = 0$$

where $\bar{t}$ is the tube wall thickness, and R is the tube radius. The boundary conditions are

$$\frac{\partial T\left(-\dfrac{\pi}{2}, t\right)}{\partial \theta} = 0 \quad \text{and} \quad \frac{\partial T\left(\dfrac{3\pi}{2}, t\right)}{\partial \theta} = 0$$

and the initial condition is

$$T(\theta, 0) = T_0$$

Radiation is permitted on the external surface but is neglected on the interior of the tube. We may characterize the problem as one dimensional heat conduction with surface radiation and model the heat transfer with an assembly of several two-node rod elements (Figure 8.10b). The problem is nonlinear because of radiation, and we obtain the transient response by time marching with Newton–Raphson iterations at each time step. The solution algorithm is

$$[J]_{n+1}^m \{\Delta T\}_{n+1}^{m+1} = -\{F\}_{n+1}^m$$

$$\{T\}_{n+1}^{m+1} = \{T\}_{n+1}^m + \{\Delta T\}_{n+1}^{m+1}$$

where the subscript n denotes the time step, and the superscript m denotes the mth iteration. The Jacobian and unbalanced load vectors are

$$[J]_{n+1}^m = [\bar{K}]_{n+1}^m + [J_r]_\theta^m$$

where

$$\{F\}_{n+1}^m = [\bar{K}]\{T\}_{n+1}^m - \{\bar{R}\}_n - \{R_\sigma(T_\sigma)\}^m - \{R_q\}$$

and

$$[\bar{K}] = \theta[K] + \frac{1}{\Delta t}[C]$$

$$\{\bar{R}\}_n = \left[(1 - \theta)[K] - \frac{1}{\Delta t}[C]\right]\{T\}_n$$

Element capacitance, conduction, and surface heat flux matrices are formed using Equation 8.21. Element radiation heat flux vectors are computed using Equation 8.49. The radiation Jacobian matrix $[J_r]_\theta^m$ is computed from both consistent and lumped formulations, equations 8.50a and 8.50b, respectively.

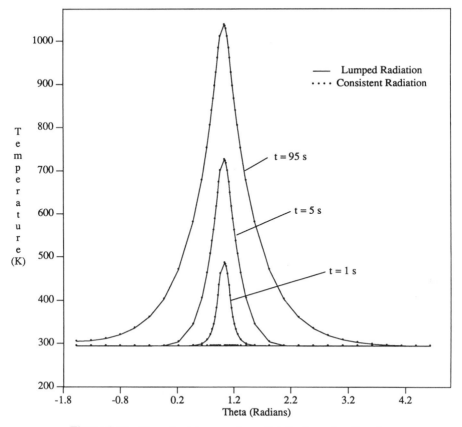

Figure 8.11 Transient temperature distributions for slit tube.

The Crank–Nicolson scheme ($\theta = \frac{1}{2}$) was used with a time step of 0.5 s. Typically from one to three Newton–Raphson iterations are required at each time step to reduce the normalized maximum nodal temperature change to less than 0.001. Spatial temperature distributions at selected times in the response are presented in Figure 8.11. The high localized heat flux induces steep temperature gradients and very high temperatures. After about 95 s temperatures approach steady-state radiation equilibrium. The results demonstrate excellent agreement between temperatures computed by consistent and lumped formulations. Since the lumped formulation requires less storage and fewer computations, the example illustrates that it can offer significant computational benefits without loss of accuracy.

Example 2: Transient Radiation With Surface Reflections A radiation exchange system (Figure 8.12) consists of a heater between two plates with dissimilar absorption and reflection characteristics [24]. The plates are at a uniform initial temperature of zero, and at time $t = 0 +$ the heater begins to generate

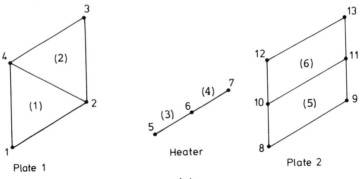

(a)

	Plate 1	Heater	Plate 2
Radiating Area	10.	2.	8.
Emissivity	0.6	0.85	0.2
Reflectivity	0.4	0.15	0.8
Thickness	0.1		0.1
Conducting Area		0.2	
Conductivity	200.	200.	200.
Mass Density	3000.	3000.	3000.
Specific Heat	0.2	0.2	0.2
Source Heat Rate		22680.	

Non-zero Viewfactors

VF13 = .04   VF14 = .02   VF15 = .2   VF16 = .1

VF23 = .02   VF24 = .04   VF25 = .1   VF26 = .2

VF31 = .2   VF32 = .1   VF35 = .16   VF36 = .16

VF41 = .1   VF42 = .2   VF45 = .16   VF46 = .16

VF51 = .25   VF52 = .125   VF53 = .04   VF54 = .04

VF61 = .125   VF62 = .25   VF63 = .04   VF64 = .04

(b)

Figure 8.12 Transient radiation example with surface heat exchanges: (a) parallel plates and heater; (b) problem data.

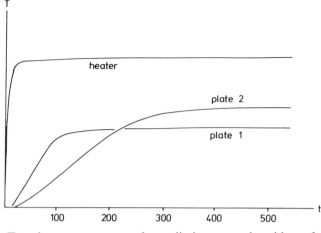

Figure 8.13 Transient temperatures for radiation example with surface heat exchanges.

heat at a constant rate. The plates are radiatively heated and a thermal transient ensues; after a long time the plates and heater approach equilibrium temperatures. Radiation properties and view factors are chosen so that plate 1 absorbs heat faster than plate 2 and most of the heat incident on plate 2 is reflected. The emitted and reflected heat not incident on either of the plates or heater is lost to space. Pertinent geometry, material, heating rate, and view factors appear in Figure 8.12.

Figure 8.13 presents the results of the analysis. The temperature-versus-time curves show that the heater temperature has a very short rise time and quickly approaches its steady-state value. Plate 1 approaches its steady-state value faster than plate 2 because plate 2 has a high reflectivity and reflects most of the incident energy. References 21–23 present other examples of finite element analyses of conduction problems with surface radiation. Applications of finite elements to the interaction of conduction and radiation in an absorbing, scattering, and emitting medium appear in references 25 and 26.

8.4 CONVECTIVE-DIFFUSION EQUATION

The energy equation for a moving fluid contains the combined effects of heat transport due to the fluid motion and heat transfer due to conduction (diffusion). If we wish to study the dispersion of a concentrated pollutant in the atmosphere or some other moving body of fluid, we must solve an equation of the same form. The governing equation known as the convective-diffusion equation is the subject of extensive research; numerous papers describing new studies appear in books and journals, [27–41]. The

convective-diffusion equation is the subject of intensive study not only because of its importance in applications, but also because the convective
transport term can introduce numerical oscillations for both steady-state and
transient solutions not present in the diffusion equation. Here we present a
finite element formulation for thermal problems governed by the
convective -diffusion equation and illustrate the solution behavior. Although
our approach takes temperature as the dependent variable, the formulation
is applicable to other field problems.

8.4.1 Problem Statement

Consider steady-state or transient heat transfer in a three-dimensional incompressible fluid flow in a solution domain Ω bounded by a surface Γ. We
use the same terminology introduced in Section 8.2 for conduction heat
transfer. The problem is governed by the fluid energy equation (see
Appendix E):

$$-\left(\frac{\partial q_x}{\partial x} + \frac{\partial q_y}{\partial y} + \frac{\partial q_z}{\partial z}\right) + Q = \rho c_v \left(\frac{\partial T}{\partial t} + u\frac{\partial T}{\partial x} + v\frac{\partial T}{\partial y} + w\frac{\partial T}{\partial z}\right) \quad (8.53)$$

where c_v is the specific heat at constant volume and u, v, and w are the flow
velocity components. The velocity components are a function of x, y, z, and
t and are known. The heat flow components are related to the temperature
gradients by Fourier's law, equation 8.2. The energy equation is the same as
for a solid, equation 8.1, except for the convective transport terms,

$$\rho c_v \left(u\frac{\partial T}{\partial x} + v\frac{\partial T}{\partial y} + w\frac{\partial T}{\partial z}\right) \quad (8.54)$$

We solve the convective-diffusion (energy) equation subject to the initial
condition, equation 8.3, and boundary conditions, equations 8.4, on all
portions of the surface Γ. For simplicity we omit the radiation boundary
condition, equation 8.4d, in the following discussion. As in the formulation of
the conduction problem, we regard thermal properties to be temperature
dependent and anticipate a nonlinear solution. We derive element equations
by the method of weighted residuals with the Petrov–Galerkin approach (see
Chapter 4), which employs weighting functions W_i different from the interpolation functions. Element matrices are the same as for conduction, with the
addition of a new matrix representing the convective transport terms, equation 8.54. In the following sections we present the general finite element
formulation and illustrate solution behavior with an example.

8.4.2 Finite Element Formulation

Consider the solution domain Ω divided into M elements of r nodes each. We express the temperature and temperature gradients within each element in terms of nodal temperatures using the $[N(x, y, z)]$ and $[B(x, y, z)]$ interpolation matrices, respectively, shown in equations 8.6. We use the method of weighted residuals with weighting functions W_i because this approach allows us to introduce the concept of upwind finite elements. Upwind finite elements are used by some authors [32–34] to eliminate numerical oscillations associated with the convective transport terms, equation 8.54. The method of weighted residual requires that

$$\int_{\Omega^{(e)}} \left[\frac{\partial q_x}{\partial x} + \frac{\partial q_y}{\partial y} + \frac{\partial q_z}{\partial z} - Q + \rho c_v \frac{\partial T}{\partial t} \right.$$
$$\left. + \rho c_v \left(u \frac{\partial T}{\partial x} + v \frac{\partial T}{\partial y} + w \frac{\partial T}{\partial z} \right) \right] W_i \, d\Omega = 0, \qquad i = 1, 2, \ldots, r \quad (8.55)$$

where $\Omega^{(e)}$ is the solution domain for element (e). Following the approach in Section 8.2, we integrate the first term on the left-hand side by the Gauss theorem and introduce the boundary conditions, equation 8.4. Next we express the heat flow components in terms of nodal temperatures via equation 8.12 and express the temperature gradients in the convective transport components in terms of the nodal temperatures by equation 8.6b. As a result, we obtain the general finite element formulation

$$[C]\{\dot{T}\} + [[K_c] + [K_h] + [K_v]]\{T\} = \{R_Q\} + \{R_q\} + \{R_h\} \quad (8.56)$$

where

$$[C] = \int_{\Omega^{(e)}} \rho c_v \{W\} \lfloor N \rfloor \, d\Omega \tag{8.57a}$$

$$[K_c] = \int_{\Omega^{(e)}} [B_w]^T [k][B] \, d\Omega \tag{8.57b}$$

$$[K_h] = \int_{S_3} h \{W\} \lfloor N \rfloor \, d\Gamma \tag{8.57c}$$

$$[K_v] = \int_{\Omega^{(e)}} \rho c_v \{W\} \lfloor u \quad v \quad w \rfloor [B] \, d\Omega \tag{8.57d}$$

$$\{R_Q\} = \int_{\Omega^{(e)}} Q \{W\} \, d\Omega \tag{8.58a}$$

$$\{R_q\} = \int_{S_2} q_s \{W\} \, d\Gamma \tag{8.58b}$$

$$\{R_h\} = \int_{S_3} h T_e \{W\} \, d\Gamma \tag{8.58c}$$

where $[B_w]$ denotes $[B]$ in equation 8.7b with N_i replaced by W_i. Element capacitance and conductance matrices and heat load vectors resemble the corresponding matrices and vectors for conduction in equation 8.14 and 8.15, except for the general weighting function W_i. The convective transport term contributes a new conductance matrix, equation 8.57d, that depends on the flow field. Note that the convective transport matrix is asymmetrical, which is characteristic of transport terms. If we use $W_i = N_i$ (the Bubnov–Galerkin approach), all matrices (except the convective transport matrix) and vectors are identical to those for a solid.

We assemble element matrices in the usual way to obtain the system equations for the solution domain. Note, however, that the convective transport matrix has "vector character" due to the velocity components. Most analysts compute these element matrices in a common global coordinate system that defines the velocity components.

8.4.3 One-Dimensional Problem

To illustrate element matrices and solution behavior for the convective-diffusion equation, consider one-dimensional flow in a channel or duct with $T(x, t)$ and $u = $ constant as the only nonzero velocity component. Then, by using Fourier's law, equation 8.53 is

$$k\frac{\partial^2 T}{\partial x^2} + Q = \rho c_v\left(\frac{\partial T}{\partial t} + u\frac{\partial T}{\partial x}\right) \tag{8.59}$$

Consider finite element matrices derived from equations 8.57 and 8.58 for two cases: conventional (Bubnov–Galerkin) weighting functions, $W_i = N_i$, and upwind (Petrov–Galerkin) weighting functions, $W_i \neq N_i$. For both cases we assume an element of length L with two nodes and use $d\Omega = dx$.

Conventional Element Matrices   We derive the conventional element matrices from equations 8.57 and 8.58 with $W_i = N_i$. The interpolation functions are the same as for the conduction rod element, equation 8.20. Element matrices are

$$[C] = \frac{\rho c_v L}{6}\begin{bmatrix} 2 & 1 \\ 1 & 2 \end{bmatrix} \tag{8.60a}$$

$$[K_c] = \frac{k}{L}\begin{bmatrix} 1 & -1 \\ -1 & 1 \end{bmatrix} \tag{8.60b}$$

$$[K_v] = \frac{\rho c_v u}{2}\begin{bmatrix} -1 & 1 \\ -1 & 1 \end{bmatrix} \tag{8.60c}$$

$$\{R_Q\} = \frac{QL}{2}\begin{Bmatrix} 1 \\ 1 \end{Bmatrix} \tag{8.60d}$$

where, except for $[K_v]$, the matrices are the same as for the conduction rod element. Note the lack of symmetry and the negative value on the diagonal in the indefinite transport matrix $[K_v]$.

Upwind Element Matrices   We derive the upwind finite element matrices [32–34] from equations 8.57 and 8.58 with the interpolation function given in equations 8.20, but use weighting functions

$$\lfloor W \rfloor = \lfloor N \rfloor + \alpha \lfloor F \rfloor \tag{8.61}$$

where α is the upwind parameter, and the upwind weighting functions are

$$\lfloor F \rfloor = \left\lfloor 3\left(\frac{x^2}{L^2} - \frac{x}{L}\right) \quad -3\left(\frac{x^2}{L^2} - \frac{x}{L}\right) \right\rfloor \tag{8.62}$$

With these weighting functions the upwind element matrices are

$$[C] = \frac{\rho c_v L}{6}\begin{bmatrix} 2 & 1 \\ 1 & 2 \end{bmatrix} + \frac{\alpha \rho c_v L}{4}\begin{bmatrix} -1 & -1 \\ 1 & 1 \end{bmatrix} \tag{8.63a}$$

$$[K_c] = \frac{k}{L}\begin{bmatrix} 1 & -1 \\ -1 & 1 \end{bmatrix} \tag{8.63b}$$

$$[K_v] = \frac{\rho c_v u}{2}\begin{bmatrix} -1 & 1 \\ -1 & 1 \end{bmatrix} + \frac{\alpha \rho c_v u}{2}\begin{bmatrix} 1 & -1 \\ -1 & 1 \end{bmatrix} \tag{8.63c}$$

$$\{R_Q\} = \frac{QL}{2}\begin{Bmatrix} 1 \\ 1 \end{Bmatrix} + \frac{\alpha QL}{2}\begin{Bmatrix} -1 \\ 1 \end{Bmatrix} \tag{8.63d}$$

In the preceding equations the first term on the right-hand side represents the conventional element formulation; the second term, which is proportional to α, represents the contribution of "upwinding." The upwind parameter α varies in the range $0 \le \alpha \le 1$. If $\alpha = 0$, the matrices reduce to the conventional elements, and if $\alpha = 1$, we have full upwinding. For steady-state problems an optimum value of α may determined that yields exact nodal temperature values [34]. As mentioned earlier, upwind elements are advantageous in eliminating spurious oscillations in fluid temperatures. The upwind finite element concept is analogous to the backward (upwind) difference approach [42] used in finite difference schemes to eliminate spurious oscillations in convective transport problems.

Solution Behavior   To evaluate conventional and upwind element behavior we consider a one-dimensional steady-state flow without internal heat generation. We write equation 8.59 as

$$\frac{d^2T}{dx^2} - \frac{\rho c_v u}{k}\frac{dT}{dx} = 0 \tag{8.64a}$$

and assume boundary conditions

$$T(0) = T_0 \tag{8.64b}$$

$$T(L) = 0 \tag{8.64c}$$

where L is the length of the solution domain. The problem has a simple exact solution that we use to evaluate the finite element performance. We obtain the finite element solution in closed form by subdividing the solution domain into a series of elements of length l, assembling the element matrices, and writing and solving the difference equation between three typical interior nodes. For example, using conventional matrices equations 8.60b and 8.60c, the assembled matrices have the form

$$\frac{k}{l}\begin{bmatrix} \ddots & & & \\ -1 & 2 & -1 \\ & & \ddots \end{bmatrix}\begin{Bmatrix} T_{i-1} \\ T_i \\ T_{i+1} \end{Bmatrix}$$

$$+ \frac{\rho c_v u}{2}\begin{bmatrix} \ddots & & & \\ -1 & 0 & 1 \\ & & \ddots \end{bmatrix}\begin{Bmatrix} T_{i-1} \\ T_i \\ T_{i+1} \end{Bmatrix} = 0$$

which yields the typical difference equation

$$-T_{i-1} + 2T_i - T_{i+1} + \frac{Pe}{2}(-T_{i-1} + T_{i+1}) = 0$$

where we introduce the grid Peclet number $Pe = \rho c_v u l/k$, which characterizes the ratio of heat transfer by convection to the heat transfer by conduction (see Appendix E). The general solution of the preceding difference equation is

$$T_i = C_1 + C_2\left(-\frac{Pe+2}{Pe-2}\right)^i \qquad \text{(conventional)} \tag{8.65}$$

where C_1 and C_2 are constants of integration. Equation 8.65 shows that if $Pe > 2$, the solution is oscillatory. Thus for a smooth temperature variation we cannot arbitrarily choose the element length but, rather, we must choose a discretization such that for each element

$$Pe = \frac{\rho c_v u l}{k} < 2 \tag{8.66}$$

This condition is well known from central differences approximations [42] to the convective-diffusion equation. If we repeat the analysis with the upwind finite elements, equations 8.63b and 8.63c, we obtain

$$T_i = C_1 + C_2 \frac{[1 + (1 + \alpha)(Pe/2)]^i}{[1 - (1 - \alpha)(Pe/2)]^i} \qquad \text{(upwind)} \qquad (8.67)$$

which reduces to the preceding result if $\alpha = 0$. However, equation 8.67 shows that the upwind elements predict temperatures that are not oscillatory for $Pe > 2$ if

$$\alpha \geq 1 - \frac{2}{Pe} \qquad (8.68)$$

Thus upwind finite elements assure an oscillation-free solutions for a given mesh if the upwind parameter α satisfies equation 8.68. Full upwinding, that is, $\alpha = 1$, always predicts a smooth temperature variation as equation 8.67 shows. The upwind approach has a clear advantage over the conventional approach, since for a given mesh it always predicts oscillation-free temperatures. The disadvantage of the upwind approach is that upwinding degrades solution accuracy by introducing artificial diffusion. Equation 8.63c shows this clearly; the additional term on the right-hand side is a conduction matrix with an equivalent conductance $\alpha \rho c_v u / 2$.

These notions are illustrated in Figure 8.14, where we compare conventional, upwind, and analytical solutions of the problem posed in equations 8.64 for two flows. For the lower flow rate the conventional finite elements with $Pe = 0.2$ predict temperatures that show excellent agreement with the exact solution, but the upwind elements ($\alpha = 1$) predict temperatures with some inaccuracy due to the effects of artificial diffusion. For the higher flow velocity the conventional finite elements with $Pe = 25$ predict temperatures with such large oscillations that the results are worthless, but the upwind elements with the same mesh show good agreement with the exact solution. However, for this same flow if the mesh is refined, the oscillations in the temperatures predicted by the conventional elements disappear and temperatures converge to the exact solution.

8.4.4   Two-Dimensional Solutions

We can compute two-dimensional, steady-state, and transient solutions of the finite element formulation, equation 8.56, of the convective-diffusion equation by using the same procedures in Section 8.2 for conduction. The principal computational difference is the convective transport matrix, equation 8.57d, which requires an asymmetric equation solver. Steady-state solution algorithms typically use Gauss elimination for linear problems and

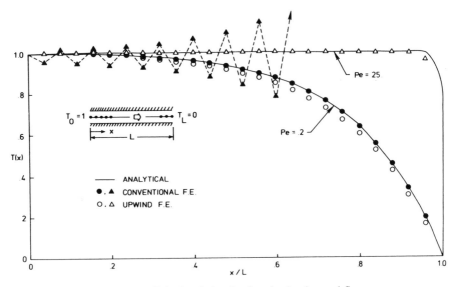

Figure 8.14 Solution behavior for single-channel flow.

Newton–Raphson iteration for nonlinear problems. Transient solutions employ either explicit or implicit time integration algorithms combined with Newton–Raphson iteration for nonlinearities. Isoparametric elements are popular, particularly the eight-node quadratic serendipity element and nine-node biquadratic Lagrangian element (Section 5.8).

The choice of consistent versus lumped capacitance matrices is an important consideration in transient solutions. A number of studies, for example, reference 10, consider the effects of lumped capacitance on the propagation of various waveforms. Conduction is neglected so that pure convection (sometimes called advection from usage in environmental problems) is used to investigate numerical diffusion and/or dispersion effects. Dispersion means that different Fourier wavelengths convect at different wave speeds, causing phase errors and a "trail of wiggles" behind the main waveforms. Studies in one- and two-dimensional convective diffusion problems [10] show that lumped capacitance matrices may introduce significant numerical dispersion, particularly for linear and quadratic elements. Thus for convective-diffusion and conduction problems (see Section 8.2.4) there is an important trade-off between the efficiency of an explicit time integration algorithm based on lumped capacitance and the accuracy of an implicit algorithm with consistent capacitance. There are, however, significant computational benefits for using explicit time-integration algorithms for three-dimensional solutions to the convective-diffusion equation, and research continues in this area.

Upwind finite elements also exist for two-dimensional convective diffusion problems [34]. The pros and cons of upwind versus conventional finite elements are the basis for lively discourse in the finite element community. The proponents of upwinding argue that upwinding offers significant benefits by removing spurious oscillations that obscure the true solution. The opponents of upwinding argue that the oscillations are an indication of an inadequate mesh or improper boundary conditions and that it is inappropriate to suppress these oscillations by upwinding with artificial diffusion. For further discussion see the references previously cited [27–41]. Thornton and Wieting [43] present an evaluation of lumped versus consistent capacitance and upwind elements for one-dimensional transient flows, including comparisons with analytical and finite difference solutions.

8.5 FREE AND FORCED CONVECTION

To analyze convection heat transfer we combine the finite element concepts introduced for conduction and the convection-diffusion equation with the concepts for viscous flow to be presented in Chapter 9. In this section we present a general finite element formulation for viscous incompressible flow with the Boussinesq approximation, discuss solution methods, and illustrate typical finite element solutions.

8.5.1 Problem Statement

Following Appendix E, consider a three-dimensional solution domain Ω. The basic conservation equations for viscous, incompressible flow are

Mass:

$$\frac{\partial u}{\partial x} + \frac{\partial v}{\partial y} + \frac{\partial w}{\partial z} = 0 \tag{8.69}$$

Momentum:

$$\rho \frac{Du}{Dt} = -\frac{\partial P}{\partial x} - \rho\beta(T - T_0)g_x + \frac{\partial \sigma_x}{\partial x} + \frac{\partial \tau_{xy}}{\partial y} + \frac{\partial \tau_{xz}}{\partial z} \tag{8.70a}$$

$$\rho \frac{Dv}{Dt} = -\frac{\partial P}{\partial y} - \rho\beta(T - T_0)g_y + \frac{\partial \tau_{xy}}{\partial x} + \frac{\partial \sigma_y}{\partial y} + \frac{\partial \tau_{yz}}{\partial z} \tag{8.70b}$$

$$\rho \frac{Dw}{Dt} = -\frac{\partial P}{\partial z} - \rho\beta(T - T_0)g_z + \frac{\partial \tau_{xz}}{\partial x} + \frac{\partial \tau_{yz}}{\partial y} + \frac{\partial \sigma_z}{\partial z} \tag{8.70c}$$

Energy:

$$\rho c_v \frac{DT}{Dt} = -\left(\frac{\partial q_x}{\partial x} + \frac{\partial q_y}{\partial y} + \frac{\partial q_z}{\partial z}\right) + Q \tag{8.71}$$

where

$$\frac{D}{Dt} = \frac{\partial}{\partial t} + u\frac{\partial}{\partial x} + v\frac{\partial}{\partial y} + w\frac{\partial}{\partial z}$$

The constitutive relations for the fluid consist of the Newtonian stress/strain-rate law, equation D.3, and Fourier's law, equation 8.11. In equations 8.70, P is the pressure, β is the coefficient of thermal expansion, T_0 is the reference temperature for which buoyant forces vanish, and g_x, g_y, and g_z are components of gravitational acceleration.

We solve the convection equations, equations 8.69–8.71, for viscous incompressible flow subject to initial conditions and an appropriate set of boundary conditions. The initial conditions consist of specifying the initial values of the velocity components, pressure, and temperature at time zero. The hydrodynamic part of the problem requires either the velocity components or the surface tractions to be specified on the boundary (see Appendix E for a statement of these conditions). The thermal part of the problem requires the temperature or heat flux to be specified on the boundary (see equation 8.4 for a statement of these conditions).

Simultaneous solution of equations 8.69–8.71 is necessary for (1) free convection, (2) combined free and forced convection, and (3) forced convection with temperature-dependent properties. For forced convection with constant fluid properties the flow problem uncouples from the thermal problem. The flow analysis that solves the mass and momentum equations is performed first, and then the energy equation is solved with the known velocity field from the flow analysis.

8.5.2   Finite Element Formulation

Finite element solutions for free and forced convection use three alternative formulations: (1) the stream function–vorticity–temperature formulation [44], (2) the velocity–penalty function–temperature formulation [45], and (3) the velocity–pressure–temperature formulation [46, 47]. Additional references illustrating these formulations appear in books [27–31] and papers [49–53].

We present the velocity–pressure–temperature formulation following the conventional Bubnov–Galerkin approach that employs weighting functions equal to the interpolation functions.

Within each finite element we approximate the velocity, pressure, and temperature distributions by

$$
\left.
\begin{aligned}
u(x, y, z, t) &= \lfloor N_v(x, y, z) \rfloor \{u(t)\} \\
v(x, y, z, t) &= \lfloor N_v(x, y, z) \rfloor \{v(t)\} \\
w(x, y, z, t) &= \lfloor N_v(x, y, z) \rfloor \{w(t)\} \\
P(x, y, z, t) &= \lfloor N_P(x, y, z) \rfloor \{P(t)\} \\
T(x, y, z, t) &= \lfloor N(x, y, z) \rfloor \{T(t)\}
\end{aligned}
\right\}
\tag{8.72}
$$

where we take the pressure interpolation functions $[N_P]$ one order lower than the velocity interpolation functions $[N_v]$ for weighted residual error consistency, as discussed in Section 9.4.2. Proceeding with the method of weighted residuals as previously described, we obtain a set of coupled matrix equations that we write in submatrix form:

$$
\begin{bmatrix}
M & 0 & 0 & 0 \\
0 & M & 0 & 0 \\
0 & 0 & M & 0 \\
0 & 0 & 0 & 0
\end{bmatrix}
\begin{Bmatrix}
\dot{u} \\ \dot{v} \\ \dot{w} \\ \dot{P}
\end{Bmatrix}
+
\begin{bmatrix}
K_A & 0 & 0 & 0 \\
0 & K_A & 0 & 0 \\
0 & 0 & K_A & 0 \\
0 & 0 & 0 & 0
\end{bmatrix}
\begin{Bmatrix}
u \\ v \\ w \\ P
\end{Bmatrix}
$$

$$
+
\begin{bmatrix}
2K_{11} + K_{22} + K_{33} & K_{12} & K_{13} & L_1 \\
K_{12}^T & K_{11} + 2K_{22} + K_{33} & K_{23} & L_2 \\
K_{13}^T & K_{23}^T & K_{11} + K_{22} + 2K_{33} & L_3 \\
L_1^T & L_2^T & L_3^T & 0
\end{bmatrix}
\begin{Bmatrix}
u \\ v \\ w \\ P
\end{Bmatrix}
$$

$$
=
\begin{Bmatrix}
R_x \\ R_y \\ R_z \\ 0
\end{Bmatrix}
\tag{8.73}
$$

$$
[C]\{\dot{T}\} + [K_v + K_c]\{T\} = \{R_Q\} + \{R_q\} + \{R_h\}
\tag{8.74}
$$

where in the hydrodynamic equations, equation 8.73, typical submatrices are

$$[M] = \int_{\Omega^{(e)}} \rho\{N_v\}\lfloor N_v \rfloor \, d\Omega \tag{8.75a}$$

$$[K_A] = \int_{\Omega^{(e)}} \rho\{N_v\}u \left\lfloor \frac{\partial N_v}{\partial x} \right\rfloor d\Omega + \int_{\Omega^{(e)}} \rho\{N_v\}v \left\lfloor \frac{\partial N_v}{\partial y} \right\rfloor d\Omega$$

$$+ \int_{\Omega^{(e)}} \rho\{N_v\}w \left\lfloor \frac{\partial N_v}{\partial z} \right\rfloor d\Omega \tag{8.75b}$$

$$[K_{11}] = \int_{\Omega^{(e)}} \mu \left\{ \frac{\partial N_v}{\partial x} \right\} \left\lfloor \frac{\partial N_v}{\partial x} \right\rfloor d\Omega \tag{8.75c}$$

$$[K_{12}] = \int_{\Omega^{(e)}} \mu \left\{ \frac{\partial N_v}{\partial y} \right\} \left\lfloor \frac{\partial N_v}{\partial x} \right\rfloor d\Omega \tag{8.75d}$$

$$[L_1] = - \int_{\Omega^{(e)}} \left\{ \frac{\partial N_v}{\partial x} \right\} \lfloor N_p \rfloor \, d\Omega \tag{8.75e}$$

$$\{R_x\} = \int_{\Gamma^{(e)}} \bar{\sigma}_x\{N_v\} \, d\Gamma - \int_{\Omega^{(e)}} \rho\beta(T - T_0)g_x\{N_v\} \, d\Omega \tag{8.76}$$

Element submatrices not shown are obtained by cyclic permutation of the subscripts 1, 2, and 3 and x, y, and z. The matrices in the energy equation, equation 8.74, appear in equations 8.57 and 8.58. In equation 8.73, $[M]$ is the fluid mass matrix, $[K_A]$ is the fluid momentum convective transport matrix, $[K]$ and $[L]$ combined represent diffusion of momentum, and $\{R\}$ is a vector of the system forcing functions. Note that $[K_A]$ depends on the flow velocity components and is asymmetric, and in equation 8.76, $\bar{\sigma}_x$ is a typical surface traction (see Appendix E). Since the momentum convective transport matrix depends on the velocity components, the hydrodynamic part of the problem is nonlinear and an iterative solution method is required. In the following sections we discuss solution techniques and several examples.

8.5.3 Solution Techniques

The fact that the matrix equations 8.73 and 8.74 may or may not be coupled must be anticipated in selecting a solution algorithm. For problems in forced convection with constant properties, the hydrodynamic problem (Navier–Stokes equations) decouples from the thermal problem, which permits the hydrodynamic problem to be solved by the methods to be discussed in Chapter 9. Normally Navier–Stokes solution algorithms employ higher-order isoparametric elements and solve the nonlinear algebraic equations iteratively with a large-capacity equation solver. When the fluid motion is influenced by buoyancy forces or the fluid properties vary with temperature, the velocity and temperature distributions are directly coupled, and equations 8.73 and 8.74 are solved simultaneously. Two approaches have been employed for steady-state solutions. Gartling [46] uses an algorithm in which the solution alternates between the two equations, and Taylor and Ijam [47] solve the equations simultaneously. These solution methods were first used to solve steady-state two-dimensional free and forced convection problems. In addition, transient two-dimensional solutions have been obtained [48]. Although solution methods have been adequate for two-dimensional solutions, fast computers with significant data storage are required. Further research in solution techniques for coupled fluid/thermal problem is necessary, particularly for three-dimensional solutions.

8.5.4 Free Convection Example

To illustrate a free convection solution [48] we present a finite element analysis of transient free convection in a right circular cylinder. Equations 8.73 and 8.74 are solved in cylindrical coordinates for the axisymmetric region, with isothermal boundary conditions shown in Figure 8.15. The computational domain was discretized with a nonuniform mesh of isoparametric quadrilateral elements concentrated near the cylinder walls in anticipation of large velocity and temperature gradients. Velocities and temperatures are approximated quadratically within each element, while the pressure varies linearly. The transient solution uses an implicit Crank–Nicolson time integration scheme that processes equations 8.73 and 8.74 sequentially for each time step. The initial conditions were taken to be a quiescent fluid maintained at the boundary temperature. Volume heating was instantaneously applied at time zero and maintained at a constant, spatially uniform value thereafter.

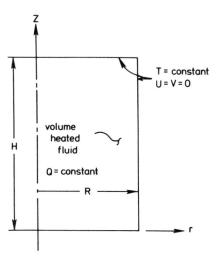

Figure 8.15 Solution domain for free convection in a right circular cylinder.

Figure 8.16 provides a series of streamline and temperature plots for various times during evaluation of the flow for a modified Grashof number $Gr = 4.39 \times 10^5$ defined as

$$Gr = \frac{\rho^2 g \beta R^5 Q}{\mu^2 k}$$

During the transient response a secondary flow is created and due to the interplay of inertia, viscous, and buoyancy forces, the secondary and main flow interact with a damped periodic motion until a steady-state configuration with two flow cells is approached for large time.

8.5.5 Forced Convection

Many of the concepts discussed previously in Section 8.4 with regard to the convective-diffusion equation apply to forced convection. Readers interested in topics such as upwind element formulations or lumped versus consistent capacitance matrices should refer to that section. Solutions of equations 8.73 and 8.74 for forced convection and combined solid–fluid convection have been obtained by several authors [46, 54–58].

Thornton and Wieting [59, 60] developed a finite element methodology for combined conduction–forced convection analyses of convectively cooled aerospace structures. In such complex structures simultaneously solving the mass, momentum, and energy conservation equations for the solid and fluid is impractical because of the great complexity of the fluid flow. For this reason a finite element engineering approach was developed based on assumptions used in practical heat transfer analysis. Finite elements with fluid mean temperature nodes and fluid–solid interface nodes were used to represent flow passages. The flow was described by a specified mass flow rate, and the fluid–solid heat exchange was expressed in terms of a convection coefficient computed from analytical–empirical equations for the Nusselt number.

8.6 CLOSURE

This chapter considered fundamental finite element approaches to the solution of conduction, convection, and radiation heat transfer problems. General formulations were presented based on the method of weighted residuals. Solution methods in current favor for the solution of nonlinear and transient problems were presented, discussed, and illustrated with examples. The exposition here is by no means complete; the finite element approaches were restricted in both scope and depth. Several important practical heat transfer problems currently of interest to finite element researchers were omitted. Yet the approaches presented illustrate the credibility of the finite element

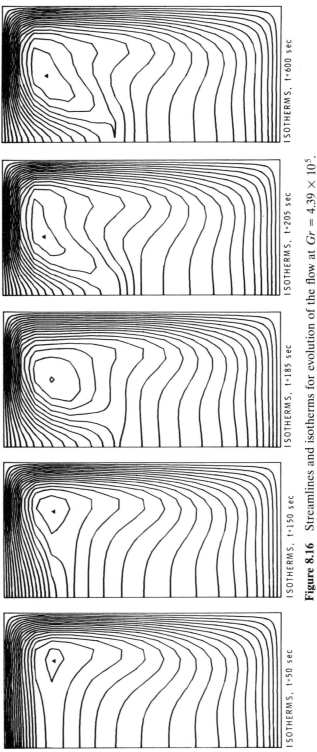

Figure 8.16 Streamlines and isotherms for evolution of the flow at $Gr = 4.39 \times 10^5$.

ISOTHERMS, t=50 sec

ISOTHERMS, t=150 sec

ISOTHERMS, t=185 sec

ISOTHERMS, t=205 sec

ISOTHERMS, t=600 sec

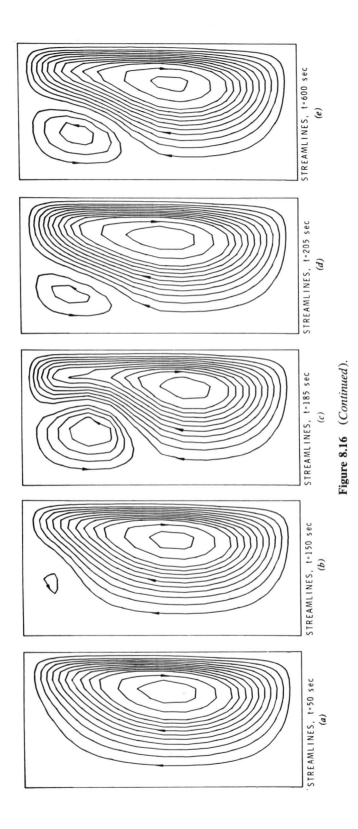

STREAMLINES, t=50 sec

(a)

STREAMLINES, t=150 sec

(b)

STREAMLINES, t=185 sec

(c)

STREAMLINES, t=205 sec

(d)

STREAMLINES, t=600 sec

(e)

Figure 8.16 (*Continued*).

method for the solution of a variety of realistic heat transfer problems. But much remains to be done before finite elements reach the same maturity for thermal analysis as finite elements have attained in other areas, such as structural analysis. There is, for example, strong motivation for more efficient nonlinear transient solutions methods for application to conduction with radiation and three-dimensional convection problems.

Readers desiring further study of finite element solutions of the heat transfer problems presented herein and other specialized problems are encouraged to see the chapter references. Also of interest for the evaluation of finite element algorithms, are the results of a comparison exercise for natural convection in a square cavity [61]. A number of contributed solutions are compared with a high-accuracy benchmark solution. Several literature surveys by T.− M. Shih, for example [62], survey recent technical papers in numerical heat transfer. Recent papers presenting application of finite elements to hyperbolic heat conduction appear in references 63–65.

REFERENCES

1. E. L. Wilson and R. E. Nickell, "Application of the Finite Element Method to Heat Conduction Analysis," *Nucl. Eng. Des.*, Vol. 4, 1966, pp. 276–286.

2. A. F. Emery and W. W. Carson, "An Evaluation of the Use of the Finite Element Method in the Computation of Temperature," *J. Heat Transfer*, Vol. 93, 1971, pp. 136–153.

3. R. H. Gallagher and R. H. Mallett, "Efficient Solution Processes for Finite Element Analysis of Transient Heat Conduction," *J. Heat Transfer*, Vol. 93, 1971, pp. 257–263.

4. J. Heuser, "Finite Element Method for Thermal Analysis," Goddard Space Flight Center, Greenbelt, Md., *NASA TN D-7274,* 1973.

5. E. L. Wilson, K. J. Bathe, and F. E. Peterson, "Finite Element Analysis of Linear and Nonlinear Heat Transfer," *Nucl. Eng. Des.*, Vol. 29, 1974, pp. 110–124.

6. O. C. Zienkiewicz, *The Finite Element Method*, 3d ed., McGraw-Hill, New York, 1977.

7. B. M. Irons and S. Ahmad, *Techniques of Finite Elements*, Ellis Horwood, Chichester, UK, 1979, p. 349.

8. G. E. Myers, "The Critical Time Step for Finite Element Solutions to Two-dimensional Heat Conduction Transients," *J. Heat Transfer*, Vol. 100, 1978, pp. 120–128.

9. A. F. Emery, K. Sugihara, and A. T. Jones, "A Comparison of Some of the Thermal Characteristics of Finite Element and Finite Difference Calculations of Transient Problems," *Numer. Heat Transfer*, Vol. 2, 1979, pp. 97–113.

10. P. M. Gresho, R. L. Lee, and R. L. Sani, "Advection Dominated Flows, with Emphasis on Consequences of Mass Lumpings," in *Finite Elements in Fluids*, Vol. 3, R. H. Gallagher, O. C. Zienkiewicz, J. T. Oden, M. Morandi Cecchi, and C. Taylor (eds.), Wiley, New York, 1978, pp. 335–350.

11. P. M. Gresho and R. L. Lee, "Don't Suppress the Wiggles—They're Telling You Something," *Comput. Fluids*, Vol. 9, 1981, pp. 223–253.

12. K. H. Hsu and J. S. Brown, "On the Verification of Finite Element Transient Temperature Analysis With Thermally Orthotropic Material Properties," in *Pressure Vessels and Piping Computer Program Evaluation and Qualification*, PVP-PB-024, D. E. Dietrich (ed.), presented at the Energy Technology Conference, American Society of Mechanical Engineers, Houston, Texas, September 18–23, 1977, pp. 83–101.

13. C. A. Anderson and O. C. Zienkiewicz, "Spontaneous Ignition: Finite Element Solutions for Steady and Transient Conditions," *J. Heat Transfer*, Vol. 96, 1974, pp. 398–404.

14. E. A. Thornton and A. R. Wieting, "A Finite Element Thermal Analysis Procedure for Several Temperature-Dependent Parameters," *J. Heat Transfer*, Vol. 100, 1978, pp. 551–553.

15. J. R. Lyness, D. R. J. Owen, and O. C. Zienkiewicz, "The Finite Element Analysis of Engineering Systems Governed by a Non-Linear Quasi-Harmonic Equation," *Comput. Struct.*, Vol. 5, 1975, pp. 65–79.

16. T. R. J. Hughes, "Unconditionally Stable Algorithm for Nonlinear Heat Conduction," *Comput. Methods Appl. Mech. Eng.*, Vol. 10, 1977, pp. 135–139.

17. M. A. Hogge, "A Survey of Direct Integration Procedures for Nonlinear Transient Heat Transfer," International Conference on Numerical Methods in Thermal Problems, University College, Swansea, UK, July 2–6, 1979. Aerospace Laboratory of the University of Liège, Liège, Belgium, *Report SA-75*; see also M. A. Hogge, "A Comparison of Two- and Three-Level Integration Schemes for Non-Linear Heat Conduction," in *Numerical Methods in Heat Transfer*, R. W. Lewis, K. Morgan, and O. C. Zienkiewicz (eds.), Wiley, New York, 1981, Chapter 4, pp. 75–90.

18. A. J. Baker and M. O. Soliman, "Utility of a Finite Element Solution Algorithm for Initial-Value Problems," *J. Comput. Phys.*, Vol. 32, 1979, pp. 289–324.

19. S. Orivuori, "Efficient Method for Solution of Nonlinear Heat Conduction Problems," *Int. J. Numer, Methods Eng.*, Vol. 14, 1979, pp. 1461–1476.

20. H. M. Adelman and R. Haftka, "On the Performance of Explicit and Implicit Transient Algorithms for Transient Thermal Analysis of Structures," *NASA Technical Memorandum 81880*, Langley Research Center, Hampton, VA, September 1980.

21. H. P. Lee, "Application of Finite Element Method in the Computation of Temperature with Emphasis on Radiative Exchanges," in *AIAA Progress in Astronautics and Aeronautics: Thermal Control and Radiation*, Vol. 31, C. L. Tien (ed.), MIT Press, Cambridge, MA, 1973, pp. 491–520.

22. H. P. Lee and C. E. Jackson, Jr., "Finite Element Solution for Radiative Conductive Analysis With Mixed Diffuse-Specular Surfaces," in *AIAA Progress in Astronautics and Aeronautics: Radiative Transfer and Thermal Control*, Vol. 49, A. M. Smith (ed.), AIAA, New York, 1976, pp. 25–46.

23. R. E. Beckett and S. C. Chu, "Finite Element Method Applied to Heat Conduction with Nonlinear Boundary Conditions," *J. Heat Transfer*, Vol. 95, 1973, pp. 126–129.

24. M. B. Marlowe, R. A. Moore, and W. D. Whetstone, "SPAR Thermal Analysis Processors Reference Manual, System Level 16," Engineering Information Systems, Inc., San Jose, CA, *NASA Contractor Report 159162*, October 1979.

25. S. T. Wu, R. E. Ferguson, and L. L. Altgilbers, "Application of Finite Element Techniques to the Interaction of Conduction and Radiation in an Absorbing, Scattering and Emitting Medium," AIAA 15th Thermophysics Conference, Snowmass, CO, July 14–16, 1980, *AIAA Paper No. 80-1486.*

26. R. Fernandes, J. Francis, and J. N. Reddy, "A Finite Element Approach to Combined Conductive and Radiative Heat Transfer in a Planar Medium," AIAA 15th Thermophysics Conference, Snowmass, CO, July 14–16, 1980, *AIAA Paper No. 80-1487.*

27. R. H. Gallagher, O. C. Zienkiewicz, J. T. Oden, M. M. Cecchi, and C. Taylor (eds.), *Finite Elements in Fluids*, Vol. 3, John Wiley and Sons, New York, 1978.

28. R. H. Gallagher, D. H. Norrie, J. T. Oden, and O. C. Zienkiewicz (eds.), *Finite Elements in Fluids*, Vol. 4, Wiley, New York, 1982.

29. R. H. Gallagher, J. T. Oden, O. C. Zienkiewicz, T. Kawai, and M. Kawahara (eds), *Finite Elements in Fluids*, Vol. 5, Wiley, New York, 1984.

30. R. H. Gallagher, G. F. Carey, J. T. Oden, and O. C. Zienkiewicz (eds.), *Finite Elements in Fluids*, Vol. 6, Wiley, New York, 1985.

31. R. H. Gallagher, R. Glowinski, P. M. Gresho, J. T. Oden, and O. C. Zienkiewicz (eds.), *Finite Elements in Fluids*, Vol. 7, Wiley, New York, 1988.

32. I. Christe, D. Griffiths, A. Mitchell, and O. C. Zienkiewicz, "Finite Element Methods for Second Order Differential Equations With Significant First Derivatives," *Int. J. Numer. Methods Eng.*, Vol. 10, 1976, pp. 1389–1396.

33. J. C. Heinrich and O. C. Zienkiewicz, "Quadratic Finite Element Schemes for Two-Dimensional Convective-Transport Problems," in *Int. J. Numer. Methods Eng.*, Vol. 11, 1977, pp. 1831–1844.

34. J. C. Heinrich, P. S. Huyakorn, O. C. Zienkiewicz, and A. R. Mitchell, "An Upwind Finite Element Scheme for Two Dimensional Convective Transport Equation," *Int. J. Numer. Methods Eng.*, Vol. 12, 1977, pp. 131–143.

35. D. K. Gartling, "Some Comments on the Paper by Heinrich, Huyakorn, Zienkiewicz and Mitchell," *Int. J. Numer. Methods Eng.*, Vol. 12, 1978, pp. 187–190.

36. D. W. Kelly, S. Nakazawa, and O. C. Zienkiewicz, "A Note on Upwinding and Anisotropic Balancing Dissipation in Finite Element Approximations to Convective Diffusion Problems," *Int. J. Numer. Methods Eng.*, Vol. 15, 1980, pp. 1705–1711.

37. A. Brooks and T. J. R. Hughes, "Streamline Upwind/Petrov–Galerkin Formulations for Convection Dominated Flows With Particular Emphasis on the Incompressible Navier–Stokes Equations," *Comput. Methods Appl. Mech. Eng.*, Vol. 32, 1982, pp. 199–259.

38. J. C. Heinrich, "A Finite Element Model for Double Diffusive Convection," Int. J. Numer. Methods Eng., Vol. 20, 1984, pp. 447–464.

39. J. Donea, S. Giuliani, H. Laval, and L. Quartapelle, "Time-Accurate Solution of Advection–Diffusion Problems by Finite Elements," *Comput. Methods Appl. Mech. Eng.* Vol. 45, 1984, pp. 123–145.

40. C.-C. Yu and J. C. Heinrich, "Petrov–Galerkin Methods for the Time-Dependent Convective Transport Equation," *Int. J. Numer. Methods Eng.*, Vol. 23, 1986, pp. 883–901.

41. C.-C. Yu and J. C. Heinrich, "Petrov–Galerkin Method for Multidimensional, Time-Dependent, Convective-Diffusion Equations," *Int. J. Numer. Methods Eng.*, Vol. 24, 1987, pp. 2201–2215.

42. P. J. Roache, *Computational Fluid Dynamics*, Hermosa, Albuquerque, NM, 1972, pp. 161–165.

43. E. A. Thornton and A. R. Wieting, "Evaluation of Finite Element Formulations for Transient Conduction Forced-Convection Analysis," *Numer. Heat Transfer*, Vol. 3, 1980, pp. 281–295.

44. B. Tabarrok and R. C. Lin, "Finite Element Analysis of Free Convection Flows," *Int. J. Heat Transfer*, Vol. 20, 1977, pp. 945–952.

45. R. S. Marshall, J. C. Heinrich, and O. C. Zienkiewicz, "Natural Convection in a Square Enclosure by a Finite Element, Penalty Function Method Using Primitive Fluid Variables," *Numer. Heat Transfer*, Vol. 1, 1978, pp. 315–330.

46. D. K. Gartling, "Convective Heat Transfer Analysis by the Finite Element Method," *Comput. Methods Appl. Mech. Eng.*, Vol. 12, 1977, pp. 365–382.

47. C. Taylor and A. Z. Ijam, "A Finite Element Numerical Solution of Natural Convection in Enclosed Cavities," *Comput. Methods Appl. Mech. Eng.*, Vol. 19, 1979, pp. 429–446.

48. D. K. Gartling, "A Finite Element Analysis of Volumetrically Heated Fluids in an Axisymmetric Enclosure," *Finite Elements in Fluids*, Vol. 4, R. H. Gallagher, D. H. Norrie, J. T. Oden and O. C. Zienkiewicz (eds.), Wiley, New York, 1982, pp. 233–250.

49. D. K. Gartling, "Finite Element Analysis of Convective Heat Transfer Problems With Change of Phase," in *Computer Methods in Fluids*, K. Morgan, C. Taylor and C. A. Brebbia (eds.), Pentech, London, UK, 1980, pp. 257–284.

50. D. K. Gartling and C. E. Hickox, "A Numerical Study of the Applicability of the Boussinesq Approximation for a Fluid-Saturated Porous Medium," *Int. J. Numer. Methods Fluids*, Vol. 5, 1985, pp. 995–1013.

51. J. C. Heinrich and C. C. Yu, "Finite Element Simulation of Buoyancy-Driven Flows With Emphasis on Natural Convection in a Horizontal Circular Cylinder," *Comput. Methods Appl. Mech. Eng.*, Vol. 69, 1988, pp. 1–27.

52. B. Ramaswamy, "Efficient Finite Element Method for Two-Dimensional Fluid Flow and Heat Transfer Problems," *in Numerical Heat Transfer, Part B*, Vol. 17, Hemisphere, Washington, DC, 1990, pp. 123–154.

53. D. W. Pepper and A. P. Singer, "Calculation of Convective Flow on the Personal Computer Using a Modified Finite-Element Method," in *Numerical Heat Transfer, Part A*, Vol. 17, Hemisphere, Washington, DC, 1990, pp. 379–400.

54. A. O. Tay and G. DeVahl Davis, "Application of the Finite Element Method to Convection Heat Transfer Between Parallel Plates," *Int. J. Heat Mass Transfer*, Vol. 14, 1971. pp. 1057–1069.

55. E. Ben-Sabar and B. Caswell, "A Stable Finite Element Simulation of Convective Transport," *Int. J. Numer. Methods Eng.*, Vol. 14, 1979, pp. 545–565.

56. M. B. Hsu and R. E. Nickell, "Coupled Convective and Conduction Heat Transfer by Finite Element Methods," in *Finite Element Methods in Flow Problems*, J. T. Oden, O. C. Zienkiewicz, R. H. Gallagher, and C. Taylor (ed.), UAH Press, University of Alabama at Huntsville, 1974, pp. 427–449.

57. C. Taylor and A. Z. Ijam, "Coupled Convective/Conduction Heat Transfer Including Velocity Field Evaluation," in *The Mathematics of Finite Elements and Applications II*, J. R. Whiteman (ed.), Academic, New York, 1976, pp. 333–348.

58. E. A. Thornton and P. Dechaumphai, "Finite Element Analyses of Plane Thermal Entry-Length Flows," in *Proceedings of the Third International Conference on Finite Elements in Flow Problems*, D. H. Norrie (ed.), held at Banff, Alberta, Canada, June 10–13, 1980, pp. 293–302.

59. E. A. Thornton and A. R. Wieting, "Finite Element Methodology for Thermal Analysis of Convectively Cooled Structures," in *Progress in Astronautics and Aeronautics: Heat Transfer and Thermal Control Systems*, Vol. 60, L. S. Fletcher (ed.), AIAA, New York, 1978, pp. 171–189.

60. E. A. Thornton and A. R. Wieting, "Finite Element Methodology for Transient Conduction/Forced-Convection Thermal Analysis," in *Progress in Astronautics and Aeronautics: Heat Transfer, Thermal Control, and Heat Pipes*, Vol. 70, W. B. Olstead (ed.). AIAA, New York, 1980, pp. 77–103.

61. G. deVahl. Davis and I. P. Jones, "Natural Convection in a Square Cavity: A Comparison Exercise," *Int. J. Numer. Methods Fluids*, Vol. 3, Wiley, 1983, pp. 227–248.

62. T.-M. Shih, "Literature Survey On Numerical Heat Transfer (1988–1989)," in *Numerical Heat Transfer, Part A*, Vol. 18, Hemisphere, Washington, DC, 1990, pp. 387–425.

63. K. K. Tamma and S. B. Railkar, "Specially Tailored Transfinite-Element Formulations for Hyperbolic Heat Conduction Involving Non-Fourier Effects," *Numer. Heat Transfer*, Part B, Vol. 15, 1989, pp. 211–226.

64. D. E. Glass, K. K. Tamma, and S. B. Railkar, "Hyperbolic Heat Conduction with Convection Boundary Conditions and Pulse Heating Effects," J. Thermophys. Heat Transfer, *Vol. 5, No. 1*, 1991, pp. 110–116.

65. K. K. Tamma and R. R. Namburu, "Hyperbolic Heat-Conduction Problems: Numerical Simulations via Explicit Lax–Wendroff-Based Finite Element Formulations," *J. Thermophys. Heat Transfer*, Vol. 5, No. 2, 1991, pp. 232–239.

PROBLEMS

1. A variational approach is an alternative to the method of weighted residuals for developing general steady-state finite element heat conduction equations. For the general solid shown in Figure 8.1 consider the functional

$$
I(T) = \int_\Omega \left[\frac{K_x}{2}\left(\frac{\partial T}{\partial x}\right)^2 + \frac{K_y}{2}\left(\frac{\partial T}{\partial y}\right)^2 + \frac{K_z}{2}\left(\frac{\partial T}{\partial z}\right)^2 - QT \right] d\Omega
$$

$$
- \int_{S_2} q_s T\, dS + \frac{1}{2}\int_{S_3} h(T - T_e)^2\, dS + \int_{S_4}\left(-\alpha q_r + \frac{1}{5}\sigma\epsilon T^5\right)T\, dS
$$

where x, y, and z are principal material axes and K_x, K_y, and K_z are principal thermal conductivities. Use the variational method of Chapter 3 to derive the general finite element equations 8.13–8.15 for steady-state heat conduction.

2. Conduction and surface convection in a fin (Figure P8.2) is governed by

$$-kA\frac{\partial^2 T}{\partial x^2} + \rho cA\frac{\partial T}{\partial t} + hp(T - T_\infty) = 0$$

$$T(0, t) = T_0 \qquad -k\frac{\partial T}{\partial x}(L, t) = 0$$

where p is the perimeter, and A is the cross-sectional area. Use the method of weighted residuals to develop a general finite element formulation with integral definitions for element capacitance, conductance, and heat load vectors. How would the formulation be modified if the boundary condition at $x = L$ was changed to $-k\partial T/\partial x = h(T - T_\infty)$?

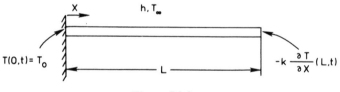

Figure P8.2

3. Consider steady one-dimensional conduction in a composite wall (Figure P8.3) with specified surface temperatures. Compute the following:

a. Temperatures at nodes 2 and 3.

b. Nodal heat flows at nodes 1 and 4.

c. The heat flux in element (1).

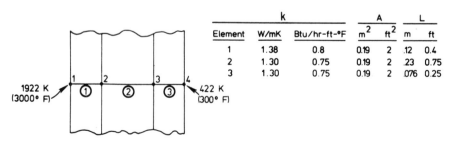

	k		A		L	
Element	W/mK	Btu/hr-ft-°F	m²	ft²	m	ft
1	1.38	0.8	0.19	2	.12	0.4
2	1.30	0.75	0.19	2	.23	0.75
3	1.30	0.75	0.19	2	.076	0.25

Figure P8.3

4. Consider heat transfer in the three-member truss shown in Figure P8.4. In a crude finite element model each member of the truss is represented by a single conduction, two-node rod element. Member 2 experiences internal heat generation, and member 3 experiences convective heat transfer to a surrounding medium. Compute the following:

a. Temperatures at nodes 2 and 3.

b. Nodal heat flow at node 1.

c. The heat fluxes in all elements.

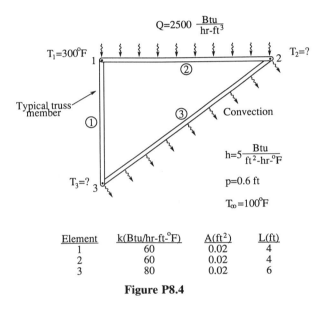

Element	k(Btu/hr-ft-°F)	A(ft²)	L(ft)
1	60	0.02	4
2	60	0.02	4
3	80	0.02	6

Figure P8.4

5. The finite element model of the truss in Problem 4 is refined to obtain a better representation of the temperature distribution in members 2 and 3. The new finite element model is shown in Figure P8.5. Using the data from the previous problem, use linear two-node elements to compute the

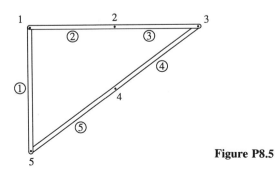

Figure P8.5

following:

a. Temperatures at nodes 2, 3, 4, and 5.

b. Nodal heat flow at node 1.

c. The heat fluxes in all elements.

Discuss how you would expect the solution to change if more elements are used to model truss members 1 and 2. Sketch expected temperature distributions in these members.

6. Heat is generated in a thick plate (Figure P8.6) at a rate of 3000 W/m² (290 Btu/h ft³). The plate is 0.2m (0.66 ft) thick and has a thermal conductivity $k = 25$ W/m K (14.5 Btu/h ft °F). The outer surfaces of the plate experience a convective heat loss to ambient air with a convection coefficient $h = 10$ W/m²K (1.76 Btu/h ft² °F) and a temperature $T_\infty = 300$ K (80°F). Assume one-half symmetry and model the plate with three nodes and two linear elements. Assuming steady heat transfer, perform the following:

a. Compute the nodal temperatures.

b. Verify a global energy balance.

c. Derive an analytical solution for $T(x)$ and compare the finite element results with the analytical predictions.

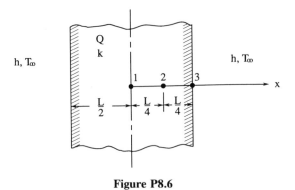

Figure P8.6

7. Consider steady two-dimensional conduction with internal heat generation for the square region shown in Figure P8.7. Compute nodal temperatures and element heat flows.

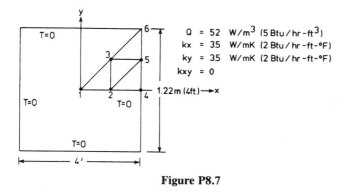

Figure P8.7

8. Consider a slab of unit thickness subjected to the boundary conditions in Figure P8.8. Using one-quarter symmetry and the finite element model shown, compute nodal temperatures and element heat flux components.

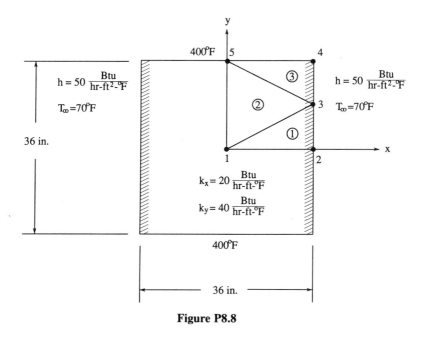

Figure P8.8

9. Read reference 5.

   a. Briefly discuss the equations used to describe transient heat transfer. Define the notation used and relate the notation to this text's notation.

  b. What nonlinearities are considered?

  c. How is the capacitance matrix computed? What justification is given?

  d. Briefly discuss the transient solution algorithm.

  e. Select and discuss one of the paper's sample problems.

10. Consider a three-node quadratic conduction/convection element (Figure P8.10. Derive the following for the element:

  a. Capacitance matrix.

  b. Conduction matrix.

  c. Convection matrix.

  d. Convection heat load vector.

  e. Heat load vector for internal heat generation.

  f. Heat load vector for surface heating.

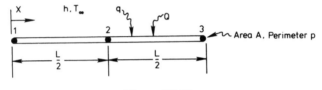

Figure P8.10

11. Consider a rectangular two-dimensional conduction element (Figure P8.11. Derive the following for the element:

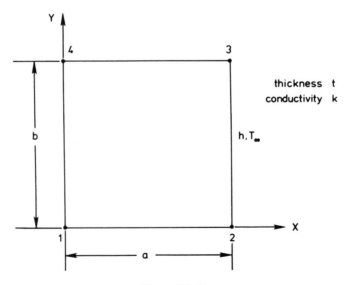

Figure P8.11

    a. Capacitance matrix.

    b. Conduction matrix.

    c. Heat load vector for surface heating q.

    d. Convection matrix and heat load vector for convection on a typical
       edge: $x = a, 0 \leq y \leq b$.

12. Consider steady heat conduction in a slab with specified wall temperatures (Figure P8.12).

    a. Write the matrix equation for computing nodal temperatures using
       the central finite difference approximation for the governing equation.

    b. Write a similar matrix equation using linear finite elements.

    c. Compare and discuss the results.

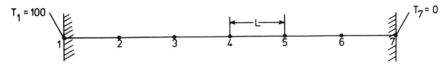

Figure P8.12

13. The geometry of a straight fin of triangular profile is shown in Figure
P8.13. In applications, the fin is used to enhance heat transfer. The heat
transfer boundary value problem may be formulated as

$$\frac{dq}{dx} + 2h(T - T_\infty) = 0$$

where q, the heat flux per unit length perpendicular to the page, is given
by

$$q = -k\frac{b}{L}x\frac{dT}{dx}$$

The boundary conditions are that $T(0)$ is finite and $T(L) = T_0$. Use the
method of weighted residuals to develop a general finite element formulation for the problem. Explain the physical significance of the boundary
terms for nodes at $x = 0$ and $x = L$.

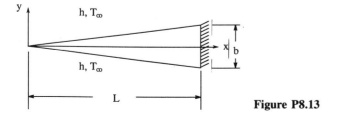

Figure P8.13

14. For the preceding fin problem derive the finite element matrices for a linear element with two nodes.

15. Consider axisymmetric heat conduction in a cylinder of inner radius $r = a$ and outer radius $r = b$. The governing differential equation is

$$-k\frac{1}{r}\frac{\partial}{\partial r}\left(r\frac{\partial T}{\partial r}\right) + \rho c\frac{\partial T}{\partial t} = Q$$

with

$$T(a,t) = T_0 \qquad -k\frac{\partial T}{\partial r}(b,t) = h[T(b,t) - T_\infty]$$

Use the method of weighted residuals to develop a general element formulation.

16. Rederive the rod element matrices, equations 8.21, for an element with linearly varying cross-sectional area. Use $A(x) = A_1 N_1(x) + A_2 N_2(x)$, where A_1 and A_2 are nodal values of the area, and write a similar equation for the perimeter p.

17. Derive the element matrices given in equation 8.25, for the axisymmetric triangle (Figure 8.4).

18. Set up the equations for Gauss–Legendre quadrature evaluation of the convection matrix for a typical edge, 1–2, of the isoparametric heat transfer element (Figure 8.5).

19. Consider a one-dimensional transient conduction problem given by

$$k\frac{\partial^2 T}{\partial x^2} + Q = \rho c\frac{\partial T}{\partial t}$$

with

$$T(x,0) = 0, \qquad 0 \le x \le L$$
$$T(0,t) = 0, \qquad T(L,t) = 0$$

and

$$Q(x,t)\begin{cases} = 0, & t < 0, \\ = Q_0, & t \ge 0, \end{cases} \qquad 0 \le x \le L$$

A two-element finite element model using symmetry leads to the matrix equations

$$\frac{1}{2}\begin{bmatrix} 4 & 1 \\ 1 & 2 \end{bmatrix}\begin{Bmatrix} \dot{T}_1 \\ \dot{T}_2 \end{Bmatrix} + \frac{1}{2}\begin{bmatrix} 2 & -1 \\ -1 & 1 \end{bmatrix}\begin{Bmatrix} T_1 \\ T_2 \end{Bmatrix} = \frac{5}{4}\begin{Bmatrix} 2 \\ 1 \end{Bmatrix}$$

Verify this equation and then compute the transient response using the implicit Crank–Nicolson method ($\theta = \frac{1}{2}$) for $0 \leq t \leq 50$. Use a time step $\Delta t = 0.5$ and sketch the nodal temperature time histories.

20. Solve for the transient response for Problem 8.19 using the explicit Euler method of time integration. First form the lumped capacitance matrix and then compute the critical time step (see Section 8.2.4). Compute and plot the transient nodal temperature histories.

21. A one-dimensional heat conduction problem may be described in some circumstances by the energy equation

$$\frac{\partial q}{\partial x} + \lambda \frac{\partial^2 T}{\partial t^2} + \rho c \frac{\partial T}{\partial t} = 0$$

and Fourier's law

$$q = -k \frac{\partial T}{\partial x}$$

where k is the thermal conductivity, λ is a known positive parameter, ρ is the density, and c is the specific heat.

  a. State appropriate boundary and initial conditions to complete the formulation.

  b. Classify the partial differential equation.

  c. Give a method of weighted residuals formulation for the problem.

  d. For a two-node linear element derive the element matrix arising from the term $\lambda \partial^2 T / \partial t^2$.

22. Read references 63–65, which discuss finite element computations for hyperbolic heat conduction.

  a. Discuss the differences between heat transfer processes modeled by the parabolic and hyperbolic heat conduction equations.

  b. Under what conditions does the hyperbolic equation predict temperatures different than the parabolic equation?

  c. Discuss differences between computational requirements for finite element solution algorithms for the parabolic and hyperbolic heat conduction equations.

  d. Discuss briefly the computational algorithms used in the references to solve the hyperbolic heat conduction equation.

23. Consider the nonlinear, steady heat conduction problem (Figure P8.23)

$$-\frac{dq}{dx} + Q = 0$$

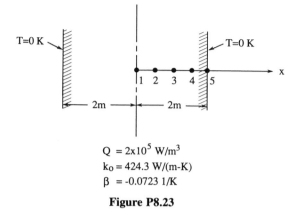

$$Q = 2\times10^5 \text{ W/m}^3$$
$$k_0 = 424.3 \text{ W/(m-K)}$$
$$\beta = -0.0723 \text{ 1/K}$$

Figure P8.23

where the heat flux is

$$q = -k(T)\frac{dT}{dx}$$

and the thermal conductivity is given by

$$k(T) = k_0(1 + \beta T)$$

where k_0 and β are given in Figure P8.23. The boundary conditions are

$$q(0) = 0 \qquad T(L) = 0$$

a. Formulate the Jacobian $[J]$ and residual load vector $[F]$ for a two-node linear conduction element.

b. Use Newton–Raphson iteration to compute the unknown temperatures for the data shown. Terminate the iterations when the value of $\Delta T_i/T_i$ at each node i is less than 10^{-3}.

c. Derive the analytical solution for the problem and assess the accuracy of your finite element solution.

24. An energy balance on a solid billet with volume V and surface area A gives the governing equation

$$\rho c V \dot{T} + hAT + \epsilon\sigma AT^4 = qA + hAT_c + \sigma\epsilon AT_r^4$$

where T_c and T_r are the exchange temperatures for surface convection and radiation, respectively. For a Newton–Raphson iteration solution for steady-state heat transfer derive the following:

a. The Jacobian J, equation 8.34a.

b. The unbalanced residual load F, equation 8.35a.

25. Use the results of Problem 24 and Newton–Raphson iteration to compute the steady-state radiation equilibrium temperature for $h = 0$, $T_c = 0$, $q = 15.8$ W/m^2 (5 Btu/h ft^2), $A = 0.186$ m^2(2 ft^2), $\sigma = 5.67 \times 10^{-8}$ W/m^2 K^4(0.171 $\times$ 10^{-8} Btu/h ft^2 °R^4), $\epsilon = 0.9$, and $T_r = 289$ K(520 °R). Use an initial temperature $T = 278$ K(500 °R) and a convergence criterion of 0.01%. Check the Newton–Raphson iteration by solving directly for the radiation equilibrium temperature.

26. Repeat problem 25, but use modified Newton–Raphson iteration. Compute the Jacobian once and hold it constant during the iterations. Compare the convergence rate with the rate obtained by updating the Jacobian at each iteration in Problem 25.

27. Derive the Jacobian and residual load vector for the nonlinear equation given in equation 8.40.

28. Compute the transient solution for the nonlinear model equation for $\lambda_0 = 1$ and $\lambda_1 = 0.03$. Use the Crank–Nicolson algorithm ($\theta = \frac{1}{2}$) and compare the results with the exact solution, equation 8.41.

29. Draw a flow chart for the following:

   a. An explicit nonlinear transient solution algorithm.

   b. An implicit nonlinear transient solution algorithm.

30. Verify the rod element consistent and lumped radiation conductance matrices, equations 8.50a and 8.50b.

31. The governing equation for one-dimensional channel flow (Figure p8.31) with surface heat exchange is

$$-kA\frac{\partial^2 T}{\partial x^2} + \rho c A u\frac{\partial T}{\partial x} + hb(T - T_\infty) + \rho c A\frac{\partial T}{\partial t} = 0$$

where b is the convection perimeter. Use the method of weighted residuals to develop a general finite element formulation for the element matrices. Use the Petrov–Galerkin approach with weighting functions $W_i \neq N_i$.

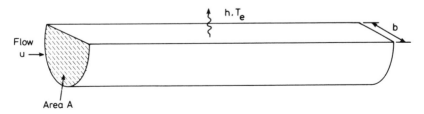

Figure P8.31

32. Derive the conventional (Bubnov–Galerkin) and upwind (Petrov–Galerkin) convection conductance matrix and convection heat load vector for the one-dimensional flow of Problem 31. Assume a linear element.

33. Show that with surface convection the condition for oscillations to develop in the conventional finite element solution for single-channel flow is

$$\frac{4}{3}\frac{L^2}{D^2}Nu + \frac{PeL}{D} > 2$$

where D is the hydraulic diameter defined by $D = 4A/b$; Nu is the Nusselt number, $Nu = hD/k$; and Pe is the Peclet number, $Pe = \rho cuD/k$.

34. Read the paper "Finite Element Analysis of Free and Forced Convection" by D. K. Gartling and R. E. Nickell, Chapter 6 in reference 27, pp. 105–121.

a. Summarize the continuum equations and list all assumptions.

b. What is the Boussinesq approximation?

c. Discuss the finite equations explaining what each term represents.

d. Explain how the authors solve the equations.

e. What is the frontal method?

f. Discuss one of the sample problems.

35. Read reference 45. Discuss the penalty function method, including advantages and disadvantages of the penalty function approach in comparison to the primitive variable approach presented in Section 8.5.

9

FLUID MECHANICS PROBLEMS

9.1 INTRODUCTION

During the last 30 years analysts relied on traditional finite difference methods to obtain computer-based solutions to difficult flow problems. The progress and success achieved in these pursuits have been, in many cases, noteworthy and remarkable. Slow viscous flow, boundary layer flows, and supersonic and hypersonic flows are just a few areas for which analysts have developed refined calculation procedures based on finite difference methods.

Yet there remain a number of problems for which finite difference methods encounter difficulties. Problems involving complex geometries, multiply connected domains, and complex boundary conditions always pose computational challenges. Finite element methods can help to alleviate these difficulties, because they offer easier ways to treat complex geometries requiring unstructured meshes, and they provide a more consistent method of using higher-order approximations. In some cases finite element methods can provide an approximate solution of the same order of accuracy as the finite difference method but at less expense. Regardless of the method used, accurate numerical solution of most viscous flow problems requires vast amounts of computer time and data storage. And, of course, problems of stability and convergence can occur with either method.

Only within the last 20 years have finite element methods been recognized as effective means for solving difficult fluid mechanics problems. In January 1974 a symposium [1, 2] on the application of finite element methods to flow problems was held at the University College of Swansea, Wales, to enable researchers to share their experiences and report their findings. The symposium established that finite element methods offer attractive advantages in a variety of fluid mechanics situations. The proceedings of the 1974 conference and the references contained therein provide a good indication of the level and scope of early finite element research in flow problems. Literature on the applications of finite element methods to fluid mechanics has rapidly increased since then.

This chapter is devoted to the description of finite element methods applied to fluid problems. For incompressible and compressible inviscid and viscous flows we lay the theoretical foundations, develop the element equations, and report promising findings of recent research efforts. Fundamental concepts and basic equations from fluid mechanics are reviewed in Appendix D for the reader's convenience.

9.2   INVISCID INCOMPRESSIBLE FLOW

Since all real fluids are viscous, an inviscid fluid is a hypothetical concept that simplifies the mathematics of fluid flow problems. Inviscid fluids experience no shearing stresses, and when they come into contact with a solid boundary they slip tangentially along it without resistance. Real fluids, of course, produce shear stresses, and they adhere to flow boundaries so that at the fluid–solid interface no slip occurs. Despite these differences between viscous and inviscid fluids, many practical problems in fluid mechanics can be analyzed with good accuracy when inviscid flow theory is used. Problems such as flow around streamlined objects, flow through converging or diverging passages, and flow over dams or weirs are just a few significant examples.

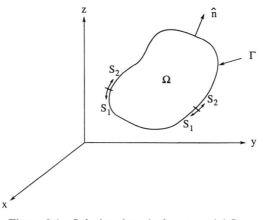

Figure 9.1   Solution domain for potential flow.

9.2.1   Problem Statement

The first attempts to extend finite element methods for the solution of fluid mechanics problems dealt with steady, inviscid, incompressible, irrotational flows [3–7]. As indicated in Appendix D, the problem of predicting details of this flow involves the solution of Laplace's equation. Finite element formulations may be developed using either the velocity potential or the stream function as the dependent variable. We present details of the finite element formulation for the velocity potential; the finite element formulation for the stream function appears as a problem at the end of the chapter.

Consider steady, incompressible, irrotational flow in a three-dimensional flow domain Ω bounded by a surface Γ (Figure 9.1). The problem is governed by the conservation of mass equation

$$\frac{\partial u}{\partial x} + \frac{\partial v}{\partial y} + \frac{\partial w}{\partial z} = 0 \tag{9.1}$$

where u, v, and w are components of the velocity vector $\mathbf{V}$. Using a potential function Φ, the velocity components are defined by

$$u = -\frac{\partial \Phi}{\partial x} \qquad v = -\frac{\partial \Phi}{\partial y} \qquad w = -\frac{\partial \Phi}{\partial z} \tag{9.2}$$

where we select the minus sign so that the positive flow direction is in the direction of decreasing Φ (analogous to the positive direction of heat conduction being defined in the direction of decreasing temperature). If we substitute equations 9.2 into the mass equation 9.1, we see that Φ satisfies Laplace's equation. But we choose to begin the method of weighted residuals finite element formulation with the mass equation so that the natural boundary conditions will appear directly in the surface integral that results from the weak (integral) form of equation 9.1. The boundary conditions specify the velocity potential on some surface segment S_1 to insure uniqueness of the solution and specify the normal derivative of the velocity potential on some surface segment S_2 where the velocity component normal to the boundary is known. Thus the boundary conditions are

$$\Phi = \Phi_1(x, y, z) \qquad \text{on } S_1 \qquad (9.3\text{a})$$

$$\frac{\partial \Phi}{\partial n} = -V_n = -(\mathbf{V} \cdot \hat{n}) \qquad \text{on } S_2 \qquad (9.3\text{b})$$

where Φ_1 is the specified velocity potential, n denotes the direction normal to the boundary, and V_n denotes a velocity component normal to the boundary. On a fixed, solid boundary, no flow penetration requires V_n to be zero. Since the flow is inviscid, there can be a velocity component tangent to the boundary. On a moving solid boundary, V_n is the normal velocity component of the moving surface.

The solution of equations 9.1–9.2 subject to the boundary conditions, equations 9.3, produces the velocity components at every point in the flow domain. For a steady, inviscid, incompressible, and irrotational flow, we may apply Bernoulli's equation between any two points in the flow. Thus, once the velocity components are known, pressure differences between points can be computed from Bernoulli's equation:

$$P + \tfrac{1}{2}\rho V^2 + \rho g z = \text{constant} \qquad (9.4)$$

where P denotes pressure, ρ is the density, V is the magnitude of the velocity ($V^2 = u^2 + v^2 + w^2$), g is the acceleration of gravity, and z is the height of a point above a reference plane.

The governing equations and boundary conditions are a special case of the equilibrium problem formulated in Chapter 7. There we develop the finite element formulation from a variational principle. However, since few variational principles exist for fluid mechanics problems, we present the finite element formulation from the more general method of weighted residuals in this chapter.

9.2.2   Finite Element Formulation

The solution domain Ω is divided into M elements of r nodes each. By the usual procedure, we express the velocity potential and velocity components within each element by

$$\Phi^{(e)}(x, y, z) = \sum_{i=1}^{r} N_i(x, y, z) \Phi_i \tag{9.5a}$$

$$u^{(e)}(x, y, z) = -\sum_{i=1}^{r} \frac{\partial N_i}{\partial x}(x, y, z) \Phi_i \tag{9.5b}$$

$$v^{(e)}(x, y, z) = -\sum_{i=1}^{r} \frac{\partial N_i}{\partial y}(x, y, z) \Phi_i \tag{9.5c}$$

$$w^{(e)}(x, y, z) = -\sum_{i=1}^{r} \frac{\partial N_i}{\partial z}(x, y, z) \Phi_i \tag{9.5d}$$

or in matrix notation,

$$\Phi^{(e)}(x, y, z) = \lfloor N(x, y, z) \rfloor \{\phi\} \tag{9.6a}$$

$$\begin{Bmatrix} u \\ v \\ w \end{Bmatrix} = -[B(x, y, z)]\{\Phi\} \tag{9.6b}$$

In the preceeding equations, $\lfloor N \rfloor$ is the velocity potential interpolation matrix, and $[B]$ is the velocity potential-gradient interpolation matrix:

$$\lfloor N(x, y, z) \rfloor = \lfloor N_1 \; N_2 \; \cdots \; N_r \rfloor \tag{9.7a}$$

$$[B(x, y, z)] = \begin{bmatrix} \dfrac{\partial N_1}{\partial x} & \dfrac{\partial N_2}{\partial x} & \cdots & \dfrac{\partial N_r}{\partial x} \\[2mm] \dfrac{\partial N_1}{\partial y} & \dfrac{\partial N_2}{\partial y} & \cdots & \dfrac{\partial N_r}{\partial y} \\[2mm] \dfrac{\partial N_1}{\partial z} & \dfrac{\partial N_2}{\partial z} & \cdots & \dfrac{\partial N_r}{\partial z} \end{bmatrix} \tag{9.7b}$$

Φ_i is the velocity potential at the ith node, and $\{\Phi\}$ is a vector of nodal velocity potentials. The second-order Laplace's equation for the velocity potential requires only C^0 continuity, and we may use velocity potential as the only nodal unknown. We focus on a single element and for simplicity omit the superscript (e). The method of weighted residuals (see Chapter 4) is used to derive the element equations starting with the mass equation 9.1. The method of weighted residuals requires

$$\int_{\Omega^{(e)}} \left(\frac{\partial u}{\partial x} + \frac{\partial v}{\partial y} + \frac{\partial w}{\partial z} \right) N_i \, d\Omega = 0, \qquad i = 1, 2, \ldots, r \tag{9.8}$$

where $\Omega^{(e)}$ is the domain for element (e). Following the procedures discussed in Chapter 4, we integrate equation 9.8 by Gauss's theorem, which introduces a surface integral on the element boundary $\Gamma^{(e)}$. Next we express the surface integral as the sum of integrals over S_1 and S_2. Thus we develop the weak form,

$$\int_{\Omega_e} \left| \frac{\partial N_i}{\partial x} \ \frac{\partial N_i}{\partial y} \ \frac{\partial N_i}{\partial z} \right| \left\{ \begin{matrix} u \\ v \\ w \end{matrix} \right\} d\Omega = \int_{S_1} (\mathbf{V} \cdot \hat{n}) N_i d\Gamma + \int_{S_2} V_n N_i \, d\Gamma,$$

$$i = 1, 2, \ldots, r \tag{9.9}$$

As the last step, we introduce the element velocities from equation 9.6b and write the set of equations in matrix form. The resulting element equations are

$$[K]\{\Phi\} = \{R_1\} + \{R_2\} \tag{9.10}$$

where

$$[K] = \int_{\Omega_e} [B]^T [B] \, d\Omega \tag{9.11a}$$

$$\{R_1\} = \int_{S_1} (\mathbf{V} \cdot \hat{n})\{N\} \, d\Gamma \tag{9.11b}$$

$$\{R_2\} = \int_{S_2} V_n\{N\} \, d\Gamma \tag{9.11c}$$

The element coefficient matrix $[K]$ is similar to the coefficient matrices encountered previously in Chapter 7; we have no difficulty in evaluating it for common element types. The vector $\{R_1\}$ represents "reactions" at nodes corresponding to specified values of Φ on S_1. The integrals are never evaluated on an element basis because the integrand is unknown. The values of the assembled global vector for $\{R_1\}$ may be computed after the global solution for $\{\Phi\}$ is obtained by the procedure described in Section 2.3.4. The vector $\{R_2\}$, however, corresponds to known values of V_n on S_2, and the vector is evaluated in the usual manner.

Assembly of the element equations to system equations follows standard procedure. Once we solve the resulting set of linear algebraic equations for the nodal values of the velocity potential, we may compute velocity components at selected points within an element using equation 9.6b. If we use the C^0 formulation described above, the velocity potential is continuous from element to element, but the velocity components are discontinuous between elements. If we use simplex elements such as triangles or tetrahedrons, velocity components are constant within an element.

9.2.3 Velocity Component Smoothing

As mentioned in the previous section, the potential formulation for steady, inviscid, incompressible, irrotational flows for elements with C^0 continuity leads to velocity components that are discontinuous from element to element. In regions of steep velocity gradients, the discontinuities may be relatively large even though they will diminish if the mesh is refined locally. To produce a velocity field that "looks better" some analysts employ a smoothing procedure that converts element velocity components to nodal velocity components. Nodal values of velocity components are useful for constructing graphical displays such as contour plots.

To illustrate a common procedure, suppose we obtain a two-dimensional potential flow solution using a mesh of linear triangles. Then a typical velocity component, say u, is constant within an element. The most simple form of velocity component smoothing is, at each node, to average the values of u from all elements surrounding the node. This procedure has the advantage that it is very simple, but it fails to take into account that contiguous elements may differ significantly in size. A more appropriate approach is to use an "area-weighted" averaging scheme. We compute the average value $\bar{u}_n$ at the nth node from

$$\bar{u}_n = \frac{\sum\limits_m u_m A_m}{\sum\limits_m A_m} \tag{9.12}$$

where the summation occurs over the m elements connected to node n. This smoothing scheme is often used, but the implementation of equation 9.12 directly in a finite element code is inconvenient, since lists of elements surrounding each node are not normally available within a program. We implement area-weighted smoothing more conveniently by looping over elements. For each element (e) we compute the element matrices defined by

$$\frac{A^{(e)}}{3}\begin{bmatrix} 1 & & \\ & 1 & \\ & & 1 \end{bmatrix}\begin{Bmatrix} u_i \\ u_j \\ u_k \end{Bmatrix} = \frac{u^{(e)}A^{(e)}}{3}\begin{Bmatrix} 1 \\ 1 \\ 1 \end{Bmatrix} \tag{9.13}$$

The left-hand side is a "lumped mass" matrix, and the right-hand side is a "load vector" based on the element velocity component. For each element, we then assemble the element contributions into global arrays using the element connectivity i, j, k. After completion of the element loop, we have assembled equations of the form

$$\lceil M \rfloor\{u\} = \{R\} \tag{9.14}$$

where $\lceil M \rfloor$ is a diagonal array. Thus equation 9.14 represents a set of uncoupled algebraic equations that we solve directly for the "smoothed" nodal values of the velocity components. The procedure based on equations 9.13–9.14 is relatively inexpensive, and it is convenient to implement. The procedure does not represent any improvement in solution accuracy, but it gives nodal velocity values that we may use to compute nodal pressures from Bernoulli's equation 9.4, or we may use the nodal values conveniently for graphical displays.

9.2.4 Example With Unstructured Mesh

The potential flow around a symmetric Joukowski airfoil was solved using the finite element potential formulation and smoothing procedure described in preceding sections. The problem statement is given in Figure 9.2. The shape of the airfoil is defined by the Joukowski transformation in the complex plane [8, 9]. The concept of adaptive remeshing was used in obtaining several solutions. The meshes and pressure contours are shown in Figures 9.3a–9.3e. The solution begins with a crude mesh that yields a relatively poor, non-smooth, solution for the pressure contours, Figure 9.3a. Based on this solution, a second mesh is generated with local mesh refinement in regions of steep pressure gradients, and a new flow solution is obtained, Figure 9.3b. This process continues through successive generation of new meshes and re-solution of the flow problem until smooth pressure contours, Figure 9.3e, indicate a converged solution. Basic concepts and further details of adaptive mesh refinement are described in Chapter 10.

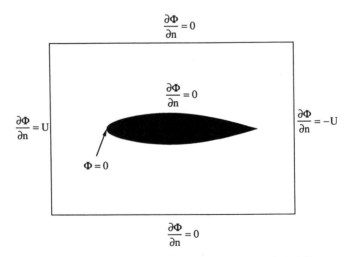

Figure 9.2 Potential flow around a Joukowski airfoil.

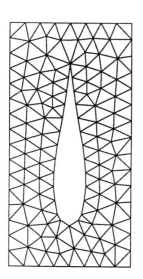

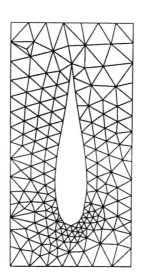

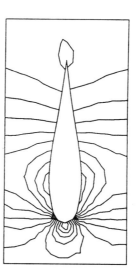

(a)   144 nodes

(b)   201 nodes

Figure 9.3   Adaptive meshes and pressure contours for flow around a Joukowski airfoil: (*a*) 144 nodes; (*b*) 201 nodes; (*c*) 315 nodes; (*d*) 990 nodes; (*e*) 2863 nodes.

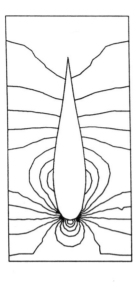

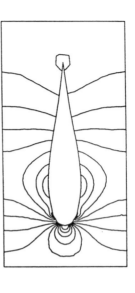

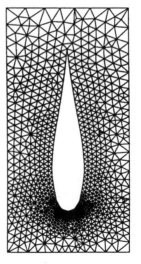

(c) 315 nodes

(d) 990 nodes

Figure 9.3 (*Continued*).

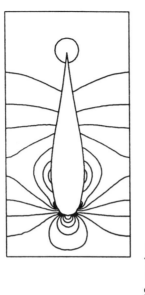

(e) 2863 nodes

Figure 9.3 (*Continued*).

9.2.5   The Kutta Condition

In the foregoing discussion we showed how to obtain a finite element solution
for the flow around an immersed body. But never did we give thought to the
physical implications of the results. The results of Figure 9.4 obtained by de
Vries and Norrie [5], serve to clarify this comment. The stagnation streamlines, which are a natural result of the solution, attach to the body at points
S_1 and S_2. However, the location of these attachment points is not physically
realistic because a real flow could not make the sharp turn around the
trailing edge as indicated. The reason for this unrealistic result is that we
have not considered the Kutta boundary condition in the solution. It is
important to employ the Kutta condition for bodies having sharp trailing
edges whenever we wish to calculate the lift of a body. Any body that is not
symmetric with respect to the approaching flow direction will have either
positive or negative lift. The Kutta condition states that the downstream
stagnation point S_2 should be at the sharp trailing edge. A procedure
suggested in reference 5 makes it possible to include the Kutta condition in
the solution using the streamline formulation. Figure 9.5 shows the correct
solution with the Kutta condition prescribed.

With a velocity potential formulation the potential field is multivalued in
the multiply connected domain due to the lift generated by the airfoil. The
circulation around the airfoil represents the difference between the multivalued potential at any point in the flow field. A unique solution can be
obtained by converting the flow region into a simply connected domain
through a "cut" or "splitting boundary" that extends from the airfoil surface
to one of the external flow field boundaries. A finite element formulation
using this approach is developed in reference 10, where the circulation
around an airfoil is computed explicitly as part of the finite element solution

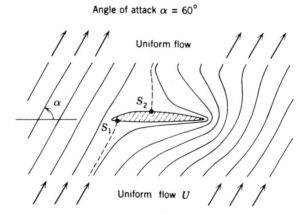

Figure 9.4   Solution for stream function (lines of constant ψ) for a NACA 4412
airfoil (Kutta condition not prescribed) [5].

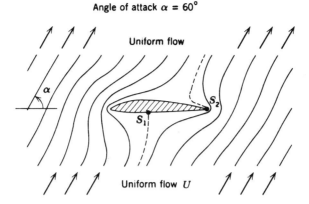

Figure 9.5   Solution for a stream function (lines of constant ψ) for a NACA 4412 airfoil (Kutta condition prescribed) [5].

by considering the circulation as a nodeless variable in the elements along the "splitting boundary." Further details of this approach and interesting applications to two- and three-dimensional flow fields appear in references 10 and 11.

9.3   VISCOUS INCOMPRESSIBLE FLOW WITHOUT INERTIA

The preceding section of this chapter centered on the finite element analysis of an important class of fluid mechanics problems involving inviscid fluids. In this section we consider the simplest of the viscous flow problems—namely, creeping flow (sometimes also called Stokes flow). If the full Navier–Stokes equations (equations D.4 in Appendix D) are made dimensionless, there results a dimensionless group known as the Reynolds number, Re, which represents the ratio of inertial forces to viscous forces in a fluid motion. When the Reynolds number is very small (usually $Re < 1$), the inertia forces are insignficant compared to the viscous forces and can be omitted from the governing momentum equations. Small Reynolds numbers characterize slow-moving flows and flows of very viscous fluids. These types of flows occur, for example, in viscometry and polymer processing.

9.3.1   Problem Statement

A steady two-dimensional, isoviscous flow is governed by the following equations:

Continuity:

$$\frac{\partial u}{\partial x} + \frac{\partial v}{\partial y} = 0 \tag{9.15a}$$

Momentum:

$$\frac{\partial}{\partial x}(\sigma_x - P) + \frac{\partial \tau_{xy}}{\partial y} = 0 \qquad (9.15b)$$

$$\frac{\partial \tau_{xy}}{\partial x} + \frac{\partial}{\partial y}(\sigma_y - P) = 0 \qquad (9.15c)$$

where the stresses are related to the velocity components by

$$\sigma_x = 2\mu \frac{\partial u}{\partial x}$$

$$\sigma_y = 2\mu \frac{\partial v}{\partial y} \qquad (9.15d)$$

$$\tau_{xy} = \mu \left(\frac{\partial u}{\partial y} + \frac{\partial v}{\partial x} \right)$$

Body forces have not been written in these equations, because they may be grouped with the pressure terms when the body force can be expressed as the gradient of a potential function. An alternative to the stress-component form of the momentum equations given above is to substitute the velocity components, equations 9.15d, into the momentum equations to express the momentum equations in terms of u and v, for example, see equations D.25 in Appendix D. We choose to start from the stress-component form because in a velocity–pressure formulation of the finite element equations the boundary conditions are handled in a consistent manner.

The boundary conditions consist of specifying the velocity components on a portion of the surface S_1 and the surface tractions on the remainder of the surface S_2. Thus

$$u = g(x, y) \qquad v = h(x, y) \qquad \text{on } S_1$$

$$\bar{\sigma}_x = (\sigma_x - P)n_x + \tau_{xy}n_y \qquad (9.16)$$

$$\bar{\sigma}_y = \tau_{xy}n_x + (\sigma_y - P)n_y \qquad \text{on } S_2$$

where $\bar{\sigma}_x$ and $\bar{\sigma}_y$ denote x and y components of the total surface traction and n_x and n_y denote direction cosines of the unit outer normal to the surface S_2.

To solve these equations by the finite element method we may use one of two different formulations: either we can introduce a stream function and work with one governing equation of the fourth order, or we can work with velocity and pressure as the field variables. Both approaches will now be outlined, because they illustrate important concepts essential for the solution of more complex problems.

9.3.2 Stream Function Formulation

As indicated in Appendix D, a stream function defined as

$$u = \frac{\partial \psi}{\partial y}, \qquad v = -\frac{\partial \psi}{\partial x}$$

can be used to combine equations 9.15b–9.15d into one relation (the biharmonic equation), which is

$$\nabla^4 \psi = \frac{\partial^4 \psi}{\partial x^4} + 2\frac{\partial^2 \psi}{\partial x^2 \partial y^2} + \frac{\partial^4 \psi}{\partial y^4} = 0 \tag{9.17}$$

and which is accompanied by the boundary conditions

$$\psi = g(x, y) \qquad \text{on } S_1 \tag{9.18a}$$

$$\nabla \psi \cdot \hat{n} = f(x, y) \qquad \text{on } S_2 \tag{9.18b}$$

The function ψ that satisfies equations 9.17 and 9.18 in domain Ω also minimizes the functional [12]

$$I(\psi) = \int_\Omega \left[\left(\frac{\partial^2 \psi}{\partial x^2}\right)^2 + \left(\frac{\partial^2 \psi}{\partial y^2}\right)^2 + 4\left(\frac{\partial^2 \psi}{\partial x \partial y}\right)^2 - 2\frac{\partial^2 \psi}{\partial x^2}\frac{\partial^2 \psi}{\partial y^2} \right] dx\, dy \tag{9.19}$$

Equation 9.19 provides a variational basis for deriving element equations. After dividing the solution domain Ω into M elements of r nodes each, we approximate $\psi(x, y)$ within each element as

$$\psi^{(e)} = \lfloor N \rfloor \{\psi_i^*\} \tag{9.20}$$

where the N_i are the interpolation functions and $\{\psi_i^*\}$ is a column vector of nodal parameters. Since $I(\psi)$ contains second-order derivatives of ψ, the interpolation functions in equation 9.20 should preserve C^1 continuity. To preserve C^1 continuity we must insure continuity of ψ as well as its normal derivatives along element interfaces. A choice of elements capable of this task is discussed in Section 5.8.2.

An intriguing aspect of formulating two-dimensional slow viscous flows in terms of a stream function is that the governing equation for ψ is identical in form to that governing Kirchhoff isotropic plate-bending theory. Furthermore, when plane elasticity problems are formulated in terms of the Airy stress function (see Appendix C), the same biharmonic equation results. This means that the same kinds of elements and, indeed, the same computer programs (with slight changes) can be used to solve all three types of problems.

To indicate the use of this analogy we refer to Section 6.5, where plate-bending problems were discussed. Comparing the functionals of equations 6.40 and 9.19, we see that the integrands are identical if the parameter ν (Poisson's ratio) in equation 6.40 is set equal to -1, and the functionals are identical if in the plate-bending problem the coefficient $Eh^3/24(1 - \nu^2)$ is taken as unity. Hence any element equations derived for plate-bending problems can also be used for the solution of creeping flow problems.

A number of authors [13–15] have used this approach to solve practical problems and to demonstrate the effectiveness of the solution procedure. Atkinson et al. [13, 14] considered both axisymmetric and planar two-dimensional flow fields of various geometries. For each of the problems considered they used three-node triangular elements with the stream function and its first two derivatives specified at each node. Atkinson et al. also studied creeping flow around a sphere, flow through a converging conical section, and developing flow in a circular pipe. Tong and Fung [15] investigated slow viscous flow in a capillary in the presence of moving particles suspended in the flow. This work has direct application to the biomedical problem of determining the influence of red blood cells on the flow in capillary blood vessels.

9.3.3   Velocity and Pressure Formulation

The stream function formulation we have just discussed offers the advantage that only one governing equation need be considered and the plate-bending analogy can be applied, but it suffers the disadvantage that elements achieving or approximating C^1 continuity must be employed. Unless such elements are used, it is impossible to specify the boundary conditions in terms of the normal and tangential derivatives of ψ.

By choosing the pressure and the velocity components as field variables, we can avoid this difficulty. The procedure is to apply the method of weighted residuals with Galerkin's criterion. Consider a two-dimensional flow domain Ω bounded by curve Γ. For a general element of this domain we select u, v, and P as nodal variables and interpolate these variables as follows:

$$u^{(e)} = \Sigma N_i^u(x, y)u_i = \lfloor N^u \rfloor\{u\} \tag{9.21a}$$

$$v^{(e)} = \Sigma N_i^v(x, y)v_i = \lfloor N^v \rfloor\{v\} \tag{9.21b}$$

$$P^{(e)} = \Sigma N_i^P(x, y)P_i = \lfloor N^P \rfloor\{P\} \tag{9.21c}$$

where N_i^u, N_i^v, and N_i^P are the interpolation functions, which need not necessarily be of the same order.

The Galerkin procedure (see Chapter 4) applied at node i of an isolated element becomes, in view of equations 9.15,

$$\int_{\Omega^{(e)}} \left[\frac{\partial}{\partial x}(\sigma_x - P) + \frac{\partial \tau_{xy}}{\partial y} \right] W_i \, d\Omega = 0 \qquad (9.22a)$$

$$\int_{\Omega^{(e)}} \left[\frac{\partial \tau_{xy}}{\partial x} + \frac{\partial}{\partial y}(\sigma_y - P) \right] W_i \, d\Omega = 0 \qquad (9.22b)$$

$$\int_{\Omega^{(e)}} \left(\frac{\partial u}{\partial x} + \frac{\partial v}{\partial y} \right) H_i \, d\Omega = 0 \qquad (9.22c)$$

where W_i and H_i are the weighting functions, which we take as

$$W_i = N_i \quad \text{and} \quad H_i = N_i^P$$

Since we have chosen the weighting functions for the momentum equations as the interpolation functions for the velocity components, we use a Bubnov–Galerkin approach. In an alternative approach some authors use a Petrov–Galerkin approach, with weighting functions $W_i \neq N_i$. The latter approach is used to introduce the concept of "upwind" finite elements. We discuss this concept for the convective-diffusion equation in Chapter 8.

If we integrate equation 9.22a and 9.22b, using the Gauss theorem given in Chapter 4, and then introduce the velocity components, equations 9.15d, and surface tractions from equations 9.16, we have

$$\int_{\Omega^{(e)}} \left[\left(2\mu \frac{\partial u}{\partial x} - P \right) \frac{\partial N_i}{\partial x} + \mu \left(\frac{\partial u}{\partial y} + \frac{\partial v}{\partial x} \right) \frac{\partial N_i}{\partial y} \right] d\Omega = \int_{S_2} \bar{\sigma}_x N_i \, d\Gamma \quad (9.23a)$$

$$\int_{\Omega^{(e)}} \left[\mu \left(\frac{\partial u}{\partial y} + \frac{\partial v}{\partial x} \right) \frac{\partial N_i}{\partial x} + \left(2\mu \frac{\partial v}{\partial y} - P \right) \frac{\partial N_i}{\partial y} \right] d\Omega = \int_{S_2} \bar{\sigma}_y N_i \, d\Gamma \quad (9.23b)$$

We note that by starting with the stress-component form of the momentum equations, the natural boundary conditions, that is, the surface tractions, appear directly in the "load" vectors on the right-hand side. Some of the first velocity and pressure formulations, for example, Yamada et al. [16] and Taylor and Hood [17], handled the boundary conditions and Gauss theorem integrations in other ways that yielded slightly different finite element formulations. The approach we follow was suggested by Gartling and Becker [18, 19].

When the approximations of equations 9.21 are substituted into equations 9.23a, 9.23b, and 9.22c, the matrix equations for node i result:

$$\left[2\int_{\Omega^{(e)}} \mu \frac{\partial N_i}{\partial x} \left\lfloor \frac{\partial N}{\partial x} \right\rfloor d\Omega + \int_{\Omega^{(e)}} \mu \frac{\partial N_i}{\partial y} \left\lfloor \frac{\partial N}{\partial y} \right\rfloor d\Omega \right] \{u\}$$

$$+ \int_{\Omega^{(e)}} \mu \frac{\partial N_i}{\partial y} \left\lfloor \frac{\partial N}{\partial x} \right\rfloor d\Omega \{v\} - \int_{\Omega^{(e)}} \frac{\partial N_i}{\partial x} \lfloor N^P \rfloor d\Omega \{P\} = \int_{S_2} \bar{\sigma}_x N_i \, d\Gamma$$

$$(9.24a)$$

$$\int_{\Omega^{(e)}} \mu \frac{\partial N_i}{\partial x} \left[\frac{\partial N}{\partial y} \right] d\Omega \{u\} + \left[2\int_{\Omega^{(e)}} \mu \frac{\partial N_i}{\partial y} \left\lfloor \frac{\partial N}{\partial y} \right\rfloor d\Omega \right.$$

$$+ \int_{\Omega^{(e)}} \mu \frac{\partial N_i}{\partial x} \left\lfloor \frac{\partial N}{\partial x} \right\rfloor d\Omega \Big] \{v\} - \int_{\Omega^{(e)}} \frac{\partial N_i}{\partial y} \lfloor N^P \rfloor d\Omega \{P\} = \int_{S_2} \bar{\sigma}_y N_i \, d\Gamma \quad (9.24b)$$

$$\int_{\Omega^{(e)}} N_i^P \left\lfloor \frac{\partial N}{\partial x} \right\rfloor d\Omega \{u\} + \int_{\Omega^{(e)}} N_i^P \left\lfloor \frac{\partial N}{\partial y} \right\rfloor d\Omega \{v\} = 0 \qquad (9.24c)$$

From these equations we can write the element matrix equations by inspection. Suppose that the velocity components are interpolated at r nodes of the element, while the pressure is interpolated at s nodes, where in general $r > s$. Then the matrix equations take the form

$$
\begin{array}{c}
r \\
r \\
s
\end{array}
\begin{bmatrix}
\overset{r}{[2K_{11} + K_{22}]} & \overset{r}{[K_{12}]} & \overset{s}{[L_1]} \\
\hline
[K_{12}]^T & [K_{11} + 2K_{22}] & [L_2] \\
\hline
[L_1]^T & [L_2]^T & [0]
\end{bmatrix}
\begin{Bmatrix}
u_1 \\
u_2 \\
\vdots \\
\underline{u_r} \\
v_1 \\
v_2 \\
\vdots \\
\underline{v_r} \\
P_1 \\
P_2 \\
\vdots \\
P_s
\end{Bmatrix}
=
\begin{Bmatrix}
R_{u_1} \\
R_{u_2} \\
\vdots \\
R_{u_r} \\
R_{v_1} \\
R_{v_2} \\
\vdots \\
R_{v_r} \\
0 \\
0 \\
\vdots \\
0
\end{Bmatrix}
\quad (9.25)
$$

where

$$[K_{11}] = \int_{\Omega^{(e)}} \mu \left\{ \frac{\partial N}{\partial x} \right\} \left\lfloor \frac{\partial N}{\partial x} \right\rfloor d\Omega \qquad (9.26\text{a})$$

$$[K_{22}] = \int_{\Omega^{(e)}} \mu \left\{ \frac{\partial N}{\partial y} \right\} \left\lfloor \frac{\partial N}{\partial y} \right\rfloor d\Omega \qquad (9.26\text{b})$$

$$[K_{12}] = \int_{\Omega^{(e)}} \mu \left\{ \frac{\partial N}{\partial y} \right\} \left\lfloor \frac{\partial N}{\partial x} \right\rfloor d\Omega \qquad (9.26\text{c})$$

$$[L_1] = -\int_{\Omega^{(e)}} \left\{ \frac{\partial N}{\partial x} \right\} \lfloor N^P \rfloor d\Omega \qquad (9.26\text{d})$$

$$[L_2] = -\int_{\Omega^{(e)}} \left\{ \frac{\partial N}{\partial y} \right\} \lfloor N^P \rfloor d\Omega \qquad (9.26\text{e})$$

$$\{R_u\} = \int_{S_2} \bar{\sigma}_x \{N\} d\Gamma \qquad (9.26\text{f})$$

$$\{R_v\} = \int_{S_2} \bar{\sigma}_y \{N\} d\Gamma \qquad (9.26\text{g})$$

We note that the complete coefficient matrix on the left-hand side of equation 9.25 is symmetric, although some of the submatrices are not symmetric.

Serious attention must be given to the choice of interpolation functions for velocity and pressure. Several different approaches have established that the interpolation functions for the velocity components should be one order higher than the pressure interpolation functions. Yamada et al. [16] arrive at this conclusion through consideration of a variational formula, Hood and Taylor [20] arrive at the same conclusion directly from considering errors in the weighted residual formulation, and Olson and Tuann [21] show that spurious rigid-body modes occur in the element coefficient matrix unless this conclusion is met. Bercovier and Pironneau [22] have confirmed the conclusion through a rigorous mathematical study of error estimates for Stokes flow. Typical finite elements for viscous flow maintain C^0 continuity for velocity components and pressure but use interpolation functions for velocity one degree higher than pressure. This means that there are more velocity unknowns than pressure unknowns.

The Stokes problem, formulated in equations 9.15–9.16, is a linear boundary value problem, and the mathematical behavior of the mixed finite element formulation shown by the interpolations given in equations 9.21 has been studied rigorously. The result is a mathematical convergence statement for the mixed finite element formulation known as the LBB stability condition after its originators—Ladyzhenskaya, Babuska, and Brezzi. The mathematics is beyond the scope of this text, but references 23–26 discuss the condition and its implications. The equations of Stokes flow are the same as the equations describing the incompressible behavior of an elastic medium. References 26–28 discuss finite element formulations for incompressible elasticity including issues closely related to Stokes flow.

For the Stokes flow, once the element equations have been evaluated, they can be assembled in the usual manner to form the system equations. The two types of boundary conditions given in equations 9.16 must be considered. On one portion of the boundary the velocity components are prescribed, and these are handled in the manner described in Chapter 2. On the remaining part of the boundary the surface tractions are prescribed, and these boundary conditions comprise the "load" vectors $\{R_u\}$ and $\{R_v\}$ via equations 9.26f and 9.26g. The assembled finite element equations are symmetric and may be solved by Gauss elimination. Care must be taken, however, so that a pressure degree of freedom does not appear first in the global equations because of the corresponding zero value in the coefficient matrix.

Numerical examples for the u, v, P formulation appear in the next section where we extend the approach for incompressible viscous flow with inertia.

9.4   VISCOUS INCOMPRESSIBLE FLOW WITH INERTIA

Finite element formulations for fluid mechanics problems classified as inviscid flows ($Re = \infty$) and slow viscous flows ($Re = 0$) were discussed in the preceding sections. In this section we discuss the more general flow problems that exist when $0 < Re < \infty$. These problems are inherently nonlinear because of the presence of the convective inertia terms.

The full Navier–Stokes equations representing a balance of inertia forces, pressure forces, and viscous forces are capable of describing some of the most interesting phenomena in fluid mechanics; unfortunately they are among the most difficult partial differential equations to solve. In very general, vague (and somewhat reckless) terms we can say that the mathematical and numerical difficulties involved in solving the governing equations for a particular laminar flow increase as the Reynolds number increases. The nature of the difficulties changes along the way, but we generally encounter stability and convergence problems when the transition Reynolds number range is approached, or even before. This seems to hold regardless of the particular computational scheme employed.

At least three fundamentally different finite element formulations appear in past research: a stream function formulation [29], a stream function and vorticity formulation [30–33], and velocity and pressure formulations [34–50]. Nowadays many researchers and engineering practitioners use software based on the latter formulation, and we develop it further in the balance of this section.

Several authors favor the velocity and pressure formulation as the most straightforward finite element procedure for solution of the nonlinear

Navier–Stokes equations. In addition to avoiding difficulties associated with the stream function or stream function and vorticity formulations, the following reasons are cited in favor of the velocity and pressure formulation:

- The formulation is readily extended to three dimensions.
- Only C^0 continuity is required of the element interpolation functions.
- Pressure, velocity, velocity gradient, and stress boundary conditions can be directly incorporated into the matrix equations.
- Free-surface problems are tractable.

There are several velocity and pressure formulations now in use. The first of these, the approach we used in our velocity and pressure formulation for Stokes flow, treats the velocity components and pressure as unknown flow variables and develops the finite element equations from simultaneous solution of the continuity and momentum equations. In a second approach the penalty function formulation eliminates the pressure as an unknown field variable through the use of a "penalty" parameter and solves modified momentum equations for the velocity components. In a third approach, upwinding is used with equal order interpolation functions to develop a velocity–pressure formulation where the momentum equations are solved separately from a pressure equation in an iterative, segregated solution scheme. We now present and discuss each of these formulations.

9.4.1 Mixed Velocity and Pressure Formulation

The approach [18, 19] to deriving the element equations relies on the Bubnov–Galerkin method. Let the velocity and pressure fields be interpolated over an element as indicated in equation 9.21 and choose these interpolation functions as the weighting functions. The development of the element equations follows that used for Stokes flow, except for the acceleration terms on the left-hand side of equations D.2 in Appendix D. The unsteady acceleration terms $\rho\,\partial u/\partial t$ and $\rho\,\partial v/\partial t$ cause no difficulty, but the nonlinear convective acceleration terms require special consideration. The Galerkin method can be used if we linearize the governing equations by approximating the nonlinear convective terms. Suppose that (u_n, v_n) is some approximate solution to the flow problem; for example, (u_n, v_n) could be the Stokes flow solution. Then the momentum equations can be written

$$\rho\left(\frac{\partial u}{\partial t} + u_n\frac{\partial u}{\partial x} + v_n\frac{\partial u}{\partial y}\right) = \frac{\partial}{\partial x}(\sigma_x - P) + \frac{\partial \tau_{xy}}{\partial y} \qquad (9.27a)$$

$$\rho\left(\frac{\partial v}{\partial t} + u_n\frac{\partial v}{\partial x} + v_n\frac{\partial v}{\partial y}\right) = \frac{\partial \tau_{xy}}{\partial x} + \frac{\partial}{\partial y}(\sigma_y - P) \qquad (9.27b)$$

The Galerkin criteria applied to equations 9.27 and the continuity equation 9.15a then leads to

$$\int_{\Omega^{(e)}}\left[-\rho\left(\frac{\partial u}{\partial t} + u_n\frac{\partial u}{\partial x} + v_n\frac{\partial u}{\partial y}\right) + \frac{\partial}{\partial x}(\sigma_x - P) + \frac{\partial \tau_{xy}}{\partial y}\right]N_i\,d\Omega = 0 \quad (9.28a)$$

$$\int_{\Omega^{(e)}}\left[-\rho\left(\frac{\partial v}{\partial t} + u_n\frac{\partial v}{\partial x} + v_n\frac{\partial v}{\partial y}\right) + \frac{\partial \tau_{xy}}{\partial x} + \frac{\partial}{\partial y}(\sigma_y - P)\right]N_i\,d\Omega = 0 \quad (9.28b)$$

$$\int_{\Omega^{(e)}}\left(\frac{\partial u}{\partial x} + \frac{\partial v}{\partial y}\right)N_i^P\,d\Omega = 0 \quad (9.28c)$$

Following the procedure of Section 9.3.2, we integrate the stress terms using the Gauss theorem and then introduce the natural boundary conditions in the resulting surface integral. The element matrices arising from the stress terms are identical to the terms we obtained for Stokes flow. New terms arise from the acceleration terms, but the derivation is straightforward and for brevity we will omit the details. The resulting element equations are

$$
\begin{array}{c}
\begin{array}{ccc} r & r & s \end{array} \\
\begin{array}{c} r \\ r \\ s \end{array}
\left[\begin{array}{c|c|c}
[M] & [0] & [0] \\ \hline
[0] & [M] & [0] \\ \hline
[0] & [0] & [0]
\end{array}\right]
\left\{\begin{array}{c} \{\dot{u}\} \\ \{\dot{v}\} \\ \{\dot{P}\} \end{array}\right\}
\end{array}
$$

$$
\begin{array}{c}
\qquad\quad\begin{array}{ccc} r & r & s \end{array} \\
+ \begin{array}{c} r \\ r \\ s \end{array}
\left[\begin{array}{c|c|c}
[C_{11} + C_{22}] & [0] & [0] \\ \hline
[0] & [C_{11} + C_{22}] & [0] \\ \hline
[0] & [0] & [0]
\end{array}\right]
\left\{\begin{array}{c} \{u\} \\ \{v\} \\ \{P\} \end{array}\right\}
\end{array}
$$

$$
\begin{array}{c}
\qquad\quad\begin{array}{ccc} r & r & s \end{array} \\
+ \begin{array}{c} r \\ r \\ s \end{array}
\left[\begin{array}{c|c|c}
[2K_{11} + K_{22}] & [K_{12}] & [L_1] \\ \hline
[K_{12}]^T & [K_{11} + 2K_{22}] & [L_2] \\ \hline
[L_1]^T & [L_2]^T & [0]
\end{array}\right]
\left\{\begin{array}{c} \{u\} \\ \{v\} \\ \{P\} \end{array}\right\}
=
\left\{\begin{array}{c} \{R_u\} \\ \{R_v\} \\ \{0\} \end{array}\right\}
\end{array}
$$

$$(9.29)$$

where

$$[M] = \int_{\Omega^{(e)}} \rho\{N\}\lfloor N \rfloor \, d\Omega \tag{9.30a}$$

$$[C_{11}] = \int_{\Omega^{(e)}} \rho u_n \{N\} \left\lfloor \frac{\partial N}{\partial x} \right\rfloor \, d\Omega \tag{9.30b}$$

$$[C_{22}] = \int_{\Omega^{(e)}} \rho v_n \{N\} \left\lfloor \frac{\partial N}{\partial y} \right\rfloor \, d\Omega \tag{9.30c}$$

and the remainder of the submatrices are defined in equations 9.26. To discuss the finite element equations we write equation 9.29 in the more compact form

$$[M]\begin{Bmatrix} \dot{u} \\ \dot{v} \\ \dot{P} \end{Bmatrix} + [C(u_n, v_n)]\begin{Bmatrix} u \\ v \\ P \end{Bmatrix} + [K]\begin{Bmatrix} u \\ v \\ P \end{Bmatrix} = \begin{Bmatrix} R_u \\ R_v \\ 0 \end{Bmatrix} \tag{9.31}$$

where the coefficient matrices $[M]$, $[C(u_n, v_n)]$, and $[K]$ are defined in equations 9.29. The coefficient matrix $[M]$ of the unsteady velocity components is the element mass matrix. We note from equation 9.29 that the element mass matrix is singular because of the null submatrix coefficient of $\dot{P}$. The mass matrix computed via equations 9.30a is known as a consistent mass matrix. As an approximation we may also use a lumped mass approach where the mass matrix is rendered diagonal by "lumping" a fraction of the element mass at the nodes. The principal advantage of a lumped mass matrix is to permit an explicit transient time integration algorithm; the disadvantage of "mass lumping" is a possible deterioration of solution accuracy. Explicit algorithms cannot be applied to equations 9.29 because $[M]$ is singular, but such algorithms can be used with the penalty function formulation to be discussed next. We discuss lumped versus consistent mass matrices for structural dynamics in Section 6.7 and for heat transfer in Section 8.2.4. Gresho et al. discuss the consequences of mass lumping in advection-dominated flows in reference 37.

The coefficient matrix $[C(u_n, v_n)]$ is a nonlinear asymmetric matrix arising from the convective acceleration terms that occur for both unsteady and steady flows. It represents the nonlinear flow character through its dependence on the velocity components (u_n, v_n) assumed known from a previous iteration. The asymmetry is characteristic of convective transport terms. Similar asymmetric matrices arise in finite element solutions of the convective-diffusion equation in heat transfer, which we discuss in Section 8.4. Physically, the matrix $[C(u_n, v_n)]$ represents the convection of momentum. The coefficient matrix $[K]$ is a linear symmetric matrix arising from the viscous terms and is common to both the Stokes and Navier–Stokes flow formulations. It represents the diffusion of momentum.

Once the element matrices are assembled to form the system equations and boundary conditions are imposed, we have the formidable task of solving the set of system equations with the form of equations 9.31. For steady flow we can drop the mass matrix term and solve the resulting nonlinear algebraic equations by iteration. For unsteady flow, we have a set of nonlinear ordinary differential equations that we can solve by combining a transient time integration algorithm with an iteration scheme at each time step. The solution of the nonlinear equations either for steady or unsteady flow is a significant computational challenge. Important considerations for high-quality solutions include the proper element choice, effective solution algorithms, correct use of the boundary conditions, and a proper discretization of the solution domain. These and other important aspects of the numerical solutions have been studied and evaluated in several papers, for example, references 34–38, which are recommended reading prior to numerical computations. Briefly stated, the computational experience of these researchers is that an element with a nine-node quadratic velocity interpolation (Lagrange) combined with a four-node pressure interpolation generally gives the best performance. Newton–Raphson iteration is generally preferred because of its superior convergence properties. Consistent use of the velocity and surface traction boundary conditions is necessary, and pressure boundary conditions are to be avoided. Finally, in regions of high velocity and/or pressure gradients refined meshes are required to avoid spatial oscillations in the flow variables.

Gresho et al. [38] demonstrated some of the capabilities of the u, v, P formulation with an unsteady solution for the flow around a circular cylinder. The finite element solution simulates the development of a Kármán vortex street and represents the first finite element primitive variable solution to this flow. Figure 9.6 shows the finite element mesh; Figure 9.6a shows the overall solution domain and boundary conditions and Figure 9.6b shows the details of the mesh near the cylinder. The nodal forces f_n at the flow inlet shown in Figure 9.6a were used to drive the flow from rest and were computed in a prior steady-state solution with a Reynolds number $Re = 100$. The lack of symmetry in these force values was attributed to a slight lack of symmetry in the nodal coordinates. The authors point out that this asymmetry was instrumental in initiating the vortex shedding. The mesh had 196 elements and 850 nodes; a total of 1929 unknowns (1700 velocities and 229 pressures) were computed in the transient solution. An implicit time integration scheme with an adaptive time-step algorithm plus Newton–Raphson iteration was used to solve equations 9.31. The paper presents several results, including time histories of the flow variables at points A, B, and C in Figure 9.6. Figure 9.7 shows a typical result, a contour plot of the relative streamlines at four times in the transient response. Figure 9.7, which represents snapshots

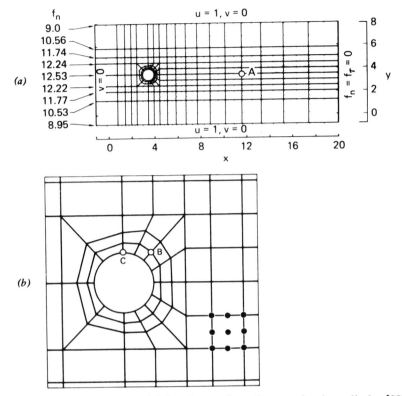

Figure 9.6   Finite element model for viscous flow about a circular cylinder [38]: (*a*) overall domain, mesh, and boundary conditions; (*b*) details of mesh near the cylinder; also shown are the nodes for a typical element.

of streamline contours as seen by an observer moving with the velocity $u = 1$ of the wall (see Figure 9.6), clearly reveals the existence of the classical Kármán vortex street.

In the velocity and pressure formulation so far we have used a Bubnov–Galerkin approach where in the momentum equations the weighting functions are chosen as the interpolation functions for the velocity components. This means that the formulation contains no "upwinding"; that is, we introduce no artificial diffusion. A consequence is that in the presence of steep velocity gradients solutions are prone to oscillate unless meshes are refined sufficiently (see reference 37). As an alternative to these possible oscillations we may introduce upwinding to stabilize the solution. The approach is to use a Petrov–Galerkin formulation and select the weighting functions so that the upwinding is introduced consistently in the formulation. A Petrov–Galerkin, streamline upwind formulation appears in reference 39. We describe upwinding for the convective-diffusion equation in Section 8.4.

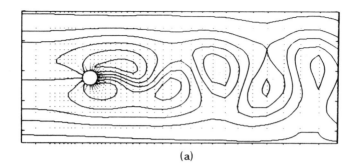

(a)

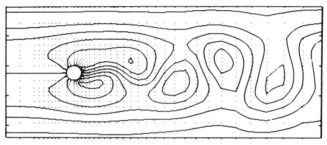

(b)

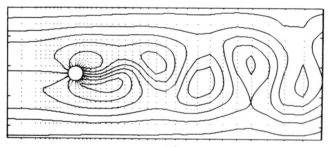

(c)

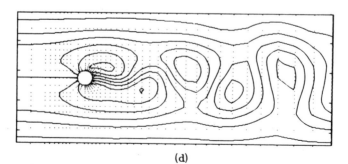

(d)

Figure 9.7   Relative streamlines for unsteady viscous flow about a circular cylinder
[38].

9.4.2 Penalty Function Formulation

In the penalty function formulation [51–54] of Navier–Stokes flow the pressure is represented by

$$P = -\lambda\left(\frac{\partial u}{\partial x} + \frac{\partial v}{\partial y}\right) \tag{9.32}$$

where $\lambda > 0$ is a parameter and the incompressibility condition, the continuity equation, is dropped. One motivation for this approach is a similar one [28] used in finite element elasticity solutions of incompressible solids where the incompressible solid is approximated as a slightly compressible one. In a viscous flow if the parameter λ is specified to have a large numerical value in the solution, the flow incompressibility condition will be approximately satisfied. The principal advantage of the penalty formulation is that the additional flow variable P is eliminated and so is the need for solving the continuity equation. To implement the concept we evaluate the element matrices associated with the penalty parameter λ in a special way. We discuss this point subsequently, but let us first formulate the element equations that we use in the penalty formulation.

We start with the momentum equations, equations 9.27, and proceed by using Galerkin's method as we did previously. The difference is that after the integration by the Gauss theorem, we eliminate the pressure as an unknown by introducing equation 9.32. For brevity we omit the details and present the final result:

$$
\begin{bmatrix} [M] & [0] \\ [0] & [M] \end{bmatrix} \begin{Bmatrix} \{\dot{u}\} \\ \{\dot{v}\} \end{Bmatrix} + \begin{bmatrix} [C_{11} + C_{22}] & [0] \\ [0] & [C_{11} + C_{22}] \end{bmatrix} \begin{Bmatrix} \{u\} \\ \{v\} \end{Bmatrix}
$$

$$
+ \begin{bmatrix} [2K_{11} + K_{22}] & [K_{12}] \\ [K_{12}]^T & [K_{11} + 2K_{22}] \end{bmatrix} \begin{Bmatrix} \{u\} \\ \{v\} \end{Bmatrix}
$$

$$
+ \lambda \begin{bmatrix} [L_{11}] & [L_{12}] \\ [L_{12}]^T & [L_{22}] \end{bmatrix} \begin{Bmatrix} \{u\} \\ \{v\} \end{Bmatrix} = \begin{Bmatrix} \{R_u\} \\ \{R_v\} \end{Bmatrix} \tag{9.33}
$$

where all terms have the same definitions as before, except for the "penalty" matrix associated with the penalty parameter λ, which is given by

$$[L_{11}] = \int_{\Omega^{(e)}} \left\{\frac{\partial N}{\partial x}\right\} \left\lfloor \frac{\partial N}{\partial x} \right\rfloor d\Omega \tag{9.34a}$$

$$[L_{12}] = \int_{\Omega^{(e)}} \left\{\frac{\partial N}{\partial x}\right\} \left\lfloor \frac{\partial N}{\partial y} \right\rfloor d\Omega \tag{9.34b}$$

$$[L_{22}] = \int_{\Omega^{(e)}} \left\{\frac{\partial N}{\partial y}\right\} \left\lfloor \frac{\partial N}{\partial y} \right\rfloor d\Omega \tag{9.34c}$$

For simplicity in considering the special treatment of the penalty matrix we consider steady flow and write equation 9.33 in the compact form

$$[[C] + [K] + \lambda[L]]\begin{Bmatrix} u \\ v \end{Bmatrix} = \begin{Bmatrix} R_u \\ R_v \end{Bmatrix} \tag{9.35}$$

where the $[C]$, $[K]$, and $[L]$ matrices are defined by inspection of equation 9.33. For an incompressible flow we must seek a solution to equation 9.35 as $\lambda \to \infty$. Since the matrices $[C]$ and $[K]$ are finite, as λ becomes large the solution tends to

$$\lambda[L]\begin{Bmatrix} u \\ v \end{Bmatrix} = \begin{Bmatrix} R_u \\ R_v \end{Bmatrix} \tag{9.36}$$

and if $[L]$ is nonsingular, then in the limit only the trivial solution for the flow velocities can result. The special consideration that is required to make the penalty function approach work is that $[L]$ must be a singular matrix. Most commonly used conforming elements produce a nonsingular $[L]$ if the integrals in equations 9.34 are evaluated exactly. The procedure used to make $[L]$ singular is to evaluate $[L]$ approximately (all other matrices are evaluated in the usual way) by using *reduced Gauss integration* [28]. For example, consider the popular four-node bilinear isoparametric quadrilateral element. For this element it is customary to employ a 2×2 Gauss–Legendre integration rule. With this level of integration the matrix $[L]$ is nonsingular. On the other hand, one-point Gauss–Legendre integration renders $[L]$ singular and makes it possible to use the penalty function approach to solve for a nontrivial velocity field via equation 9.35. The four-node quadrilateral element with reduced integration on the $[L]$ matrix is a popular and effective element in the penalty function formulation. Hughes et al. [51] recommend the elements and integration rules shown in Figure 9.8.

Other than the reduced integration rule on the $[L]$ matrix the finite element computations proceed in the usual way. Numerical studies [51] show that λ should be picked according to the rule

$$\lambda = c \max\{\mu, \mu Re\}$$

where c is a constant that depends on the computer word length, μ is the viscosity, and Re is a Reynolds number. For a floating point word length of 60–64 bits a recommended choice of c is 10^7.

The solutions of numerous steady- and unsteady-state problems [51–54] by the penalty function formulation show the approach to be very cost effective in solving the Navier–Stokes equations. The principal advantage of the penalty formulation is that the approach is easier to program, less storage is required, and there are fewer equations to solve. These benefits result

Element		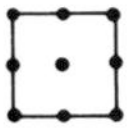	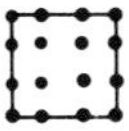
Shape Functions	Bilinear	Biquadratic	Bicubic
[L]	1 point	2×2	3×3
[C],[K]	2 × 2	3 × 3	4×4

Figure 9.8 Selective Gauss–Legendre integration rules for two-dimensional penalty function elements [51].

because the pressure has been eliminated. In addition, after the velocity components have been computed the pressures can be easily computed on an element basis via equation 9.32.

9.4.3 Equal-Order Velocity and Pressure Formulation

The additional complexity introduced in the velocity and pressure formulation by mixed interpolation led some researchers [55–59] to investigate equal-order interpolation formulations. Additionally, there was the motivation to improve efficiency of finite element incompressible flow solutions, particularly for three-dimensional problems, by reducing storage requirements and computational effort. Often alternative approaches are motivated by schemes that have been successful with finite different approximations on structured meshes. One such approach [57–59] we describe here is based on utilizing equal-order velocity and pressure interpolation and a segregated solution scheme.

The steady-flow formulation begins with the momentum equations written as

$$\rho u \frac{\partial u}{\partial x} + \rho v \frac{\partial u}{\partial y} = \mu \frac{\partial}{\partial x}\left(\frac{\partial u}{\partial x}\right) + \mu \frac{\partial}{\partial y}\left(\frac{\partial u}{\partial y}\right) - \frac{\partial P}{\partial x} \qquad (9.37a)$$

$$\rho u \frac{\partial v}{\partial x} + \rho v \frac{\partial v}{\partial y} = \mu \frac{\partial}{\partial x}\left(\frac{\partial v}{\partial x}\right) + \mu \frac{\partial}{\partial y}\left(\frac{\partial v}{\partial y}\right) - \frac{\partial P}{\partial y} \qquad (9.37b)$$

For a typical element we select u, v, and P as nodal variables and interpolate these variables as

$$u = \lfloor N \rfloor \{u\} \qquad (9.38a)$$

$$v = \lfloor N \rfloor \{v\} \qquad (9.38b)$$

$$P = \lfloor N \rfloor \{P\} \qquad (9.38c)$$

where the interpolation functions for velocities and pressures are the same.

We discretize the momentum equations using the Bubnov–Galerkin approach. However, a special treatment for the convection terms on the left-hand side of equations 9.37 is introduced. These terms are approximated by a monotone streamline upwind formulation [57]. In this approach, we rewrite the convection terms for a transported variable ϕ

$$u \frac{\partial \phi}{\partial x} + v \frac{\partial \phi}{\partial y}$$

in streamline coordinates as

$$u_s \frac{\partial \phi}{\partial s}$$

where u_s is the velocity in the streamline direction, and $\partial/\partial s$ is the gradient in the streamline direction. For pure convection (sometimes called advection), the last equation is constant along a streamline. These terms are treated as constants in the finite element formulation and are evaluated by a streamline tracing method. Details are given in reference 57.

Anticipating an iterative solution, we assume that nodal values for the velocities $\{u_n\}$, $\{v_n\}$ and pressure $\{P_n\}$ are known from a previous iteration. These values are used to linearize the nonlinear convective terms on the left-hand side and to define the pressure gradients on the right-hand side of equation 9.37. We multiply each momentum equation by a weighting function N_i, and integrate the viscous diffusion terms by parts using the Gauss theorem from Chapter 4. After writing the results in matrix form we obtain for a single element,

$$[A_u]\{u\} = \{R_u\} + \{R_{Px}\} \tag{9.39a}$$

$$[A_v]\{v\} = \{R_v\} + \{R_{Py}\} \tag{9.39b}$$

where the coefficient matrices $[A_u]$ and $[A_v]$ contain known contributions from convection and diffusion, and the load vectors are defined by

$$\{R_u\} = \int_{S_2} \{N\} \left(\frac{\partial u}{\partial x} n_x + \frac{\partial u}{\partial y} n_y \right) d\Gamma \tag{9.40a}$$

$$\{R_v\} = \int_{S_2} \{N\} \left(\frac{\partial v}{\partial x} n_x + \frac{\partial v}{\partial y} n_y \right) d\Gamma \tag{9.40b}$$

$$\{R_{Px}\} = -\int_{\Omega^{(e)}} \{N\} \frac{\partial P}{\partial x} d\Omega \tag{9.40c}$$

$$\{R_{Py}\} = -\int_{\Omega^{(e)}} \{N\} \frac{\partial P}{\partial y} d\Omega \tag{9.40d}$$

In equation 9.40, the vectors $\{R_u\}$ and $\{R_v\}$ are boundary terms, which we will not have to evaluate since typically they are "reactions" corresponding to specified nodal velocities, or they correspond to specified velocity gradients. These boundary terms will be dropped in subsequent discussion since they normally are not of interest and on outflow boundaries these vectors will be zero because velocity gradients are zero. The vectors $\{R_{Px}\}$ and $\{R_{Py}\}$ are load vectors from the pressure gradients. Using equation 9.37c, the pressure gradients can be evaluated for an element since nodal pressures $\{P_n\}$ are known from a previous iteration. From the element contributions, we assemble the global equations for the velocity components, modify the matrices for specified velocities on the boundary S_1, and solve each set of equations for the new velocity components.

We obtain the equation for updating the pressures by enforcing conservation of mass; that is, we discretize the continuity equation. Multiplying equation 9.15a by a weighting function N_i and integrating by parts yields for an element,

$$\int_{\Omega^{(e)}} N_i \left(\frac{\partial u}{\partial x} + \frac{\partial v}{\partial y} \right) d\Omega = - \int_{\Omega^{(e)}} \left(\frac{\partial N_i}{\partial x} u + \frac{\partial N_i}{\partial y} v \right) d\Omega$$

$$+ \int_{S_2} N_i (un_x + vn_y) \, d\Gamma = 0 \tag{9.41}$$

The surface integral in equation 9.41 represents the mass flow across the boundary and is zero on all slip or no-slip boundaries. Typically the integral is evaluated for element surfaces on the inflow plane.

To express the discretized continuity equation in terms of pressure, we return to the element momentum equations, 9.39. Ignoring the boundary terms, we write typical equations at the ith node as

$$A_{ii}^u u_i = - \sum_{j \neq i} A_{ij}^u u_j - \int_{\Omega^{(e)}} N_i \frac{\partial P}{\partial x} \, d\Omega \tag{9.42a}$$

$$A_{ii}^v v_i = - \sum_{j \neq i} A_{ij}^v u_j - \int_{\Omega^{(e)}} N_i \frac{\partial P}{\partial y} \, d\Omega \tag{9.42b}$$

Assuming the pressure gradient to be constant over an element, we can then write

$$u_i = \hat{u}_i - K_{Pi}^u \frac{\partial P}{\partial x} \tag{9.43a}$$

$$v_i = \hat{v}_i - K_{Pi}^v \frac{\partial P}{\partial y} \tag{9.43b}$$

where

$$\hat{u}_i = -\frac{\sum\limits_{j \neq i} A^u_{ij} u_j}{A^u_{ij}} \tag{9.44a}$$

$$\hat{v}_i = -\frac{\sum\limits_{j \neq i} A^v_{ij} v_j}{A^v_{ij}} \tag{9.44b}$$

$$K^u_{Pi} = \frac{\int_{\Omega^{(e)}} N_i \, dA}{A^u_{ii}} \tag{9.44c}$$

$$K^v_{Pi} = \frac{\int_{\Omega^{(e)}} N_i \, dA}{A^v_{ii}} \tag{9.44d}$$

Equations 9.43 represent coupling between velocities and pressure gradients.

We return now to the continuity equation, 9.41. With the interpolation equations 9.38, we may first write

$$-\int_{\Omega^{(e)}} \frac{\partial N_i}{\partial x} \left(\sum_j N_j u_j \right) d\Omega - \int_{\Omega^{(e)}} \frac{\partial N_i}{\partial y} \left(\sum_j N_j v_j \right) d\Omega$$

$$+ \int_{S_2} N_i (un_x + vn_y) \, d\Gamma = 0$$

and then introduce the nodal velocities u_j and v_j from equations 9.43. After the substitution and some rearrangement we obtain

$$\int_{\Omega^{(e)}} \frac{\partial N_i}{\partial x} \left(\sum_j N_j K^u_{Pj} \right) \frac{\partial P}{\partial x} \, d\Omega + \int_{\Omega^{(e)}} \frac{\partial N_i}{\partial y} \left(\sum_j N_j K^v_{Pj} \right) \frac{\partial P}{\partial x} \, d\Omega$$

$$= \int_{\Omega^{(e)}} \frac{\partial N_i}{\partial x} \left(\sum_j N_j \hat{u}_j \right) d\Omega + \int_{\Omega^{(e)}} \frac{\partial N_i}{\partial y} \left(\sum_j N_j \hat{v}_j \right) d\Omega$$

$$- \int_{S_2} N_i (un_x + vn_y) \, d\Gamma$$

Finally, after introducing the interpolation equation 9.38c for the pressure gradients we may write the element matrix equation for nodal pressures:

$$[K_x + K_y]\{P\} = \{R_u\} + \{R_v\} + \{R_b\} \tag{9.45}$$

where

$$[K_x] = \int_{\Omega^{(e)}} \left\{\frac{\partial N}{\partial x}\right\} \left(\sum_j N_j K^u_{Pj}\right) \left[\frac{\partial N}{\partial x}\right] d\Omega \qquad (9.46a)$$

$$[K_y] = \int_{\Omega^{(e)}} \left\{\frac{\partial N}{\partial y}\right\} \left(\sum_j N_j K^v_{Pj}\right) \left[\frac{\partial N}{\partial x}\right] d\Omega \qquad (9.46b)$$

$$\{R_u\} = \int_{\Omega^{(e)}} \left(\sum_j N_j \hat{u}_j\right) \left\{\frac{\partial N}{\partial x}\right\} d\Omega \qquad (9.46c)$$

$$\{R_v\} = \int_{\Omega^{(e)}} \left(\sum_j N_j \hat{v}_j\right) \left\{\frac{\partial N}{\partial y}\right\} d\Omega \qquad (9.46d)$$

$$\{R_b\} = -\int_{S_2} (un_x + vn_y)\{N\} \, d\Gamma \qquad (9.46e)$$

The element coefficient matrix $[K_x + K_y]$ has the form of a standard, symmetric diffusion matrix with orthogonal diffusion constants that depend on K^u_{Pj} and K^v_{Pj}. The load vectors $\{R_u\}$ and $\{R_v\}$ depend on the element velocity components. The load vector $\{R_b\}$ will be evaluated only on inflow boundaries with known velocities. The element pressure equations are assembled into global arrays, nodal pressure boundary conditions are imposed, and the resulting equations are solved directly for nodal pressures. Further discussion of boundary conditions appears in reference 58.

Figure 9.9 summarizes the iterative solution algorithm. As shown, we assume an initial velocity and pressure field. Based on this estimate, first the nodal velocities are computed using equation 9.39, then the nodal pressures are computed using equation 9.45. Finally, using the new pressure field, the velocities are updated using equation 9.42. This process continues until convergence is achieved. The segregated solution approach results in savings of computer storage because the equations for the velocities and pressure components are solved separately; the number of unknowns in each set of equations is less than or equal the number of nodes.

To illustrate the effectiveness of the iterative algorithm, we consider incompressible flow over a backward-facing step. Reference 45 describes a thorough study of the problem using a finite element algorithm based on mixed velocity and pressure interpolation. As shown in Figure 9.10a, a parabolic inflow velocity is specified over the upper one-half of the channel. All other channel surfaces have the no-slip velocity specification except the outflow plane where the boundary condition specifies the pressure to be zero. The Reynolds number Re is based on the channel height H, and we present results for $Re = 800$. For this Reynolds number, the flow has a large

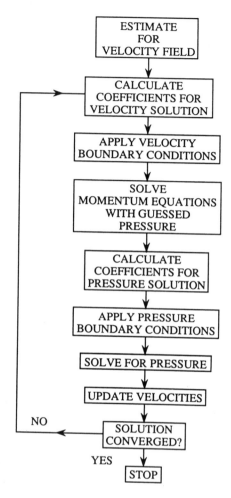

Figure 9.9 Iterative solution algorithm [58].

recirculation eddy on the lower wall for $0 < x < 6$ and a second recirculation eddy on the upper wall for $5 < x < 10$.

Reference 45 analyzes the problem with nine-node biquadratic Lagrange interpolation for the velocity components and a linear pressure distribution. A mesh refinement study varies the number of elements from 720 to 32,000 and the number of unknowns from 8426 to 355,362. A typical mesh was uniform across the channel and graded in the downstream direction. For the iterative, equal-order interpolation analysis conducted at the University of Virginia, bilinear quadrilateral elements were used. Figure 9.10b shows the graded mesh employed, which has 8000 elements and about 24,700 unknowns.

Figure 9.11 compares the u velocity components computed by the two algorithms at $x = 7$ and $x = 15$. The velocity profiles at $x = 7$ (Figure 9.11a) show the negative velocities that occur in the eddy on the upper wall. The

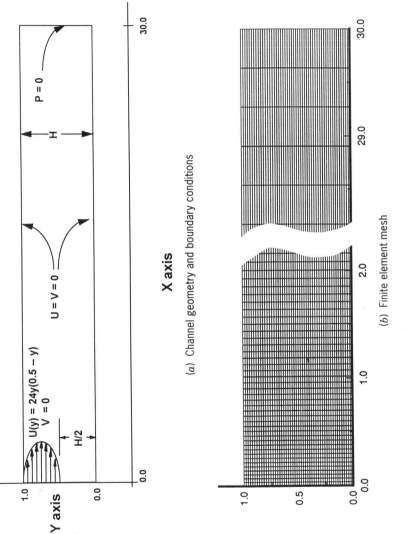

X axis

(a) Channel geometry and boundary conditions

(b) Finite element mesh

Figure 9.10 Incompressible flow over backward-facing step. (a) channel geometry and boundary conditions; (b) finite element mesh.

421

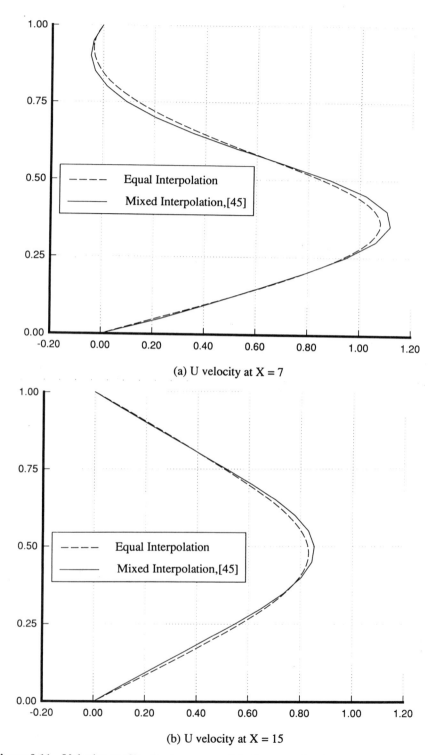

Figure 9.11 Velocity profiles for incompressible flow over backward step, $Re = 800$: (a) U velocity at $x = 7$; (b) U velocity at $x = 15$.

velocity distributions at $x = 15$ (Figure 9.11b) show that after the second eddy the flow is becoming fully developed on the way to the exit plane. Agreement between the analyses is very good with a maximum difference in the u velocities of 3%. Reference 45 observes that for $Re = 800$ experimental studies indicate that three-dimensional effects are likely to occur. Thus, the finite element analyses cannot be validated by comparisons with experimental data.

9.5 COMPRESSIBLE FLOW

The preceding sections of this chapter focus on finite element analysis of fluid mechanics problems involving incompressible fluids. In this section we consider compressible flow problems. The application of finite elements to compressible flows is a relatively new development; you will notice the references we cite describe research that began in the 1980s. A strong motivation for the research is applications to high-speed flow problems encountered in supersonic and hypersonic flight. Until the recent finite element methodology developments, the principal computational fluid dynamics algorithms were finite difference and finite volume schemes [60]. These algorithms have a dominant role in CFD, but recent advances in finite elements provide approaches that have significant benefits for some complex compressible flow problems.

Advances in finite element methodology for compressible flows include the development of new solution algorithms and adaptive mesh refinement strategies. The remarkably complex character of compressible flows places very strong demands on solution algorithms. High-speed flows around flight vehicles generate complex shock structures, thin boundary layers, shock–shock interactions, thin shear layers, and shock–boundary layer interactions. Resolution of flow details is required to predict aerodynamic pressure and skin friction distributions as well as aerodynamic heat fluxes that are critical for high-speed vehicle design. Many of these flow features are characterized by steep gradients that mandate extremely dense mesh refinement. These requirements mean that finite element analyses typically involve very large-scale computations with super-computers.

The conservation equations for unsteady compressible fluid flow offer one of the greatest computational challenges in all of mathematical physics. The highly nonlinear equations for unsteady compressible flow—often called the Navier–Stokes equations—constitute a mixed set of hyperbolic–parabolic equations. In comparison, the unsteady incompressible conservation equations considered in sections 9.3–9.4 are a mixed set of elliptic-parabolic equations. Thus analysis of compressible flow requires different solution algorithms than the algorithms we previously describe for incompressible flow.

This section presents basic finite element algorithms for compressible flow and lists references where further details of more advanced algorithms appear. We first consider low-speed flows with variable density and then high-speed flows. Examples illustrate applications of adaptive mesh refinement. Chapter 10 presents further details and additional references describing this important new development. The reader may refer to Appendix D for a brief review of the basic equations of compressible flow. The texts by John D. Anderson, Jr. [61, 62] provide an excellent description of the broad field of supersonic and hypersonic fluid mechanics.

9.5.1   Problem Statement

We consider a transient, laminar compressible flow in a two-dimensional flow domain Ω bounded by a surface Γ (Figure 9.12). The flow is governed by the equations for conservation of mass, momentum, and energy. Written in conservation form using matrix notation, the equations are

$$\frac{\partial}{\partial t}\{U\} + \frac{\partial}{\partial x}\{E_I - E_V\} + \frac{\partial}{\partial y}\{F_I - F_V\} = 0 \qquad (9.47)$$

where the vector $\{U\}$ denotes the conservation variables,

$$\{U\}^T = \begin{vmatrix} \rho & \rho u & \rho v & \rho E_t \end{vmatrix} \qquad (9.48a)$$

the vectors $\{E_I\}$ and $\{F_I\}$ denote the inviscid flux components,

$$\{E_I\} = \begin{vmatrix} \rho u & \rho u^2 + P & \rho u v & (\rho E_t + P)u \end{vmatrix} \qquad (9.48b)$$

$$\{F_I\} = \begin{vmatrix} \rho v & \rho u v & \rho v^2 + P & (\rho E_t + P)v \end{vmatrix} \qquad (9.48c)$$

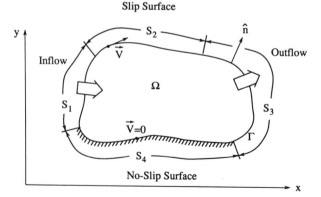

Figure 9.12   Solution domain for compressible flow.

and the vectors $\{E_V\}$ and $\{F_V\}$ denote the viscous flux components,

$$\{E_V\} = \begin{bmatrix} 0 & \sigma_x & \tau_{xy} & (u\sigma_x + v\tau_{xy} + q_x) \end{bmatrix} \tag{9.47d}$$

$$\{F_V\} = \begin{bmatrix} 0 & \tau_{xy} & \sigma_y & (u\tau_{xy} + v\sigma_y + q_y) \end{bmatrix} \tag{9.48e}$$

where ρ is the fluid density, u and v are velocity components, P is the pressure, a function of density and temperature $P = P(\rho, T)$, and E_t represents the total energy (per unit mass) given by

$$E_t = e + \tfrac{1}{2}(u^2 + v^2) \tag{9.49a}$$

$$e = c_v T \tag{9.49b}$$

where e is the internal energy, c_v is the specific heat at constant volume, and T is temperature. The stress components for a Newtonian fluid are

$$\sigma_x = \tfrac{2}{3}\mu(T)\left(2\frac{\partial u}{\partial x} - \frac{\partial v}{\partial y}\right) \tag{9.50a}$$

$$\sigma_y = \tfrac{2}{3}\mu(T)\left(2\frac{\partial v}{\partial y} - \frac{\partial u}{\partial x}\right) \tag{9.50b}$$

$$\tau_{xy} = \mu(T)\left(\frac{\partial u}{\partial y} + \frac{\partial v}{\partial x}\right) \tag{9.50c}$$

where $\mu(T)$ denotes the temperature dependent viscosity. The heat fluxes are

$$q_x = -k(T)\frac{\partial T}{\partial x} \tag{9.51a}$$

$$q_y = -k(T)\frac{\partial T}{\partial y} \tag{9.51b}$$

where $k(T)$ is the temperature-dependent thermal conductivity.

Equation 9.47 is solved subject to appropriate initial and boundary conditions. The initial conditions consist of specifying the spatial distributions of the conservation variables at an initial time. There are several boundary conditions that we may use. With reference to Figure 9.12, typical conditions include (1) a subsonic or supersonic inflow boundary, S_1; a slip surface (or symmetry plane), S_2; a subsonic or supersonic outflow boundary, S_3; and (4) a no-slip isothermal or adiabatic surface, S_4. Later examples illustrate these boundary conditions.

9.5.2 Low-Speed Flow With Variable Density

For sustained high-speed flight in the atmosphere, convectively cooled structures will be required for future hypersonic vehicles. Figure 9.13 depicts a typical cross section of a convectively cooled structure. An aerodynamic skin and coolant passage protect the primary aircraft structure from intense aerodynamic heating. The skin transfers the energy from the external aerodynamic heating to a low-temperature coolant flow. In the National Aero-Space Plane engine structure the coolant is cryogenic hydrogen later used as the propulsion system fuel. A typical Mach number of cryogenic hydrogen flow through a convectively cooled structure is about 0.1.

Normally for isothermal flows of compressible fluids if the Mach number is smaller than about 0.3, then pressure gradients are small enough so that the density is nearly constant, and the flow may be assumed incompressible. However, for convectively cooled structures with significant heat transfer, large density changes occur due to high local fluid temperatures, and the flow must be assumed compressible.

We begin our discussion of compressible flow with development of a finite element algorithm for low-speed but variable density flow. The flow is described by equation 9.47, and we anticipate that no shocks will be developed because the flow is subsonic. The algorithm [63] computes the steady flow solution by time marching the transient equations to steady state. A similar approach is used for supersonic inviscid flows in the next section, but there we include additional features in solution algorithms to capture the shocks encountered in supersonic flows.

In our previous formulations for incompressible flow, the dependent variables at the nodes are the primitive variables u, v, P, T. However, for the flow equations in conservation form a different approach is taken. At a node we take as basic unknowns the components of the vector of conservation variables $\{U\}$; that is, the nodal unknowns are U_i, $i = 1, 2, 3, 4$. Once we compute these unknowns, we use the definitions of the conservation variables in equation 9.48a and the total energy in equation 9.49 to solve for the

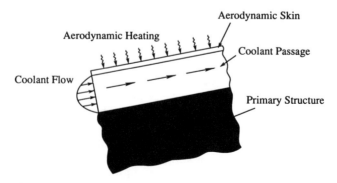

Figure 9.13 Cross section of convectively cooled structure.

TABLE 9.1. Conservation and Primitive Variables

Conservation Variable	Primitive Variables
$U_1 = \rho$	$\rho = U_1$
$U_2 = \rho u$	$u = \dfrac{U_2}{U_1}$
$U_3 = \rho v$	$v = \dfrac{U_3}{U_1}$
$U_4 = \rho E_t$	$E_t = \dfrac{U_4}{U_1}$
	$T = \dfrac{U_4}{c_v U_1} - \dfrac{u^2 + v^2}{2}$
	$P = P(\rho, T)$

primitive variables. Table 9.1 defines the conservation variables and gives the relationships between the primitive and conservation variables. Representing the unknowns in this way is called a group variable approach [64].

For simplicity, we develop the finite element equations for a single conservation equation of the form

$$\frac{\partial U}{\partial t} + \frac{\partial}{\partial x}(E_I - E_V) + \frac{\partial}{\partial y}(F_I - F_V) = 0 \qquad (9.52)$$

where the terms are analogous to the vector form of the conservation equations given in equation 9.47. The first step in the discretization is to multiply equation 9.52 by a weighting function W_i and integrate over the element domain $\Omega^{(e)}$:

$$\int_{\Omega^{(e)}} W_i \left[\frac{\partial U}{\partial t} + \frac{\partial}{\partial x}(E_I - E_V) + \frac{\partial}{\partial y}(F_I - F_V) \right] d\Omega = 0, \qquad i = 1, 2, \ldots, r$$

$$(9.53)$$

where r denotes the number of element nodes. The second step is to integrate the viscous flux components in equation 9.53 by parts using the Gauss theorem of Chapter 4. After rearrangement, the result is

$$\int_{\Omega^{(e)}} W_i \frac{\partial U}{\partial t} \, d\Omega + \int_{\Omega^{(e)}} W_i \left(\frac{\partial E_I}{\partial x} + \frac{\partial F_I}{\partial y} \right) d\Omega + \int_{\Omega^{(e)}} \left(\frac{\partial W_i}{\partial x} E_V + \frac{\partial W_i}{\partial y} F_V \right) d\Omega$$

$$- \int_{\Gamma^{(e)}} W_i (E_V n_x + F_V n_y) \, d\Gamma = 0 \qquad (9.54)$$

where n_x and n_y are the components of a unit vector normal to the element boundary $\Gamma^{(e)}$.

Next, we interpolate the conservation variable as well as the inviscid and viscous fluxes over each element in terms of the nodal values of these quantities:

$$U = \lfloor N \rfloor \{U\}$$

$$E_I = \lfloor N \rfloor \{E_I\}$$

$$F_I = \lfloor N \rfloor \{F_I\} \qquad (9.55)$$

$$E_V = \lfloor N \rfloor \{E_V\}$$

$$F_v = \lfloor N \rfloor \{F_V\}$$

where $\lfloor N \rfloor$ represents the element interpolation functions. After substituting the element interpolation functions from equation 9.55 into equation 9.54, we write the element matrix equation:

$$\lfloor M \rfloor \left\{ \frac{dU}{dt} \right\} + [L_x]\{E_I\} + [L_y]\{F_I\}$$

$$+ [K_x]\{E_V\} + [K_y]\{F_V\} = \{R_b\} \quad (9.56)$$

If we use the Bubnov–Galerkin approach with $W_i = N_i$, the element matrices are given by

$$[M] = \int_{\Omega^{(e)}} \{N\} \lfloor N \rfloor \, d\Omega \qquad (9.57a)$$

$$[L_x] = \int_{\Omega^{(e)}} \{N\} \left\lfloor \frac{\partial N}{\partial x} \right\rfloor d\Omega \qquad (9.57b)$$

$$[L_y] = \int_{\Omega^{(e)}} \{N\} \left\lfloor \frac{\partial N}{\partial y} \right\rfloor d\Omega \qquad (9.57c)$$

$$[K_x] = \int_{\Omega^{(e)}} \left\{ \frac{\partial N}{\partial x} \right\} \lfloor N \rfloor \, d\Omega \qquad (9.57d)$$

$$[K_y] = \int_{\Omega^{(e)}} \left\{ \frac{\partial N}{\partial y} \right\} \lfloor N \rfloor \, d\Omega \qquad (9.57e)$$

$$\{R_b\} = \int_{\Gamma^{(e)}} \{N\} (E_V n_x + F_V n_y) \, d\Gamma \qquad (9.57f)$$

Note that element matrices in equations 9.57a–9.57e are independent of flow parameters and need be evaluated only once at the beginning of the computations. Equations 9.56 and 9.57 represent the finite element spatial discretization. In the balance of the formulation, we develop the time-marching scheme.

For time marching, we use the recurrence relations developed for general field problems in Section 7.5.3. We let t_n denote a typical time in the response and seek the response at $t_{n+1} = t_n + \Delta t$, where Δt is the time step. We evaluate the equation at $t_\theta = t_n + \theta \Delta t$ where θ is a parameter such that $0 \le \theta\theta \le 1$. We write equation 9.56 at time t_θ and introduce the following approximations:

$$\left\{ \frac{dU}{dt} \right\}_\theta = \frac{1}{\Delta t}(\{U\}_{n+1} - \{U\}_n) \tag{9.58a}$$

$$\{E_I\}_\theta = (1 - \theta)\{E_I\}_n + \theta\{E_I\}_{n+1} \tag{9.58b}$$

$$\{F_I\}_\theta = (1 - \theta)\{F_I\}_n + \theta\{F_I\}_{n+1} \tag{9.58c}$$

$$\{E_V\}_\theta = (1 - \theta)\{F_V\}_n + \theta\{F_V\}_{n+1} \tag{9.58d}$$

$$\{F_V\}_\theta = (1 - \theta)\{F_V\}_n + \theta\{F_V\}_{n+1} \tag{9.58e}$$

$$\{R_b\}_\theta = (1 - \theta)\{R_b\}_n + \theta\{R_b\}_{n+1} \tag{9.58f}$$

Substituting equations 9.58 into equation 9.56 gives

$$[M]\{U\}_{n+1} = [M]\{U\}_n - \theta \Delta t\{R\}_{n+1} - (1 - \theta) \Delta t\{R\}_n \tag{9.59}$$

where

$$\{R\}_n = [L_x]\{E_I\}_n + [L_y]\{F_I\}_n + [K_x]\{E_V\}_n + [K_y]\{F_V\}_n - \{R_b\}_n \tag{9.60a}$$

$$\{R\}_{n+1} = [L_x]\{E_I\}_{n+1} + [L_y]\{F_I\}_{n+1} + [K_x]\{E_V\}_{n+1}$$
$$+ [K_y]\{F_V\}_{n+1} - \{R_b\}_{n+1} \tag{9.60b}$$

Equation 9.59 represents the contributions of a single element to the global equations. From such element contributions, global matrices are assembled in the usual way. For $\theta \ne 0$, the assembled global equations are nonlinear because of the $\{R\}_{n+1}$ vector on the right-hand side of equation 9.59; equation 9.60b shows that $\{R\}_{n+1}$ depends on unknown flux components at t_{n+1}.

At each time step the nonlinear equation 9.59 is solved by Newton–Raphson iteration. Because the solution of the Newton–Raphson equations is computationally expensive, the most efficient way to use the algorithm is to use $\theta = 1$ and take the largest possible time step to march to steady state. For some applications, best results are obtained by using a variable time step, starting with a small step and increasing the step size as the solution approaches steady state. Further details of algorithm implementation appear in reference 63.

To illustrate the algorithm performance, we present the solution for flow and heat transfer in a convectively cooled plate experiment. The experiment is part of a study at the University of Virginia to investigate flow–thermal–structural interactions for convectively cooled structures subjected to localized heating. The problem analyzed consists of air cooling a thin, aluminum, flat plate subjected to localized heating by a specified heat flux. Figure 9.14 shows the geometry of the flow domain and the boundary conditions.

To obtain the steady-state temperature distribution in the plate and coolant, the plate conservation of energy equation is solved simultaneously with the coolant equations. The walls of the channel are treated as insulated no-slip boundaries with the exception of the interface between the aluminum plate and the air coolant. At the interface, the flow velocity is zero, the temperatures of the plate and coolant are continuous, and the normal heat flux between the plate and coolant is continuous. The inflow and outflow boundary conditions are shown in Figure 9.14. The air coolant is assumed to be a perfect gas.

The flow is analyzed for a mesh of triangular elements (Figure 9.15) consisting of 2900 nodes and 5500 elements in the plate and 8700 nodes and 17,000 elements in the coolant. The mesh is concentrated heavily in regions of steep temperature gradients and within thin boundary layers. Boundary

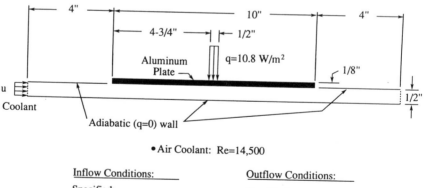

• Air Coolant: Re=14,500

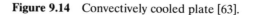

Inflow Conditions:	Outflow Conditions:
Specified:	Specified:
u=7.6 m/s	P=120 kN/m^2
v=0	
Density=1.4 kg/m^3	Unspecified:
Unspecified:	Density, u, v
Total Energy	

Figure 9.14   Convectively cooled plate [63].

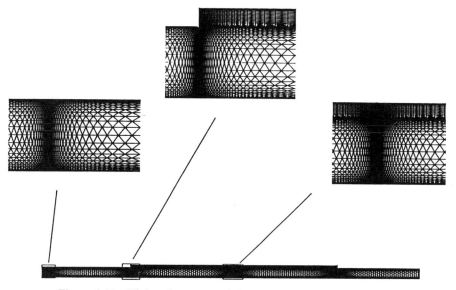

Figure 9.15   Finite element mesh for convectively cooled plate.

layers are quite thin because of the high Reynolds number of 14,500 (based on hydraulic diameter equal to twice the flow channel height).

Solution convergence was obtained by starting with small time steps at early times, and the time step is ramped up slowly until full convergence is achieved. The thermal response of the plate is much slower than for the air coolant, so the time step must become very large to attain convergence in a reasonable number of time steps. The nondimensional time steps ranged from $\nu = 400$ to 400×10^6 over 30 steps where ν is the Courant number defined by

$$\nu = \frac{\left(|u|_{\max} + c\right)\Delta t}{\Delta x_{\min}} \tag{9.61}$$

In the Courant number, $|u|_{\max}$ is the maximum absolute flow velocity, c is the speed of sound of air (for a perfect gas $c = \sqrt{\gamma R T}$ where γ is the ratio of specific heats, $\gamma = c_p/c_v$, R is the gas constant, and T is temperature), and $\Delta x_{\min}$ is the minimum mesh size. Convergence was measured by reducing temperature changes from one time step to the next by 5 or more orders of magnitude.

Figures 9.16 and 9.17 present computed coolant temperature and density profiles for the upper one-half of the channel. The figures show rather intense temperature and density gradients in the cross-flow planes of the coolant, but there is almost no temperature gradient through the thickness of

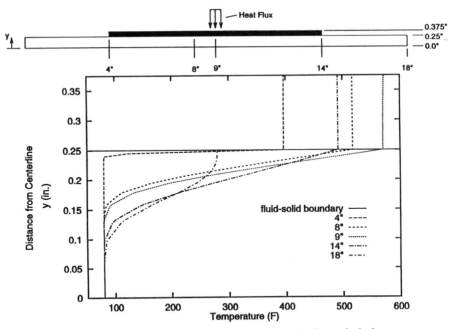

Figure 9.16 Temperature profiles for convectively cooled plate.

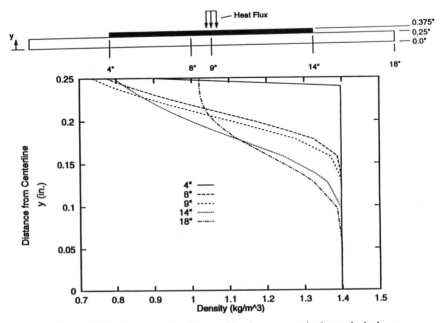

Figure 9.17 Coolant density profiles for convectively cooled plate.

the thin aluminum plate. There is, however, a significant variation of temperature along the plate in the flow direction due to the localized heating. Due to the increase in coolant temperature there is about a 50% reduction in the cooling density in the region of the localized heating.

9.5.3 High-Speed Flow

Since the early 1980s, a variety of finite element algorithms and adaptive refinement schemes have been under development for high-speed flow. Popular algorithms fall roughly into three classes: (1) Taylor–Galerkin and Runge–Kutta schemes [65–83], streamline upwind Petrov–Galerkin (SUPG) schemes [84–90], and least-squares schemes [91–93]. In comparison to algorithms based on finite differences and finite volumes for high-speed compressible flow, the development of finite element algorithms is in a relatively nascent state.

We begin by describing a basic finite element algorithm for inviscid flow and present some illustrative results. Then we extend our discussion with recent developments for viscous compressible flow.

Inviscid Flow   For inviscid, non-heat-conducting equations we drop the viscous flux vectors from equations 9.47 to obtain the set of conservations equations known as the Euler equations:

$$\frac{\partial}{\partial t}\{U\} + \frac{\partial}{\partial x}\{E_I\} + \frac{\partial}{\partial y}\{F_I\} = 0 \qquad (9.62)$$

where equations 9.48a–9.48c define the conservation variables and inviscid fluxes. The unsteady Euler equations are hyperbolic for all Mach numbers. In contrast, the steady equations have mixed elliptic–hyperbolic character; for subsonic flow regions the flow is elliptic, but for supersonic flow regions the flow is hyperbolic. Thus the most effective solution technique for steady flows is to time march the unsteady, hyperbolic equations to steady state. In the process, the solution algorithm captures shock discontinuities that occur in the flow. A measure of the mathematical effectiveness of an algorithm is the "crispness" of the captured shocks. Relatively unsophisticated algorithms tend to "smear" the extremely thin shock discontinuities over finite regions in the flow. More sophisticated (and computationally expensive) algorithms produce crisper shocks. We present an unsophisticated but effective algorithm and provide references to the more mathematically sophisticated schemes.

We describe an explicit time-marching scheme known as the Taylor–Galerkin algorithm. The Taylor-Galerkin algorithm was first proposed for convective transport problems [65] and then applied to compressible flows [66, 67, 70]. The basic concept of Taylor–Galerkin algorithms is to

(1) use Taylor series expansions in time to establish recurrence relations for time marching, and (2) use the method of weighted residuals with Galerkin's criteria to develop the finite element matrix equations describing the spatial distribution of the flow variables. For simplicity, we develop the Taylor–Galerkin algorithm for a single scalar equation:

$$\frac{\partial U}{\partial t} + \frac{\partial E}{\partial x} + \frac{\partial F}{\partial y} + 0 \qquad (9.63)$$

where the variables U, E, and F are analogous to the corresponding vector quantities in equation 9.62. Let $\{U\}^n$ denote the element nodal values of the flow variables $U(x, y, t)$ at time t_n. The time step Δt spans two typical times t_n and t_{n+1} in the transient response. The computation proceeds through two time levels, $t_{n+1/2}$ and t_{n+1}. At time level $t_{n+1/2}$, values for U that are constant within each element are computed explicitly. At time level t_{n+1}, the constant element values computed at the first time level are used to compute nodal values for U. In the time level t_{n+1} computations, element contributions are assembled to yield the global equations for nodal unknowns. The resulting equations are approximately diagonalized to yield an explicit algorithm.

Time Level $t_{n+1/2}$ To advance the solution to time level $t_{n+1/2}$, a truncated Taylor series yields

$$U(x, y, t_{n+1/2}) = U(x, y, t_n) + \frac{1}{2} \Delta t \frac{\partial U}{\partial t}(x, y, t_n) \qquad (9.64)$$

where terms of the order Δt^2 have been neglected. Then equation 9.63 is introduced on the right-hand side of equation 9.64 so that

$$U(x, y, t_{n+1/2}) = U(x, y, t_n) - \frac{1}{2} \Delta t \left[\frac{\partial E}{\partial x}(x, y, t_n) + \frac{\partial F}{\partial y}(x, y, t_n) \right]$$

$$(9.65)$$

At time level $t_{n+1/2}$, we assume the dependent variable $U(x, y, t_{n+1/2})$ to have a constant value $U_D^{n+1/2}$ within an element. Since the dependent variable has a constant value within an element, we assume that the flux components $E^{n+1/2}$ and $F^{n+1/2}$ are constant within an element as well.

At time level t_n in the response U, E, and F vary within an element and are interpolated from nodal values. Thus, we take the following spatial approximation within an element:

$$U^{(e)}(x, y, t_{n+1/2}) = U_D^{n+1/2} \qquad (9.66a)$$

$$U^{(e)}(x, y, t_n) = \lfloor N(x, y) \rfloor \{U\}^n \qquad (9.66b)$$

$$E^{(e)}(x, y, t_n) = \lfloor N(x, y) \rfloor \{E\}^n \qquad (9.66c)$$

$$F^{(e)}(x, y, t_n) = \lfloor N(x, y) \rfloor \{F\}^n \qquad (9.66d)$$

where $\lfloor N(x, y) \rfloor$ denotes element interpolation functions, and $\{U\}^n$ is a vector of the element nodal quantities. We now use the method of weighted residuals to derive the equation for $U_D^{n+1/2}$. First we substitute the spatial approximations equations 9.66 into equation 9.65 to give a residual; next we multiply by a weighting function of unity; and finally we integrate over the element domain $\Omega^{(e)}$. The result is

$$A^{(e)} U_D^{n+1/2} = \int_{\Omega^{(e)}} \lfloor N \rfloor \, d\Omega^{(e)} \{U\}^n$$

$$- \frac{\Delta t}{2} \left[\int_{\Omega^{(e)}} \left\lfloor \frac{\partial N}{\partial x} \right\rfloor d\Omega \{E\}^n + \int_{\Omega^{(e)}} \left\lfloor \frac{\partial N}{\partial y} \right\rfloor d\Omega \{F\}^n \right] \quad (9.67)$$

where $A^{(e)}$ denotes the area of an element. With equation 9.67, the dependent variable $U_D^{n+1/2}$ for each element can be computed explicitly using known nodal values for U, E, and F from the previous time t_n. We may interpret the constant element variable $U_D^{n+1/2}$ as a weighted average of an element's nodal values at time t_n. In advancing the solution to the next time level, the values of the dependent variables on the outflow surface are required also. Let $U_S^{n+1/2}$ denote the surface value on a typical element edge on the outflow surface S_3. Following the procedure used in developing equation 9.67, we may write

$$L^{(e)} U_S^{n+1/2} = \int_{\Gamma^{(e)}} \lfloor N \rfloor \, d\Gamma$$

$$- \frac{\Delta t}{2} \left[\int_{\Gamma^{(e)}} \left\lfloor \frac{\partial N}{\partial x} \right\rfloor d\Gamma \{E\}^n + \int_{\Gamma^{(e)}} \left\lfloor \frac{\partial N}{\partial y} \right\rfloor d\Gamma \{F\}^n \right] \quad (9.68)$$

where $L^{(e)}$ denotes the length of an element's edge $\Gamma^{(e)}$ on S_3. Thus, by using equations 9.67 and 9.68, we may advance explicitly element and surface values of the dependent variables to $t_{n+1/2}$. Beginning with nodal values $\{U\}^n$, $\{E\}^n$, and $\{F\}^n$ at time t_n, we use equation 9.67 to compute constant values $U_D^{n+1/2}$ for each element. In a similar way, we use equation 9.68 to compute constant surface values $U_S^{n+1/2}$ for element edges on the outflow surface S_3. These values are computed explicitly by looping through all elements and appropriate element edges.

Time Level t_{n+1}   To advance the solution to t_{n+1}, we use forward and backward truncated Taylor series expansions at $t_{n+1/2}$ to write the approximation

$$U(x, y, t_{n+1}) = U(x, y, t_n) + \Delta t \frac{\partial U}{\partial t}(x, y, t_{n+1/2}) \qquad (9.69)$$

Then following the approach used previously, we introduce equation 9.63 on the right-hand side to obtain

$$U(x, y, t_{n+1}) = U(x, y, t_n)$$
$$- \Delta t \left[\frac{\partial E}{\partial x}(x, y, t_{n+1/2}) + \frac{\partial F}{\partial y}(x, y, t_{n+1/2}) \right] \qquad (9.70)$$

At time levels t_n and t_{n+1}, we interpolate the variables U, E, and F using equations 9.66. We obtain the equations for the nodal values for $\{U\}^{n+1}$ by using the method of weighted residuals with weighting functions N_i. In the process, we integrate the last term in equation 9.70 by parts. These operations yield

$$[M]^{(e)}\{U\}^{n+1} = [M]^{(e)}\{U\}^n$$
$$+ \Delta t \left[\int_{\Omega^{(e)}} \left\{ \frac{\partial N}{\partial x} \right\} d\Omega \, E(t_{n+1/2}) + \int_{\Omega^{(e)}} \left\{ \frac{\partial N}{\partial y} \right\} d\Omega \, F(t_{n+1/2}) \right]$$
$$- \Delta t \int_{\Gamma^{(e)}} \left(n_x E_S^{n+1/2} + n_y F_S^{n+1/2} \right) d\Gamma \qquad (9.71)$$

where n_x and n_y are components of a unit normal vector on S_3. The element mass matrix $[M]^{(e)}$ is

$$[M]^{(e)} = \int_{\Omega^{(e)}} \{N\} \lfloor N \rfloor \, d\Omega \qquad (9.72)$$

To yield an explicit algorithm, the element mass matrix is diagonalized (or lumped) where the diagonal elements are computed from

$$M_i^{(e)} = \int_{\Omega^{(e)}} N_i \, d\Omega, \qquad i = 1, 2, \ldots, r \qquad (9.73)$$

See Sections 7.5.3 and 8.2.4 for a discussion of "lumped" mass matrices. Following usual finite element procedures we use element contributions to assemble global equations in the form

$$[^\backprime M_\backsim]\{U\}^{n+1} = [^\backprime M_\backsim]\{U\}^n + \{R\}^{n+1/2} \tag{9.74}$$

where $[^\backprime M_\backsim]$ denotes a diagonal system mass matrix and $\{R\}^{n+1/2}$ denotes a load vector computed from element contributions defined by equation 9.71. Since the mass matrices $[^\backprime M_\backsim]$ are diagonal, equation 9.74 defines an explicit computation where we advance the solution from time step t_n to time step t_{n+1} using the intermediate values computed at $t_{n+1/2}$.

When implementing the Taylor–Galerkin algorithm we need to consider several computational issues. These issues include (1) algorithm stability, (2) global or local time steps, and (3) artificial dissipation.

Algorithm Stability   The stability of the Taylor–Galerkin algorithm is studied in reference 79. Using a von Neumann stability analysis [60], the stability criteria for the Taylor–Galerkin algorithm are determined for the advection equation in one and two dimensions. The approach is to use a structured, uniform mesh to generate a finite difference relation corresponding to the finite element algorithm (see the problems at the end of the chapter). For the one-dimensional advection equation, the Taylor–Galerkin algorithm is identical to a two-step Lax–Wendroff finite difference scheme [60]. The stability analysis shows that the algorithm is stable provided $|\nu| < 1$, where ν is the Courant number defined by $\nu = a\,\Delta t/\Delta x$, where a is the speed of sound, and Δx is the mesh spacing. (CFD literature also calls the Courant number the CFL number, giving credit to three authors: Courant, Friedrichs, and Lewy.) For a general two-dimensional mesh the analysis is more complicated, and general conclusions are difficult to make. However, for a structured mesh of square, four-noded quadrilateral elements the stability criteria is also $|\nu| < 1$. If the quadrilateral elements are bisected by parallel diagonals, then six triangles surround a node. For stability of this triangular configuration, $|\nu| < 0.5$.

In the analysis of a practical problem with an unstructured mesh of quadrilaterals or triangles, the stability is usually determined by the smallest element. The allowable time step Δt_{CFL} is typically estimated by

$$\Delta t_{\mathrm{CFL}} = \sigma \frac{h}{a} \tag{9.75}$$

where h is a characteristic element dimension, and σ is a safety factor, $0 < \sigma < 1$.

Global or Local Time Steps   When finite difference algorithms are applied to graded meshes, the concept of local time steps is often used to accelerate convergence of time-marching solutions to steady state. The approach is to use a local time step at each node to avoid the limitation of having the smallest mesh spacing determine solution convergence for the entire mesh. If local time steps are used, the transient solution will not be time accurate, but convergence to steady state will be accelerated. With the Taylor–Galerkin algorithm, at time level $t_{n+1/2}$ where element quantities are computed, a local element time step is estimated by equation 9.75. At t_{n+1}, a local time step at each node is computed by using the minimum time step from the surrounding elements. Typically, the use of local time steps reduces the number of time steps required for convergence by a factor of two.

Artificial Dissipation   The Taylor–Galerkin algorithm is first-order accurate in time and second-order accurate in the spatial variables x and y. This second-order accuracy occurs because the finite element spatial discretization is equivalent to central difference approximations. In the presence of steep gradients, numerical solutions of flow problems based on central difference approximations tend to oscillate (see Section 8.4). For supersonic flow problems with shocks, artificial dissipation (sometimes called artificial viscosity) is used to reduce oscillations typical of second-order approximations. Several forms of artificial dissipation have been used in CFD, but a popular form used with the Taylor–Galerkin algorithm is due to Lapidus. Reference 69 describes the original Lapidus artificial viscosity concept and provides a multidimensional extension. We present an artificial viscosity approach based on the original Lapidus model.

The approach consists of introducing artificial viscous flux components $\overline{E}$ and $\overline{F}$ where

$$\overline{E} = \lambda A^{(e)} \left| \frac{\partial u}{\partial x} \right| \frac{\partial U}{\partial x} \tag{9.76a}$$

$$\overline{F} = \lambda A^{(e)} \left| \frac{\partial v}{\partial x} \right| \frac{\partial U}{\partial y} \tag{9.75b}$$

where λ is the artificial viscosity constant, and $A^{(e)}$ is the element area. The magnitudes of the velocity gradients serve as indicators to increase the artificial dissipation in regions of steep gradients. In the implementation, we correct (or smooth) the nodal values $\{U\}^{n+1}$ at time level t_{n+1} by incremental values $\{\Delta U\}$ computed from

$$[^{\smile}M_{\smile}]\{\Delta U\} = \{\overline{R}\} \tag{9.77a}$$

where $[^{\smile}M_{\smile}]$ is the diagonal mass matrix, and $\{\overline{R}\}$ is a load vector assembled from element contributions:

$$\{\overline{R}\}^{(e)} = -\Delta t \int_{\Omega^{(e)}} \left[\left\{ \frac{\partial N}{\partial x} \right\} \overline{E}^{n+1} + \left\{ \frac{\partial N}{\partial y} \right\} \overline{F}^{N+1} \right] d\Omega \tag{9.77b}$$

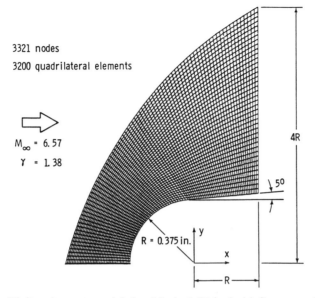

3321 nodes

3200 quadrilateral elements

M_∞ = 6. 57

γ = 1. 38

4R

5^0

R = 0.375 in.

y

x

R

Figure 9.18 Finite element model for Mach 6.57 inviscid flow over blunt leading edge [70].

where $\bar{E}^{n+1}$ and $\bar{F}^{n+1}$ are the artificial viscous flux components evaluated at time level t_{n+1}. Computational experience shows that $1 < \lambda < 2$ provides acceptable solution convergence without excessive smearing of shocks.

An early application of the Taylor–Galerkin algorithm was the inviscid analysis of a Mach 6.57 flow over a blunt leading edge of a wind tunnel panel holder [70]. The finite element model shown in Figure 9.18 consists of a structured mesh of 3200 quadrilateral elements and 3321 nodes. For an initial condition of uniform flow and a time step of 0.5×10^{-6} s, the steady-state solution was obtained in 3250 time steps. Convergence was achieved when the L_2 norm of the density change decreased three orders of magnitude. Density contours, Figure 9.19 show the bow shock that forms ahead of the body and the flow expansion. Solutions were also obtained from two other numerical techniques. Comparison of the three predictions for the velocity distributions at the exit plane, Figure 9.20, show good agreement. Other applications of the two-step Taylor–Galerkin scheme with triangular elements appear in references 66–68.

The results shown in Figures 9.18–9.20 are for the basic Taylor–Galerkin algorithm with Lapidus artificial dissipation. Solution methods based on higher-resolution schemes give sharper definition of flow discontinuities. Reference 71 describes a higher-resolution Taylor–Galerkin scheme based on a concept known as flux-corrected transport. Later applications [72] of the algorithm employ adaptive mesh refinement.

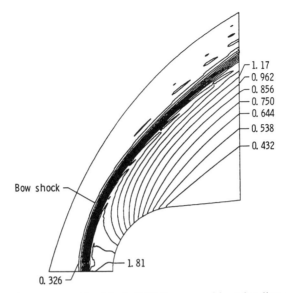

Figure 9.19   Density contours for Mach 6.57 flow over blunt leading edge [70].

Other time-marching schemes have also been developed. Reference 79 presents a Galerkin–Runge–Kutta algorithm. The algorithm is formulated by first performing the method of weighted residuals spatial discretization and then using a Runge–Kutta multistep time integration scheme. The algorithm was implemented with an adaptive quadrilateral and triangular mesh refinement algorithm.

Some of the most challenging high-speed flow problems are encountered on hypersonic vehicles such as the National AeroSpace Plane. One such problem is a shock–shock interaction that occurs on the engine cowl leading edge. Figure 9.21 depicts the hypersonic vehicle flow field. Note that the shock from the vehicle intersects the engine cowl bow shock. The flow field behind the bow shock contains a supersonic "jet" that impinges on the cowl surface with intense local pressure and heat flux.

Reference 79 uses the Galerkin–Runge–Kutta algorithm and adaptive mesh refinement to capture the complex shock interaction. Figure 9.22 presents a schematic of an experimental study of the shock interaction problem. (Note that for the experimental study, the configuration is upside-down relative to the actual flow shown in Figure 9.21.) The finite element mesh (Figure 9.23) obtained after five levels of adaptation contains 7380 nodes and 7513 elements. The density contours that result from this mesh (Figure 9.24) agree well with schlieren photographs (not shown). A comparison of predicted surface pressures with experimental data (Figure 9.25) shows good agreement. This problem shows the capability of the finite element approach to model complex flows with good resolution.

— Finite element shock capturing
--- Finite volume shock capturing
○   Method of lines shock fitting

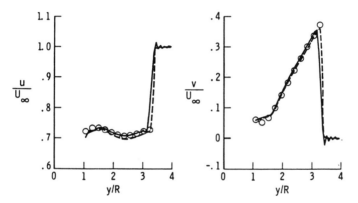

Figure 9.20 Comparative outflow velocity distributions for Mach 6.57 flow over blunt leading edge [70].

In our discussion so far, we describe explicit time-marching algorithms for solving the hyperbolic Euler conservation equations. Explicit algorithms have also been extended for viscous flows, [78, 81]. Using structured and unstructured meshes the algorithms have been used to solve several complex flows successfully. However, the drawback of explicit algorithms for viscous flow is the poor convergence rate. For high-speed flows, steep gradients at the boundary surfaces must be resolved with very small elements. Moreover,

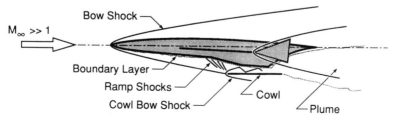
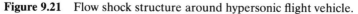

Figure 9.21 Flow shock structure around hypersonic flight vehicle.

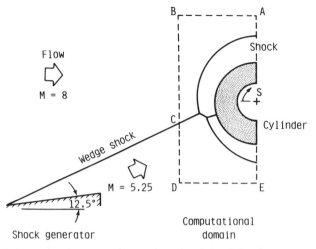

Figure 9.22   Schematic of an experimental study of shock impingement problem [79].

small element sizes imply small time steps because of an explicit algorithm's stability limit. Thus, explicit algorithms are not computationally effective for viscous flows.

The alternative to explicit schemes is to develop implicit, unconditionally stable algorithms. This is an area of current research and remains a challenge. Reference 82 describes a family of implicit/explicit Taylor–Galerkin

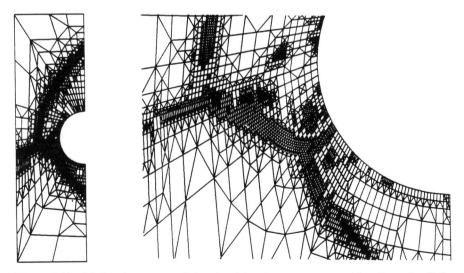

Figure 9.23   Finite element mesh for shock impingement on cowl leading edge [74].

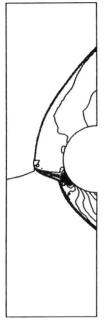

Figure 9.24 Density contours for shock impingement on cowl leading edge [79].

algorithms. The stability of several algorithms is established, and combined implicit/explicit algorithms are implemented. Several difficult inviscid flow problems are solved successfully, but only relatively simple viscous flow problems are considered. Additional study will be required to evaluate the implicit/explicit algorithm's performance for viscous flows. Reference 83 presents an implicit finite element algorithm for high-speed flows in which

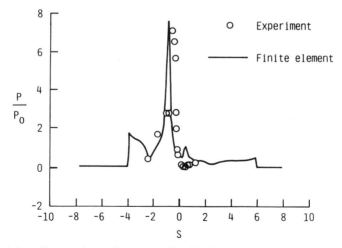

Figure 9.25 Comparison of pressure distributions on cowl leading edge [79].

the equation system is solved by line relaxation on an unstructured mesh. Hypersonic viscous flow solutions in two and three dimensions are illustrated, although no comparative validation solutions are given. Another approach under development is the streamline upwind Petrov–Galerkin [84–90] algorithm. This algorithm shows considerable promise but tends to be complex mathematically and computationally expensive [90]. Thus a promising beginning has been made on implicit algorithm development, but there is a need for further implicit algorithm evaluation and validation for high-speed viscous flows.

9.6 CLOSURE

This chapter considered fundamental finite element approaches to the solution of incompressible and compressible, inviscid and viscous Newtonian fluid flows. For inviscid, incompressible flows we discussed a velocity potential formulation for general three-dimensional problems. For viscous, incompressible flow without inertia, a stream function formulation and a formulation using velocity and pressure as field variables were presented. The velocity and pressure formulation has become widely used. For viscous incompressible flow with inertia, three different velocity and pressure algorithms were presented. For compressible flows we described solution algorithms for low-speed and high-speed flows. The role of adaptive mesh refinement was discussed and illustrated with an example. We describe this important new development more fully in Chapter 10.

The exposition here is by no means exhaustive. The brief examples we have presented are intended only to illustrate that a number of reliable and accurate approaches for some fundamental flow problems are presently available. In comparison to other CFD methods, developments in the analysis of general fluid flows by the finite element method are still in an early stage. Much research remains to be done before all avenues are explored fully. Readers desiring further study of flow problems are encouraged to examine the references given as well as recent conference proceedings and journal articles. Finite element fluid mechanics research is expanding rapidly and new research results are becoming available frequently. For example, a recent volume in the AIAA Progress in Astronautics and Aeronautics Series [94] titled *Computational Nonlinear Mechanics in Aerospace Engineering* devotes five of 15 chapters to finite element analyses of fluid mechanics problems.

REFERENCES

1. J. T. Oden, O. C. Zienkiewicz, R. H. Gallagher, and C. Taylor (eds.), *Finite Element Methods in Flow Problems*, Proceedings of the International Symposium

on Finite Element Methods in Flow Problems held at University College of Swansea, Wales, January 1974, University of Alabama Press, Huntsville, 1974.

2. R. H. Gallagher, J. T. Oden, C. Taylor, and O. C. Zienkiewicz (eds.), *Finite Elements in Fluids*, Vols. 1 and 2, selected papers from the Proceedings of the Symposium on Finite Element Methods in Flow Problems held at the University College of Swansea, Wales, 1974, Wiley, New York, 1975.

3. H. C. Martin, "Finite Element Analysis of Fluid Flows," Proceedings of Second Conference on Matrix Methods in Structural Mechanics, *AFFDL-TR-68-150*, Wright-Patterson Air Force Base, Dayton, OH, October, 1968.

4. T. Fujino, "Analyses of Hydrodynamic and Plate Structures Problems by Finite Element Methods," in *Recent Advances in Matrix Methods of Structural Analysis and Design*, R. H. Gallagher, Y. Yamada, and J. T. Oden (eds.), University of Alabama Press, Huntsville, 1971.

5. G. de Vries and D. H. Norrie, "The Application of the Finite-Element Technique to Potential Flow Problems," *Trans. ASME*, Series E: *J. Appl. Mech.*, Vol. 38, 1971, pp. 798–802.

6. J. H. Argyris and G. Mareczek, "Potential Flow Analysis by Finite Elements," *Ingenier-Archiv*, Vol. 42, No. 1, December 12, 1972, pp. 1–25.

7. V. Meissner, "A Mixed Finite Element Model for Use in Potential Flow Problems," *Int. J. Numer. Methods Eng.*, Vol. 6, No. 4, 1973, pp. 467–473.

8. V. L. Streeter, *Fluid Dynamics*, McGraw-Hill, New York, 1948.

9. R. H. F. Pao, *Fluid Dynamics*, Charles E. Merrill, Columbus, OH, 1967.

10. E. Baskharone and A. Hamed, "A New Approach in Cascade Flow Analysis Using the Finite Element Method," *AIAA J.*, Vol. 19, No. 1, January 1981, pp. 65–71.

11. A. Hamed and E. Baskharone, "Analysis of the Three-Dimensional Flow in a Turbine Scroll," *J. Fluids Eng.*, Vol. 102, September 1980, pp. 297–301.

12. L. V. Kantorovich and V. I. Krylov, *Approximate Methods of Higher Analysis* (translated by C. D. Benster), Wiley-Interscience, New York, 1958.

13. B. Atkinson, M. P. Brocklebank, C. C. H. Card, and J. M. Smith, "Low Reynolds Number Developing Flows," *Am. Inst. Chem. Eng. J.*, Vol. 15, July 1969, pp. 548–553.

14. B. Atkinson, C. C. H. Card, and B. M. Irons, "Application of the Finite Element Method to Creeping Flow Problems," *Trans. Inst. Chem. Eng.*, Vol. 48, 1970, pp. T265–T284.

15. P. Tong and Y. C. Fung, "Slow Particulate Viscous Flow in Channels and Tubes: Applications to Biomechanics," *Trans. ASME*, Series E: *J. Appl. Mech.*, Vol. 38, December 1971, pp. 721–728.

16. Y. Yamada, K. Ito, Y. Yokouchi, T. Tamano, and T. Ohtsubo, "Finite Element Analysis of Steady Fluid and Metal Flow," Chapter 4 of reference 2, pp. 73–94.

17. C. Taylor and P. Hood, "A Numerical Solution to the Navier–Stokes Equations Using the Finite Element Technique," *Comput. Fluids*, Vol. 1, No. 1, 1973, pp. 73–100.

18. D. K. Gartling and E. B. Becker, "Finite Element Analysis of Viscous, Incompressible Fluid Flow, Part 1: Basic Methodology," *Comput. Methods Appl. Mech. Eng.*, Vol. 8, 1960, pp. 51–60.

19. D. K. Gartling and E. B. Becker, "Finite Element Analysis of Viscous, Incompressible Fluid Flow, Part 2: Applications," *Comput. Methods Appl. Mech. Eng.*, Vol. 8, 1960, pp. 127–138.

20. P. Hood and C. Taylor, "Navier Stokes Equations Using Mixed Interpolation," reference 1, pp. 121–132.

21. M. D. Olson and S. Y. Tuann, "Primitive Variables versus Stream Function Finite Element Solutions of the Navier–Stokes Equations," Chapter 4 of *Finite Elements in Fluids*, Vol. 3, R. H. Gallagher, O. C. Zienkiewicz, J. T. Oden, M. Morandi Cecchi and C. Taylor (eds.), selected papers from the Second International Conference on Finite Elements in Flow Problems held at St. Margherita, Italy, 1976, Wiley, New York, 1978, pp. 73–87.

22. M. Bercovier and O. Pironneau, "Error Estimates for Finite Element Method Solution of the Stokes Problem in Primitive Variables," *Numer. Math.*, Vol. 33, 1979, pp. 211–224.

23. J. T. Oden and G. F. Carey, *Finite Elements: Mathematical Aspects*, Volume IV, Prentice-Hall, Englewood Cliffs, NJ, 1983; see also G. C. Carey and J. T. Oden, *Finite Elements: Fluid Mechanics*, Vol. VI, Prentice-Hall, Englewood Cliffs, NJ, 1986.

24. M. D. Gunsburger, *Finite Element Methods for Viscous Incompressible Flows*, Academic, San Diego, CA, 1989.

25. O. Pironneau, *Finite Element Methods for Fluids*, Wiley, New York, 1989.

26. T. J. R. Hughes, *The Finite Element Method: Linear Static and Dynamic Finite Element Analysis*, Prentice-Hall, Englewood Cliffs, NJ, 1987.

27. R. D. Cook, D. S. Malkus and M. E. Plesha, *Concepts and Applications of Finite Element Analysis*, 3d ed., Wiley, New York, 1989.

28. O. C. Zienkiewicz and R. L. Taylor, *The Finite Element Method*, Vol. 1, *Basic Formulation and Linear Problems*, 4th ed., McGraw Hill, New York, 1989.

29. M. D. Olson, "Variational-Finite Element Methods for Two-Dimensional and Axisymmetric Navier–Stokes Equations," Chapter 3 of reference 2, pp. 57–72.

30. R. T. Cheng, "Numerical Solution of the Navier–Stokes Equations by the Finite Element Method," *Phys. Fluids*, Vol. 15, No. 12, 1972, pp. 2098–2105.

31. A. J. Baker, "Finite Element Solution Algorithm for Viscous Incompressible Fluid Dynamics," *Int. J. Numer. Methods Eng.*, Vol. 6, No. 1, 1973, pp. 89–103.

32. T. Bratanow, A. Ecer, and M. Kobiske, "Finite-Element Analysis of Unsteady Incompressible Flow Around an Oscillating Obstacle of Arbitrary Shape," *AIAA J.*, Vol. 11, No. 11, November 1973, pp. 1471–1477.

33. M. F. Peeters, W. G. Habashi, and E. G. Dueck, "Finite Element Stream Function–Vorticity Solutions of the Incompressible Navier–Stokes Equations," *Int. J. Numer. Methods Fluids*, Vol. 7, 1987, pp. 17–27.

34. D. K. Gartling, R. E. Nickell, and R. I. Tanner, "A Finite Element Convergence Study for Accelerating Flow," *Int. J. Numer. Methods Eng.*, Vol. 11, 1977, pp. 1155–1174.

35. E. Ben-Sabar and B. Caswell, "A Stable Finite Element Simulation of Convective Transport," *Int. J. Numer. Methods Eng.*, Vol. 14, 1979, pp. 545–565.

36. P. M. Gresho, R. L. Lee and R. L. Sani, "On the Time-Dependent Solution of the Incompressible Navier–Stokes Equations in Two and Three Dimensions,"

Recent Advances in Numerical Methods in Fluids, Vol. 1, C. Taylor and K. Morgan, eds., Pineridge Press, Swansea, UK, 1980, pp. 27–79.

37. P. M. Gresho and R. L. Lee, "Don't Suppress the Wiggles—They're Telling You Something!" *Comput. Fluids*, Vol. 9, 1981, pp. 223–253.

38. P. M. Gresho, R. L. Lee, and C. D. Upson, "FEM Solution of the Navier–Stokes Equations for Vortex Shedding Behind a Cylinder: Experiments with the Four-Node Element," *Adv. Water Resources*, Vol. 4, December 1981, pp. 175–184.

39. A. N. Brooks and T. J. R. Hughes, "Streamline Upwind/Petrov–Galerkin Formulations for Convection Dominated Flows With Particular Emphasis on the Incompressible Navier–Stokes Equations," *Comput. Methods Appl. Mech. Eng.*, Vol. 32, 1982, pp. 199–259.

40. P. M. Gresho, S. T. Chan, R. L. Lee, and C. D. Upson, "A Modified Finite Element Method for Solving the Time-Dependent, Incompressible Navier–Stokes Equations, Part 1: Theory," *Int. J. Numer. Methods Fluids*, Vol. 4, 1984, pp. 557–598.

41. P. M. Gresho and R. L. Sani, "On Pressure Boundary Conditions for the Incompressible Navier–Stokes Equations," *Int. J. Numer. Methods Fluids*, Vol. 7, 1987, pp. 1111–1145.

42. J. L. Sohn, "Evaluation of FIDAP on Some Classical Laminar and Turbulent Benchmarks," *Int. J. Numer. Methods Fluids*, Vol. 8, 1988, pp. 1469–1490.

43. P. M. Gresho, "On the Theory of Semi-Implicit Projection Methods for Viscous Incompressible Flow and Its Implementation via a Finite Element Method That Also Introduces a Nearly Consistent Mass Matrix, Part 1: Theory," *Int. J. Numer. Methods Fluids*, Vol. 11, 1990, pp. 587–620.

44. P. M. Gresho and S. T. Chan, "On the Theory of Semi-Implicit Projection Methods for Viscous Incompressible Flow and Its Implementation via a Finite Element Method That Also Introduces a Nearly Consistent Mass Matrix, Part 2: Implementation," *Int. J. Numer. Methods Fluids*, Vol. 11, 1990, pp. 621–659.

45. D. K. Gartling, "A Test Problem for Outflow Boundary Conditions: Flow Over a Backward-Facing Step," *Int. J. Numer. Methods in Fluids*, Vol. 11, 1990, pp. 953–967.

46. P. M. Gresho, "Incompressible Fluid Dynamics: Some Fundamental Formulation Issues," *Annu. Rev. Fluid Mech.*, Vol. 23, 1991, pp. 413–453.

47. R. C. Givler, D. K. Gartling, M. S. Engelman, and V. Haroutunian, "Navier–Stokes Simulations of Flow Past Three-Dimensional Submarine Models," *Comput. Methods Appl. Mech. Eng.*, Vol. 87, 1991, pp. 175–200.

48. B. Ramaswamy, T. C. Jue, and J. E. Akin, "Semi-implicit and Explicit Finite Element Schemes for Coupled Fluid/Thermal Problems," *Int. J. Numer. Methods Eng.*, Vol. 34, 1992, pp. 675–696.

49. R. Glowinski and O. Pironneau, "Finite Element Methods for Navier–Stokes Equations," *Annu. Rev. Fluid Mech.*, Vol. 24, 1992, pp. 167–204.

50. B-N. Jiang, "A Least-Squares Finite Element Method for Incompressible Navier–Stokes Problems," *Int. J. Numer. Methods Fluids*, Vol. 14, 1992, pp. 843–859.

51. T. J. R. Hughes, W. K. Liu, and A. Brooks, "Finite Element Analysis of Incompressible Viscous Flows by the Penalty Function Formulation," *J. Comput. Phys.*, Vol. 30, No. 1, January 1979, pp. 1–60.

52. M. Bercovier and M. Engelman, "A Finite Element for the Numerical Solution of Viscous Incompressible Flows," *J. Comput. Phys.*, Vol. 30, No. 2, February 1979, pp. 181–201.

53. J. N. Reddy, "On Penalty Function Methods in the Finite-Element Analysis of Flow Problems," *Int. J. Numer. Methods Fluids*, Vol. 2, 1982, pp. 151–171.

54. J. N. Reddy, "Penalty-Finite-Element Analysis of 3-D Navier–Stokes Equations," *Comput. Methods Appl. Eng.*, Vol. 35, 1982, pp. 87–106.

55. G. Comini and S. Del Giudice, "Finite-Element Solution of the Incompressible Navier–Stokes Equations," *Numerical Heat Transfer*, Vol. 5, Hemisphere, Washington, DC, 1982, pp. 463–478.

56. C. Prakash and S. V. Patankar, "A Control Volume-Based Finite-Element Method for Solving the Navier–Stokes Equations Using Equal-Order Velocity-Pressure Interpolation," *Numerical Heat Transfer*, Vol. 8, Hemisphere, Washington, DC, 1985, pp. 259–280.

57. J. G. Rice and R. J. Schnipke, "A Monotone Streamline Upwind Finite Element Method for Convection-Dominated Flows," *Comput. Methods Appl. Mech. Eng.*, Vol. 48, 1985, pp. 313–327.

58. J. G. Rice and R. J. Schnipke, "An Equal-Order Velocity–Pressure Formulation That Does Not Exhibit Spurious Pressure Modes," *Comput. Methods in Appl. Mech. Eng.*, Vol. 58, 1986, pp. 135–149.

59. R. J. Schnipke and J. G. Rice, "A Finite Element Method for Free and Forced Convection Heat Transfer," *Int. J. Numer. Methods Eng.*, Vol. 24, 1987, pp. 117–128.

60. D. A. Anderson, J. C. Tannehill, and R. H. Pletcher, *Computational Fluid Mechanics and Heat Transfer*, Hemisphere, Washington, DC, 1984.

61. J. D. Anderson, Jr., *Modern Compressible Flow With Historical Prospective*, McGraw-Hill, New York, 1982.

62. J. D. Anderson, Jr., *Hypersonic and High Temperature Gas Dynamics*, McGraw-Hill, New York, 1989.

63. P. W. Yarrington, "Finite Element Analysis of Low-Speed Compressible Flows Within Convectively Cooled Structures," M.S. Thesis, University of Virginia, May 1993.

64. C. A. J. Fletcher, *Computational Galerkin Methods*, Springer, New York, 1984, p. 158.

65. J. Donea, "A Taylor–Galerkin Method for Convective Transport Problems," *Int. J. Numer. Methods Eng.*, Vol. 20, 1984, pp. 101–119.

66. R. Löhner, K. Morgan, and O. C. Zienkiewicz, "The Solution of Non-linear Systems of Hyperbolic Equations by the Finite Element Method," *Int. J. Numer. Method Fluids*, Vol. 4, 1984, pp. 1043–1063.

67. R. Löhner, K. Morgan, and O. C. Zienkiewicz, "The Use of Domain Splitting With an Explicit Hyperbolic Solver," *Comput. Method Appl. Mech. Eng.*, Vol. 45, 1984, pp. 313–329.

68. R. Löhner, K. Morgan, and O. C. Zienkiewicz, "An Adaptive Finite Element Procedure for Compressible High Speed Flows," *Comput. Methods Appl. Mech. Eng.*, Vol. 51, 1985, pp. 441–465.

69. R. Löhner, K. Morgan, and J. Peraire, "A Simple Extension to Multidimensional Problems of the Artificial Viscosity Due to Lapidus," *Commun. Appl. Numer. Methods*, Vol. 1, 1985, pp. 141–147.

70. K. S. Bey, E. A. Thornton, P. Dechaumphai, and R. Ramakrishnan, "A New Finite Element Approach for Prediction of Aerothermal Loads: Progress in Inviscid Flow Computations," *AIAA 7th Computational Fluids Dynamics Conference*, Cincinnati, OH, July 15–17, 1985, pp. 411–424.

71. R. Löhner, K. Morgan, J. Peraire, and M. Vahdati, "Finite Element Flux-Corrected Transport (FEM-FCT) for the Euler and Navier–Stokes Equations," *Int. J. Numer. Methods Fluids*, Vol. 7, 1987, pp. 1093–1109.

72. R. Löhner, "An Adaptive Finite Element Scheme for Transient Problems in CFD," *Comput. Methods Appl. Mech. Eng.*, Vol. 61, 1987, pp. 323–338.

73. J. T. Oden, T. Strouboulis, and P. Devloo, "Adaptive Finite Element Methods for High-Speed Compressible Flows," *Int. J. Numer. Methods Fluids*, Vol. 7, 1987, pp. 1211–1228.

74. J. Donea, L. Quartapelle, and V. Selmin, "An Analysis of Time Discretization in the Finite Element Solution of Hyperbolic Problems," *J. Comput. Phys.*, Vol. 70, 1987, pp. 463–499.

75. J. Peraire, M. Vahdati, K. Morgan, and O. C. Zienkiewicz, "Adaptive Remeshing for Compressible Flow Computations," *J. Comput. Phys.*, Vol. 72, 1987, pp. 449–466.

76. J. Peraire, J. Peiro, L. Formaggia, K. Morgan, and O. C. Zienkiewicz, "Finite Element Euler Computations in Three Dimensions," *Int. J. Numer. Methods Eng.*, Vol. 26, 1988, pp. 2135–2159.

77. L. Formaggia, J. Peraire, and K. Morgan, "Simulation of a Store Separation Using the Finite Element Method," *Appl. Math. Modelling*, Vol. 12, April 1988, pp. 175–181.

78. E. A. Thornton and P. Dechaumphai, "Coupled Flow, Thermal, and Structural Analysis of Aerodynamically Heated Panels," *J. Aircraft*, Vol. 25, No. 11, November 1988, pp. 1052–1059.

79. R. Ramakrishnan, K. S. Bey, and E. A. Thornton, "Adaptive Quadrilateral and Triangular Finite-Element Scheme for Compressible Flows," *AIAA J.*, Vol. 28, No. 1, January 1990, pp. 51–59.

80. J. Probert, O. Hassan, J. Peraire, and K. Morgan, "An Adaptive Finite Element Method for Transient Compressible Flows," *Int. J. Numer. Methods Eng.*, Vol. 32, 1991, pp. 1145–1159.

81. R. Ramakrishnan, E. A. Thornton, and A. R. Wieting, "Adaptive Finite Element Analysis of Hypersonic Laminar Flows for Aerothermal Loads Predictions," *J. Thermophys. Heat Transfer*, Vol. 5, No. 3, July–September 1991, pp. 308–317.

82. W. W. Tworzydlo, J. T. Oden, and E. A. Thornton, "Adaptive Implicit/Explicit Finite Element Method for Compressible Viscous Flows," *Comput. Methods Appl. Mech. Eng.*, Vol. 95, 1992, pp. 397–440.

83. O. Hassan, K. Morgan, and J. Peraire, "An Implicit Finite Element Method for High Speed Flows," *Int. J. Numer. Methods Eng.*, Vol. 32, 1991, pp. 183–205.

84. T. J. R. Hughes, "Recent Progress in the Development and Understanding of SUPG Methods With Special Reference to the Compressible Euler and

Navier–Stokes Equations," *Int. J. Numer. Methods Fluids*, Vol. 7, 1987, pp. 1261–1275.

85. T. J. R. Hughes and M. Mallet, "A New Finite Element Formulation for Computational Fluid Dynamics, IV: A Discontinuity-Capturing Operator for Multidimensional Advective-Diffusive Systems," *Comput. Methods Appl. Mech. Eng.*, Vol. 58, 1986, pp. 329–336.

86. T. J. R. Hughes and M. Mallet, "A New Finite Element Formulation for Computational Fluid Dynamics, III: The Generalized Streamline Operator for Multidimensional Advective-Diffusive Systems," *Comput. Methods. Appl. Mech. Eng.*, Vol. 58, 1986, pp. 305–328.

87. T. J. R. Hughes, M. Mallet, and A. Mizukami, "A New Finite Element Formulation for Computational Fluid Dynamics, II: Beyond SUPG," *Comput. Methods Appl. Mech. Eng.*, Vol. 54, 1986, pp. 341–355.

88. T. J. R. Hughes, L. P. Franca, and M. Mallet, "A New Finite Element Formulation for Computational Fluid Dynamics, I: Symmetric Forms of the Compressible Euler and Navier–Stokes Equations and the Second Law of Thermodynamics," *Comput. Methods Appl. Mech. Eng.*, Vol. 54, 1986, pp. 223–234.

89. F. P. Brueckner and J. C. Heinrich, "Petrov–Galerkin Finite Element Model for Compressible Flows," *Int. J. Numer. Methods Eng.*, Vol. 32, 1991, pp. 255–274.

90. G. R. Vemaganti, E. A. Thornton, and A. R. Wieting, "A Structured and Unstructured Remeshing Method for High Speed Flows," *J. Spacecraft Rockets*, Vol. 28, No. 2, March–April 1991, pp. 158–164.

91. B-N. Jiang and G. F. Carey, "A Stable Least-Squares Finite Element Method for Non-Linear Hyperbolic Problems," *Int. J. Numer. Methods Fluids*, Vol. 8, 1988, pp. 933–942.

92. B-N. Jiang and L. A. Povinelli, "Least-Squares Finite Element Method for Fluid Dynamics," *Comput. Methods Appl. Mech. Eng.*, Vol. 81, 1990, pp. 13–37.

93. T. J. R. Hughes, L. P. Franca, and G. M. Hulbert, "A New Finite Element Formulation for Computational Fluid Dynamics, VIII: The Galerkin/Least-Squares Method for Advective-Diffusive Equations," *Comput. Methods Appl. Mech. Eng.*, Vol. 73, 1989, pp. 173–189.

94. S. N. Atluri (ed.), *Computational Nonlinear Mechanics in Aerospace Engineering*, Vol. 146, Progress in Astronautics and Aeronautics, AIAA, Washington, DC, 1992.

PROBLEMS

1. Consider a stream function formulation of inviscid incompressible flow. Starting from the boundary value problem (Figure 9.1),

$$\nabla^2 \psi = 0 \qquad \text{in } \Omega$$
$$\psi = g(x, y) \qquad \text{on } S_1$$
$$\psi = 0 \qquad \text{on } S_2$$

use Galerkin's method of weighted residuals to derive the element matrices $[K]^{(e)}$ and $\{R\}^{(e)}$ for the flow. Discuss how the boundary conditions are implemented.

2. Three-node triangular elements are to be used to solve the problem shown in Figure 9.1 via a two-dimensional potential function formulation. Evaluate the element matrices $[K]^{(e)}$ and $\{R\}^{(e)}$ for a typical triangular element. Derive equations for calculating the element velocity components u and v.

3. Four-node tetrahedron elements are to be used to solve the problem shown in Figure 9.1 via a three-dimensional potential function formulation. Evaluate the element matrices $[K]^{(e)}$ and $\{R\}^{(e)}$ for a typical tetrahedron element. Derive equations for calculating the element velocity components u, v, and w.

4. Heat conduction in a two-dimensional slab of unit thickness is described by Laplace's equation $\nabla^2 T = 0$, where $T(x, y)$ denotes the temperature. The heat flux components are $q_x = -k \, \partial T/\partial x$ and $q_y = -k \, \partial T/\partial y$, where k is the thermal conductivity. Develop an analogy between heat conduction and inviscid, incompressible flow with the potential function formulation. Explain how you would use a finite element heat conduction computer program to solve the analogous potential flow problem.

5. Consider the problem of steady, fully developed viscous flow between two walls (Figure P9.5). We may take $u = u(y)$ and $v = 0$. Show that the conservation equations reduce to

$$\mu \frac{d^2 u}{dy^2} = \frac{dP}{dx}$$

for constant viscosity. We may assume that the pressure gradient is known. The lower wall is stationary and the upper wall has constant

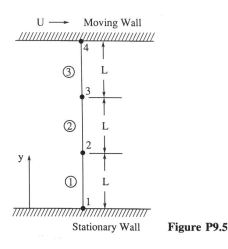

Stationary Wall   **Figure P9.5**

velocity U; thus the boundary conditions are

$$u(0) = 0$$
$$u(3L) = U$$

a. Develop the finite element equations that can be used to solve for u_i, $i = 1, 2, \ldots, N$.

b. Solve for the velocities u_2 and u_3 for the mesh shown, assuming that μ and dP/dx are known constants.

c. Solve for the shearing stress $\tau = \mu(du/dy)$ in element 1.

d. Determine the reactions R_1 and R_4 at the walls. What is the physical significance of these quantities?

e. Compare your finite element results to the analytical solution to the problem.

6. Consider the problem of steady, fully developed flow between two fixed walls. We may take $u = u(y)$ and $v = 0$. Show that the conservation equations reduce to

$$\frac{d\tau_{xy}}{dy} = \frac{dP}{dx}$$

where

$$\tau_{xy} = \mu(T)\frac{du}{dy}$$

where the viscosity is temperature dependent, and the pressure gradient is known. By symmetry, only one-half of the flow region must be modeled (Figure P9.6).

a. What are the boundary conditions at $y = 0$ and $y = 2L$?

b. Develop the finite element equations that can be used to solve for u_i, $i = 1, 2, \ldots, N$.

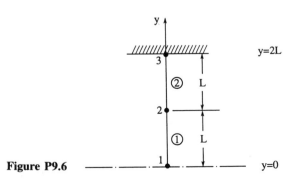

Figure P9.6

c. Solve for the unknown nodal velocities for the mesh shown assuming dP/dx is a known constant and

$$\mu(T) = \mu_0, \qquad 0 < y < L$$

$$\mu(T) = 2\mu_0, \qquad L < y < 2L$$

d. Compute the shearing stress in the element adjacent to the wall.

7. Consider the steady laminar flow of a thin sheet of viscous liquid with a free surface down a plane with inclination angle θ (Figure P9.7). Assuming that $u = u(y)$ and $v = 0$, show that the conservation equations reduce to the x momentum equation

$$\rho g \sin \theta + \frac{d\tau_{xy}}{dy} = 0$$

where

$$\tau_{xy} = \mu \frac{du}{dy}$$

and the viscosity μ is assumed constant. A two-element finite element model is to be used to obtain an approximate solution.

a. Write appropriate finite element boundary conditions for nodes 1 and 3.

b. Develop finite element equations that can be used to solve for u_i, $i = 1, 2, \ldots, N$.

c. Solve for u_2 and u_3 for the mesh shown.

d. Solve for the shearing stress at the wall.

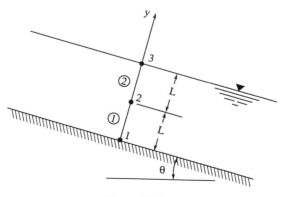

Figure P9.7

8. Consider a viscous flow between two parallel flat plates (Figure P9.8). The top plate is suddenly accelerated from rest and moves in its own plane with constant velocity U. Assuming that $u = u(y, t)$, $v = 0$, $P = P(y)$, the conservation equations reduce to the x momentum equation

$$\rho \frac{\partial u}{\partial t} = \frac{\partial \tau_{xy}}{\partial y}$$

where

$$\tau_{xy} = \mu \frac{\partial u}{\partial y}$$

and the viscosity μ is constant. The initial condition is $u(y, 0) = 0$, and the boundary conditions are $u(0, t) = 0$ and $u(3L, t) = U$.

a. Develop a finite element formulation to determine $u_i(t)$, $i = 1, 2, \ldots, N$.

b. Form an explicit time marching algorithm for determining $u_i(t)$.

c. Compute the time histories for $u_2(t)$ and $u_3(t)$ using an explicit time-marching algorithm.

9. Consider natural convection between two heated vertical plates, Figure P9.9. The vertical plates are maintained at steady but different temperatures T_L and T_R. Assuming zero pressure gradient, $u = u(y)$, $v = 0$, and $T = T(y)$, show that the conservation equations are

$$\rho g \beta (T - T_m) + \frac{d\tau_{xy}}{dy} = 0$$

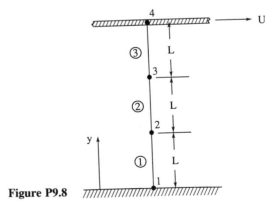

Figure P9.8

where

$$\tau_{xy} = \mu \frac{du}{dy}$$

$$T_m = \frac{T_L + T_R}{2}$$

and

$$\frac{dq_y}{dy} + \tau_{xy} \frac{du}{dy} = 0$$

where

$$q_y = -k \frac{dT}{dy}$$

The viscosity and thermal conductivity are assumed constant. The nonlinear term in the energy equation is viscous dissipation.

a. Develop a finite element formulation to determine u_i, T_i, $i = 1, 2, \ldots, N$.

b. For a two-node, linear element develop the matrices that are needed for a Newton–Raphson solution algorithm. Work out the details for the Jacobian.

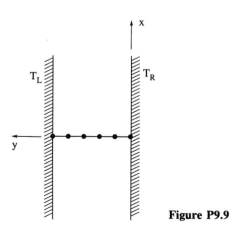

Figure P9.9

10. Reference 46 describes fundamental formulation issues for incompressible fluid dynamics. Read this paper and summarize the basic formulation issues.

11. Reference 41 discusses pressure boundary conditions for the incompressible Navier–Stokes equations. Read this paper and briefly discuss the issue and the author's conclusions.

12. The title of reference 37 is "Don't Suppress the Wiggles—They're Telling You Something!" Read this paper and summarize what the authors say the "wiggles" are telling. Illustrate the wiggles message with examples.

13. Modify the velocity–pressure formulation for the Stokes problem for unsteady flow and derive the additional element matrices that must be included.

14. The one-dimensional Burgers equation is sometimes used as a simple model to study solution algorithms for the incompressible Navier–Stokes equations. Consider the unsteady Burgers equation:

$$\frac{\partial u}{\partial t} + u\frac{\partial u}{\partial x} - \nu\frac{\partial^2 u}{\partial x^2} = 0, \qquad 0 < x < L$$

Write appropriate boundary and initial conditions for the equation. Use the Bubnov–Galerkin method to derive general finite element matrix equations for Burgers' equation.

15. Evaluate the finite element matrices for Burgers' equation for a two node one-dimensional element with linear interpolation.

16. Develop a Newton–Raphson iteration algorithm for the finite element solution to the steady Burgers equation discussed in the two preceding problems. Evaluate the element Jacobian matrix.

17. Read reference 51 on the penalty function formulation for viscous incompressible flow. Discuss the motivation cited by the authors for the penalty function formulation. What are the advantages and disadvantages of the approach?

18. Read reference 57, which describes a monotone streamline upwind finite element method for convection-dominated flows. What does "monotone" mean? Describe how the upwinding approach works and explain how it differs from streamline upwind Petrov–Galerkin formulations (SUPG) [39].

19. With reference to Section 9.5.2 briefly discuss how low-speed compressible flows differ from flows described by the incompressible Navier–Stokes equations. Illustrate your discussion with specific examples.

20. Section 9.5.3 states that for the one-dimensional advection equation

$$\frac{\partial u}{\partial t} + a\frac{\partial u}{\partial x} = 0$$

the two-step Taylor-Galerkin algorithm is identical to a two-step Lax–Wendroff finite difference scheme. Using a CFD textbook, reference 60, or other references, describe the two-step Lax–Wendroff finite difference algorithm giving key equations.

21. For the one-dimensional advection equation consider a finite difference representation where u_j^{n+1} represents the solution at node j at time step $n + 1$. Node j is located at $x_j = j\Delta x$, where Δx denotes the nodal spacing. Show that the Lax–Wendroff and Taylor–Galerkin algorithms lead to the same difference equation [79],

$$u_j^{n+1} = u_j^n - \tfrac{1}{2}\nu\left(u_{j+1}^n - u_{j-1}^n\right) + \tfrac{1}{2}\nu^2\left(u_{j+1}^n - 2u_j^n + u_{j-1}^n\right)$$

where $\nu = a\,\Delta t/\Delta x$ is the Courant number.

22. Using the difference equation developed in the previous problem show that the numerical stability requirement is $|\nu| < 1$. Use the von Neumann stability analysis method [60].

23. Reference 79 describes a stability analysis for the Taylor–Galerkin algorithm applied to a mesh of rectangular finite elements with triangular transition elements. Discuss the analysis and the conclusion reached.

24. Acoustic waves in a fluid are described by the nondimensional linear hyperbolic set of equations

$$\frac{\partial \rho}{\partial t} + \frac{\partial u}{\partial x} = 0$$

$$\frac{\partial u}{\partial t} + c^2\frac{\partial \rho}{\partial x} = 0$$

where ρ and u represent small changes in the density and velocity, respectively; c is the speed of sound. Use the Taylor–Galerkin approach to develop a finite element solution algorithm. Evaluate all element matrices for a linear two-node element.

25. The linear shock tube is governed by the hyperbolic set of equations described in the previous problem. The initial and boundary conditions are shown in Figure P9.25. The tube, closed at one end and open at the other, is divided by a diaphragm at $x = 0.3$, across which a density difference exists initially. The diaphragm is suddenly broken and an expansion wave propagates to the left and a compression wave propa-

Linear Shock Tube

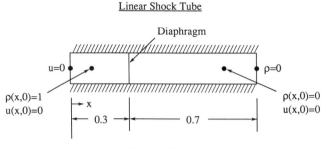

Figure P9.25

gates to the right. Reflections of these waves at the closed and open ends determine the subsequent response. For a sound speed $c = 1$, compute the transient response using the Taylor–Galerkin algorithm developed in the previous problem. Plot the density distribution for $0 \le t \le 1.50$ at intervals $\Delta t = 0.25$.

26. An important application of unsteady wave motion is the shock tube [61]. This is a tube closed at both ends with a diaphragm separating a region of high-pressure gas on the left from a region of low-pressure gas on the right. When the diaphragm is broken, a shockwave propagates to the right, and an expansion wave propagates to the left. The problem is described by the unsteady, one-dimensional, hyperbolic Euler equations. In CFD literature, the problem is often used to evaluate time-marching algorithms, and it is called the Riemann shock tube problem. The initial conditions for a Riemann shock tube problem are given in Figure P9.26. Use the Taylor–Galerkin algorithm with Lapidus dissipation to compute the transient response. Compare your results with the solution presented in reference 66.

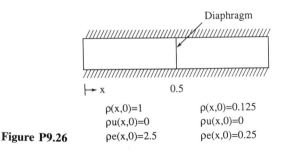

Figure P9.26

27. Review the finite element implicit compressible flow algorithms described in references 82–90. Briefly explain the differences in the three approaches. Discuss some of the problems that have been solved with the algorithms.

10

A SAMPLE COMPUTER CODE
AND OTHER PRACTICAL
CONSIDERATIONS

10.1   INTRODUCTION

The major portion of this chapter is written for readers who have never used a computer to solve a continuum problem by the finite element method. Having an understanding of the underlying mathematics of the finite element method and knowing how to derive element equations for a given problem are not sufficient to comprehend how problems are actually solved on a computer. We need to understand how the element equations are evaluated and assembled into global equations, and how boundary conditions are imposed and the system equations are solved. Today, a significant number of commercially available finite element programs give powerful finite element solution capability. These problems may be, and often are, executed by individuals without a clear understanding of the operations that the programs actually perform. For the best usage of these tools, however, we should have a deeper understanding of how finite element programs actually work. In this regard, we discuss details of a computer program for solving a heat conduction problem. A reader familiar with the FORTRAN language should be able to follow the coding with the aid of comments appearing throughout the program. The program is not the only way, nor even the optimum way, that the finite element method can be implemented for heat conduction problems. But the program illustrates the basic finite element operations, and we believe that you will benefit from studying the program. After describing the program structure, we present input and output data for two example problems.

A major consideration in solving real problems is development of high-quality finite element meshes. By a high-quality mesh, we mean a mesh that is refined sufficiently so that details of the important physical behavior are captured accurately. Over the last 10 years, researchers have invested significant effort in mesh generation methods. We describe basic concepts for mesh generation and give references where algorithms may be found. A closely related and important new area for improving the quality of finite element solutions is adaptive mesh refinement. Adaptive mesh refinement is developing rapidly, and we present an overview of the approach with illustrative examples.

An important part of finite element analysis is the evaluation of the integrals in the element equations. When these integrals cannot be evaluated in closed form, we must resort to approximate numerical methods. In Section 10.6 we summarize a number of useful numerical integration formulas for this purpose.

Although a detailed discussion of methods for solving matrix equations is beyond the scope of this text, we briefly illustrate methods for solving linear and nonlinear matrix equations and suggest references where additional information may be found.

10.2   SETTING UP A SIMPLE HEAT CONDUCTION PROBLEM

To illustrate a finite element computer code we present a FORTRAN computer program in Section 10.3 that solves steady-state heat conduction problems for bodies of arbitrary shape that can be represented by plane or axisymmetric elements. The program is not necessarily the most general or the most efficient, because it is limited to linear steady heat transfer and employs simple solution methods. The program, however, illustrates the essential features of finite element solutions. Also, the program is modular, with logical operations grouped in clearly defined subroutines to permit convenient modifications or additions to the program.

10.2.1   Problem Formulation

We begin with a statement of the problem and the equations that must be programmed. General forms of the equations and boundary conditions are given in Cartesian coordinates in Chapter 8, but here we summarize the equations solved by the program. The steady-state temperature distribution in a two-dimensional solid of thickness t satisfies the thermal energy equation

$$-\left(\frac{\partial q_x}{\partial x} + \frac{\partial q_y}{\partial y}\right) + Q = 0 \tag{10.1a}$$

where

$$\begin{Bmatrix} q_x \\ q_y \end{Bmatrix} = -\begin{bmatrix} k_{xx} & k_{xy} \\ k_{xy} & k_{yy} \end{bmatrix} \begin{Bmatrix} \dfrac{\partial T}{\partial x} \\ \dfrac{\partial T}{\partial y} \end{Bmatrix} \tag{10.1b}$$

and boundary conditions

$$T = T(x, y) \qquad \text{on } S_1 \tag{10.2a}$$

$$q_x n_x + q_y n_y = 0 \qquad \text{on } S_2 \tag{10.2b}$$

For a three-dimensional axisymmetric solid the temperature distribution satisfies

$$-\left[\frac{1}{r}\frac{\partial}{\partial r}(rq_r) + \frac{\partial q_z}{\partial z}\right] + Q = 0 \tag{10.3a}$$

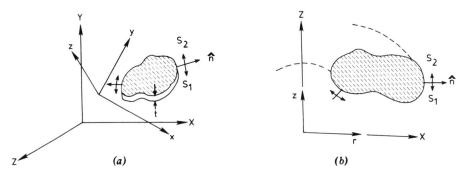

Figure 10.1 Definition of coordinate systems: (*a*) coordinate systems for two-dimensional body; (*b*) coordinate system for axisymmetric solid.

where

$$
\left\{ \begin{matrix} q_r \\ \\ q_z \end{matrix} \right\} = - \begin{bmatrix} k_{rr} & k_{rz} \\ \\ k_{rz} & k_{zz} \end{bmatrix} \left\{ \begin{matrix} \dfrac{\partial T}{\partial r} \\ \\ \dfrac{\partial T}{\partial z} \end{matrix} \right\}
\tag{10.3b}
$$

and boundary conditions

$$
T = T(r, z) \qquad \text{on } S_1
\tag{10.4a}
$$

$$
q_r n_r + q_z n_z = 0 \qquad \text{on } S_2
\tag{10.4b}
$$

We assume that the thermal conductivity tensors and internal heat generation rates are known as functions of position. Figure 10.1 shows the coordinate systems for the two-dimensional solid and the three-dimensional axisymmetric solid. Note that the two-dimensional solid is located in the x–y plane of a local coordinate system, and this local coordinate system may have a general orientation with respect to the global X, Y, Z coordinate system. The axisymmetric solid is located in a local r–z plane that coincides with the X–Z plane of the global coordinate system. These coordinate systems permit us to solve a variety of three-dimensional heat conduction problems with two-dimensional plane elements. The problem statement is as follows: *Given* (1) the geometry of a body, (2) the data describing the thermal properties of the heat-conducting medium, and (3) the thermal loading conditions (boundary conditions), *find* (1) the temperature distribution throughout the body and (2) the heat flows (fluxes) at locations of interest.

10.2.2 The Finite Element Equations

In Chapter 8 we presented heat conduction element matrices for general boundary conditions, evaluated these matrices for simple element shapes such as rods and triangles, and described the computation of isoparametric element matrices. To simplify the computer program, in this chapter we consider only conduction elements with specified temperature and/or adiabatic (zero heat flow) boundary conditions, equations 10.2 and 10.4. Additional elements and other boundary conditions may easily be included in the program, and they appear as exercises at the end of this chapter.

For convenience in the use of FORTRAN real variables, some variables in the program use a slightly different notation than in Chapter 8, but the changes are limited to a few terms and should be clear from the context. Using the program notation, element thermal equilibrium equations are

$$[S]^{(e)} \ \{T\}^{(e)} = \{Q\}^{(e)} \tag{10.5}$$
$$\underset{\text{ND}\times\text{ND}}{} \ \underset{\text{ND}\times1}{} \ \underset{\text{ND}\times1}{}$$

where $[S]^{(e)}$ is the element conductance matrix, $\{T\}^{(e)}$ is the vector of nodal temperatures, $\{Q\}^{(e)}$ is the element equivalent nodal heat load vector due to the internal heat generation rate Q in equations 10.1 and 10.3, and ND is the number of element nodes. The basic element in the program is a four-node isoparametric quadrilateral. Four forms of the element (Figure 10.2) are available as program options: (1) plane quadrilateral, (2) plane triangle, (3) axisymmetric quadrilateral, and (4) axisymmetric triangle. Element matrices are computed by four-point Gauss–Legendre integration. For example, the element conductance matrix from equation 8.29 is

$$[S]^{(e)} = \sum_{i=1}^{2} \sum_{j=1}^{2} W_i W_j t \big[B(\xi_i, \eta_j) \big]^T [k] \big[B(\xi_i, \eta_j) \big] |J(\xi_i, \eta_j)| \tag{8.29}$$

where the Gauss weights and points are given in Table 10.5. The triangular element results from coalescing two nodes of the quadrilateral, which is done automatically within the program. The axisymmetric elements are computed by modifying the Jacobian in equation 8.29 to account for the axisymmetric element volume. Note in Figure 10.2 that elements may have orthotropic thermal conductivities. The plane quadrilateral and triangular plane elements may be located arbitrarily in the X, Y, Z global coordinates, but their

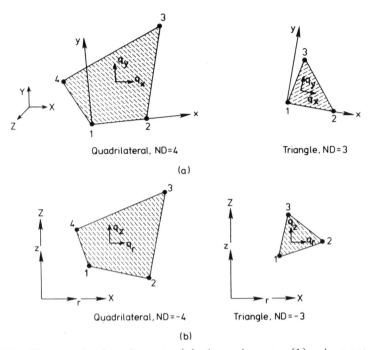

Figure 10.2 Heat conduction elements: (*a*) plane elements; (*b*) axisymmetric elements.

nodes must define a plane. The quadrilateral and triangular axisymmetric elements must lie in the global *X–Z* plane.

For convenience in the program we assume that the thermal conductivities and volumetric heat generation rate are constant within each element. This assumption does not preclude the variation of these quantities throughout the solution region, since they may be assigned different values for each element and the values need not be continuous from one element to another. Thus nonhomogeneous materials may be treated.

10.2.3 The Overall Program Logic

Figure 10.3*a* shows the key operations in the main driving program. Subroutines perform typical input operations such as reading and storing nodal data or elements. Figure 10.3*b* shows the linear steady-state solution module that consists of a set of subroutines to use the element data to carry out the solution. For other solutions such as a transient solution we may use similar modules, often with calls to many of the same subroutines used by the steady-state solution module.

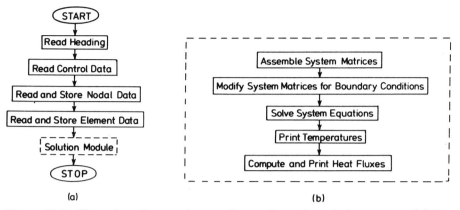

Figure 10.3   Flow chart for steady-state linear thermal analysis program: (*a*) key operations in main program; (*b*) linear steady-state solution module.

10.3   THE COMPUTER PROGRAM AND ITS EXPLANATION

Now we examine in detail the computer program that performs the steps just mentioned. Throughout this section we refer to the FORTRAN listing given in Section 10.3.5. The listing contains numerous COMMENT statements to explain the various subroutine operations.

10.3.1   Structure of the Program

The main program consists of subroutines that are called sequentially in a normal program execution. Dynamic storage allocation stores nodal data, system matrices, and vectors in a blank common designated in the main program as A. The dimension of the blank common is the only restriction on the amount of the input data and the size of the problem that can be solved. The dimension of A and the variable MTOT denote the available storage; there are no other limitations on the number of nodes, elements, and thermal data. To increase program capacity the user need change only the dimension of A and the value of MTOT. The program uses low-speed disk storage (File 1) to store element data, but all assembly operations and equation solutions are performed with system matrices completely within the central memory. To give further insight into the program we now briefly describe some of the important subroutines.

Subroutine NODES   The nodal data are read by the subroutine NODES called by the main program. A nodal point is described by the node number, a boundary condition code, the nodal coordinates, and a specified nodal temperature. Nodal data are specified in the global X, Y, Z Cartesian system. All nodal point data are retained in memory during the complete solution.

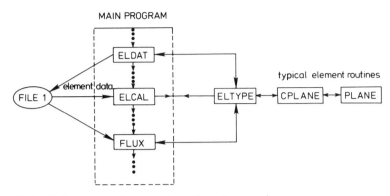

Figure 10.4   Calls to typical element routines from main program.

Subroutines ELDAT, ELCAL, and FLUX   The main program calls these subroutines, which in turn call other subroutines to input element data and perform element matrix and heat flux calculations.

Elements enter the program in groups that consist of sequentially numbered elements of the same type. There may be more than one group of the same element type. The data for all elements follow the same general scheme: (1) one line of control data, (2) one or more lines of thermal property data, and (3) one or more lines of element data. Each element may have different properties.

Figure 10.4 shows the sequence of calls to ELDAT, ELCAL, and FLUX from the main program. Observe that each subroutine calls ELTYPE, which serves as a "switching" routine to call appropriate element routines. In this version of the program there is only one element type available, so there is only one option available in ELTYPE. However, new elements may be added by inserting calls in ELTYPE to new element routines. After the first call to ELTYPE and to its subordinate routines PLANE and CPLANE, basic element data are subsequently returned to ELDAT, which writes the data on low-speed storage File 1. Later in the execution the element data are read into the central memory by ELCAL and FLUX as they are called by the main program.

Subroutine PLANE   The element routines such as CPLANE perform the fundamental element operations. Figure 10.5 shows that the element routine performs three basic functions, depending on the value of ISW: (1) If ISW = 1, the subroutine processes the input data; (2) if ISW = 2, the subroutine computes the element matrices; and (3) if ISW = 3, the subroutine computes the element heat fluxes. Subordinate routines perform the various functions. The comments in the FORTRAN listing explain the functions of each of these subroutines.

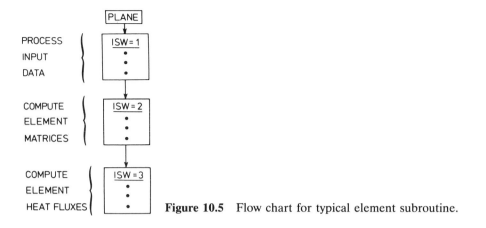

Figure 10.5   Flow chart for typical element subroutine.

Subroutines ADDSTF and ADDLD   The subroutines ADDSTF and ADDLD called by element routines assemble the system matrices. The system conductance matrix is assembled and stored in banded form as shown in Figure 10.6. We recognize that the conduction matrix is symmetrical, and we could choose to store only terms on and above the main diagonal as is done in linear structural analysis programs; however, we store the complete banded matrix because asymmetrical matrices can arise from a Newton–Raphson Jacobian in nonlinear problems or from convective transport matrices and upwind element formulations (see Chapter 8). Use of an asymmetrical storage scheme permits these techniques to be added to the program.

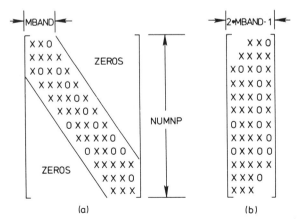

Figure 10.6   Program storage of system matrices: (*a*) actual assembled system matrix; (*b*) banded storage of system matrix.

Subroutine TEMPBC   The only boundary conditions required are specified nodal temperatures that enter with the nodal input data. For an adiabatic boundary, no boundary condition need be specified. The program handles the specified temperature boundary conditions in subroutine TEMPBC, using the first method described in Section 2.3.4. This method consists of modifying the conductance matrix and heat load vector such that the size of the matrix is unchanged. The advantage of this approach is the ease of indexing the equations; that is, the node numbers and equation numbers are the same. A disadvantage is that extra equations are carried in the solution process. The penalty is usually not large, because typically only a few nodes have specified temperatures.

Subroutines FACTOR and SOLVE   The general banded simultaneous equations are solved using Gauss elimination by the subroutines FACTOR and SOLVE called from the main program. FACTOR reduces the conductance matrix to upper triangular form. SOLVE first reduces the load vector and then solves for the nodal temperatures by back-substitution. These operations are performed in separate subroutines to facilitate solutions where the set of equations needs to be solved for multiple load vectors. One example of such use is in linear transient analysis with an implicit algorithm.

Having discussed the general structure of the program and the main program subroutines, we now turn to a description of how the program is used in actual problem-solving situations. The first step is to describe the geometry of the solution domain and then represent it by a finite element model. The second step is to prepare the input data containing the model description.

10.3.2   Preparation of the Finite Element Model

Finite element modeling of realistic problems is difficult to learn from a textbook because of the judgmental factors that contribute to a bonafide finite element model. Good judgment in finite element modeling, as in other areas of engineering, can best be acquired through experience. There is a growing awareness in the engineering community of the importance of this often-neglected aspect of solving problems with finite elements. We take this opportunity to suggest that readers prepare models carefully, giving thoughtful consideration to the assumptions and limitations of the theory, model, and computer program.

With modern software, the construction of finite element models is highly automated. Finite element meshes are often generated from a geometric model developed in a CAD system. A basic principle is to locate nodes on the boundary of the body so that we define the geometry accurately. Node points should be placed in regions wherever values of temperature are specified or sought. And, of course, nodes are more closely spaced in regions where large temperature gradients are expected. Nodes are then connected by appropriate elements, often triangles or quadrilateral elements or a mixture of the two element types.

One main rule governs the construction of a mesh from a set of nodes: each node must connect to a node of an element it touches; that is, nodes may not be placed along the sides of an adjacent element. Figure 10.7 illustrates this point by considering elements formed from a group of typical nodes. In addition, elements that approach squares or equilateral triangles are better than long, narrow elements, which should be avoided if possible. When working with bodies composed of different materials, the parting lines of the materials should be represented by the boundaries of elements. In fact, no difficulty arises if every element is chosen to be a different material; thus almost any heterogenous body can be handled easily.

Once the solution region has been properly discretized, we are ready to establish the system topology or numbering scheme. This is done by numbering the nodes consecutively starting from 1. The computational effort for the solution of the system equations is proportional to the bandwidth squared times the number of equations. Thus for large problems consideration should be given to numbering the nodes to minimize the conductance matrix bandwidth. Computer algorithms [1–7] have been developed to renumber nodes to give an optimum bandwidth, and these schemes can be effective in reducing computer time for the solution of complex systems. To minimize the bandwidth we should number the nodes so as to minimize the difference between connected nodes on an element. For instance, when numbering nodes on a two-dimensional rectangular body, we should first number the nodes in the narrow direction. The elements must also be numbered consecu-

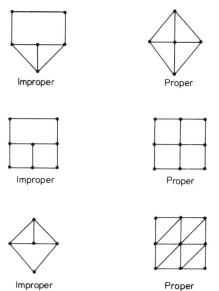

Improper   Proper

Improper   Proper

Improper   Proper   **Figure 10.7** Possible ways to form a mesh.

tively starting from 1, but except for the requirement that elements be consecutively numbered, the element numbering scheme is arbitrary. Once the node and element numbers have been assigned, they must remain fixed, because interpretation of the final output relies on the numbering scheme.

10.3.3   Preparation of Input Data

In this section we describe the preparation of input data for a given problem. Input data are *free field* and may be entered in any column on a line separated by blanks or commas. All parameters on a line must be entered; a blank cannot be used for zero. Any consistent set of units may be used; example units are given by illustrative purposes only.

The data are arranged in four different classifications, and they must be entered in the order listed below:

Classification	Data Type
I	Problem identification
II	Master control parameters
III	Nodal point data
IV	Material and element data

A detailed description of these classifications follows.

Classification I

Variable	Entry
HED	Problem identification

Explanation: Any desired heading information to be printed with the output.

Classification II

Variable	Entry
NUMNP	Total number of nodal points in the model
NELGRP	Number of element groups

Explanation: Nodes are labeled with integers ranging from 1 to the total number of nodes in the system, NUMNP. For selected elements a new element group may be defined. Elements within groups are assigned integer labels ranging from 1 to the total number of elements in the group.

Classification III

Variable	Entry
N	Node number
ID(N)	Boundary condition code
	.EQ.0 Temperature unknown
	.EQ.1 Temperature specified
X(N)	X coordinate
Y(N)	Y coordinate
Z(N)	Z coordinate
T(N)	Nodal temperature

Explanation: Nodal point data must be defined for all NUMNP nodes. Nodal data are entered consecutively for all nodes on lines from 1 to NUMNP. The boundary condition code is used to designate these nodes, which will have fixed values of temperature in the solution process. The fixed value of temperature is entered in the T(N) array.

Classification IV

A. Control Line

Variable	Entry
MTYPE	1 to denote the plane element
NUME	Number of elements in this group
NUMAT	Number of material property data lines
ND	Number of element nodes
	.EQ. 3 Plane triangle
	.EQ. − 3 Axisymmetric triangle
	.EQ. 4 Plane quadrilateral
	.EQ. − 4 Axisymmetric quadrilateral

B. Typical Material Property Data Line

Variable	Entry
N	Material type
TH(N)	Element thickness
	.EQ.t Plane element
	.EQ.1 Axisymmetric element
KXX(N)	Component of conductivity tensor, k_{xx}(real)
KXY(N)	Component of conductivity tensor, k_{xy}(real)
KYY(N)	Component of conductivity tensor, k_{yy}(real)

C. Typical Element Data Line

Variable	Entry
M	Element number
IE(I),I = 1,|ND|	Element node numbers (I, J, K, L)
MTYP	Material type
VOL	Volumetric heat generation rate Q (e.g., Btu/h ft^3)

Explanation: See Figure 10.2 for a definition of element coordinate systems and Section 10.2.2 for a discussion of the element formulation. For an isotropic material the conductivity value k should be entered for K_{xx} and K_{yy}, and K_{xy} should be entered as zero. Within each element group, the element data are entered consecutively for all elements on lines from 1 to NUME.

Element heat flows are computed at element centroids and are defined by

$$\begin{Bmatrix} q_x \\ q_y \end{Bmatrix} = -t \begin{bmatrix} k_{xx} & k_{xy} \\ k_{xy} & k_{yy} \end{bmatrix} \begin{Bmatrix} \dfrac{\partial T}{\partial x} \\ \dfrac{\partial T}{\partial y} \end{Bmatrix}$$

10.3.4 Description of Output

The first line of output is the problem identification, which is printed exactly as entered under Classification I. Next a brief table of control information is printed using the data entered under Classification II. Then all nodal point data are printed. Next the element data (Classification IV) are printed showing material properties and element connections. After the element data, a short table is printed showing the solution parameters, the number of equations, and matrix bandwidth. The output continues with a table displaying the computed nodal temperatures and concludes with lists of computed heat flows for each element group. Note that heat flows are computed in each element's local coordinate system (see Figure 10.2).

10.3.5 FORTRAN Listing

```
      PROGRAM TAP                                              TAP0010
                                                               TAP0020
      INTEGER INPUT,OUTPUT,TAPE5,TAPE6, TAPE1                  TAP0030
                                                               TAP0040
      PARAMETER (INPUT = 64,OUTPUT = 64,TAPE5 = INPUT,TAPE6 = OUTPUT,  TAP0050
     *           TAPE1 = 512)                                  TAP0060
C     **  **  **  **  **  **  **  **  **  **  **  **  **       TAP0070
C                                                              TAP0080
```

```
C                                                                TAP0090
C        TAP: A FINITE ELEMENT THERMAL ANALYSIS PROGRAM          TAP0100
C                                                                TAP0110
C     EARL A. THORNTON    UNIVERSITY OF VIRGINIA    1993          TAP0120
C                                                                TAP0130
C     **  **  **  **  **  **  **  **  **  **  **  **  **          TAP0140
C                                                                TAP0150
C         TAPE      CONTENTS                                     TAP0160
C                                                                TAP0170
C         1         ELEMENT DATA                                 TAP0180
C         5         INPUT                                        TAP0190
C         6         OUTPUT                                       TAP0200
C                                                                TAP0210
      COMMON / SOL / NUMNP,NELGRP,MTOT,NANA                      TAP0220
      COMMON / EL / ISW,NPAR(4),NFIRST,NLAST,LELST,MAXEST        TAP0230
      COMMON / DYN / N1,N2,N3,N4,N5,N6,N7,N8,N9,N10              TAP0240
      COMMON / EQS / NEQ,NCOLS,MBAND                             TAP0250
      COMMON / HEAD / HED(18)                                    TAP0260
      COMMON / IO / IN,IT                                        TAP0270
C                                                                TAP0280
      COMMON A(100000)                                          TAP0290
      MTOT = 100000                                             TAP0300
C                                                                TAP0310
      IN = 5                                                    TAP0320
      IT = 6                                                    TAP0330
C                                                                TAP0340
C     OPEN FILES                                                TAP0350
C                                                                TAP0360
      OPEN (FILE = 'INPUT.DAT ', UNIT = IN)                     TAP0370
      OPEN (FILE = 'OUTPUT.DAT ', UNIT = IT)                    TAP0380
                                                                TAP0390
      READ (IN,10) HED                                          TAP0400
      READ (IN,*) NUMNP,NELGRP                                  TAP0410
      WRITE (IT,20) HED,NUMNP,NELGRP                            TAP0420
C                                                                TAP0430
C     INPUT NODE DATA                                           TAP0440
C                                                                TAP0450
      N1 = 1                                                    TAP0460
      N2 = N1 + NUMNP                                           TAP0470
      N3 = N2 + NUMNP                                           TAP0480
      N4 = N3 + NUMNP                                           TAP0490
      N5 = N4 + NUMNP                                           TAP0500
      N6 = N5 + NUMNP                                           TAP0510
C                                                                TAP0520
      IF (N6.GT.MTOT) CALL ERROR (N6-MTOT)                      TAP0530
C                                                                TAP0540
      CALL NODES (A(N1,A(N2),A(N3),A(N4),A(N5),NUMNP)           TAP0550
C                                                                TAP0560
C              CURRENT CONTENTS OF COMMON AREA A(I)             TAP0570
                                                                TAP0580
C              LOCATION    VARIABLE    DEFINITION               TAP0590
C                                                                TAP0600
C     STARTING AT N1.........ID(1)      BOUNDARY CONDITION ID   TAP0610
C                 N2......... T(1)      NODAL TEMPERATURES       TAP0620
C                 N3......... X(1)      X NODAL COORDINATES      TAP0630
C                 N4......... Y(1)      Y NODAL COORDINATES      TAP0640
C                 N5......... Z(1)      Z NODAL COORDINATES      TAP0650
C                 N6.....TEMPORARY STORAGE FOR ELEMENT INPUT    TAP0660
C                                                                TAP0670
C     INPUT,GENERATE AND STORE ELEMENT DATA                     TAP0680
```

```
C                                                                TAP0690
      CALL ELDAT                                                 TAP0700
C                                                                TAP0710
      NEQ = NUMNP                                                TAP0720
      NCOLS = 2*MBAND- 1                                         TAP0730
      NWK = NEQ*NCOLS                                            TAP0740
      N7 = N6 + NWK                                              TAP0750
      N8 = N7 + NEQ                                              TAP0760
      N9 = N8 + NEQ                                              TAP0770
      N10 = HN9 + MAXEST                                         TAP0780
      WRITE (IT,30)                                              TAP0790
      WRITE (IT,40) NEQ,MBAND                                    TAP0800
      IF (N10.GT.MTOT) CALL ERROR (N10- MTOT)                    TAP0810
C                                                                TAP0820
C     NOW USE.....A(N6)  -  A(1,1)  CONDUCTANCE MATRIX           TAP0830
C                 A(N7)  -  Q(1)    LOAD VECTOR FROM BOUND. COND. TAP0840
C                 A(N8)  -  B(1)    TOTAL LOAD VECTOR, OR        TAP0850
C                 A(N8)  -  B(1)    TEMP. FROM LINEAR ANALYSIS   TAP0860
C                 A(N9)  -  ELEMENT DATA                         TAP0870
C                                                                TAP0880
C     LINEAR STEADY ANALYSIS                                     TAP0890
C                                                                TAP0900
      CALL ZERO (A(N6),NWK)                                      TAP0910
      CALL ZERO (A(N7),NEQ)                                      TAP0920
      CALL ZERO (A(N8),NEQ)                                      TAP0930
      CALL ELCAL (A(N9))                                         TAP0940
      CALL TEMPBC (NEQ,NCOLS,MBAND,A(N1),A(N2),A(N6),A(N7))      TAP0950
      CALL FACTOR (NEG,NCOLS,MBAND,A(N6),IPUNT)                  TAP0960
      IF (IPUNT.EQ.1) STOP                                       TAP0970
      CALL ADDBC (NEQ,A(N7),A(N8))                               TAP0980
      CALL SOLVE (NEQ,NCOLS,MBAND,A(N6),A(N8),IPUNT)             TAP0990
      IF (IPUNT.EQ.1) STOP                                       TAP1000
      CALL PRINTT (A(N8),NUMNP)                                  TAP1010
      CALL FLUX (A(N9))                                          TAP1020
C                                                                TAP1030
C                                                                TAP1040
*     CLOSE FILES                                                TAP1050
*                                                                TAP1060
      CLOSE (IN)                                                 TAP1070
      CLOSE (IT)                                                 TAP1080
                                                                 TAP1090
      STOP                                                       TAP1100
C                                                                TAP1110
10    FORMAT (18A4)                                              TAP1120
20    FORMAT (18A4 /// 38H C O N T R O L  I N F O R M A T I O N,// 4X,  TAP1130
     127H NUMBER OF NODAL POINTS   = ,I5 / 4X,27H NUMBER OF      TAP1140
     ELEMENT GROUPS 2 = ,I5 / 4X)                                TAP1150
30    FORMAT (38H S O L U T I O N  P A R A M E T E R S,// )      TAP1160
40    FORMAT (5X,34H TOTAL NUMBER OF EQUATIONS     = ,I5,/,5X,34H TAP1170
      SEMI 1 BANDWIDTH            = ,IT)                         TAP1180
      END                                                        TAP1190
      SUBROUTINE ZERO (A,N)                                      ZER0010
C                                                                ZER0020
C     ZERO AN ARRAY A(I)                                         ZER0030
C                                                                ZER0040
      DIMENSION A(1)                                             ZER0050
      DO 10 I = 1,N                                              ZER0060
      A(I) = 0.0                                                 ZER0070
10    CONTINUE                                                   ZER0080
      RETURN                                                     ZER0090
```

```
      END                                                    ZERO100
      SUBROUTINE ERROR (N)                                   ERR0010
      COMMON / IO / IN,IT                                    ERR0020
      WRITE (IT,10) N                                        ERR0030
      STOP                                                   ERR0040
C                                                            ERR0050
10    FORMAT ( / / ,20H STORAGE EXCEEDED BY16)               ERR0060
      END                                                    ERR0070
      SUBROUTINE NODES (ID,T,X,Y,Z,NUMNP)                    NOD0010
      DIMENSION ID(1),X(1),Y(1),Z(1),T(1)                    NOD0020
      COMMON / IO / IN,IT                                    NOD0030
C                                                            NOD0040
C    READ NODAL POINT DATA                                   NOD0050
C                                                            NOD0060
      WRITE (IT,40)                                          NOD0070
      DO 10 N = 1 , NUMNP                                    NOD0080
      READ (IN,*) I,ID(N),X(N),Y(N),Z(N),T(N)               NOD0090
10    CONTINUE                                               NOD0100
C                                                            NOD0110
C    PRINT ALL NODAL POINT DATA                              NOD0120
C                                                            NOD0130
      WRITE (IT,50)                                          NOD0140
      WRITE (IT,60) (N,ID(N),X(N),Y(N),Z(N),T(N),N = 1,NUMNP)  NOD0150
C                                                            NOD0160
      RETURN                                                 NOD0170
C                                                            NOD0180
40    FORMAT ( / /23H NODAL POINT INPUT DATA / )             NOD0190
50    FORMAT ( / ,1X,4HNODE3X,12H  B.C. CODE 11X,23HNODAL POINT  NOD0200
      COORDINATE 1S / 7H NUMBER22X,HX12X,1HY12X,1HZ8X,       NOD0210
      11HTEMPERATURE, / )
60    FORMAT (I5,5X,I5,6X,3E13.5,2X,E13.5)                   NOD0220
      END                                                    NOD0230
      SUBROUTINE ELDAT                                       ELD0010
C                                                            ELD0020
C    CALLS ELTYPE TO BRANCH TO ELEMENT ROUTINES              ELD0030
C     FOR ELEMENT INPUT DATA                                 ELD0040
C                                                            ELD0050
      COMMON / SOL / NUMNP,NELGRP,MTOT,NANA                  ELD0060
      COMMON / EL / ISW,NPAR(4),NFIRST,NLAST,LELST,MAXEST    ELD0070
      COMMON / EQS / NEQ,NCOLS,MBAND                         ELD0080
      COMMON / IO / IN,IT                                    ELD0090
      COMMON A(100000)                                       ELD0100
      NUMEL = 0                                              ELD0110
      MBAND = 0                                              ELD0120
      MAXEST = 0                                             ELD0130
      ISW = 1                                                ELD0140
      REWIND 1                                               ELD0150
      DO 10 M = 1,NELGRP                                     ELD0160
      WRITE (IT,20) M                                        ELD0170
      READ (IN,*) (NPAR(I),I = 1,4)                          ELD0180
      NUMEL = NUMEL + NPAR(2)                                ELD0190
      MTYPE = NPAR(1)                                        ELD0200
      CALL ELTYPE (MYTPE)                                    ELD0210
      IF (LELST.GT.MAXEST) MAXEST = LELST                    ELD0220
      WRITE (1) LELST,NPAR,(A(I),I = NFIRST,NLAST)           ELD0230
10    CONTINUE                                               ELD0240
      RETURN                                                 ELD0250
C                                                            ELD0260
20    FORMAT ( / / ,1X,13HELEMENT GROUP,I3)                  ELD0270
      END                                                    ELD0280
      SUBROUTINE ELCAL (AA)                                  ELC0010
```

```
C                                                                   ELC0020
C      CALLS ELTYPE TO BRANCH TO ELEMENT ROUTINES                   ELC0030
C      TO FORM EQUIVALENT SYSTEM MATRICES                           ELC0040
C                                                                   ELC0050
       COMMON / SOL / NUMNP,NELGRP,MTOT,NANA                        ELC0060
       COMMON / EL / ISW,NPAR(4),NFIRST,NLAST,LELST,MAXEST          ELC0070
       DIMENSION AA(1)                                              ELC0080
       ISW = 2                                                      ELC0090
       REWIND = 1                                                   ELC0100
       DO 10 M = 1,NELGRP                                           ELC0110
       READ (1) LRD,NPAR,(AA(I),I = 1,LRD)                         ELC0120
       MTYPE = NPAR(1)                                              ELC0130
       CALL ELTYPE (MTYPE)                                          ELC0140
10     CONTINUE                                                     ELC0150
       RETURN                                                       ELC0160
       END                                                          ELC0170
       SUBROUTINE ELTYPE (MTYPE)                                    ELT0010
       COMMON / IO / IN,IT                                          ELT0020
C                                                                   ELT0030
C      CALLED BY MAIN AND FLUX                                      ELT0040
C                                                                   ELT0050
       GO TO (10,20,30), MTYPE                                      ELT0060
10     CALL PLANE                                                   ELT0070
       RETURN                                                       ELT0080
20     CONTINUE                                                     ELT0090
30     WRITE (IT,40) MTYPE                                          ELT0100
       RETURN                                                       ELT0110
C                                                                   ELT0120
40     FORMAT (8H ELEMENT,I4,26H HAS NOT BEEN IMPLEMENTED ,/ )     ELT0130
       END                                                          ELT0140
       SUBROUTINE CALBAN (ND,LM)                                    CAL0010
C                                                                   CAL0020
C      CALLED BY ELEMENT SUBROUTINES                                CAL0030
C                                                                   CAL0040
C      CALCULATES SEMI-BANDWIDTH                                    CAL0050
C                                                                   CAL0060
       COMMON / EQS / NEQ,NCOLS,MBAND                               CAL0070
       DIMENSION LM(ND)                                             CAL0080
       MIN = 100000                                                 CAL0090
       MAX = 0                                                      CAL0100
       DO 10 L = 1,ND                                               CAL0110
       IF (LM(L).GT.MAX) MAX = LM(L)                               CAL0120
       IF (LM(L).LT.MIN) MIN = LM(L)                               CAL0130
10     CONTINUE                                                     CAL0140
       NDIF = MAX - MIN + 1                                         CAL0150
       IF (NDIF.GT.MBAND) MBAND = NDIF                             CAL0160
       RETURN                                                       CAL0170
       END                                                          CAL0180
       SUBROUTINE ADDSTF (ND,LM,NEQ,NCOLS,MBAND,S,A)               ADD0010
                                                                    ADD0020
C      FORMS SYSTEM CONDUCTANCE MATRIX                              ADD0030
C                                                                   ADD0040
       DIMENSION A(NEQ,NCOLS),S(ND,ND),LM(ND)                      ADD0050
       DO 20 I = 1,ND                                               ADD0060
       IT = LM(I)                                                   ADD0070
       DO 10 J = 1,ND                                               ADD0080
       JJ = MBAND - LM(I) + LM(J)                                   ADD0090
       A(II,JJ) = A(II,JJ) + S(I,"J)                                ADD0100
10     CONTINUE                                                     ADD0110
20     CONTINUE                                                     ADD0120
```

```
      RETURN                                              ADD0130
      END                                                 ADD0140
      SUBROUTINE ADDLD (ND,LM,NEQ,ID,Q,B)                 ADD0010
C                                                         ADD0020
C    FORMS LOAD VECTOR                                    ADD0030
C                                                         ADD0040
      DIMENSION ID(NEQ),B(NEQ),Q(ND),LM(ND)               ADD0050
      DO 10 I=1,ND                                        ADD0060
      II=LM(I)                                            ADD0070
      IF (ID(II).NE.0) GO TO 10                           ADD0080
      B(II)=B(II)+Q(I)                                    ADD0090
10    CONTINUE                                            ADD0100
      RETURN                                              ADD0110
      END                                                 ADD0120
      SUBROUTINE ADDBC (NEQ,Q,B)                          ADD0010
C                                                         ADD0020
C    ADDS CONTRIBUTION OF TEMPERATURE BOUNDARY CONDITION  ADD0030
C    TO LOAD VECTOR                                       ADD0040
C                                                         ADD0050
      DIMENSION Q(NEQ),B(NEQ)                             ADD0060
      DO 10 L=1,NEQ                                       ADD0070
      B(L)=B(L)+Q(L)                                      ADD0080
10    CONTINUE                                            ADD0090
      RETURN                                              ADD0100
      END                                                 ADD0110
      SUBROUTINE TEMPID (T,NTP,LTP,TND)                   TEM0010
C                                                         TEM0020
C    DETERMINES ELEMENT TEMPERATURES FOR AN INPUT NODE LIST.  TEM0030
C                                                         TEM0040
      DIMENSION T(1),LTP(NTP),TND(NTP)                    TEM0050
      DO 10 I=1,NTP                                       TEM0060
      IDUM=LTP(I)                                         TEM0070
      TND(I)=T(IDUM)                                      TEM0080
10    CONTINUE                                            TEM0090
      END                                                 TEM0100
      SUBROUTINE TEMPBC (NEQ,NCOLS,MBAND,ID,T,A,B)        TEM0010
C                                                         TEM0020
C      IMPOSES BOUNDARY CONDITIONS ON THERMAL EQUILIBRIUM EQUATIONS  TEM0030
C                                                         TEM0040
      DIMENSION A(NEQ,NCOLS),B(NEQ),ID(NEQ),T(NEQ)        TEM0050
      DO 30 I=1,NEQ                                       TEM0060
      IF (ID(I).EQ.0) TO TO 30                            TEM0070
      DO 20 JJ=1,NCOLS                                    TEM0080
      II=I+MBAND-JJ                                       TEM0090
      IF (II.LT.1) GO TO 10                               TEM0100
      IF (II.GT.NEQ) GO TO 10                             TEM0110
      IF (A(II,JJ).EQ.0.0) GO TO 10                       TEM0120
      B(II)=B(II)-T(I)*A(II,JJ)                           TEM0130
      A(II,JJ)=0.0                                        TEM0140
10    CONTINUE                                            TEM0150
      IF (A(I,JJ).EQ.0.0) GO TO 20                        TEM0160
      A(I,JJ)=0.0                                         TEM0170
20    CONTINUE                                            TEM0180
      B(I)=T(I)                                           TEM0190
      A(I,MBAND)=1.0                                      TEM0200
30    CONTINUE                                            TEM0210
      RETURN                                              TEM0220
      END                                                 TEM0230
      SUBROUTINE FACTOR (NEQ,NCOLS,MBAND,A,IPUNT)         FAC0010
                                                          FAC0020
```

```
C     REDUCE MATRIX BY GAUSS ELIMINATION                         FAC0030
C                                                                FAC0040
C     A(I,J) HAS SEMI-BANDWIDTH MBAND AND MAY BE ASYMMETRIC.     FAC0050
                                                                 FAC0060
      COMMON / IO / IN,IT                                        FAC0070
      DIMENSION A(NEQ,NCOLS)                                     FAC0080
      IPUNT = 0                                                  FAC0090
      KMIN = MBAND + 1                                           FAC0100
      DO 50 N = 1,NEQ                                            FAC0110
      IF (A(N,MBAND).EQ.0.0) GO TO 60                            FAC0120
      C = 1./A(N,MBAND)                                          FAC0130
      DO 10 K = KMIN,NCOLS                                       FAC0140
      IF (A(N,K).EQ.0.0) GO TO 10                                FAC0150
      A(N,K) = C*A(N,K)                                          FAC0160
10    CONTINUE                                                   FAC0170
      DO 40 L = 2,MBAND                                          FAC0180
      JJ = MBAND-L + 1                                           FAC0190
      I = N + L-1                                                FAC0200
      IF (I.GT.NEQ) GO TO 40                                     FAC0210
      IF (A(I,JJ).EQ.0.0) GO TO 40                               FAC0220
      KI = MBAND + 2-L                                           FAC0230
      KF = NCOLS + 1-L                                           FAC0240
      J = MBAND                                                  FAC0250
      DO 30 K = KI,KF                                            FAC0260
      J = J + 1                                                  FAC0270
      IF (A(N,J).EQ.0.0) TO TO 30                                FAC0280
      A(I,K) = A(I,K)-A(I,JJ)*A(N,J)                             FAC0290
30    CONTINUE                                                   FAC0300
40    CONTINUE                                                   FAC0310
50    CONTIUNE                                                   FAC0320
      RETURN                                                     FAC0330
60    CONTINUE                                                   FAC0340
      IPUNT = 1                                                  FAC0350
      WRITE (IT,70) N,A,(N,MBAND)                                FAC0360
      RETURN                                                     FAC0370
C                                                                FAC0380
70    FORMAT (,5X,34H SET OF EQUATIONS MAY BE SINGULAR,//,5X,25  FAC0390
     1HDIAGONAL TERM OF EQUATION,I5,13H IS EQUAL TO ,E15.8)      FAC0400
      END                                                        FAC0410
      SUBROUTINE SOLVE (NEQ,NCOLS,MBAND,A,B,IPUNT)               SOL0010
C                                                                SOL0020
C     REDUCTION OF A LOAD VECTOR B(I)                            SOL0030
C                                                                SOL0040
      COMMON / IO / IN,IT                                        SOL0050
      DIMENSION A(NEQ,NCOLS),B(NEQ)                              SOL0060
      IPUNT = 0                                                  SOL0070
      DO 30 N = 1,NEQ                                            SOL0080
      IF (A(N,MBAND).EQ.0.0) GO TO 60                            SOL0090
      B(N) = B(N) / A(N,MBAND)                                   SOL0100
      DO 20 L = 2,MBAND                                          SOL0110
      JJ = MBAND-L + 1                                           SOL0120
      I = N + L-1                                                SOL0130
      IF (I.GT.NEQ) GO TO 20                                     SOL0140
      IF (A(I,JJ).EQ.0.0) GO TO 20                               SOL0150
      B(I) = B(I)-A(I,JJ)*B(N)                                   SOL0160
20    CONTINUE                                                   SOL0170
30    CONTINUE                                                   SOL0180
C                                                                SOL0190
C     BACKSUBSTITUTION                                           SOL0200
C                                                                SOL0210
```

```
      LL = MBAND + 1                                            SOL0220
        DO 50 M = 1, NEQ                                        SOL0230
      N = NEQ + 1 - M                                           SOL0240
      DO 40 L = LL, NCOLS                                       SOL0250
      IF (A(N,L).EQ.0) GO TO 40                                 SOL0260
      K = N + L - MBAND                                         SOL0270
      B(N) = B(N) - A(N,L)*B(K)                                 SOL0280
      CONTINUE                                                  SOL0290
      CONTINUE                                                  SOL0300
      RETURN                                                    SOL0310
60    CONTINUE                                                  SOL0320
      IPUNT = 1                                                 SOL0330
      WRITE (IT,70) N,A(N,MBAND)                                SOL0340
      RETURN                                                    SOL0350
C                                                               SOL0360
70    FORMAT (,5X,31H SET OF EQUATIONS ARE SINGULAR ,//,5X,25   SOL0370
     1HDIAGONAL TERM OF EQUATION,I5,13H IS EQUAL TO ,E15.8)     SOL0380
      END                                                       SOL0390
      SUBROUTINE PRINTT (T,NUMNP)                               PRI0010
C                                                               PRI0020
C     PRINTS TEMPERATURE SOLUTION                               PRI0030
C                                                               PRI0040
      COMMON / IO / IN,IT                                       PRI0050
      DIMENSION T(NUMNP)                                        PRI0060
      WRITE (IT,20)                                             PRI0070
      WRITE (IT,30)                                             PRI0080
      NLINES = NUMNP / 5 + 1                                    PRI0090
      J1 = 1                                                    PRI0100
      DO 10 I = 1, NLINES                                       PRI0110
      NO = (I - 1)*5 + 1                                        PRI0120
      IF (NO.GT.NUMNP) GO TO 10                                 PRI0130
      J2 = J1 + 4                                               PRI0140
      IF (J2.GT.NUMNP) J2 = NUMNP                               PRI0150
      WRITE (IT,40) NO,(T(J),J = J1,J2)                         PRI0160
      J1 = J1 + 5                                               PRI0170
10    CONTINUE                                                  PRI0180
      RETURN                                                    PRI0190
C                                                               PRI0200
20    FORMAT (//,36H T E M P E R A T U R E   V E C T O R,/)     PRI0210
30    FORMAT (80H NODE NO.     NO    VALUE    NO + 1 VALUE      PRI0220
     1 NO + 2 VALUE     NO + 3 VALUE     NO + 4 VALUE,/ )       PRI0230
40    FORMAT (I6,4X,5E14.6)                                     PRI0240
      END                                                       PRI0250
      SUBROUTINE FLUX (AA)                                      FLU0010
C                                                               FLU0020
C     CALCULATE FLUXES FOR ALL ELEMENTS                         FLU0030
C                                                               FLU0040
      COMMON / SOL / NUMNP,NELGRP,MTOT,NANA                     FLU0050
      COMMON / EL / ISW,NPAR(4),NFIRST,NLAST,LELST,MAXEST       FLU0060
      COMMON / IO / IN,IT                                       FLU0070
      DIMENSION AA(1)                                           FLU0080
      ISW = 3                                                   FLO0090
      REWIND 1                                                  FLU0100
      DO 10 M = 1, NELGRP                                       FLU0110
      WRITE (IT,20) M                                           FLU0120
      READ (1) LRD,NPAR,(AA(I),I = 1,LRD)                       FLU0130
      MTYPE = NPAR(1)                                           FLU0140
      CALL ELTYPE (MTYPE)                                       FLU0150
10    CONTINUE                                                  FLU0160
      RETURN                                                    FLU0170
```

```
C                                                                      FLU0180
20   FORMAT ( / / ,1X,13HELEMENT GROUP,I3)                             FLU0190
     END                                                               FLU0200
     FUNCTION DOT(A,B)                                                 FLU0210
C                                                                      FLU0220
C    COMPUTES THE DOT PRODUCT OF TWO VECTORS A,B                       FLU0230
C                                                                      FLU0240
     DIMENSION A(4),B(4)                                               FLU0250
     DOT = A(1)*B(1) + A(2)*B(2) + A(3)*B(3)                           FLU0260
C                                                                      FLU0270
     RETURN                                                            FLU0280
     END                                                               FLU0290
     SUBROUTINE CROSS (A,B,C)                                          CRO0010
C                                                                      CRO0020
C     COMPUTES....THE MAGNITUDE C(4) OF A VECTOR C = A CROSS B,        CRO0030
C                 AND COMPONENTS OF A UNIT VECTOR IN THE               CRO0040
C                 C DIRECTION.                                         CRO0050
     COMMON / IO / IN,IT                                               CRO0060
     DIMENSION A(4),B(4),C(4)                                          CRO0070
     X = A(2)*B(3) - A(3)*B(2)                                         CRO0080
     Y = A(3)*B(1) - A(1)*B(3)                                         CRO0090
     Z = A(1)*B(2) - A(2)*B(1)                                         CRO0100
     C(40) = SQRT(X*X + Y*Y + Z*Z)                                     CRO0110
     IF (C(4).GT.0.1E-8) GO TO 10                                      CRO0120
     WRITE (IT,20)                                                     CRO0130
     STOP                                                              CRO0140
10   CONTINUE                                                          CRO0150
     C(3) = Z / C(4)                                                   CRO0160
     C(2) = Y / C(4)                                                   CRO0170
     C(1) = X / C(4)                                                   CRO0180
     RETURN                                                            CRO0190
C                                                                      CRO0200
20   FORMAT ( / / ,5X,20HFATAL ERROR IN CROSS,16H, CHECK GEOMETRY / / ) CRO0210
     END                                                               CRO0220
     SUBROUTINE VECTOR (V,XI,YI,ZI,XJ,YJ,ZJ)                           VEC0010
C                                                                      VEC0020
C     COMPUTES....THE MAGNITUDE V(4) OF A VECTOR FROM I TO J,AND       VEC0030
C                 COMPONENTS V(1),V(2),V(3) OF A UNIT VECTOR           VEC0040
C                 FROM I TO J.                                         VEC0050
C                                                                      VEC0060
     COMMON / IO / IN,IT                                               VEC0070
     DIMENSION V(4)                                                    VEC0080
     X = JX - XI                                                       VEC0090
     Y = YJ - YI                                                       VEC0100
     Z = ZJ - ZI                                                       VEC0110
     V(4) = SQRT(X*X + Y*Y + Z*Z)                                      VEC0120
     IF (V(4).GT.0.1E-8) GO TO 10                                      VEC0130
     WRITE (IT,20)                                                     VEC0140
     STOP                                                              VEC0150
10   CONTINUE                                                          VEC0160
     V(3) = Z / V(4)                                                   VEC0170
     V(2) = Y / V(4)                                                   VEC0180
     V(1) = X / V(4)                                                   VEC0190
     RETURN                                                            VEC0200
C                                                                      VEC0210
20   FORMAT ( / / ,5X,21HFATAL ERROR IN VECTOR,16H, CHECK GEOMETRY / / ) VEC0220
     END                                                               VEC0230
     SUBROUTINE PLANE                                                  PLA0010
C                                                                      PLA0020
C    SETS UP STORAGE FOR PLANE ELEMENT                                 PLA0030
```

```
C                                                          PLA0040
      COMMON / SOL / NUMNP,NELGRP,MTOT,NANA                 PLA0050
      COMMON / EL / ISW,NPAR(4),NFIRST,NLAST,LELST,MAXEST   PLA0060
      COMMON / DYN / N1,N2,N3,N4,N5,N6,N7,N8,N9,N10         PLA0070
      COMMON A(100000)                                      PLA0080
      NFIRST = N6                                           PLA0090
      IF (ISW.GT.1) NFIRST = N9                             PLA0100
      L1 = NFIRST                                           PLA0110
      L2 = L1 + NPAR(3)                                     PLA0120
      L3 = L2 + NPAR(3)                                     PLA0130
      L4 = L3 + NPAR(3)                                     PLA0140
      L5 = L4 + NPAR(3)                                     PLA0150
      L6 = L5 + NPAR(2)                                     PLA0160
      L7 = L6 + NPAR(2)                                     PLA0170
      NLAST = L7 + 4*NPAR(2)                                PLA0180
      LELST = NLAST- NFIRST                                 PLA0190
      MM = NLAST- MTOT                                      PLA0200
      IF (MM.GT.0) CALL ERROR (MM)                          PLA0210
      CALL CPLANE (A(N1),A(N2),A(N3),A(N4),A(N5),A(L1),A(L2),A(L3)  PLA0220
     1,A(L4),A(L5),A(L6),A(L7))                             PLA0230
      RETURN                                                PLA0240
      END                                                   PLA0250
      SUBROUTINE CPLANE (ID,T,X,Y,Z,TH,KXX,KXY,KYY,MTYPE,VOLQ,LM)  CPL0010
C                                                          CPL0020
C     COMPUTES ELEMENT MATRICES FOR PLANE ELEMENT          CPL0030
C                                                          CPL0040
      COMMON / SOL / NUMNP,NELGRP,MTOT,NANA                 CPL0050
      COMMON / DYN / N1,N2,N3,N4,N5,N6,N7,N8,N9,N10         CPL0060
      COMMON / IO / IN,IT                                   CPL0070
      COMMON / EQS / NEQ,NCOLS,MBAND                        CPL0080
      COMMON / EL / ISW,NPAR(4),NFIRST,NLAST,LELST,MAXEST   CPL0090
      COMMON A(100000)                                      CPL0100
      DIMENSION ID(1),T(1),X(1),Y(1),Z(1)                  CPL0110
      DIMENSION MTYPE(1),VOLQ(1),LM(4,1)                    CPL0120
      DIMENSION S(4,4),PVOL(4),TH(1)                        CPL0130
      DIMENSION TND(4),ST(2,4),IE(4),IX(4)                  CPL0140
      DIMENSION XL(4),YL(4)                                 CPL0150
      REAL KXX(1),KXY(1),KYY)1)                             CPL0160
      NUME = NPAR(2)                                        CPL0170
      NUMAT = NPAR(3)                                       CPL0180
      ND = NPAR(4)                                          CPL0190
      KAT = 3                                               CPL0200
      IF (ND.LT.0) KAT = 2                                  CPL0210
      IF (ND.LT.0) ND = -ND                                 CPL0220
      IF (ISW.EQ.1) GO TO 10                                CPL0230
      IF (ISW.EQ.2) GO TO 100                               CPL0240
      IF (ISW.EQ.3) GO TO 120                               CPL0250
C                                                          CPL0260
C     ELEMENT DATA INPUT                                   CPL0270
C                                                          CPL0280
10    CONTINUE                                             CPL0290
      WRITE (IT,140) NUME,NUMAT,ND                          CPL0300
      IF (KAT.EQ.2) WRITE (IT,150)                          CPL0310
      DO 20 I = 1,NUMAT                                     CPL0320
      READ (IN,*) N,TH(N),KXX(N),KXY(N),KYY(N)              CPL0330
20    CONTINUE                                             CPL0340
C                                                          CPL0350
      WRITE (IT,210)                                        CPL0360
      DO 30 I = 1,NUMAT                                     CPL0370
      WRITE (IT,220) I,TH(I),KXX(I),KXY(I),KYY(I)           CPL0380
```

```
30  CONTINUE                                                     CPL0390
    WRITE (IT,170)                                               CPL0400
    DO 40 I = 1 , NUME                                           CPL0410
    READ (IN,*) N,(IE(II),II=1,ND),MTYP,VOL                      CPL0420
    IF (ND.EQ.3) IE(4)=0                                         CPL0430
    DO 60 II=1,4                                                 CPL0440
    IX(II)=IE(II)                                                CPL0450
60  CONTINUE                                                     CPL0460
C                                                                CPL0470
C     SAVE ELEMENT INFORMATION                                   CPL0480
C                                                                CPL0490
    MTYPE(N)=MTYP                                                CPL0500
    VOLQ(N)=VOL                                                  CPL0510
C                                                                CPL0520
C     FORM LOCATION MATRIX AND COMPUTE BANDWIDTH                 CPL0530
                                                                 CPL0540
    DO 80 II=1,4                                                 CPL0550
    LM(II,N)=IX(II)                                              CPL0560
80  CONTINUE                                                     CPL0570
    CALL CALBAN (ND,LM(1,N))                                     CPL0580
    WRITE (IT,180) N,IX,MTYPE(N),VOLQ(N)                         CPL0590
40  CONTINUE                                                     CPL0600
    RETURN                                                       CPL0610
C                                                                CPL0620
C     ELEMENT CONDUCTION MATRIX AND HEAT LOAD VECTORS            CPL0630
C                                                                CPL0640
100 CONTINUE                                                     CPL0650
                                                                 CPL0660
    DO 110 N = 1 , NUME                                          CPL0670
    MTYP = MTYPE(N)                                              CPL0680
    CXX = KXX(MTYP)                                              CPL0690
    CXY = KXY(MTYP)                                              CPL0700
    CYY = KYY(MTYP)                                              CPL0710
    IF (KAT.NE.2) CALL QCORD (ND,LM(1,N),X,Y,Z,XL,YL,AS)        CPL0720
    IF (KAT.EQ.2) CALL AXCORD (ND,LM(1,N),X,Y,Z,XL,YL)          CPL0730
    CALL KISO4 (KAT,0,ND,TH(MTYP),XL,YL,CXX,CXY,CYY,VOLQ(N),S,   CPL0740
    ST,PVOL)
    CALL ADDSTF (ND,LM(1,N),NEQ,NCOLS,MBAND,S,A(N6))            CPL0750
    CALL ADDLD (ND,LM(1,N),NEQ,A(N1),PVOL,A(N8))                CPL0760
110 CONTINUE                                                     CPL0770
    RETURN                                                       CPL0780
C                                                                CPL0790
C     FLUX COMPUTATION                                           CPL0800
C                                                                CPL0810
120 CONTINUE                                                     CPL0820
    WRITE (IT,160)                                               CPL0830
    WRITE (IT,190)                                               CPL0840
    DO 130 N=1,NUME                                              CPL0850
    MTYP=MTYPE(N)                                                CPL0860
    CXX=KXX(MTYP)                                                CPL0870
    CXY=KXY(MTYP)                                                CPL0880
    CYY=KYY(MTYP)                                                CPL0890
    IF (KAT.NE.2) CALL QCORD (ND,LM(1,N),X,Y,Z,XL,YL,AS)        CPL0900
    IF (KAT.EQ.2) CALL AXCORD (ND,LM(1,N),X,Y,Z,XL,YL)          CPL0910
    CALL KISO4 (KAT,1,ND,TH(MTYP),XL,YL,CXX,CXY,CYY,VOLQ(N),     CPL0920
    S,ST,PVOL)
    CALL TEMPID (A(N8),ND,LM(1,N),TND)                          CPL0930
    CALL QCOND4 (ND,TND,ST,QX,QY)                               CPL0940
    WRITE (IT,200) N,QX,QY                                       CPL0950
130 CONTINUE                                                     CPL0960
```

```
      RETURN                                                        CPL0970
C                                                                   CPL0980
C                                                                   CPL0990
140 FORMAT (//,1X,27HP L A N E   E L E M E N T S,///,2X,26HNUMBER  CPL1000
   1OF PLANE ELEMENTS =,I3,/2X,26HNUMBER OF MATERIALS               CPL1010
   2=,I3,/2X,26HNUMBER OF NODES           =,I3,///)                 CPL1020
150 FORMAT (2X,17HAXISYMMETRIC CASE,///)                           CPL1030
160 FORMAT (//,1X,26HP L A N E   E L E M E N T  ,2X,20H            CPL1040
   1H E A T   F L U X E S,//)                                       CPL1050
170 FORMAT (//,37H     N     I     J     K     L     MATID ,5X,1HQ/) CPL1060
180 FORMAT (5I5,2X,I5,5X,E11.4)                                     CPL1070
190 FORMAT (9X,7HELEMENT,6X,2HQX,8X,2HQY,/)                        CPL1080
200 FORMAT (9X,I4,1X,E11.4,2X,E11.4)                                CPL1090
210 FORMAT (//,1X,8HMATERIAL,2X,9HTHICKNESS,9X,                    CPL1100
   119HCONDUCTIVITY TENSOR/27X,3HKXX,8X,3HKXY,8X,3HKYY)             CPL1110
220 FORMAT (1X,I4,4X,4(E11.4))                                      CPL1120
      END                                                           CPL1130
      SUBROUTINE AXCORD (ND,IX,X,Y,Z,XL,YL)                         AXC0010
C                                                                   AXC0020
C    FINDS LOCAL COORDINATES OF AXISYMMETRIC ELEMENTS              AXC0030
C    IN GLOBAL REFERENCE FRAME. NOTE ELEMENTS MUST                AXC0040
C    LIE IN GLOBAL X- Z PLANE.                                     AXC0050
C                                                                   AXC0060
      DIMENSION X(1),Y(1),Z(1)                                      AXC0070
      DIMENSION IX(4),XL(4),YL(4)                                   AXC0080
      I = IX(1)                                                     AXC0090
      J = IX(2)                                                     AXC0100
      K = IX(3)                                                     AXC0110
      XL(1) = X(I)                                                  AXC0120
      XL(2) = X(J)                                                  AXC0130
      XL(3) = X(K)                                                  AXC0140
      YL(1) = Z(I)                                                  AXC0150
      YL(2) = Z(J)                                                  AXC0160
      YL(3) = Z(K)                                                  AXC0170
      IF (ND.EQ.3) RETURN                                           AXC0180
      L = IX(4)                                                     AXC0190
      XL(4) = X(L)                                                  AXC0200
      YL(4) = Z(L)                                                  AXC0210
      RETURN                                                        AXC0220
      END                                                           AXC0230
      SUBROUTINE QCORD (ND,IX,X,Y,Z,XL,YL,AS)                       QCO0010
C                                                                   QCO0020
C    COMPUTES LOCAL COORDINATES AND AREA FOR TRIANGLE OR QUAD      QCO0030
C    WITH LOCAL X AXIS FROM NODE I TO J                           QCO0040
C                                                                   QCO0050
      DIMENSION X(1),Y(1),Z(1)                                      QCO0060
      DIMENSION IX(4),XL(4),YL(4),G(4),U(4),V(4),W(4)               QCO0070
      I = IX(1)                                                     QCO0080
      J = IX(2)                                                     QCO0090
      K = IX(3)                                                     QCO0100
      CALL VECTOR (V,X(I),Y(I),Z(I),X(J),Y(J),Z(J))                 QCO0110
      CALL VECTOR (G,X(I),Y(I),Z(I),X(K),Y(K),Z(K))                 QCO0120
      CALL CROSS (V,G,W)                                            QCO0130
      CALL CROSS (W,V,U)                                            QCO0140
      XL(1) = 0.0                                                   QCO0150
      YL(1) = 0.0                                                   QCO0160
      XL(2) = V(4)                                                  QCO0170
      YL(2) = 0.0                                                   QCO0180
      XL(3) = G(4)*DOT(G,V)                                         QCO0190
      YL(3) = G(4)*DOT(G,U)                                         QCO0200
```

```
      XL(4) = XL(3)                                               QCO0210
      YL(4) = YL(3)                                               QCO0220
      AS = 0.0                                                    QCO0230
      AS = AS + 0.5*V(4)*G(4)*W(4)                                QCO0240
      IF (ND.EQ.3) RETURN                                         QCO0250
      L = IX(4)                                                   QCO0260
      CALL VECTOR (W,X(I),Y(I),Z(I),X(L),Y(L),Z(L))              QCO0270
      XL(4) = W(4)*DOT(W,V)                                       QCO0280
      YL(4) = W(4)*DOT(W,U)                                       QCO0290
      CALL CROSS (G,W,U)                                          QCO0300
      AS = AS + 0.5*W(4)*G(4)*U(4)                                QCO0310
      RETURN                                                      QCO0320
      END                                                         QCO0330
      SUBROUTINE KISO4 (KAT,KODE,ND,THICK,XL,YL,KXX,KXY,KYY,Q,   KIS0010
C     S,ST,P)                                                     KIS0020
C     COMPUTES CONDUCTANCE MATRIX,LOAD VECTOR,AND                KIS0030
C     FLUX RECOVERY MATRIX FOR A 3 NODE TRIANGLE AND A 4 NODE    KIS0040
C     ISOPARAMETRIC QUADILATERAL                                  KIS0050
C                                                                 KIS0060
C     CALLED BY CISO4                                             KIS0070
C     CALLS     SHAPE                                             KIS0080
C                                                                 KIS0090
C     IF KODE.EQ.O COMPUTES S,P                                   KIS0100
C                 1 COMPUTES ST                                   KIS0110
C                                                                 KIS0120
      DIMENSION S(ND,ND),ST(2,ND)P(ND),B(2,4),E(2,2),C(2,2)      KIS0130
      DIMENSION XL(ND),YL(ND)                                     KIS0140
      REAL N(4),KXX,KXY,KYY                                       KIS0150
      COMMON / IO / IN,IT                                         KIS0160
      CXX = KXX*THICK                                             KIS0170
      IF (CXX.EQ.0.0) GO TO 130                                   KIS0180
      CXY = KXY*THICK                                             KIS0190
      CYY = KYY*THICK                                             KIS0200
      IF (KODE.EQ.1) GO TO 90                                     KIS0210
      DO 10 I = 1,ND                                              KIS0220
      P(I) = 0.0                                                  KIS0230
      DO 10 J = 1,ND                                              KIS0240
      S(I,J) = 0.0                                                KIS0250
   10 CONTINUE                                                    KIS0260
C                                                                 KIS0270
C     COMPUTE UPPER TRIANGLE OF CHOLESKY FACTOR OF CONDUCTION    KIS0280
C     MATRIX.                                                     KIS0290
      E(1,1) = SQRT(CXX)                                          KIS0300
      E(1,2) = CXY / E(1,1)                                       KIS0310
      E(2,1) = 0.0                                                KIS0320
      E(2,2) = SQRT(CYY- E(1,2)*E(1,2))                           KIS0330
C                                                                 KIS0340
       START GAUSS QUADRATURE LOOP. FOUR POINT INTEGRATION.      KIS0350
C                                                                 KIS0360
      DO 70 II = 1,2                                              KIS0370
      DO 70 JJ = 1,2                                              KIS0380
      CALL SHAPE (ND,II,JJ,XL,YL,N,B,DETJAC)                     KIS0390
      IF (KAT.EQ.2) CALL AXI (ND,N,XL,DETJAC)                    KIS0400
C                                                                 KIS0410
C      MULTIPLY CHOLESKY FACTOR OF E TIMES B.                     KIS0420
C      OVERWRITE RESULT IN B. FORM THERMAL CONDUCTANCE MATRIX.   KIS0430
C                                                                 KIS0440
      DO 30 K = 1,2                                               KIS0450
      DO 30 L = 1,ND                                              KIS0460
      DUMY = 0.0                                                  KIS0470
```

```
        DO 20 M=K,2                                          KIS0480
        DUMY=DUMY+E(K,M)*B(M,L)                              KIS0490
20      CONTINUE                                             KIS0500
        B(K,L)=DUMY                                          KIS0510
30      CONTINUE                                             KIS0520
C                                                            KIS0530
C       ADD CONTRIBUTIONS. GAUSS WEIGHTS ARE 1.0             KIS0540
C                                                            KIS0550
C       MULTIPLY E*B TRANSPOSE * E*B                         KIS0560
C                                                            KIS0570
        DO 50 NROW=1,ND                                      KIS0580
        DO 50 NCOL=NROW,ND                                   KIS0590
        DUMY=0.0                                             KIS0600
        DO 40 L=1,2                                          KIS0610
        DUMY=DUMY+B(L,NROW)*B(L,NCOL)                        KIS0620
40      CONTINUE                                             KIS0630
        S(NROW,NCOL)=(NROW,NCOL)+DUMY*DETJAC                 KIS0640
50      CONTINUE                                             KIS0650
C                                                            KIS0660
C       COMPUTE THERMAL LOAD VECTOR                          KIS0670
C                                                            KIS0680
        DO 60 I=1,ND                                         KIS0690
        P(I)=P(I)+Q*THICK*N(I)*DETJAC                        KIS0700
60      CONTINUE                                             KIS0710
70      CONTINUE                                             KIS0720
C                                                            KIS0730
C       QUADRATURE COMPLETE. COMPLETE CONDUCTANCE MATRIX BY  KIS0740
C       SYMMETRY.                                            KIS0750
        DO 80 K=2,ND                                         KIS0760
        DO 80 L=1,K                                          KIS0770
        S(K,L)=S(L,K)                                        KIS0780
80      CONTINUE                                             KIS0790
        RETURN                                               KIS0800
C                                                            KIS0810
C       COMPUTE FLUX RECOVERY MATRIX                         KIS0820
C                                                            KIS0830
90      CONTINUE                                             KIS0840
        DO 100 I=1,2                                         KIS0850
        DO 100 J=1,ND                                        KIS0860
        ST(I,J)=0.0                                          KIS0870
100     CONTINUE                                             KIS0880
        C(1,1)=CXX                                           KIS0890
        C(1,2)=CXY                                           KIS0900
        C(2,1)=CXY                                           KIS0910
        C(2,2)=CYY                                           KIS0920
C                                                            KIS0930
C       EVALUATE B MATRIX AT CENTROID FOR FLUX RECOVERY      KIS0940
C                                                            KIS0950
        CALL SHAPE (ND,3,3,XL,YL,N,B,DETJAC)                 KIS0960
        IF (KAT.EQ.2) CALL AXI (ND,N,XL,DETJAC)              KIS0970
        DO 120 K=1,2                                         KIS0980
        DO 120 L=1,ND                                        KIS0990
        DUMY=0.0                                             KIS1000
        DO 110 M=1,2                                         KIS1010
        DUMY=DUMY+C(K,M)*B(M,L)                              KIS1020
110     CONTINUE                                             KIS1030
        ST(K,L)=ST(K,L)-DUMY                                 KIS1040
120     CONTINUE                                             KIS1050
        RETURN                                               KIS1060
130     CONTINUE                                             KIS1070
```

```
      WRITE (IT,140)                                            KIS1080
      STOP                                                      KIS1090
C                                                               KIS1100
140   FORMAT ( / ,5X,28HFATAL ERROR, KXX EQUALS ZERO,/ )        KIS1110
      END                                                       KIS1120
      SUBROUTINE SHAPE (ND,II,JJ,XL,YL,N,B,DETJAC)              SHA0010
      DIMENSION B(2,ND),XL(ND),YL(ND),XII(4),ETI(4),AA(3)       SHA0020
      REAL N(4),NXI(4),JAC(2,2),NET(4)                          SHA0030
      COMMON / IO / IN,IT                                       SHA0040
      DATA   AA /-0.57735026918963 , 0.57735026918963 , 0.0  /  SHA0050
      DATA XII /-1.,1.,1.,-1. /                                 SHA0060
      DATA ETI /-1.,-1.,1.,1. /                                 SHA0070
      DO 10 I=1,4                                               SHA0080
      DUM1=(1.+XII(I)*AA(II))*0.25                              SHA0090
      DUM2=(1.+ETI(I)*AA(JJ)) 4.25                              SHA0100
      N(I)=4.*DUM1*DUM2                                         SHA0110
      NXI(I)=XII(I)*DUM2                                        SHA0120
      NET(I)=ETI(I)*DUM1                                        SHA0130
10    CONTINUE                                                  SHA0140
      IF (ND.EQ.4) GO TO 20                                     SHA0150
C                                                               SHA0160
C     FORM TRIANGLE BY ADDING THIRD AND FOURTH TOGETHER         SHA0170
C                                                               SHA0180
      N(3)=N(3)+N(4)                                            SHA0190
      NXI(3)=NXI(3)+NXI(4)                                      SHA0200
      NET(3)=NET(3)+NET(4)                                      SHA0210
20    CONTINUE                                                  SHA0220
C                                                               SHA0230
C     FIND JACOBIAN,ITS INVERSE AND ITS DETERMINANT             SHA0240
C                                                               SHA0250
      DO 30 I=1,2                                               SHA0260
      DO 30 J=1,2                                               SHA0270
      JAC(I,J)=0.0                                              SHA0280
30    CONTINUE                                                  SHA0290
      DO 40 I=1,ND                                              SHA0300
      JAC(1,1)=JAC(1,1)+NXI(I)*XL(I)                            SHA0310
      JAC(1,2)=JAC(1,2)+NXI(I)*YL(I)                            SHA0320
      JAC(2,1)=JAC(2,1)+NET(I)*XL(I)                            SHA0330
      JAC(2,2)=JAC(2,2)+NET(I)*YL(I)                            SHA0340
40    CONTINUE                                                  SHA0350
      DETJAC=JAC(1,1)*JAC(2,2)-JAC(2,1)*JAC(1,2)                SHA0360
      DUMY=JAC(1,1) / DETJAC                                    SHA0370
      JAC(1,1)=JAC(2,2) / DETJAC                                SHA0380
      JAC(1,2)=-JAC(1,2) / DETJAC                               SHA0390
      JAC(2,1)=-JAC(2,1) / DETJAC                               SHA0400
      JAC(2,2)=DUMY                                             SHA0410
C                                                               SHA0420
C     FORM THE GRADIENT - TEMPERATURE MATRIX...B(2,ND)          SHA0430
C                                                               SHA0440
      DO 50 I=1,2                                               SHA0450
      DO 50 J=1,ND                                              SHA0460
      B(I,J)=JAC(I,1)*NXI(J)+JAC(I,2)*NET(J)                    SHA0470
50    CONTINUE                                                  SHA0480
      RETURN                                                    SHA0490
      END                                                       SHA0500
      SUBROUTINE AXI (ND,N,XL,DETJAC)                           AXI0010
C                                                               AXI0020
C     MODIFIES JACOBIAN DETERMINANT FOR AXISYMMETRY             AXI0030
C                                                               AXI0040
      REAL N(4)                                                 AXI0050
```

```
      DIMENSION XL(4)                               AXI0060
      RR = 0.0                                      AXI0070
      DO 10 I = 1,ND                                AXI0080
      RR = RR + N(I)*XL(I)                          AXI0090
 10   CONTINUE                                      AXI0100
      DETJAC = RR*DETJAC                            AXI0110
      RETURN                                        AXI0120
      END                                           AXI0130
      SUBROUTINE QCOND4 (ND,TND,ST,QX,QY)           QCO0010
C                                                   QCO0020
C     CALLED BY CPLANE                              QCO0030
C     COMPUTES TRIANGLE AND QUAD. CONDUCTION HEAT FLUXES   QCO0040
C                                                   QCO0050
      DIMENSION TND(ND),ST(2,ND)                    QCO0060
      QX = 0.0                                      QCO0070
      QY = 0.0                                      QCO0080
      DO 10 JJ = 1,ND                               QCO0090
      QX = QX + ST(1,JJ)*TND(JJ)                    QCO0100
      QY = QY + ST(2,JJ)*TND(JJ)                    QCO0110
 10   CONTINUE                                      QCO0120
      RETURN                                        QCO0130
      END                                           QCO0140
```

10.4 EXAMPLE PROBLEMS WITH INPUT AND OUTPUT

10.4.1 Example 1: Axisymmetric Heat Flow in an Insulated Hollow Cylinder

The problem of heat flow in an axisymmetric insulated hollow cylinder is a simple example with which to illustrate the use of the program. The inner surface of the cylinder is held at $T = 200$, while the outer surface of the low-conductivity insulation is maintained at $T = 100$. We want to determine the temperature distribution and heat flux in the cylinder and insulation. A typical cross section of the cylinder wall and insulation is shown in Figure 10.8.

We decide to use the finite element model shown in Figure 10.9, with specified boundary temperatures and heat loads as illustrated. We enter the specified boundary temperatures as part of the nodal data, but the program automatically takes care of the zero nodal heat loads on the interior.

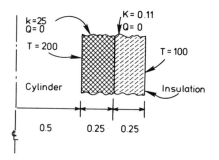

Figure 10.8 Example 1, axisymmetric heat flow.

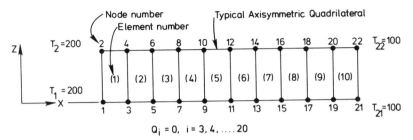

Figure 10.9   Finite element mesh for Example 1, Figure 10.8.

After numbering the nodes and elements as shown in Figure 10.9, we prepare the input data file as shown in Table 10.1. Observe that the nodes are numbered in the narrow direction to give a minimum bandwidth.

Inspection of the output in Table 10.2 shows that the temperatures on the top and the bottom of the elements are equal. This result is to be expected since the problem is one dimensional. Note that in the high-conductivity cylinder (nodes 1–12) there is only a slight temperature drop, but in the low-conductivity insulation (nodes 13–22) there is a large temperature drop. In the element heat fluxes QX is the radial flux and QY is the vertical flux; QY should be zero because of the one-dimensional heat flow, but it is nonzero because of round-off error. Temperatures and radial heat flows computed from the finite element analysis are compared with an exact analytical solution in Figures 10.10 and 10.11, respectively. Figure 10.10 shows the excellent agreement between the temperatures computed from the finite element model and the exact solution. Figure 10.10 shows that the constant element heat flows gives excellent agreement with the analytical solution at element centroids, but there is some error at the element boundaries, as indicated by the step discontinuities in the finite element heat flux distribution. From Figure 10.11 we see that nodal averages would be in excellent agreement with the analytical solution. In Section 9.2.3, we present a "smoothing" technique for computing average nodal velocities components in potential flow. The same technique may be applied here to compute continuous nodal heat flux distributions.

10.4.2   Example 2: Heat Transfer With Steep Local Gradients

As a second example, we consider the irregular, planar region of unit thickness shown in Figure 10.12. The region is heated externally so that it experiences local temperatures up to 2000 °F on its bottom surface. On the U-shaped top surface, temperatures are specified to be zero. Surfaces without specified temperatures are assumed to be perfectly insulated so that the heat flux normal to the surface is zero. The problem roughly simulates a thin-walled segment of a hypersonic aircraft subjected to intense local heating due to a shock impingement.

TABLE 10.1.   Example 1 Input Data

Example	1	Axisymmetric Hollow Cylinder of Two Materials				
22	1	1				
1	1	.5	0.	0.	200.	
2	1	.5	0.	.1	200.	
3	0	.55	0.	0.	0.	
4	0	.55	0.	.1	0.	
5	0	.60	0.	0.	0.	
6	0	.60	0.	.1	0.	
7	0	.65	0.	0.	0.	
8	0	.65	0.	.1	0.	
9	0	.70	0.	0.	0.	
10	0	.70	0.	.1	0.	
11	0	.75	0.	0.	0.	
12	0	.75	0.	.1	0.	
13	0	.80	0.	0.	0.	
14	0	.80	0.	.1	0.	
15	0	.85	0.	0.	0.	
16	0	.85	0.	.1	0.	
17	0	.90	0.	0.	0.	
18	0	.90	0.	.1	0.	
19	0	.95	0.	0.	0.	
20	0	.95	0.	.1	0.	
21	1	1.0	0.	0.	100	
22	1	1.0	0.	.1	100	
1	10	2	− 4			
1	1.	25.	0.	25.		
2	1.	.11	0.	.11		
1	1	3	4	2	1	0.
2	3	5	6	4	1	0.
3	5	7	8	6	1	0.
4	7	9	10	8	1	0.
5	9	11	12	10	1	0.
6	11	13	14	12	2	0.
7	13	15	16	14	2	0.
8	15	17	18	16	2	0.
9	17	19	20	18	2	0.
10	19	21	22	20	2	0.

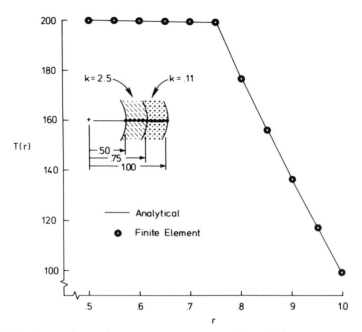

Figure 10.10 Comparison of temperatures computed from finite element model with analytical solution, Example 1.

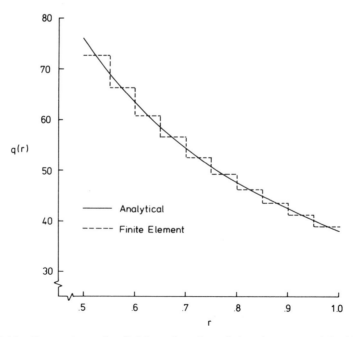

Figure 10.11 Comparison of radial heat flow from finite element model with analytical solution, Example 1.

TABLE 10.2.   Example 1 Output Data

Example 1   Axisymmetric Hollow Cylinder of Two Materials
C O N T R O L   I N F O R M A T I O N
     NUMBER OF NODAL POINTS   =   22
     NUMBER OF ELEMENT GROUPS =   1

NODAL POINT INPUT DATA

NODE NUMBER	B.C. CODE	NODAL POINT COORDINATES			
		X	Y	Z	TEMPERATURE
1	1	.50000E+00	.00000E+00	.00000E+00	.20000E+03
2	1	.50000E+00	.00000E+00	.10000E+00	.20000E+03
3	0	.55000E+00	.00000E+00	.00000E+00	.00000E+00
4	0	.55000E+00	.00000E+00	.10000E+00	.00000E+00
5	0	.60000E+00	.00000E+00	.00000E+00	.00000E+00
6	0	.60000E+00	.00000E+00	.10000E+00	.00000E+00
7	0	.65000E+00	.00000E+00	.00000E+00	.00000E+00
8	0	.65000E+00	.00000E+00	.10000E+00	.00000E+00
9	0	.70000E+00	.00000E+00	.00000E+00	.00000E+00
10	0	.70000E+00	.00000E+00	.10000E+00	.00000E+00
11	0	.75000E+00	.00000E+00	.00000E+00	.00000E+00
12	0	.75000E+00	.00000E+00	.10000E+00	.00000E+00
13	0	.80000E+00	.00000E+00	.00000E+00	.00000E+00
14	0	.80000E+00	.00000E+00	.10000E+00	.00000E+00
15	0	.85000E+00	.00000E+00	.00000E+00	.00000E+00
16	0	.85000E+00	.00000E+00	.10000E+00	.00000E+00
17	0	.90000E+00	.00000E+00	.00000E+00	.00000E+00
18	0	.90000E+00	.00000E+00	.10000E+00	.00000E+00
19	0	.95000E+00	.00000E+00	.00000E+00	.00000E+00
20	0	.95000E+00	.00000E+00	.10000E+00	.00000E+00
21	1	.10000E+01	.00000E+00	.00000E+00	.10000E+03
22	1	.10000E+01	.00000E+00	.10000E+00	.10000E+03

ELEMENT GROUP   1

P L A N E   E L E M E N T S

     NUMBER OF PLANE ELEMENTS = 10
     NUMBER OF MATERIALS   =   2
     NUMBER OF NODES   =   4

AXISYMMETRIC CASE

MATERIAL	THICKNESS	CONDUCTIVITY TENSOR		
		KXX	KXY	KYY
1	.1000E+01	.2500E+02	.000E+00	.2500E+02
2	.1000E+01	.1100E+00	.000E+00	.1100E+00

N	I	J	K	L	MATID	Q
1	1	3	4	2	1	.0000E+00
2	3	5	6	4	1	.0000E+00
3	5	7	8	6	1	.0000E+00
4	7	9	10	8	1	.0000E+00

TABLE 10.2. **Example Output Data** (*Continued*)

N	I	J	K	L	MATID	Q
5	9	11	12	10	1	.0000E + 00
6	11	13	14	12	2	.0000E + 00
7	13	15	16	14	2	.0000E + 00
8	15	17	18	16	2	.0000E + 00
9	17	19	20	18	2	.0000E + 00
10	19	21	22	20	2	.0000E + 00

S O L U T I O N   P A R A M E T E R S

```
        TOTAL NUMBER OF EQUATIONS    = 22
        SEMI BANDWIDTH               =  4
```

T E M P E R A T U R E   V E C T O R

NODE NO.	NO VALUE	NO + 1 VALUE	NO + 2 VALUE	NO + 3 VALUE	NO + VALUE
1	.200000E + 03	.200000E + 03	.199855E + 03	.199855E + 03	.199723E + 03
6	.199723E + 03	.199601E + 03	.199601E + 03	.199489E + 03	.199489E + 03
11	.199384E + 03	.199384E + 03	.177089E + 03	.177089E + 03	.156146E + 03
16	.156146E + 03	.136400E + 03	.136400E + 03	.117721E + 03	.117721E + 03
21	.100000E + 03	.100000E + 03			

ELEMENT GROUP   1

P L A N E   E L E M E N T   H E A T   F L U X E S

ELEMENT	QX	QY
1	.7243E + 02	-.2014E- 02
2	.6613E + 02	.1043E- 02
3	.6083E + 02	.2594E- 02
4	.5633E + 02	-.2994E- 02
5	.5244E + 02	-.4964E- 02
6	.4905E + 02	-.4730E- 04
7	.4607E + 02	.4901E- 04
8	.4344E + 02	.3857E- 04
9	.4109E + 02	.7315E- 05
10	.3899E + 02	.1933E- 04

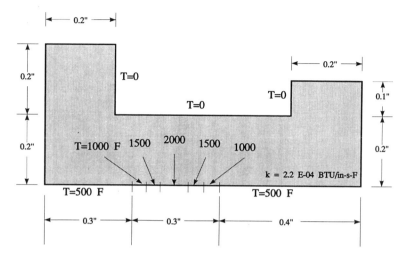

Figure 10.12   Example 2, planar region with steep local temperature gradients.

We use this example for two purposes: (1) to illustrate data preparation and a solution for a planar problem, and (2) for illustrating the convergence of the temperature distribution as we refine the mesh adaptively. Table 10.3 presents the corresponding data file; the program output file is shown in Table 10.4. The unstructured mesh of triangular elements and temperature contours are shown in Figure 10.13*a*. For this relatively coarse mesh, the temperature contours are not very smooth, especially in the region of high local temperatures on the bottom surface. The lack of smoothness is an indication that the mesh is too coarse to capture the steep temperature gradients near the boundary. Figures 10.13*b* and *c* show two subsequent meshes and the corresponding temperature contours. Each of these meshes was generated in succession from the previous temperature solution. Note that as the mesh is refined, smaller elements are generated in the steep gradient region, but larger elements are generated in the regions of the solution domain where temperature gradients are smaller. We describe the adaptive remeshing approach in Section 10.6. Examination of the temperature contours in Figure 10.13 shows that as the mesh is refined, the contours become smoother, indicating improving resolution of the steep temperature gradients.

This example illustrates the effectiveness of finite elements in analyzing heat conduction for an irregular region with highly unstructured meshes. The same basic program, without modification, is able to handle nonuniform boundary conditions, irregular geometry, and vastly different meshes. For an industrial or research application, the analysis would be more sophisticated with different boundary conditions, such as convection and radiation, as well as temperature-dependent properties. And yet, the fundamental approach

TABLE 10.3. Example 2 Input Data

Example 2	Heat Transfer With Steep Local Gradients			
136		1		
1	1	.00000E+00	.00000E+00	.00000E+00	.50000E+03
2	1	.50000E-01	.00000E+00	.00000E+00	.50000E+03
3	1	.10000E+00	.00000E+00	.00000E+00	.50000E+03
4	1	.15000E+00	.00000E+00	.00000E+00	.50000E+03
5	1	.20000E+00	.00000E+00	.00000E+00	.50000E+03
6	1	.25000E+00	.00000E+00	.00000E+00	.50000E+03
7	1	.30000E+00	.00000E+00	.00000E+00	.10000E+04
8	1	.32500E+00	.00000E+00	.00000E+00	.10000E+04
9	1	.35000E+00	.00000E+00	.00000E+00	.15000E+04
10	1	.37500E+00	.00000E+00	.00000E+00	.15000E+04
11	1	.40000E+00	.00000E+00	.00000E+00	.20000E+04
12	1	.45000E+00	.00000E+00	.00000E+00	.20000E+04
13	1	.50000E+00	.00000E+00	.00000E+00	.15000E+04
14	1	.52500E+00	.00000E+00	.00000E+00	.15000E+04
15	1	.55000E+00	.00000E+00	.00000E+00	.10000E+04
16	1	.57500E+00	.00000E+00	.00000E+00	.10000E+04
17	1	.60000E+00	.00000E+00	.00000E+00	.50000E+03
18	1	.65000E+00	.00000E+00	.00000E+00	.50000E+03
19	1	.70000E+00	.00000E+00	.00000E+00	.50000E+03
20	1	.75000E+00	.00000E+00	.00000E+00	.50000E+03
21	1	.80000E+00	.00000E+00	.00000E+00	.50000E+03
22	1	.85000E+00	.00000E+00	.00000E+00	.50000E+03
23	1	.90000E+00	.00000E+00	.00000E+00	.50000E+03
24	1	.95000E+00	.00000E+00	.00000E+00	.50000E+03
25	0	.10000E+01	.00000E+00	.00000E+00	.00000E+00
26	0	.10000E+01	.50000E+01	.00000E+00	.00000E+00
27	0	.10000E+01	.10000E+00	.00000E+00	.00000E+00
28	0	.10000E+01	.15000E+00	.00000E+00	.00000E+00
29	0	.10000E+01	.20000E+00	.00000E+00	.00000E+00
30	0	.10000E+01	.25000E+00	.00000E+00	.00000E+00
31	0	.10000E+01	.30000E+00	.00000E+00	.00000E+00
32	0	.95000E+00	.30000E+00	.00000E+00	.00000E+00
33	0	.90000E+00	.30000E+00	.00000E+00	.00000E+00
34	0	.85000E+00	.30000E+00	.00000E+00	.00000E+00
35	1	.80000E+00	.30000E+00	.00000E+00	.00000E+00
36	1	.80000E+00	.25000E+00	.00000E+00	.00000E+00
37	1	.80000E+00	.20000E+00	.00000E+00	.00000E+00
38	1	.75000E+00	.20000E+00	.00000E+00	.00000E+00
39	1	.70000E+00	.20000E+00	.00000E+00	.00000E+00
40	1	.65000E+00	.20000E+00	.00000E+00	.00000E+00
41	1	.60000E+00	.20000E+00	.00000E+00	.00000E+00
42	1	.55000E+00	.20000E+00	.00000E+00	.00000E+00
43	1	.50000E+00	.20000E+00	.00000E+00	.00000E+00
44	1	.45000E+00	.20000E+00	.00000E+00	.00000E+00
45	1	.40000E+00	.20000E+00	.00000E+00	.00000E+00
46	1	.35000E+00	.20000E+00	.00000E+00	.00000E+00
47	1	.30000E+00	.20000E+00	.00000E+00	.00000E+00
48	1	.25000E+00	.20000E+00	.00000E+00	.00000E+00
49	1	.20000E+00	.20000E+00	.00000E+00	.00000E+00
50	1	.20000E+00	.25000E+00	.00000E+00	.00000E+00

51	1	.20000E+00	.30000E+00	.00000E+00	.00000E+00
52	1	.20000E+00	.35000E+00	.00000E+00	.00000E+00
53	0	.20000E+00	.40000E+00	.00000E+00	.00000E+00
54	0	.15000E+00	.40000E+00	.00000E+00	.00000E+00
55	0	.10000E+00	.40000E+00	.00000E+00	.00000E+00
56	0	.50000E-01	.40000E+00	.00000E+00	.00000E+00
57	0	.000000+00	.40000E+00	.00000E+00	.00000E+00
58	0	.00000E+00	.35000E+00	.00000E+00	.00000E+00
59	0	.00000E+00	.30000E+00	.00000E+00	.00000E+00
60	0	.00000E+00	.25000E+00	.00000E+00	.00000E+00
61	0	.00000E+00	.20000E+00	.00000E+00	.00000E+00
62	0	.00000E+00	.15000E+00	.00000E+00	.00000E+00
63	0	.00000E+00	.10000E+00	.00000E+00	.00000E+00
64	0	.00000E+00	.50000E-01	.00000E+00	.00000E+00
.	.	.	.	.	.
.	.	.	.	.	.
.	.	.	.	.	.
131	0	.15475E+00	.89214E-01	.00000E+00	.00000E+00
132	0	.40817E-01	.82464E-01	.00000E+00	.00000E+00
133	0	.20251E+00	.83624E-01	.00000E+00	.00000E+00
134	0	.23021E+00	.45002E-01	.00000E+00	.00000E+00
135	0	.25170E+00	.95695E-01	.00000E+00	.00000E+00
136	0	.75373E+00	.97049E-01	.00000E+00	.00000E+00

```
    1    206           1     3
    1 .100000E+01      .22000E-.00000E+00   .22000E-03   .10000E+01
    1      16      17    65         1   .00000E+00
    2       7       8    66         1   .00000E+00
    3      10      11    67         1   .00000E+00
    4       8       9    66         1   .00000E+00
    5      13      14    68         1   .00000E+00
    6      15      16    65         1   .00000E+00
    7      14      15    68         1   .00000E+00
    8       9      10    67         1   .00000E+00
    9      65      17    18         1   .00000E+00
   10      28      29    69         1   .00000E+00
   11       4       5    70         1   .00000E+00
   12      38      39    71         1   .00000E+00
   13      18      19    72         1   .00000E+00
   14      59      60    73         1   .00000E+00
   15      60      61    74         1   .00000E+00
   16      32      33    75         1   .00000E+00
   17      32      75    30         1   .00000E+00
   18      75      33    76         1   .00000E+00
   19      75      76    77         1   .00000E+00
   20      76      33    34         1   .00000E+00
   21      19      20    78         1   .00000E+00
   22      57      58    56         1   .00000E+00
   23      37      38    79         1   .00000E+00
   24      42      43    80         1   .00000E+00
   25      40      41    81         1   .00000E+00
   26       7      66     6         1   .00000E+00
   27      77      76    82         1   .00000E+00
   28      69      29    83         1   .00000E+00
   29      72      19    78         1   .00000E+00
```

TABLE 10.3. **Example 2 Input Data** (*Continued*)

30	72	78	84	1	.00000E + 00
31	3	4	85	1	.00000E + 00
32	71	39	86	1	.00000E + 00
33	78	20	87	1	.00000E + 00
34	42	80	88	1	.00000E + 00
35	81	41	88	1	.00000E + 00
36	81	88	89	1	.00000E + 00
37	88	41	42	1	.00000E + 00
38	37	79	90	1	.00000E + 00
39	28	69	91	1	.00000E + 00
40	72	84	92	1	.00000E + 00
41	92	84	86	1	.00000E + 00
42	72	92	65	1	.00000E + 00
43	25	26	24	1	.00000E + 00
44	1	2	64	1	.00000E + 00
45	45	46	93	1	.00000E + 00
46	11	12	67	1	.00000E + 00
47	36	37	82	1	.00000E + 00
48	52	53	54	1	.00000E + 00
49	69	83	77	1	.00000E + 00
50	35	36	34	1	.00000E + 00
51	12	13	68	1	.00000E + 00
52	27	28	91	1	.00000E + 00
53	27	91	94	1	.00000E + 00
54	94	91	95	1	.00000E + 00
55	94	95	96	1	.00000E + 00
56	95	91	97	1	.00000E + 00
57	58	59	98	1	.00000E + 00
58	95	97	99	1	.00000E + 00
.	.	.	.	.	.
.	.	.	.	.	.
.	.	.	.	.	.
191	5	6	134	1	.00000E + 00
192	108	102	109	1	.00000E + 00
193	102	103	107	1	.00000E + 00
194	134	6	66	1	.00000E + 00
195	76	34	36	1	.00000E + 00
196	136	87	128	1	.00000E + 00
197	127	48	106	1	.00000E + 00
198	92	89	65	1	.00000E + 00
199	126	135	66	1	.00000E + 00
200	15	65	68	1	.00000E + 00
201	9	67	66	1	.00000E + 00
202	108	109	54	1	.00000E + 00
203	135	127	106	1	.00000E + 00
204	79	136	128	1	.00000E + 00
205	66	67	124	1	.00000E + 00
206	68	65	111	1	.00000E + 00

TABLE 10.4.   Example 2 Output Data

Example 2   Heat Transfer With Steep Local Gradients
 C O N T R O L   I N F O R M A T I O N
       NUMBER OF NODAL POINTS   = 136
       NUMBER OF ELEMENT GROUPS =   1
NODAL POINT INPUT DATA

NODE NUMBER	B.C. CODE	NODAL POINT COORDINATES X	Y	Z	TEMPERATURE
1	1	.00000E + 00	.00000E + 00	.00000E + 00	.50000E + 03
2	1	.50000E- 01	.00000E + 00	.00000E + 00	.50000E + 03
3	1	.10000E + 00	.00000E + 00	.00000E + 00	.50000E + 03
4	1	.15000E + 00	.00000E + 00	.00000E + 00	.50000E + 03
5	1	.20000E + 00	.00000E + 00	.00000E + 00	.50000E + 03
6	1	.25000E + 00	.00000E + 00	.00000E + 00	.50000E + 03
7	1	.30000E + 00	.00000E + 00	.00000E + 00	.10000E + 04
8	1	.32500E + 00	.00000E + 00	.00000E + 00	.10000E + 04
9	1	.35000E + 00	.00000E + 00	.00000E + 00	.15000E + 04
10	1	.37500E + 00	.00000E + 00	.00000E + 00	.15000E + 04
11	1	.40000E + 00	.00000E + 00	.00000E + 00	.20000E + 04
12	1	.45000E + 00	.00000E + 00	.00000E + 00	.20000E + 04
13	1	.50000E + 00	.00000E + 00	.00000E + 00	.15000E + 04
14	1	.52500E + 00	.00000E + 00	.00000E + 00	.15000E + 04
15	1	.55000E + 00	.00000E + 00	.00000E + 00	.10000E + 04
16	1	.57500E + 00	.00000E + 00	.00000E + 00	.10000E + 04
17	1	.60000E + 00	.00000E + 00	.00000E + 00	.50000E + 03
18	1	.65000E + 00	.00000E + 00	.00000E + 00	.50000E + 03
19	1	.70000E + 00	.00000E + 00	.00000E + 00	.50000E + 03
20	1	.75000E + 00	.00000E + 00	.00000E + 00	.50000E + 03
21	1	.80000E + 00	.00000E + 00	.00000E + 00	.50000E + 03
22	1	.85000E + 00	.00000E + 00	.00000E + 00	.50000E + 03
23	1	.90000E + 00	.00000E + 00	.00000E + 00	.50000E + 03
24	1	.95000E + 00	.00000E + 00	.00000E + 00	.50000E + 03
25	0	.10000E + 01	.00000E + 00	.00000E + 00	.00000E + 00
26	0	.10000E + 01	.50000E- 01	.00000E + 00	.00000E + 00
27	0	.10000E + 01	.10000E + 00	.00000E + 00	.00000E + 00
28	0	.10000E + 01	.15000E + 00	.00000E + 00	.00000E + 00
29	0	.10000E + 01	.20000E + 00	.00000E + 00	.00000E + 00
30	0	.10000E + 01	.25000E + 00	.00000E + 00	.00000E + 00
31	0	.10000E + 01	.30000E + 00	.00000E + 00	.00000E + 00
32	0	.95000E + 00	.30000E + 00	.00000E + 00	.00000E + 00
33	0	.90000E + 00	.30000E + 00	.00000E + 00	.00000E + 00
34	0	.85000E + 00	.30000E + 00	.00000E + 00	.00000E + 00
35	1	.80000E + 00	.30000E + 00	.00000E + 00	.00000E + 00
36	1	.80000E + 00	.25000E + 00	.00000E + 00	.00000E + 00
37	1	.80000E + 00	.20000E + 00	.00000E + 00	.00000E + 00
38	1	.75000E + 00	.20000E + 00	.00000E + 00	.00000E + 00
39	1	.70000E + 00	.20000E + 00	.00000E + 00	.00000E + 00
40	1	.65000E + 00	.20000E + 00	.00000E + 00	.00000E + 00
41	1	.60000E + 00	.20000E + 00	.00000E + 00	.00000E + 00
42	1	.55000E + 00	.20000E + 00	.00000E + 00	.00000E + 00
43	1	.50000E + 00	.20000E + 00	.00000E + 00	.00000E + 00
44	1	.45000E + 00	.20000E + 00	.00000E + 00	.00000E + 00
.	.	.	.	.	.
.	.	.	.	.	.
.	.	.	.	.	.
117	0	.46952E + 00	.67081E- 01	.00000E + 00	.00000E + 00
118	0	.47465E + 00	.15610E + 00	.00000E + 00	.00000E + 00

TABLE 10.4. Example 2 Output Data (*Continued*)

119	0	.83429E-01	.19500E+00	.00000E+00	.00000E+00
120	0	.43495E-01	.17417E+00	.00000E+00	.00000E+00
121	0	.89065E-01	.15094E+00	.00000E+00	.00000E+00
122	0	.44775E-01	.12521E+00	.00000E+00	.00000E+00
123	0	.93476E-01	.96086E-01	.00000E+00	.00000E+00
124	0	.34859E+00	.10395E+00	.00000E+00	.00000E+00
125	0	.32472E+00	.15439E+00	.00000E+00	.00000E+00
126	0	.30042E+00	.11175E+00	.00000E+00	.00000E+00
127	0	.27184E+00	.15139E+00	.00000E+00	.00000E+00
128	0	.81523E+00	.99530E-01	.00000E+00	.00000E+00
129	0	.82193E+00	.43305E-01	.00000E+00	.00000E+00
130	0	.13475E+00	.13185E+00	.00000E+00	.00000E+00
131	0	.15475E+00	.89214E-01	.00000E+00	.00000E+00
132	0	.40817E-01	.82464E-01	.00000E+00	.00000E+00
133	0	.20251E+00	.83624E-01	.00000E+00	.00000E+00
134	0	.23021E+00	.45002E-01	.00000E+00	.00000E+00
135	0	.25170E+00	.95695E-01	.00000E+00	.00000E+00
136	0	.75373E+00	.97049E-01	.00000E+00	.00000E+00

ELEMENT GROUP   1

P L A N E   E L E M E N T S

NUMBER OF PLANE ELEMENTS = 206
NUMBER OF MATERIALS   =   1
NUMBER OF NODES   =   3

MATERIAL	THICKNESS	CONDUCTIVITY TENSOR		
		KXX	KXY	KYY
1	.1000E+01	.2200E-03	0000E+00	.2200E-03

N	I	J	K	L	MATID	Q
1	16	17	65	0	1	.0000E+00
2	7	8	66	0	1	.0000E+00
3	10	11	67	0	1	.0000E+00
4	8	9	66	0	1	.0000E+00
5	13	14	68	0	1	.0000E+00
6	15	16	65	0	1	.0000E+00
7	14	15	68	0	1	.0000E+00
8	9	10	67	0	1	.0000E+00
9	65	17	18	0	1	.0000E+00
10	28	29	69	0	1	.0000E+00
11	4	5	70	0	1	.0000E+00
12	38	39	71	0	1	.0000E+00
13	18	19	72	0	1	.0000E+00
14	59	60	73	0	1	.0000E+00
15	60	61	74	0	1	.0000E+00
16	32	33	75	0	1	.0000E+00
17	32	75	30	0	1	.0000E+00
18	75	33	76	0	1	.0000E+00
19	75	76	77	0	1	.0000E+00

20	76	33	34	0	1	.0000E + 00
21	19	20	56	0	1	.0000E + 00
22	57	58	56	0	1	.0000E + 00
23	37	38	79	0	1	.0000E + 00
.	.	.	.	.	.	.
.	.	.	.	.	.	.
.	.	.	.	.	.	.
156	58	98	56	0	1	.0000E + 00
157	87	20	21	0	1	.0000E + 00
158	87	21	129	0	1	.0000E + 00
159	18	72	65	0	1	.0000E + 00
160	55	56	108	0	1	.0000E + 00
161	30	31	32	0	1	.0000E + 00
162	104	107	103	0	1	.0000E + 00
163	83	29	30	0	1	.0000E + 00
164	38	71	79	0	1	.0000E + 00
165	80	43	118	0	1	.0000E + 00
166	106	105	130	0	1	.0000E + 00
167	84	78	136	0	1	.0000E + 00
168	119	121	105	0	1	.0000E + 00
169	83	30	75	0	1	.0000E + 00
170	83	75	77	0	1	.0000E + 00
171	107	50	51	0	1	.0000E + 00
172	98	108	56	0	1	.0000E + 00
173	2	110	64	0	1	.0000E + 00
174	136	78	87	0	1	.0000E + 00
175	109	52	54	0	1	.0000E + 00
176	54	55	108	0	1	.0000E + 00
177	107	51	109	0	1	.0000E + 00
178	82	76	36	0	1	.0000E + 00
179	81	89	92	0	1	.0000E + 00
180	90	79	128	0	1	.0000E + 00
181	79	71	136	0	1	.0000E + 00
182	71	86	84	0	1	.0000E + 00
183	84	136	71	0	1	.0000E + 00
184	44	118	43	0	1	.0000E + 00
185	23	24	96	0	1	.0000E + 00
186	126	127	135	0	1	.0000E + 00
187	107	109	102	0	1	.0000E + 00
188	111	112	68	0	1	.0000E + 00
189	22	23	96	0	1	.0000E + 00
190	21	22	129	0	1	.0000E + 00
191	5	6	134	0	1	.0000E + 00
192	108	102	109	0	1	.0000E + 00
193	102	103	107	0	1	.0000E + 00
194	134	6	66	0	1	.0000E + 00
195	76	34	36	0	1	.0000E + 00
196	136	87	128	0	1	.0000E + 00
197	127	48	106	0	1	.0000E + 00
198	92	89	65	0	1	.0000E + 00

TABLE 10.4.   Example 2 Output Data (*Continued*)

199	126	135	66	0	1	.0000E+00
200	15	65	68	0	1	.0000E+00
201	9	67	66	0	1	.0000E+00
202	108	109	54	0	1	.0000E+00
203	135	127	106	0	1	.0000E+00
204	79	136	128	0	1	.0000E+00
205	66	67	124	0	1	.0000E+00
206	68	65	111	0	1	.0000E+00

S O L U T I O N S   P A R A M E T E R S

```
TOTAL NUMBER OF EQUATIONS    = 136
SEMI BANDWIDTH               = 130
```

T E M P E R A T U R E   V E C T O R

NODE NO.	NO VALUE	NO+1 VALUE	NO+2 VALUE	NO+3 VALUE	NO+4 VALUE
1	.500000E+03	.500000E+03	.500000E+03	.500000E+03	.500000E+03
6	.500000E+03	.100000E+04	.100000E+04	.150000E+04	.150000E+04
11	.200000E+04	.500000E+03	.500000E+03	.500000E+03	.500000E+03
16	.100000E+04	.500000E+03	.500000E+03	.500000E+03	.500000E+03
21	.500000E+03	.500000E+03	.500000E+03	.500000E+03	.450404E+03
26	.400809E+03	.333964E+03	.269808E+03	.218474E+03	.187544E+03
31	.175521E+03	.163499E+03	.129658E+03	.728784E+02	.000000E+00
36	.000000E+00	.000000E+00	.000000E+00	.000000E+00	.000000E+00
41	.000000E+00	.000000E+00	.000000E+00	.000000E+00	.000000E+00
46	.000000E+00	.000000E+00	.000000E+00	.000000E+00	.000000E+00
51	.000000E+00	.000000E+00	.205300E+02	.410599E+02	.661722E+02
56	.827495E+02	.894768E+02	.962042E+02	.115595E+03	.149802E+03
61	.199877E+03	.264523E+03	.339178E+03	.418812E+03	.607962E+03
66	.790592E+03	.127554E+04	.105992E+04	.235039E+04	.437760E+03
71	.155042E+03	.434718E+03	.123928E+03	.168868E+03	.164002E+03
76	.111059E+03	.177807E+03	.406654E+03	.151583E+03	.275594E+03
81	.187523E+03	.104764E+03	.197079E+03	.304696E+03	.426064E+03
86	.167778E+03	.401442E+03	.223907E+03	.380175E+03	.165284E+03
91	.296754E+03	.348491E+03	.307027E+03	.378038E+03	.327495E+03
96	.426587E+03	.245822E+03	.950342E+02	.262062E+03	.368799E+03
101	.189298E+03	.859194E+02	.121772E+03	.713192E+02	.156569E+03
106	.216309E+03	.548320E+02	.698813E+02	.424578E+02	.425608E+03
111	.527587E+03	.635907E+03	.319556E+03	.691210E+03	.663065E+03
116	.108566E+04	.103746E+04	.307851E+03	.184834E+03	.226916E+03
121	.247327E+03	.297817E+03	.338753E+03	.621560E+03	.262683E+03
126	.456557E+03	.225263E+03	.292099E+03	.412557E+03	.264903E+03
131	.350721E+03	.365756E+03	.380830E+03	.478120E+03	.415348E+03
136	.288868E+03				

ELEMENT GROUP   1

P L A N E   E L E M E N T   H E A T   F L U X E S

ELEMENT	QX	QY
1	.4400E+01	−.8581E−02
2	.4396E−06	.1022E+01
3	−.4400E+01	.2047E+01

4	-.4400E + 01	-.9451E + 00
5	-.2943E- 06	.2347E + 01
6	.4396E- 06	.1909E + 01
7	.4400E + 01	.3197E + 01
8	.2570E- 06	.1197E + 01
9	.5220E + 00	-.6134E- 01
10	.2259E + 00	.4158E- 01
11	.5619E- 07	.3124E + 00
12	.0000E + 00	-.6624E + 00
13	.4843E- 07	.3027E + 00
14	-.1505E + 00	.4369E- 01
15	-.2203E + 00	.4138E- 01
16	.1489E + 00	-.6698E- 01
17	-.2276E- 02	-.7955E- 01
18	.1395E + 00	.1461E + 00
19	.1985E + 00	-.1688E + 00
20	-.7589E- 01	.2462E + 00
21	.4843E- 07	.4337E + 00
22	-.2960E- 01	.2960E- 01
23	.0000E + 00	-.6811E + 00
24	.0000E + 00	-.1344E + 01
25	.0000E + 00	-.8768E + 00
26	.1016E + 01	.2322E + 01
27	.2507E + 00	.1535E + 00
28	.71338- 01	.1853E + 00
29	-.2536E + 00	.2966E + 00
30	.1095E + 00	.5123E + 00
31	.5402E- 07	.3588E + 00
32	.5891E + 00	-.3763E + 00
.	.	.
.	.	.
.	.	.
165	.1179E + 01	-.8460E + 00
166	.1806E + 00	-.4230E + 00
167	-.3978E + 00	.2861E + 00
168	-.3095E + 00	.2194E + 00
169	.4010E- 01	.1576E + 00
170	.1578E + 00	.1683E- 01
171	.2200E + 00	.1113E + 00
172	.9560E- 01	-.1404E- 01
173	.3362E + 00	.1293E + 00
174	-.4534E + 00	-.2317E + 00
175	.1697E + 00	-.6350E- 01
176	-.1105E + 00	-.1413E- 01
177	.2191E + 00	-.4704E- 01
178	-.2633E- 01	.3819E + 00
179	-.8804E + 00	-.1103E + 00
180	.5863E- 01	-.5176E + 00
181	-.1445E- 01	-.5653E + 00

TABLE 10.4.   Example 2 Output Data (*Continued*)

182	−.5333E− 01	−.6051E + 00
183	.6323E− 01	.6048E + 00
184	−.1345E + 01	.7553E + 00
185	.4843E− 07	.4010E + 00
186	.1041E + 01	−.1440E + 00
187	.4872E− 01	−.1570E + 00
188	−.4407E + 00	−.1601E + 01
189	.1099E− 06	.4010E + 00
190	.4284E− 07	.4442E + 00
191	.6519E− 07	.1070E + 00
192	−.6008E− 1	.1524E + 00
193	−.1308E + 00	.2063E + 00
194	−.9791E− 01	−.9628E + 00
195	.1645E + 00	.2987E + 00
196	−.4480E + 00	.1825E + 00
197	.9300E + 00	−.2630E + 00
198	−.1468E + 00	−.7991E + 00
199	.1767E + 00	−.1181E + 01
200	.1357E + 01	−.4104E + 00
201	.9351E + 00	.2358E + 01
202	.1031E + 00	.6138E− 01
203	.7061E + 00	.3851E + 00
204	−.5063E− 00	−.2763E + 00
205	−.1364E + 01	.1524E + 01
206	.1279E + 01	.1136E + 01

would be the same. Note also that after the model is used for the thermal analysis, the same nodes and element topology could be used for a thermal stress analysis with a structural analysis program.

These example problems, though simple, illustrate the essential features of the program. Problems with nonuniform, orthotropic thermal conductivity, distributed internal heat generation, and arbitrary specification of boundary or interior temperatures are all easily handled. Transient problems and nonlinear problems such as those with temperature-dependent thermal conductivity are beyond the program's scope. However, the program is modular and is organized to permit the addition of other solution modules and elements. The program is in standard FORTRAN in single precision to permit it to operate on a variety of computer systems.

One difficulty that new users of finite element methodology experience is the apparent complexity of finite element programs. The authors hope the

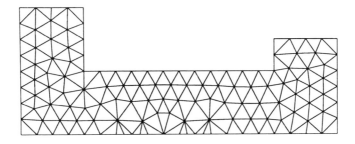

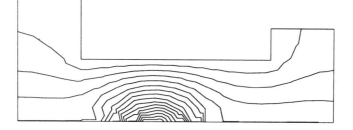

(a) 136 nodes and 206 elements

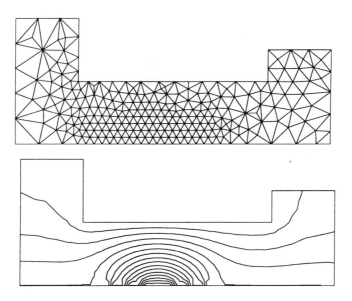

(b) 251 nodes and 436 elements

Figure 10.13 Example 2, finite element meshes and temperature contours for planar region: (a) 135 nodes and 206 elements; (b) 251 nodes and 438 elements; (c) 894 nodes and 1661 elements.

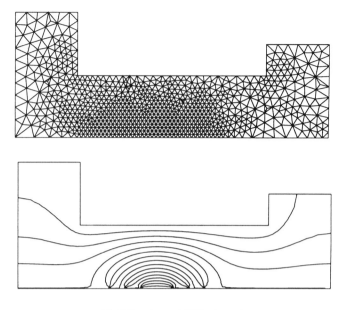

(c) 894 nodes and 1661 elements

Figure 10.13   (*Continued*)

program that we present here will help overcome these difficulties. Excellent introductory finite element programs appear in the texts by Bathe and Wilson [8], Hughes [9], Zienkiewicz and Taylor [10], and Reddy [11]. The book by Hinton and Owen [12] is devoted to finite element programming and describes several programs in detail. The text by Cook et al. [13] describes several subroutines and other aspects of finite element programming.

10.5   MESH GENERATION

A cumbersome aspect of obtaining a finite element solution is the preparation of input data. Most of the input data consist of a description of the element mesh topology. We must, by some means, provide the computer program with the node numbers and the coordinates of nodal points along with the element numbers and node numbers associated with each element. When our finite element mesh contains hundreds or even thousands of nodes, this task takes on major proportions. If all data had to be prepared by hand, the job would be very tedious and time-consuming indeed. And if the input data contain errors, these are, of course, reflected in an erroneous solution, which is simply a waste of an engineer's time.

Because of this situation, commercially available software has been developed for mesh generation. A wide variety of software is available based on several different algorithms. A description of mesh-generation methods and their algorithms are far beyond our purpose here. Instead, we suggest a number of references where various schemes are presented, and we give an example of what some mesh generation schemes can do. Samples of different approaches for mesh generation are found in references 14–20.

Mesh generation always involves some engineering judgment to decide on the effective placement of the nodes and elements within the solution region. Hence an automatic mesh generation scheme usually requires the user to specify some information about the mesh desired. Often one selects zones in which a certain node-placement density is desired, and one specifies some of the important boundary node locations. Then the mesh generator places additional nodes on the zone boundaries and within the zones and assembles a consistent network of elements from them. The only other data that the user need specify are those defining the connectivity of the zones. Figure 10.14 illustrates the general idea.

When a problem calls for the use of three-dimensional elements (such as tetrahedra or rectangular prisms), automatic mesh generation is almost mandatory, because the spatial visualization of a body divided into such elements is appallingly difficult.

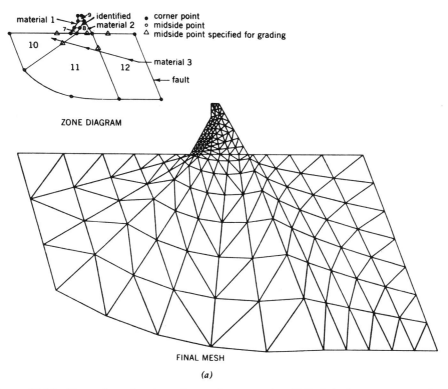

Figure 10.14 Examples of automatic mesh generation [14]: (*a*) dam on an earth foundation; (*b*) hyperbolic cooling tower.

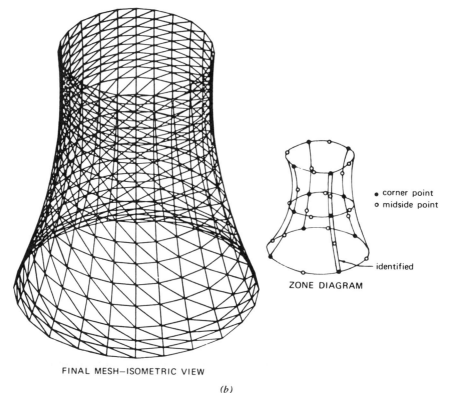

FINAL MESH—ISOMETRIC VIEW

ZONE DIAGRAM

• corner point
o midside point

identified

(*b*)

Figure 10.14   (*Continued*).

Computer graphics facilities are also an invaluable aid to the finite element analyst. After the element mesh has been generated (either by hand or automatically), it is most helpful to have the mesh displayed graphically by the computer. Then a quick visual check is all that is needed to detect unwanted errors. Computer plotting of the output data is also most useful in interpreting the results of a finite element solution. In our simple heat conduction problems we saw that the output consisted of nodal values of temperature and heat flux. For problems with hundreds of nodes, tabulation of these results could result in pages and pages of printed numbers. Moreover, in a more complicated problem where we have multiple nodal variables (such as the velocity components and pressure in a fluid problem) or in transient problems involving time histories, the output data become even more voluminous. Interpretation of these output data is easier if computer graphics is used to produce solution surfaces or contour plots.

10.6 ADAPTIVE MESH REFINEMENT

One of the most significant advances in finite element methodology in recent years has been the development of adaptive mesh refinement. An objective of adaptive refinement is to obtain the best resolution of physical phenomena for a given computational effort. In the preceding section, 10.5, we discussed the practical importance of using computers to generate meshes. This is a priori mesh generation, since we seek a mesh to obtain our initial solution. Adaptive mesh refinement is a posteriori mesh generation, since we use the initial mesh and the solution obtained on the initial mesh to generate a new mesh for the next solution.

10.6.1 Mesh Refinement Methods

The approach is to obtain a solution on an initial, sometimes coarse, mesh. Based on this solution and error indicators, we create a new mesh. We use error indicators to quantify the relative element discretization error for the entire mesh. If in our initial coarse mesh solution we have used relatively large elements in regions of steep gradients, the error indicator will show larger errors there than in other areas where gradients are less steep. In areas of large relative error, we need to refine the mesh. The ultimate objective of adaptive refinement is to obtain an equal distribution of error over all elements in the mesh.

We have a choice between several adaptive refinement methods. Among the most popular approaches are the h method, the p method, the r method, and combined h/p method. In the h method [21–25], the elements of the initial mesh are refined into smaller elements or de-refined into larger elements. In the p method [26–30], the order of the polynomial used for the element interpolation function is increased (or decreased) while keeping the geometry of the element constant. The r method [31] keeps the number of elements and their connectivity the same but moves the nodes. There are also methods of mesh refinement that use combinations of the above three methods; the h/p method [32–35] refines or de-refines some elements while increasing or decreasing the order of interpolation polynomials in other elements. Another alternative is adaptive remeshing, where the solution on the initial mesh and error indicators are used to generate a completely new mesh.

Figure 10.15 presents three examples of adaptive mesh refinement. The meshes are for an inviscid supersonic flow over an inclined plane. The meshes are refined to capture an oblique shock emanating from the corner and making an angle above the inclined plane. Figures 10.15a and 10.15b are two examples of h refinement. In Figure 15a the mesh consists of all triangles, but the mesh is locally refined to capture the oblique shock. Figure 15b shows a mesh of quadrilateral elements where the transition from the

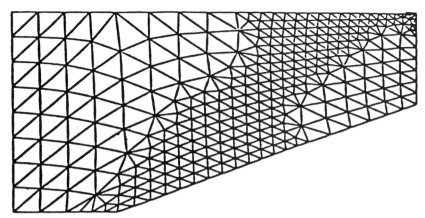

(a) h method with triangles

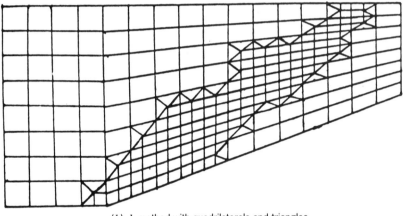

(b) h method with quadrilaterals and triangles
for transition

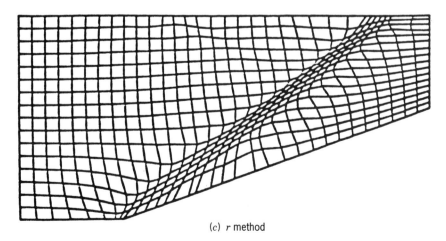

(c) r method

Figure 10.15 Examples of adaptive mesh refinement: (a) h method with triangles; (b) h method with quadrilaterals and triangles for transition; (c) r method.

larger elements to the smaller elements is accomplished with triangular transition elements. Figure 15c is a mesh illustrating the r method where nodes have been relocated to give small elements along the oblique shock.

The papers cited above show that adaptive refinement/de-refinement methods are being applied to a wide range of applications in structures, heat transfer, and fluid mechanics. Each method has limitations. In the neighborhood of strictly one-dimensional features like shocks and thin boundary layers in fluid mechanics, the h method of mesh enrichment is often not very efficient, since, from one refinement to another, the number of elements increases significantly. Another disadvantage of mesh-enrichment methods is that the original location of the nodes does not alter through successive refinements. Though new nodes are added and old ones are deleted at each refinement stage, the initial orientation of the elements does not change. Though the r method does not increase the number of elements, it may give rise to highly distorted elements. Distorted elements can degrade solution quality with some algorithms and should be avoided. Implementation of the p method is more complicated than the h method because extensive modifications of the analysis programs are required. However, convergence rates for the p method are higher than for h refinement. The combination of the h and p methods is also very complicated to program, but it has excellent convergence rates.

10.6.2 Error Indicators

For engineers it is intuitive that as we refine the mesh, the finite element solution converges to the exact solution. But for applied mathematicians, proof of convergence of finite element solutions is a fundamental issue. Intrinsic to convergence and convergence rates is the notion of approximation errors. Texts [36–38] on the mathematical aspects of finite elements discuss convergence and approximation errors with rigor. We mention here some very basic concepts of error estimation and convergence as an aid in learning about error indicators.

The error in the finite element approximation is a function E, which we define as the difference between the exact solution and the finite element solution. For the two-dimensional heat conduction problems we solve in this chapter, the error is

$$E(x, y) = T(x, y) - T_{\text{FE}}(x, y) \tag{10.6}$$

where $T(x, y)$ is the exact solution, and $T_{\text{FE}}(x, y)$ is the finite element solution produced by our computer program. The error varies over the solution domain. To measure how the global error decreases as we refine the mesh, we use the norm of the error. A widely used norm measures the root-mean-square error over the domain, the L_2 (pronounced "ell-two") norm. The norm of the error E is denoted by $\|E\|$, and the L_2 norm of the error is defined by

$$\|E\|_{L_2} = \left[\int_\Omega E^2 \, dA\right]^{1/2} \tag{10.7}$$

where the integral is evaluated over the solution domain Ω. If we are solving a problem for which an exact analytical solution exists, we can evaluate equation 10.7 by evaluating the integral over each element and then summing over all elements. We then use the L_2 norm as a global measure of the solution convergence as we refine the mesh. For elliptic boundary value problems such as our heat conduction problem, mathematical analyses [36, 38] show that as the mesh is refined, the L_2 norm of the error is estimated in the limit by

$$\|E\|_{L_2} = Ch^p \tag{10.8}$$

where C is a constant, h is a mesh parameter that characterizes element size, and p is an integer that depends on the interpolation functions. For linear interpolation functions, $p = 2$. Thus for the linear elements used in our computer program the solution error decreases with h^2. Thus, for a uniform mesh, if we reduce h by two, the solution error as measured by the L_2 norm decreases by four.

The error as measured by the L_2 norm depends on our knowing the exact solution. Moreover, it is a *global* measure of the total solution error. In solving engineering problems the exact solution is unknown, and we need a practical way of estimating the error. In addition, we wish to estimate errors on a *local* level in elements of the mesh where steep gradients occur. Although several mathematical approaches have been developed for *local error estimation*, a straight forward method is based on the notion of interpolation error [37].

For simplicity, let us consider first a one-dimensional thermal problem. Figure 10.16 shows a one-dimensional element of length h connecting nodes i and $i + 1$. Let the error $E(x) = T(x) - T_{FE}(x)$ and consider a point within the element denoted by $\bar{x}$. With a Taylor series, we can write

$$E(x) = E(\bar{x}) + \frac{dE}{dx}\bigg|_{\bar{x}} (x - \bar{x}) + \frac{1}{2}\frac{d^2E}{dx^2}\bigg|_{\bar{x}} (x - \bar{x})^2 + \cdots$$

where we assume that E has bounded second derivatives. Now we select $\bar{x}$ to be the point x_M where the error is a maximum, hence

$$\frac{dE}{dx}(x_M) = 0$$

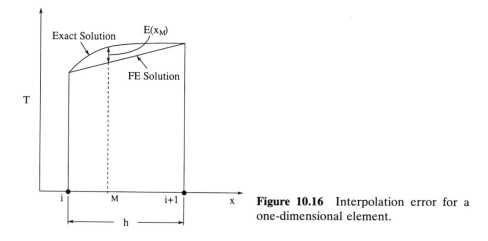

Figure 10.16 Interpolation error for a one-dimensional element.

Then after neglecting higher-order terms we may write

$$E(x) = E(x_M) + \frac{1}{2} \frac{d^2E}{dx^2}\bigg|_{x_M} (x - x_M)^2$$

Assuming that the finite element solution is exact at the nodes means that $E(x_i) = E(x_{i+1}) = 0$. (The finite element solution will be exact at the nodes for equations of the form $d^2T/dx^2 = Q$, where Q is constant. For other equations, the finite element nodal values will, of course, approximate the solution. We assume exact nodal values here as a practical means to derive a simple mathematical expression to indicate solution error.) Thus at x_i we may write

$$0 = E(x_M) + \frac{1}{2} \frac{d^2E}{dx^2}\bigg|_{x_M} (x_i - x_M)^2$$

or

$$E(x_M) = -\frac{1}{2} \frac{d^2E}{dx^2}\bigg|_{x_M} (x_i - x_m)^2$$

Similarly from x_{i+1},

$$E(x_M) = -\frac{1}{2} \frac{d^2E}{dx^2}\bigg|_{x_m} (x_{i+1} - x_M)^2$$

In the last two equations we will have either

$$|x_i - x_m| < \frac{h}{2}$$

or

$$|x_{i+1} - x_m| < \frac{h}{2}$$

and thus we may write an equation for the magnitude of the error as

$$|E| \le \frac{1}{8}\left|\frac{d^2E}{dx^2}\right|_{x_M} h^2 \qquad (10.9)$$

Equation 10.9 provides an estimate for the element interpolation error. For engineering problems, we cannot evaluate the second derivative because E depends on the unknown exact solution. Researchers have found, however, that by approximating this derivative as

$$\left.\frac{d^2E}{dx^2}\right|_{x_M} \cong \left.\frac{d^2T_{FE}}{dx^2}\right|_{(e)}$$

we may write for a typical element

$$|E|_{(e)} \le \frac{1}{8}\left|\frac{d^2T_{FE}}{dx^2}\right|_{(e)} h^2 \qquad (10.10)$$

Experience shows that although equation 10.10 is not a rigorous expression for estimating the element error, it is useful as an error *indicator* for establishing relative solution errors throughout a mesh. In a similar way for two-dimensional elements, we may use a Taylor series expansion to develop an error indicator as

$$|E|_{(e)} \le \frac{1}{8}\left(\left|\frac{\partial^2T_{FE}}{\partial x^2}\right| + 2\left|\frac{\partial^2T_{FE}}{\partial x\,\partial y}\right| + \left|\frac{\partial^2T_{FE}}{\partial y^2}\right|\right) h^2 \qquad (10.11)$$

where h is a characteristic element dimension.

Our first reaction upon consideration of equations 10.10 and 10.11 is that the indicators cannot be computed for the finite element solution with linear elements because the second derivatives do not exist. As a practical matter, credible values of the second derivatives can be computed with the following approach. For each element, we compute the element first derivatives in the usual way. For example,

$$\frac{\partial T^{(e)}}{\partial x} = \left\lfloor \frac{\partial N}{\partial x} \right\rfloor \{T\}^{(e)} \tag{10.12}$$

Then values of the first derivatives are computed at nodal points by assembling system equations from element contributions of the form

$$\int_{\Omega^{(e)}} \{N\} \lfloor N \rfloor \, dA \left\{\frac{\partial T}{\partial x}\right\} = \int_{\Omega^{(e)}} \{N\} \, dA \frac{\partial T^{(e)}}{\partial x} \tag{10.13}$$

where $\{\partial T/\partial x\}$ denotes nodal derivatives. After the system equations are assembled from equations 10.13, the "mass matrix" on the left-hand side is diagonalized to yield an explicit set of equations that we can solve for the nodal values of the derivatives. We may interpret the procedure based on equation 10.13 as a method for computing a nodal derivative from a weighted average of element derivatives surrounding the node. Element areas serve as weighting factors. (We describe a similar technique in Section 9.2.3 as a method for "smoothing" velocities computed in potential flow.)

The computation of second derivatives follows the same steps. First, element second derivatives are computed from nodal first derivatives from

$$\frac{\partial^2 T^{(e)}}{\partial x^2} = \left\lfloor \frac{\partial N}{\partial x} \right\rfloor \left\{ \frac{\partial T}{\partial x} \right\}^{(e)} \tag{10.14}$$

Then, if needed, nodal second derivatives are computed using system equations assembled from

$$\int_{\Omega^{(e)}} \{N\} \lfloor N \rfloor \, dA \left\{\frac{\partial^2 T}{\partial x^2}\right\} = \int_{\Omega^{(e)}} \{N\} \, dA \frac{\partial^2 T^{(e)}}{\partial x^2} \tag{10.15}$$

The procedure described by equations 10.12–10.15 has proven adequate for computing the derivatives needed for error indicators. The second derivatives computed on the boundaries are typically less accurate than those computed at interior nodes.

The computational procedure based on the error indicators given in equations 10.10 and 10.11 with the preceding derivative computational method has been used successfully in both refinement/de-refinement and adaptive remeshing schemes. Other error indicators have also been employed. A popular error indicator, particularly for structural analysis, appears in references 39–45. In the following section, we describe briefly how error indicators are used in adaptive remeshing.

10.6.3 Adaptive Remeshing

We present here some basic concepts of adaptive remeshing [46–53]. The approach generates an entirely new mesh based on the information provided by the solution on an earlier mesh. The basic approach was developed in reference 46 using triangular elements, and other variants are described in references 47–53, including extensions to transient and three-dimensional problems. We present only highlights of the approach; the details are found in the references. Figure 10.13 presents a series of adaptive meshes that we used in the solution of Example 2.

Mesh generation parameters for the construction of a new mesh are shown in Figure 10.17. The parameters are (1) two components of a vector $\boldsymbol{\alpha}$ (in global coordinates) along which an element is to be stretched, (2) a spacing h_1 normal to this vector, and (3) a spacing h_2 parallel to this vector. The key step between a solution on a previous mesh and the generation of an adaptive remesh is the computation of the mesh generation parameters. The mesh generation parameters are computed from an error indicator based on a dependent variable. If we are solving a heat conduction problem the dependent variable is the single scalar temperature, but if we are solving a fluid mechanics problem with several unknowns the dependent variable may be a scalar such as the magnitude of the velocity vector for an incompressible flow or density for a compressible flow. An optimum mesh is achieved when the local error is distributed uniformly throughout the mesh. In one dimension if we use the error indicator shown in equation 10.10, we will obtain an optimum mesh by requiring

$$h^2 \left| \frac{\partial^2 T}{\partial x^2} \right| = \text{const.} \tag{10.16}$$

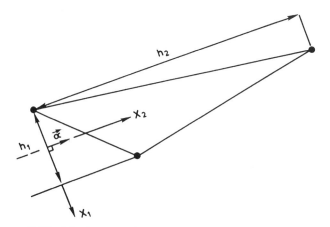

Figure 10.17 Mesh generation parameters for adaptive remeshing.

In application of remeshing to a two-dimensional problem, the second derivatives at a point are given by the matrix

$$\begin{bmatrix} \dfrac{\partial^2 T}{\partial x^2} & \dfrac{\partial^2 T}{\partial x \, \partial y} \\[2ex] \dfrac{\partial^2 T}{\partial y \, \partial x} & \dfrac{\partial^2 T}{\partial y^2} \end{bmatrix} \tag{10.17}$$

For remeshing, principal directions are computed along which the cross derivatives vanish. (The approach is analogous to the prediction of planes of principal stress in elasticity.) Principal eigenvalues λ_1 and λ_2 and principal directions X_1 and X_2 are computed. Equal distribution of solution error at nodes is achieved by requiring that in each direction

$$h_1^2 |\lambda_1| = h_2^2 |\lambda_2| = \text{const.} \tag{10.18}$$

where h_1 refers to an element dimension in the X_1 direction, and h_2 refers to the element dimension in the X_2 direction (Figure 10.17). Thus

$$\frac{h_2}{h_1} = \sqrt{\left| \frac{\lambda_1}{\lambda_2} \right|} \tag{10.19}$$

The eigenvalue problem is solved at each nodal point of a previous mesh, including the boundaries. Using the maximum eigenvalue and a user-specified minimum h, the constant in equation 10.16 is determined. Then at each point, h_1 and h_2 can be computed using equations 10.18 and 10.19. These dimensions determine the size of an element to be created. The shape of the element is constrained by limits on the internal angles so that a distorted angle is not created.

To illustrate the basic steps in remeshing, an example is presented in Figure 10.18. The figure illustrates one of the meshes used in Example 2 being created from a solution obtained on a previous mesh. In Figure 10.18a, the initial mesh (sometimes called the background mesh) is shown. Using this mesh a finite element solution has been obtained. Figures 10.18b–d show the evolution of the new mesh. Figure 10.18b shows the boundary points that are created first. Figures 10.18c and d show the mesh at two stages of development, and Figure 10.18e shows the final mesh. Notice that the mesh is created one element at a time, and the remeshing proceeds inward from the boundary.

This description of adaptive remeshing concludes our discussion of adaptive mesh refinement. The discussion, by necessity, has been brief, and readers are encouraged to read the papers cited. Adaptive mesh refinement clearly improves the quality of finite element solutions. Recent and current research is focused on developing the most effective approaches for implementing the basic concept in software for solving practical problems. Un-

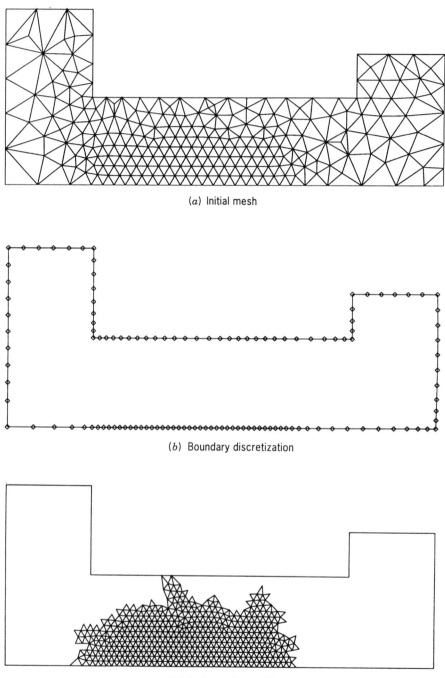

(a) Initial mesh

(b) Boundary discretization

(c) A stage of remeshing

Figure 10.18 Evolution of a mesh during adaptive remeshing: (a) initial mesh; (b) boundary discretization; (c) a stage of remeshing; (d) later stage of remeshing; (e) final mesh.

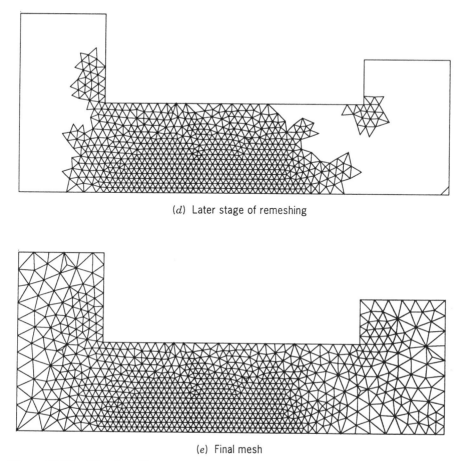

(d) Later stage of remeshing

(e) Final mesh

Figure 10.18   (*Continued*)

doubtedly, in the future adaptive refinement will become a standard step for obtaining high-quality solutions to real-world problems.

10.7 NUMERICAL INTEGRATION FORMULAS

Often the evaluation of the element equations for a particular problem involves the evaluation of integrals over an element. We face the problem of evaluating

$$J_1 = \int f(x)\, dx \tag{10.20}$$

$$J_2 = \int\int f(x, y)\, dx\, dy \tag{10.21}$$

$$J_3 = \int\int\int f(x, y, z)\, dx\, dy\, dz \tag{10.22}$$

where f is a known function and the region of integration is defined by the element boundaries. As we have seen, it is sometimes possible to obtain exact expressions for these integrals when natural coordinates are used. But when the form of f or the shape of the element does not permit closed-form integration, we must resort to numerical integration. A number of effective formulas are available for this task, but we shall mention here just a few of them.

10.7.1 Newton–Cotes

We consider first the one-dimensional case—equation 10.20. A class of numerical integration formulas known as *Newton–Cotes (closed) formulas* assumes that the function f is evaluated at equally spaced points along the x axis (Figure 10.19). These integration formulas are derived by integrating exactly a Lagrangian interpolation polynomial of order n fit to $n + 1$ values of $f(x)$. The first eight formulas in the infinite series that can be derived are as follows [54, pp. 885–887]:

$n = 1$:

$$\int_{x_0}^{x_1} f(x)\, dx = \frac{h}{2}(f_0 + f_1) - \frac{h^2}{12} f^{(2)}(\xi) \tag{10.23a}$$

$n = 2$:

$$\int_{x_0}^{x_1} f(x)\, dx = \frac{h}{3}(f_0 + 4f_1 + f_2) - \frac{h^5}{90} f^{(4)}(\xi) \tag{10.23b}$$

$n = 3$:

$$\int_{x_0}^{x_3} f(x)\, dx = \frac{3h}{8}(f_0 + 3f_1 + 3f_2 + f_3) - \frac{3h^5}{80} f^{(4)}(\xi) \tag{10.23c}$$

$n = 4$:

$$\int_{x_0}^{x_4} f(x)\, dx = \frac{2h}{45}(7f_0 + 32f_1 + 12f_2 + 32f_3 + 7f_4) - \frac{8h^7}{945} f^{(6)}(\xi) \tag{10.23d}$$

$n = 5$:

$$\int_{x_0}^{x_5} f(x)\, dx = \frac{5h}{288}(19f_0 + 75f_1 + 50f_2 + 50f_3 + 75f_4 + 19f_5)$$

$$- \frac{275h^7}{12{,}096} f^{(6)}(\xi) \tag{10.23e}$$

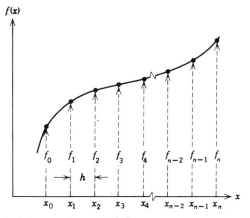

Figure 10.19 Numerical integration of $f(x)$ via the Newton–Cotes formulas. The sample points are equally spaced along the x axis.

$n = 6$:

$$\int_{x_0}^{x_6} f(x)\, dx = \frac{h}{140}(41f_0 + 216f_1 + 27f_2 + 272f_3 + 27f_4 + 216f_5 + 41f_6)$$

$$-\frac{9h^9}{1,400} f^{(8)}(\xi) \tag{10.23f}$$

$n = 7$:

$$\int_{x_0}^{x_7} f(x)\, dx = \frac{7h}{17,280}(751f_0 + 3577f_1 + 1323f_2 + 2989f_3 + 2989f_4$$

$$+ 1323f_5 + 3577f_6 + 751f_7) - \frac{8183h^9}{518,400} f^{(8)}(\xi)$$

$$\tag{10.23g}$$

$n = 8$:

$$\int_{x_0}^{x_8} f(x)\, dx = \frac{4h}{14,175}(989f_0 + 5888f_1 - 928f_2 + 10,496f_3 - 4540f_4$$

$$+ 10,496f_5 - 928f_6 + 5888f_7 + 989f_8) - \frac{2368h^{11}}{467,775} f^{(10)}(\xi)$$

$$\tag{10.23h}$$

where $x_0 < \xi < x_n$ and $f^{(k)}$ designates the kth derivative of f.

All of the formulas in this series can be written generally as

$$\int_{x_a}^{x_b} f(x)\, dx = \sum_{i=0}^{n} W_i^x f(x_i) + E_x^n \tag{10.24}$$

where W_i^x are the weight coefficients and E_x^n is the error. As an example, for $n = 4$ we have

$$W_0^x = \frac{14h}{45} \qquad W_1^x = \frac{64h}{45} \qquad W_2^x = \frac{24h}{45}$$

$$W_3^x = \frac{64h}{45} \qquad W_4^x = \frac{14h}{45}$$

$$E_x^4 = \frac{-8h^7}{945} f^{(6)}(\xi)$$

The Newton–Cotes formulas for one-dimensional integration can be easily extended to evaluate the multiple integrals of equations 10.21 and 10.22 when the integration is taken over a rectangle or a right prism. We can develop formulas for the multiple-dimensional integrals by applying successively the formulas for one-dimensional integration. Suppose that we select equally spaced integration points (n in the x direction, l in the y direction, and m in the z direction). Then for the double integral, equation 10.21, we can write, from equation

$$J_2 = \int_{y_a}^{y_b}\int_{x_a}^{x_b} f(x, y)\, dx\, dy = \int_{y_a}^{y_b}\left[\sum_{i=0}^{n} W_i^x f(x_i, y) + E_x^n\right] dy$$

$$= \sum_{j=0}^{l} W_j^y \left[\sum_{i=0}^{n} W_i^x f(x_i, y_j) + E_x^n\right] + E_y^l$$

$$= \sum_{j=0}^{l}\sum_{i=0}^{n} W_j^y W_i^x f(x_i, y_j) + E_{xy}^{ln} \tag{10.25}$$

where E_{xy}^{ln} is the combined integration error, and W_j^y are the weight coefficients for the y direction of integration. Similarly, for the triple integral, equation 10.22, we can write

$$J_3 = \int_{z_a}^{z_b}\int_{y_a}^{y_b}\int_{x_a}^{x_b} f(x, y, z)\, dx\, dy\, dz$$

$$= \sum_{k=0}^{m}\sum_{j=0}^{l}\sum_{i=0}^{n} W_k^z W_j^y W_i^x f(x_i, y_j, z_k) + E_{xyz}^{min} \tag{10.26}$$

10.7.2   Legendre–Gauss

Another class of numerical integration formulas, known as *Gaussian-type formulas*, does not rely on equally spaced integration but, rather, uses abscissas that are the zeros of the particular interpolation polynomial employed. For a given number of abscissas this presumably leads to better accuracy. If Legendre polynomials are used, the resulting formulas are known as Legendre–Gauss formulas. Integration in one dimension is then accomplished by

$$J_1 = \int_{-1}^{1} f(x)\, dx \approx \sum_{i=1}^{n} W_i f(x_i) \tag{10.27}$$

where the weights W_i and the abscissas x_i are given in Table 10.5. Multiple integration can be handled in the same way as indicated in equations 10.25 and 10.26.

Similar integration formulas have been developed for evaluating area and volume integrals when the regions of integration are triangles or tetrahedra and the integrands are expressed in terms of natural coordinates. For triangles, we have from equation 5.15

$$\begin{aligned} x &= L_1 x_1 + L_2 x_2 + L_3 x_3 \\ y &= L_1 y_1 + L_2 y_2 + L_3 y_3 \end{aligned} \tag{5.15}$$

where x_1, y_1, $i = 1, 2, 3$, are the coordinates of the nodes of the triangle, and L_i, $i = 1, 2, 3$, are the natural coordinates (see Section 5.5.1) for the three-node triangle. Thus we may express J_2 as

$$J_2 = \iint_{\Delta} f(x, y)\, dA = \iint_{\Delta} f(L_1, L_2, L_3)\, dA$$

and then by Gauss integration

$$J_2 \approx \Delta \sum_{i=1}^{n} W_i f(L_{1i}, L_{2i}, L_{3i}) \tag{10.28}$$

where Δ is the area of the triangle, W_i are the Gauss weights, and L_{1i}, L_{2i}, L_{3i} are natural coordinates of the Gauss integration points. Table 10.6 gives the values of the weights and the natural coordinates of the integration points [56]. Corresponding formula for tetrahedra may be found in reference 10.

10.8   SOLVING ALGEBRAIC EQUATIONS

A chapter on coding techniques and other practical considerations associated with the finite method would be incomplete if it did not mention the subject

TABLE 10.5. Abscissas and Weight Factors for Legendre–Gauss Integration

$$\int_{-1}^{1} f(x)\,dx \approx \sum_{i=1}^{n} W_i^x f(x_i)$$

Abscissas $= \pm x_i$ (zeros of Legendre polynomials)
Weight factors $= W_i^x$

$\pm x_i$			W_i^x		
		$n = 1$			
0.0			2.0		
		$n = 2$			
0.57735	02691	89626	1.00000	00000	00000
		$n = 3$			
0.00000	00000	00000	0.88888	88888	88889
0.77459	66692	41483	0.55555	55555	55556
		$n = 4$			
0.33998	10435	84856	0.65214	51548	62546
0.86113	63115	94053	0.34785	48451	37454
		$n = 5$			
0.00000	00000	00000	0.56888	88888	88889
0.53846	93101	05683	0.47862	96704	99366
0.90617	98459	38664	0.23692	68850	56189
		$n = 6$			
0.23861	91860	83197	0.46791	39345	72691
0.66120	93864	66265	0.36076	15730	48139
0.93246	95142	03152	0.17132	44923	79170
		$n = 7$			
0.00000	00000	00000	0.41795	91836	73469
0.40584	51513	77397	0.38183	00505	05119
0.74153	11855	99394	0.27970	53914	89277
0.94910	79123	42759	0.12948	49661	68870
		$n = 8$			
0.18343	46424	95650	0.36268	37833	78362
0.52553	24099	16329	0.31370	66458	77887
0.79666	64774	13627	0.22238	10344	53374
0.96028	98564	97536	0.10122	85362	90376

Source. From Davis and Rabinowitz [55].

of algebraic equation-solving procedure. Many finite element analyses ultimately reduce to solving a set of linear or nonlinear algebraic equations of the standard form

$$[K]\{x\} = \{R\} \qquad (10.29)$$

Improving a given finite element analysis often means increasing the number of nodes and the number of degrees of freedom in the solution region and hence enlarging the system matrix $[K]$. Consequently, highly accurate finite

TABLE 10.6.   Numerical Integration Formulas for Triangles

No.	Order	Figure	Rem.	Points	Triangular Coordinates	Weights W_k
1	Linear		$R = 0(h^2)$	a	1/3, 1/3, 1/3	1
2	Quadratic		$R = 0(h^3)$	a b c	1/2, 1/2, 0 0, 1/2, 1/2 1/2, 0, 1/2	1/3 1/3 1/3
3	Cubic		$R = 0(h^4)$	a b c d	1/3, 1/3, 1/3 0.6, 0.2, 0.2 0.2, 0.6, 0.2 0.2, 0.2, 0.6	−27/48 25/48 25/48 25/48
5	Quintic		$R = 0(h^6)$	a b c d e f g	1/3, 1/3, 1/3 $\alpha_1, \beta_1, \beta_1$ $\beta_1, \alpha_1, \beta_1$ $\beta_1, \beta_2, \beta_2$ $\alpha_2, \beta_2, \beta_2$ $\beta_2, \alpha_2, \beta_2$ $\beta_2, \beta_2, \alpha_2$	0.225 0.13239415 0.12593918

With:
$\alpha_1 = 0.05961587$
$\beta_1 = 0.47014206$
$\alpha_2 = 0.79742699$
$\beta_2 = 0.10128651$

element analyses usually involve the solution of a large number of equations, and the analyst desires an efficient scheme for solving these large-order systems.

Associated with efficient schemes of solving the large-order algebraic system encountered in finite element analysis is the method of storing the system coefficient matrix. We have noted (Figure 10.6) that the assembled matrix often has nonzero terms clustered about the main diagonal, while locations distant from the main diagonal contain zero terms. As we note in Section 2.3, finite element programs exploit the banded character of the matrix by storing and manipulating only the terms within the band. This approach, while straightforward and effective, does not achieve optimum efficiency, because, as Figure 10.6a shows, there are nonzero terms contained within the band. Two storage schemes used to take advantage of these zero terms are the envelope (also called the skyline or profile) method and sparse matrix methods. In the envelope method we store only the terms between the diagonal and the last nonzero term in a column (or row); this storage scheme results in a "ragged" edge to the band called the envelope or the skyline of the matrix. In sparse matrix methods we store and operate on only the actual nonzeros occurring in the matrix. The penalty for utilization of envelope or sparse matrix storage methods is the requirement for more complex programming to keep track of the terms stored. Envelope methods, however, offer potentially large savings over band methods and justify the additional programming detail. Envelope methods are discussed in detail in references 8–10 and are illustrated by the programs contained therein.

Also related to the efficient solving of large-order algebraic systems is the method of employing the computer mass-storage media. A solution scheme (whether a banded, envelope or sparse matrix method) that retains the entire system matrix in the computer's central memory is an *in-core* algorithm. A solution scheme that uses additional back-up storage such as a disk file is an *out-of-core* algorithm. Since the amount of computer central memory (core) is limited, production-type finite element programs with the capacity to solve very large-order systems employ out-of-core algorithms. Smaller programs such as we present in Section 10.3 retain the simplicity of an in-core algorithm.

Further discussion of these programming aspects of finite element analysis is beyond the scope of the present discussion, but in the next two sections we briefly describe popular approaches used in finite element programs for solving linear and nonlinear algebraic equations.

10.8.1 Linear Matrix Equations

Linear matrix equation-solving algorithms can be generally classified as *direct* or *iterative*. Gauss elimination and Cholesky decomposition are examples of direct schemes, while the Gauss–Seidel and Givens methods are two of the many iterative schemes available. The direct schemes are the most popular methods for finite element analysis, and while it is impractical to discuss these methods in depth in this book, we illustrate the Gauss elimination and Cholesky schemes with examples. Readers interested in further details of the various techniques as employed in finite element analysis should consult references 57–67.

Gauss Elimination   To illustrate the Gauss elimination method of solving equations 10.29 we solve the set of equations

$$\begin{bmatrix} 1 & 1 & 1 \\ 1 & 5 & -3 \\ 1 & -3 & 9 \end{bmatrix} \begin{Bmatrix} x_1 \\ x_2 \\ x_3 \end{Bmatrix} = \begin{Bmatrix} 6 \\ 2 \\ 22 \end{Bmatrix} \qquad (10.30)$$

where we note that the coefficient matrix is symmetric and nonsingular. To solve a set of equations using Gauss elimination we proceed in the following systematic steps:

Step 1. Subtract a multiple of the first equation from the second and third equations to obtain zeros below the diagonal in the first column. In this simple example the values in the first column are 1, so we simply subtract the first equation from the second and third equations. Then equations 10.30 are

$$\begin{bmatrix} 1 & 1 & 1 \\ 0 & 4 & -4 \\ 0 & -4 & 8 \end{bmatrix} \begin{Bmatrix} x_1 \\ x_2 \\ x_3 \end{Bmatrix} = \begin{Bmatrix} 6 \\ -4 \\ 16 \end{Bmatrix}$$

Step 2. We continue by subtracting a multiple of the second equation from the third equation to obtain zeros below the diagonal in the second column. Thus we subtract -1 times the second equations from the third equation to yield

$$\begin{bmatrix} 1 & 1 & 1 \\ 0 & 4 & -4 \\ 0 & 0 & 4 \end{bmatrix} \begin{Bmatrix} x_1 \\ x_2 \\ x_3 \end{Bmatrix} = \begin{Bmatrix} 6 \\ -4 \\ 12 \end{Bmatrix}$$

If we have a larger system of equations, we continue in this fashion, reducing all elements below the diagonal to zero until we reach the final equation. The result of this procedure is to reduce the original system matrix to an upper triangular matrix.

Step 3. Finally, we solve for the unknowns by back-substitution, working from the last equation to the first equation:

$$x_3 = \frac{12}{4} = 3$$

$$x_2 = \frac{-4 + 4(3)}{4} = 2$$

$$x_1 = \frac{67 - 1(2) - 1(3)}{1} = 1$$

In the computer implementation of Gauss elimination we may choose to perform the reduction (also called factoring) of the coefficient matrix separately from the reduction of the load vector and the back-substitution. Separation of the operations permits efficient solution for nodal unknowns due to multiple load vectors, since the coefficient matrix needs to be reduced only once. The subroutines FACTOR and SOLVE listed in Section 10.3 are written in this manner. FACTOR uses Gauss elimination to reduce an unsymmetric matrix stored in banded form to an upper triangular matrix, and SOLVE reduces a load vector and then performs the back-substitution to solve for the nodal unknowns. Note that Gauss elimination can be used to solve equations with either a symmetric or an unsymmetric coefficient matrix.

Cholesky Decomposition   In Cholesky decomposition we decompose the coefficient matrix into the product of an upper triangular matrix $[U]$ and its transpose:

$$[K] = [U]^T[U] \tag{10.31}$$

where we assume that $[K]$ is symmetric and nonsingular. Substitution of equation 10.31 into equation 10.29 gives

$$[U]^T[U]\{x\} = \{R\}$$

which we rewrite as

$$[U]^T\{S\} = \{R\} \tag{10.32a}$$

where

$$[U]\{x\} = \{S\} \tag{10.32b}$$

With $[U]$ known, we first solve for the elements of the vector $\{S\}$ from equation 10.32a by forward-substitution, and then solve for the nodal unknowns from equation 10.32b by back-substitution. To illustrate the procedure for deriving general equations for the elements of $[U]$, we write a 3×3 system matrix in the factored form of equation 10.31:

$$\begin{bmatrix} k_{11} & k_{12} & k_{13} \\ k_{12} & k_{22} & k_{23} \\ k_{13} & k_{23} & k_{33} \end{bmatrix} = \begin{bmatrix} u_{11} & 0 & 0 \\ u_{12} & u_{22} & 0 \\ u_{13} & u_{23} & u_{33} \end{bmatrix} \begin{bmatrix} u_{11} & u_{12} & u_{13} \\ 0 & u_{22} & u_{23} \\ 0 & 0 & u_{33} \end{bmatrix}$$

Then, by multiplying the right-hand side and equating the results term by term to the left-hand side, we may solve for the elements of $[U]$. In general,

$$u_{11} = \sqrt{k_{11}} \tag{10.33a}$$

$$u_{1j} = \frac{k_{1j}}{u_{11}}, \quad j > 1 \tag{10.33b}$$

$$u_{ii} = \left(k_{ii} - \sum_{l=1}^{i-1} u_{li}^2 \right)^{1/2} \tag{10.33c}$$

$$u_{ij} = \frac{k_{ij} - \sum_{l=1}^{i-1} u_{li} u_{lj}}{u_{ii}}, \quad j > i > 1 \tag{10.33d}$$

To illustrate the scheme for equations 10.30 we first solve for the elements of $[U]$ using equation 10.33:

$$u_{11} = \sqrt{k_{11}} = 1$$

$$u_{12} = \frac{k_{12}}{u_{11}} = 1$$

$$u_{13} = \frac{k_{13}}{u_{11}} = 1$$

$$u_{22} = \left(k_{22} - u_{12}^2 \right)^{1/2} = 2$$

$$u_{23} = \frac{k_{23} - u_{12} u_{13}}{u_{22}} = -2$$

$$u_{33} = \left(k_{33} - u_{13}^2 - u_{23}^2 \right)^{1/2} = 2$$

Then, using these values in equation 10.32a, we solve for $\{S\}$,

$$\{S\}^T = \lfloor 6 \quad -2 \quad 6 \rfloor$$

and finally solve for $\{x\}$ from equation 10.32b,

$$\{x\}^T = \lfloor 1 \quad 2 \quad 3 \rfloor$$

Note that this scheme is suitable only for the solution of symmetric positive-definite systems of equations. Cholesky decomposition is used not only as a method for solving algebraic equations but in other matrix operations that employ a factored matrix. One example appears in subroutine KIS04 in Section 10.3 where the conductivity tensor is factored by Cholesky decomposition to permit efficient formation of the symmetric isoparametric conductance matrix; other applications of Cholesky decomposition are discussed in reference 8.

10.8.2 Nonlinear Matrix Equations

Nonlinear algebraic equations are usually solved by an iteration procedure. We discuss here the popular Newton–Raphson iteration method, but interested readers can find a more general discussion in reference 68. Consider a typical equation from a nonlinear set of equations of the form of equation 10.29. We write

$$F_i(x_1, x_2, \ldots, x_n) = \sum_{j=1}^{n} k_{ij}(x_1, x_2, \ldots, x_n) x_j - R_i(x_1, x_2, \ldots, x_n),$$

$$i = 1, 2, \ldots, n \quad (10.34)$$

where $F_i(x_1, x_2, \ldots, x_n)$ is the unbalanced load in the ith equation, and n is the number of equations. If the vector of nodal variables $x_1, x_2, \ldots, x_n$ is the exact solution, then $F_i(x_1, x_2, \ldots, x_n) \equiv 0$ for $i = 1, 2, \ldots, n$. Usually, however, we cannot compute the exact solution, but we can obtain an approximate solution such that the unbalance in a typical equation is smaller than a specified tolerance. We do this with Newton–Raphson iteration, which is often introduced in elementary textbooks as an effective method for computing the roots of a transcendental equation. The basic idea used here is the same, but we extend it to apply to n variables. We begin with an approximate solution vector and use a Taylor series expansion of F_i in n variables to obtain an improved solution vector. Thus

$$F_i(x_1 + \Delta x_1, x_2 + \Delta x_2, \ldots, x_n + \Delta x_n)$$

$$= F_i(x_1, x_2, \ldots, x_n) + \sum_{j=1}^{n} \frac{\partial F_i}{\partial x_j}(x_1, x_2, \ldots, x_n) \Delta x_j + \cdots,$$

$$i = 1, 2, \ldots, n$$

Now as an approximation we neglect all higher-order terms and require for a solution that the left-hand side vanish. Then we obtain a set of linear algebraic equations with unknowns Δx_j:

$$\sum_{j=1}^{n} \frac{\partial F_i}{\partial x_j}(x_1, x_2, \ldots, x_n) \Delta x_j = -F_i(x_1, x_2, \ldots, x_n)$$

This set of equations is the basis for the Newton–Raphson iteration algorithm, which we write in matrix notation as

$$[J]^m \{\Delta x\}^{m+1} = -\{F\}^m \tag{10.35a}$$

$$\{x\}^{m+1} = \{x\}^m + \{\Delta x\}^{m+1} \tag{10.35b}$$

where the superscript m denotes the mth iteration, and $[J]^m$ is the system Jacobian at the mth iteration. The elements of $[J]^m$ are

$$J_{ij}^m = \frac{\partial F_i}{\partial x_j}(x_1^m, x_2^m, \ldots, x_n^m) \tag{10.36}$$

We proceed by computing the Jacobian and unbalanced nodal load vector and solving a set of linear algebraic equations at each iteration. When used with a reasonable initial guess, the Newton–Raphson method shows an

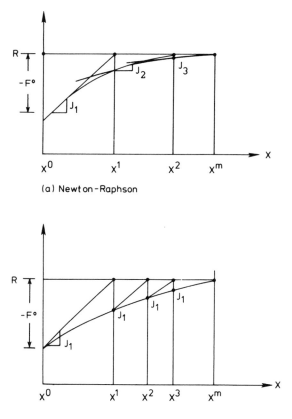

(a) Newton–Raphson

(b) Modified Newton–Raphson

Figure 10.20 Comparison of Newton–Raphson methods for a single-degree-of-freedom system.

excellent convergence rate, but in some problems if the initial guess is a poor approximation, the method may diverge. Since the Jacobian has to be reformed and factored at every iteration, the method may be computationally expensive for large sets of equations. Large computational costs have led some analysts to use a modified Newton–Raphson method wherein the Jacobian is formed and factored only once and is held constant throughout the balance of the iteration process. More iterations are required with the modified Newton–Raphson method, but usually net computational costs are reduced. Graphical interpretations of the Newton–Raphson method and modified schemes are shown in Figure 10.20 for a single-degree-of-freedom system.

10.9   CLOSURE

The sample computer program and the other practical considerations presented in this final chapter illustrate important computational operations involved in implementation of the finite element method. Once the reader has gained an understanding of the underlying fundamentals presented in the preceding chapters and has learned how these fundamentals are implemented on a computer, he or she will have command of a powerful engineering analysis tool.

As this book concludes, the authors would like to make a final observation about the evolution of finite element technology. The development of the technology has occurred since we received our undergraduate degrees. An entire industry for development of finite element software, software support, and marketing has evolved. As the method has become essential in a vast array of applications, thousands of engineering jobs have been created. At the same time a significant array of new university courses have been developed. The evolutionary process continues as new advancements make the technology more effective in design and research. The growth of this technology is just one of many examples of how our education as engineers does not end—but only begins—when we leave the university. We hope that this text helps you to begin and continue your education in finite elements, and that you will find the rewards of your study as beneficial as have the authors.

REFERENCES

1. E. Cuthill, "Several Strategies for Reducing the Bandwidth of Matrices," in *Sparse Matrices and Their Applications*, D. J. Rose and R. A. Willoughby (eds.), Plenum, New York, 1972.
2. R. J. Collins, "Bandwidth Reduction by Automatic Renumbering," *Int. J. Numer. Methods Eng.*, Vol. 6, 1973, pp. 345–356.

3. W.-H. Liu and A. H. Sherman, "Comparative Analysis of the Cuthill–McKee and the Reversed Cuthill–McKee Ordering Algorithms for Sparse Matrices," *SIAM J. Numer. Anal.*, Vol. 13, 1976, pp. 198–213.

4. N. E. Gibbs, W. G. Poole, Jr., and P. K. Stockmeyer, "An Algorithm for Reducing the Bandwidth and Profile of a Sparse Matrix," *SIAM J. Numer. Anal.*, Vol. 13, 1976, pp. 236–250.

5. G. C. Everstine, "A Comparison of Three Resequencing Algorithms for the Reduction of Matrix Profile and Wavefront," *Int. J. Numer. Methods Eng.*, Vol. 14, No. 6, 1979, pp. 837–853.

6. M. Hoit and E. L. Wilson, "An Equation Numbering Algorithm Based on a Minimum Front Criteria," *Comput. Structures*, Vol. 16, 1983, pp. 225–239.

7. S. W. Sloan, "An Algorithm for Profile and Wavefront Reduction of Sparse Matrices," *Int. J. Numer. Methods Eng.*, Vol. 23, No. 2, 1986, pp. 239–251.

8. K. J. Bathe and E. L. Wilson, *Numerical Methods in Finite Element Analysis*, Prentice-Hall, Englewood Cliffs, NJ, 1976.

9. T. J. R. Hughes, *The Finite Element Method: Linear Static and Dynamic Finite Element Analysis*, Prentice-Hall, Englewood Cliffs, NJ, 1987.

10. O. C. Zienkiewicz and R. L. Taylor, *The Finite Element Method*, 4th ed., Vol. 1, McGraw-Hill, London, 1989.

11. J. N. Reddy, *An Introduction to the Finite Element Method*, 2d ed., McGraw-Hill, New York, 1993.

12. E. Hinton and D. R. J. Owen, *Finite Element Programming*, Academic, New York, 1977.

13. R. D. Cook, D. S. Malkus, and M. E. Plesha, *Concepts and Applications of Finite Element Analysis*, 3d ed., John Wiley & Sons, New York, 1989.

14. O. C. Zienkiewicz and D. V. Phillips, "An Automatic Mesh Generation Scheme for Plane and Curved Surfaces by 'Isoparametric' Co-Ordinates," *Int. J. Numer. Methods Eng.*, Vol. 3, 1971, pp. 519–528.

15. W. C. Thacker, "A Brief Review of Techniques for Generating Irregular Computational Grids," *Int. J. Numer. Methods in Eng.*, Vol. 15, 1980, pp. 1335–1341.

16. M. A. Yerry and M. S. Shephard, "Automatic Three-Dimensional Mesh Generation by the Modified-Octree Technique," *Int. J. Numer. Methods Eng.*, Vol. 20, 1984, 1965–1990.

17. J. C. Cavendish, D. A. Field, and W. H. Frey, "An Approach to Automatic Three-Dimensional Finite Element Mesh Generation," *Int. J. Numer. Methods Eng.*, Vol. 21, 1985, pp. 329–347.

18. S. H. Lo, "A New Mesh Generation Scheme for Arbitrary Planar Domains," *Int. J. Numer. Methods Eng.*, Vol. 21, 1985, pp. 1403–1426.

19. P. L. Baehmann, S. L. Wittchen, M. S. Shephard, K. R. Grice, and M. A. Yerry, "Robust, Geometrically Based, Automatic Two-Dimensional Mesh Generation," *Int. J. Numer. Methods Eng.*, Vol. 24, 1987, pp. 1043–1078.

20. S. H. Lo, "Delaunay Triangulation of Non-convex Planar Domains," *Int. J. Numer. Methods Eng.*, Vol. 28, 1989, pp. 2695–2707.

21. L. Demkowicz, P. Devloo, and J. T. Oden, "On an h-Type Mesh-Refinement Strategy Based on Minimization of Interpolation Errors," *Comput. Methods Appl. Mech. Eng.*, Vol. 53, 1985, pp. 67–89.

22. R. Löhner, K. Morgan, and O. C. Zienkiewicz, "An Adaptive Finite Element Procedure for Compressible High Speed Flows," *Comput. Methods Appl. Mech. Eng.*, Vol. 51, 1985, pp. 441–465.

23. R. Löhner, "An Adaptive Finite Element Scheme for Transient Problems in CFD," *Comput. Methods Appl. Mech. Eng.*, Vol. 61, 1987, pp. 323–338.

24. R. Ramakrishnam, K. S. Bey, and E. A. Thornton, "Adaptive Quadrilateral and Triangular Finite Element Scheme for Compressible Flows," *AIAA J.*, Vol. 28, No. 1, 1990, pp. 51–59.

25. R. Ramakrishnan, A. R. Wieting, and E. A. Thornton, "A Transient Finite Element Adaptation Scheme for Thermal Problems with Steep Gradients," *Int. J. Numer. Methods Heat Fluid Flow*, Vol. 2, 1992, pp. 517–535.

26. O. C. Zienkiewicz, J. P. de S. R. Gago, and D. W. Kelly, "The Hierarchical Concept in Finite Element Analysis," *Comput. Struct.*, Vol. 16, No. 1–4, 1983, pp. 53–65.

27. A. G. Peano, "Hierarchies of Conforming Finite Elements for Plane Elasticity and Plate Bending," *Comput. Math. Appl.*, Vol. 2, 1976, pp. 211–224.

28. A. G. Peano, A. Pasini, R. Riccioni, and L. Sardella, "Adaptive Approximation in Finite Element Structural Analysis," *Comput. Struct.*, Vol. 10, 1979, pp. 332–342.

29. B. A. Szabo', "Some Recent Developments in Finite Element Analysis," *Comput. Math. Appl.*, Vol. 5, 1979, pp. 99–115.

30. B. A. Szabo', "Mesh Design for the *p*-Version of the Finite Element Method," *Comput. Methods Appl. Mech. Eng.*, Vol. 55, 1986, pp. 181–197.

31. J. T. Oden, T. Strouboulis, and P. Devloo, "Adaptive Finite Element Methods for the Analysis of Inviscid Compressible Flow, I: Fast Refinement/Unrefinement and Moving Mesh Methods for Unstructured Meshes," *Comput. Methods Appl. Mech. Eng.*, Vol. 59, 1986, pp. 327–362.

32. L. Demkowicz, J. T. Oden, W. Rachowicz, and O. Hardy, "Toward a Universal *h–p* Adaptive Finite Element Strategy, 1: Constrained Approximation and Data Structures," *Comput. Methods Appl. Mech. Eng.*, Vol. 77, 1989, pp. 79–112.

33. J. T. Oden, L. Demkowicz, W. Rachowicz, and T. A. Westermann, "Toward a Universal *h–p* Adaptive Finite Element Strategy, 2: A Posteriori Error Estimation," *Comput. Methods Appl. Mech. Eng.*, Vol. 77, 1989, pp. 113–180.

34. W. Rachowicz, J. T. Oden, and L. Demkowicz, "Toward a Universal *h–p* Adaptive Finite Element Strategy, 3: Design of *h–p* Meshes," *Comput. Methods Appl. Mech. Eng.*, Vol. 77, 1989, pp. 181–212.

35. W. W. Tworzydlo, J. T. Oden, and E. A. Thornton, "Adaptive Implicit/Explicit Finite Element Method for Compressible Viscous Flows," *Comput. Methods Appl. Mech. Eng.*, Vol. 95, 1992, pp. 397–440.

36. G. Strang and G. F. Fix, *An Analysis of the Finite Element Method*, Prentice-Hall, Englewood Cliffs, NJ, 1973.

37. E. B. Becker, G. F. Carey and J. T. Oden, *Finite Elements*, Vol. I, *An Introduction*, Prentice-Hall, Englewood Cliffs, NJ, 1981.

38. J. T. Oden and G. F. Carey, *Finite Elements*, Vol. IV, *Mathematical Aspects*, Prentice-Hall, Englewood Cliffs, NJ, 1983.

39. O. C. Zienkiewicz and J. Z. Zhu, "A Simple Error Estimator and Adaptive Procedure for Practical Engineering Analysis," *Int. J. Numer. Methods Eng.*, Vol. 24, 1987, pp. 337–357.

40. J. Z. Zhu and O. C. Zienkiewicz, "Adaptive Techniques in the Finite Element Method," *Commun. Appl. Numer. Methods*, Vol. 4, 1988, pp. 197–204.

41. M. Ainsworth, J. Z. Zhu, A. W. Craig, and O. C. Zienkiewicz, "Analysis of the Zienkiewicz–Zhu *a Posteriori* Error Estimator in the Finite Element Method," *Int. J. Numer. Methods Eng.*, Vol. 28, 1989, pp. 2161–2174.

42. O. C. Zienkiewicz and J. Z. Zhu, "Error Estimates and Adaptive Refinement for Plate Bending Problems," *Int. J. Numer. Methods Eng.*, Vol. 28, 1989, pp. 2839–2853.

43. J. Z. Zhu and O. C. Zienkiewicz, "Superconvergence Recovery Technique and a Posteriori Error Estimators," *Int. J. Numer. Methods Eng.*, Vol. 30, 1990, pp. 1321–1339.

44. O. C. Zienkiewicz and J. Z. Zhu, "The Three R's of Engineering Analysis and Error Estimation and Adaptivity," *Comput. Methods Appl. Mech. Eng.*, Vol. 82, 1990, pp. 95–13.

45. O. C. Zienkiewicz and J. Z. Zhu, "Adaptivity and Mesh Generation," *Int. J. Numer. Methods Eng.*, Vol. 32, 1991, pp. 783–810.

46. J. Peraire, M. Vahdati, K. Morgan, and O. C. Zienkiewicz, "Adaptive Remeshing for Compressible Flow Computations," *J. Comput. Phys.*, Vol. 72, 1987, pp. 449–466.

47. L. Formaggia, J. Peraire, and K. Morgan, "Simulation of a Store Separation Using the Finite Element Method," *Appl. Math. Modelling*, Vol. 12, April 1988, pp. 175–181.

48. J. Peraire, J. Peiro, L. Formaggia, K. Morgan, and O. C. Zienkiewicz, "Finite Element Euler Computations in Three Dimensions," *Int. J. Numer. Methods Eng.*, Vol. 26, 1988, pp. 2135–2159.

49. R. Löhner, "Adaptive Remeshing for Transient Problems," *Comput. Methods Appl. Mech. Eng.*, Vol. 75, 1989, pp. 195–214.

50. E. A. Thornton and G. Vemaganti, "Adaptive Remeshing Method for Finite Element Thermal Analysis," *J. Thermophys. Heat Transfer*, Vol. 4, No. 2, April 1990, pp. 212–220.

51. H. Jin and N.-E. Wiberg, "Two-Dimensional Mesh Generation, Adaptive Remeshing and Refinement," *Int. J. Numer. Methods Eng.*, Vol. 29, 1990, pp. 1501–1526.

52. J. Probert, O. Hassan, J. Peraire, and K. Morgan, "An Adaptive Finite Element Method for Transient Compressible Flows," *Int. J. Numer. Method Eng.*, Vol. 32, 1991, pp. 1145–1159.

53. C. J. Hwang and S. J. Wu, "Global and Local Remeshing Algorithms for Compressible Flows," *J. Comput. Phys.*, Vol. 102, 1992, pp. 98–113.

54. M. Abramowitz and I. A. Stegun (ed.), *Handbook of Mathematical Functions*, National Bureau of Standards Applied Mathematics Series 55, U.S. Government Printing Office, Washington, DC, June 1964.

55. P. Davis and P. Rabinowitz, "Abscissas and Weights for Gaussian Quadratures of High Order," *J. Res. Nat. Bur. Stand.*, Vol. 56, RP2645, 1956.

56. G. R. Cowper, "Gaussian Quadrature Formulas for Triangles," *Int. J. Numer. Methods Eng.*, Vol. 7, 1973, pp. 405–408.

57. B. M. Irons, "A Frontal Solution Program for Finite Element Analysis," *Int. J. Numer. Methods Eng.*, Vol. 2, No. 1, 1970.

58. H. G. Jensen and G. A. Parks, "Efficient Solutions for Linear Matrix Equations," *Proc. ASCE*, Vol. 96, No. ST-1, January 1970.

59. G. Cantin, "An Equation Solver of Very Large Capacity," *Int. J. Numer. Methods Eng.*, Vol. 3, 1971, pp. 379–388.

60. W. T. Segui, "Computer Programs for the Solution of Systems of Linear Algebraic Equations," *Int. J. Numer. Methods Eng.*, Vol. 7, 1973, pp. 479–490.

61. D. P. Mondkar and G. H. Powell, "Towards Optimal in-Core Equation Solving," *Comput. Struct.*, Vol. 4, 1974, pp. 531–548.

62. D. P. Mondkar and G. H. Powell, "Large Capacity Equation Solver for Structural Analysis," *Comput. Struct.*, Vol. 4, 1974, pp. 699–728.

63. C. A. Felippa, "Solution of Linear Equations with Skyline-Stored Symmetrix Matrix," *Comput. Struct.*, Vol. 5, 1975, pp. 13–30.

64. P. Hood, "Frontal Solution Program for Unsymmetric Matrices," *Int. J. Numer. Methods Eng.*, Vol. 10, 1976, pp. 379–400.

65. E. L. Wilson and H. H. Dovey, "Solution or Reduction of Equilibrium Equations for Large Complex Structural Systems," *Adv. Eng. Software*, Vol. 1, No. 1, 1978, pp. 19–25.

66. D. J. Rose, G. F. Whitten, A. H. Sherman, and R. E. Tarjan, "Algorithms and Software for in-Core Factorization of Sparse Symmetric Positive Definite Matrices," *Comput. Struct.*, Vol. 10, 1979, pp. 411–418.

67. E. Mandelssohn and M. Baruch, "Solution of Linear Equations With a Symmetrically Skyline Stored Nonsymmetric Matrix," *Comput. Struct.*, Vol. 18, 1984, pp. 215–246.

68. J. M. Ortega and W. C. Rheinboldt, *Iterative Solution of Nonlinear Equations in Several Variables*, Academic, New York, 1970.

PROBLEMS

1. Draw a flow chart of the thermal analysis program presented in Section 10.3 showing "calls" to all subroutines. Prepare a list and a brief description of the subroutines.

2. The thermal analysis program currently contains only one element. Outline the steps that are necessary to add a new element to the program. Give a list of the required new subroutines, their purpose, and where they would be called.

3. Add a two-node rod element to the program. Include the following in the element:

   a. Conduction

   b. Surface convection

   c. Internal heat generation

   d. Surface heating

4. Add a family of surface convection elements to the program to supplement the plane conduction elements. Include the following in the family:

   a. A two-node edge convection element

   b. A two-node axisymmetric surface convection element

5. Outline the steps that are necessary to modify the program for transient analysis. Give a list of required modifications and new subroutines. Give a flow chart for a transient solution module.

6. Use the computer program to solve Problem 3, Chapter 8.

7. Use the computer program to solve Problem 7, Chapter 8. Repeat the analysis with a refined mesh with a minimum of 15 nodes. Plot the computed temperature distributions $T(x, 0)$ for $0 < x < 2$.

8. Use the computer program to determine the temperature distribution in a wall constructed of wooden beams with wood exterior siding, sheet rock on the interior, and rock wool insulation in the space between the beams (Figure P10.8). Assume that the inside wall is at 292 K (65 °F) and that the outside wall is at 272 K (30 °F).

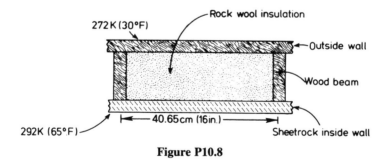

Figure P10.8

Data. Beam dimensions: 4.14 cm (1.63 in.) × 8.89 cm (3.5 in.)
Interior and exterior wall thickness = 1.91 cm (0.75 in.)
Wood conductivity = 0.173 W/mK (0.1 Btu/h ft °F)
Sheet rock conductivity = 0.519 W/m K (0.3 Btu/h ft °F)
Rock wool conductivity = 0.0519 W/mK (0.03 Btu/h ft °F).

9. A long, hollow cylinder is subjected to an inside temperature of $T_i = 100$ °F and an outside temperature of $T_0 = 0$ °F. The cylinder has an inside radius $r_i = 1$ in. and an outside radius $r_0 = 2$ in. Compute the radial temperature distribution in the cylinder wall and compare the computed

temperatures with the analytical solution

$$T(r) = T_i - \frac{T_i - T_0}{\ln\left(\dfrac{r_0}{r_i}\right)} \ln\frac{r}{r_i}$$

Use 10 axisymmetric quadrilateral elements in your finite element model.

10. A 2-ft-thick concrete slab is heated by hot water flowing through 6-in. pipes located on 2-ft centers, Figure P10.10. The top surface of the slab is maintained at 40 °F, and the lower surface of the slab is maintained at 50 °F. If the average inside surface of the pipes is maintained at 100 °F, determine the temperature distribution in the slab. Use symmetry in your finite element model. For concrete, $k_x = k_y = 0.7$ Btu/h ft °F and $k_{xy} = 0$.

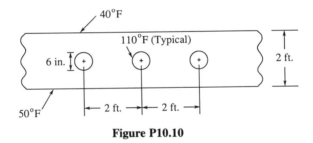

Figure P10.10

11. Determine the temperature distribution in the molded pipe insulation shown in Figure P10.11. Use symmetry in your finite element model. Use $k = 0.173$ W/mK (0.1 Btu/h ft °F).

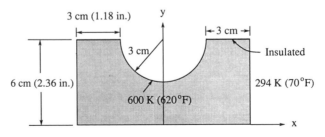

Figure P10.11

12. A schedule 40 iron pipe is covered with insulation. The inside temperature of the pipe wall is measured to be 422 K (300 °F) and the outside
   temperature of the insulation is 303 K (85 °F). Compute the outside
   surface temperature of the pipe.

   Data. Pipe inside radius = 5.13 cm (2.02 in.)
   Pipe outside radius = 5.73 cm (2.25 in.)
   Pipe thermal conductivity = 55.4 W/mK (32 Btu/h ft °F)
   Insulation thickness = 2.54 cm (1.0 in.)
   Insulation thermal conductivity = 0.071 W/mK (0.041 Btu/h ft
   °F).

13. To verify the error estimate given in equation 10.8, consider the following
   one-dimensional heat conduction problem:

$$k\frac{d^2T}{dx^2} + Q = 0$$

$$T(0) = T(L) = 0$$

The analytical solution is

$$T = \frac{QL^2}{2k}\frac{x}{L}\left(1 - \frac{x}{L}\right)$$

For convenience, make the problem nondimensional by using $Q = 2$ and
$k = L = 1$. Using the thermal analysis program, compute temperature
distributions for three uniform meshes with $h = 0.2, 0.1$, and 0.05, where
h is the nondimensional mesh spacing. Then compute the error as
measured in the L_2 norm, equation 10.7, for each solution. Note that
from equation 10.8 we may write

$$\ln\|E\|_{L_2} = p\ln(h) + \ln C$$

For your three solutions plot $\ln\|E\|_{L_2}$ versus $\ln(h)$ and determine the
slope p. Discuss how your value of p compares with the theoretical
estimate of $p = 2$.

14. For the one-dimensional heat conduction problem given in the preceding
   problem use the thermal analysis program to compute nondimensional
   nodal temperatures and element heat fluxes for $h = 0.1$. Then use the
   method described in Section 10.6.2 to compute nodal values of the first
   and second derivatives. Compare your results with derivatives obtained
   from the analytical solution.

15. Using results from Problems 13 and 14, evaluate the element's error
   indicators given in equation 10.10 for $h = 0.1$. Compare the actual
   element errors with errors indicated by the error indicator. Discuss your
   results in terms of adaptive refinement.

16. Legendre–Gauss numerical integration of a polynomial $f(x) = x^{2n-1}$ is exact if we use n Gauss points. Verify this statement by integrating the polynomials x, x^2, x^3, x^4, x^5, and x^6 using three Gauss points, $n = 3$.

17. For a rectangular finite element with sides a, b, one term of the element consistent mass matrix is given by

$$M_{23} = \int_0^a \int_0^b N_2(x, y) N_3(x, y) \, dx \, dy$$

where N_2 and N_3 are the element interpolation functions (Section 5.8.1). Derive the interpolation functions and evaluate the integral analytically. Now compare this result with values obtained by Legendre–Gauss integration. First transform the integral into natural coordinates ξ, η to obtain the integral

$$M_{23} = \int_{-1}^1 \int_{-1}^1 N_2(\xi, \eta) N_3(\xi, \eta) |J(\xi, \eta)| \, d\xi \, d\eta$$

where $|J(\xi, \eta)|$ is the determinant of the Jacobian of the coordinate transformation matrix (Section 5.10.1). Evaluate the integral by using (a) one-point integration, (b) four-point integration, and (c) nine-point integration. Evaluate and discuss the accuracy of your results.

18. Evaluate the integral of the function $f(x, y) = x^2 y$ over the triangle defined by the nodes $(1, 1)$, $(3, 2)$, and $(2, 3)$. Evaluate the integral exactly using equation 5.22 and then approximately using Legrendre–Gauss integration. Use one, three, and four Gauss points and discuss the accuracy of your results.

19. Use a 3×3 matrix to derive equations 10.33 for Cholesky decomposition. Given the following coefficient matrix and load vectors

$$[K] = \begin{bmatrix} 4 & 2 & 0 & 0 \\ 2 & 8 & 2 & 0 \\ 0 & 2 & 8 & 2 \\ 0 & 0 & 2 & 4 \end{bmatrix} \quad \{R_1\} = \begin{Bmatrix} 4 \\ 8 \\ 8 \\ 4 \end{Bmatrix} \quad \{R_2\} = \begin{Bmatrix} 6 \\ 12 \\ 6 \\ 12 \end{Bmatrix}$$

solve for the nodal unknowns for the two load vectors by (a) Gauss elimination and (b) Cholesky decomposition.

20. Consider a nonlinear single-degree-of-freedom system of the form of equation 10.29 where

$$K = x^{-1/2} \quad \text{and} \quad R = 2.5$$

For an initial value of $x^0 = 0.3$ use an iterative solution to solve for x by computing the Jacobian and unbalanced load vector. First use

Newton–Raphson iteration and then modified Newton–Raphson. Do three iterations by each method and compare your results with the exact solution. Draw figures similar to Figure 10.20 to show the convergence rates.

21. Consider a one-dimensional heat conduction problem where the thermal conductivity varies linearly with temperature. The problem may be stated as

$$\frac{d}{dx}\left(k_0 T \frac{dT}{dx}\right) + Q = 0$$

where k_0 is a constant, and the internal heat generation is a constant. For steady heat transfer the boundary conditions are

$$T(0) = 0$$

$$\frac{dT}{dx}(L) = 0$$

For a two-node, linear element of length l connecting nodes i and $i + 1$ the element Jacobian matrix is

$$[J]^{(e)} = \frac{k_0}{l}\begin{bmatrix} T_i & -T_{i+1} \\ -T_i & T_{i+1} \end{bmatrix}$$

and the unbalanced element load vector is

$$\{F\}^{(e)} = \begin{Bmatrix} \dfrac{k_0}{2l}(T_i^2 - T_{i+1}^2) - \dfrac{Ql}{2} \\[3mm] \dfrac{k_0}{2l}(-T_i^2 + T_{i+1}^2) - \dfrac{Ql}{2} \end{Bmatrix}$$

For convenience, take $Q = 2$ and $k_0 = L = 1$. Use the element matrices to assemble global equations of the form of equation 10.35. Use four elements and compute the unknown nodal temperatures with Newton–Raphson iteration. Compare your finite element results with the analytical solution to the problem.

APPENDIX A

MATRICES

In the 1850s Cayley introduced matrix notation as an effective shorthand scheme for dealing with systems of linear algebraic equations. Without this convenient notation the labor of writing out every term of a large system of simultaneous equations would be enough to discourage anyone.

Matrix notation is intimately associated with the finite element method because ultimately the application of finite element techniques leads to sets of simultaneous equations. In many ways matrix notation is like a parachute. To carry and store the parachute conveniently we keep it folded, but to use it we must unfold it. Similarly, to discuss or manipulate ordered sets of numbers or functions conveniently we represent them compactly with matrix notation. But at the calculation stage the notation must be unraveled by hand or by computer so that individual terms can be treated explicitly.

Here[1] we present the fundamentals of matrix notation and matrix algebra necessary for the finite element method.

A.1 DEFINITIONS

Suppose that we have a rectangular array of numbers or functions arranged as follows:

$$\begin{bmatrix} a_{11} & a_{12} & a_{13} & \cdots & a_{1n} \\ a_{21} & a_{22} & a_{23} & \cdots & a_{2n} \\ \vdots & & & & \\ a_{m1} & a_{m2} & a_{m3} & \cdots & a_{mn} \end{bmatrix}$$

Such an array is called an $m \times n$ *matrix* when certain laws of manipulation, to be given later, are specified. The entries in the matrix are themselves called elements and these elements may be real or complex.[2] The elements appearing in any given horizontal line comprise a row of the matrix, whereas those appearing in any vertical line comprise a column. Each element of a matrix carries a double subscript label. The first subscript indicates its row position, and the second subscript its column position. Thus a_{ij} is the element in the ith row and jth column: The matrix we illustrated has m rows and n columns. If $m = n$, the matrix is a *square matrix* of *order n*.

In the application of finite element techniques we also encounter row matrices ($m = 1$) and column matrices ($n = 1$). A matrix with 1 row and n columns is called a row matrix or row vector. A matrix with m rows and 1 column is called a column matrix or column vector. The convention that we use to distinguish the different types of matrices is as follows.

We denote a *rectangular $m \times n$ matrix A* by enclosing A in brackets: $[A]$. For the special case where $m = 1$ *and* $n = 1$, $[A]$ is simply a scalar:

If $m = 1$ and $n > 1$, $[A]$ is a *row matrix*, and we denote it as $\lfloor A \rfloor$.

If $m > 1$ and $n = 1$, $[A]$ is a *column matrix*, and we denote it as $\{A\}$.

---

[1]A more extensive treatment can be found in standard reference books such as *Elementary Linear Algebra* by B. Kolman, Macmillan, New York, 1986.

[2]In this text we assume that the elements of matrices are real numbers or functions. However, the results we present here hold also for matrices with complex elements.

A.2   SPECIAL TYPES OF SQUARE MATRICES

1. A *null matrix* or *zero matrix* is one whose elements are all zero.
2. A *diagonal matrix* is a square matrix with zero elements everywhere except on its main diagonal. The main diagonal runs from the upper left corner to the lower right corner.
3. An *identity matrix* is a diagonal matrix whose elements on the main diagonal are all unity. The elements of an identity matrix are often symbolized by the Kronecker delta, δ_{ij}, which is defined as

$$\delta_{ij} \equiv \begin{cases} 1, & i = j \\ 0, & i \neq j \end{cases}$$

4. A *symmetric matrix* is one whose elements satisfy the condition $a_{ij} = a_{ji}$. If all corresponding rows and columns of a symmetric matrix are interchanged, the matrix remains unchanged.

A.3   MATRIX OPERATIONS

The definitions that we have given thus far describe certain kinds of arrays of numbers or matrices. Before we can hope to manipulate these matrices, we first must define some rules for manipulations. Hence we consider next some definitions for matrix operations.

Two matrices are *equal* to one another only if they are identical; that is, the elements of one matrix must be the same as the corresponding elements of the other matrix. For example, if $[A] = [B]$, then $a_{ij} = b_{ij}$ for all i and j. The *sum* of $[A]$ and $[B]$ is defined only when $[B]$ has the same number of rows and the same number of columns as $[A]$. When these conditions hold, the addition of $[A]$ and $[B]$ results in a matrix, $[C]$, whose elements are the sum of the corresponding elements of $[A]$ and $[B]$. Thus if $[C] = [A] + [B]$, then $c_{ij} = a_{ij} + b_{ij}$. Since the addition of matrices reduces to the addition of elements, matrix addition follows the associative and commutative laws of real numbers. Therefore

$$[A] + [B] = [B] + [A]$$

and

$$[A] + ([B] + [C]) = ([A] + [B]) + [C] = [A] + [B] + [C]$$

Multiplication of matrix $[A]$ *by a scalar* β results in a matrix $[B]$ whose elements are $b_{ij} = \beta a_{ij}$. When we write $[B] = \beta[A]$, we imply that every element of $[A]$ is multiplied by the constant β. The rule for *matrix subtraction* now follows from the other rules we have established, namely,

$$[A] + (-1)[B] = [A] - [B]$$

Thus, if $[C] = [A] - [B]$, then $c_{ij} = a_{ij} - b_{ij}$. We can see from this rule that the restrictions on the number of rows and columns are the same for matrix addition and subtraction.

There are many feasible ways to define matrix multiplication. The definition that we give here is the most common because it leads to an efficient means for dealing with systems of simultaneous linear equations. Suppose that we have an $m \times n$ matrix $[A]$ and a $p \times q$ matrix $[B]$. If $n = p$, the matrix product $[A][B]$ is defined and is an $m \times q$ matrix $[C]$ whose elements are

$$c_{ij} = \sum_{k=1}^{n} a_{ik}b_{kj}, \qquad i = 1, 2, \dots, m, \quad j = 1, 2, \dots, q$$

If $n \neq p$, the matrix product is undefined.

In the matrix product $[A][B]$ matrix $[A]$ is the premultiplier and matrix $[B]$ is the postmultiplier. Matrix multiplication is defined only when the number of columns of the premultiplier equals the number of rows of the postmultiplier. The following equation illustrates the procedure for matrix multiplication:

$$
\overset{2 \times 3}{\begin{bmatrix} a_{11} & a_{12} & a_{13} \\ a_{21} & a_{22} & a_{23} \end{bmatrix}}
\overset{3 \times 3}{\begin{bmatrix} b_{11} & b_{12} & b_{13} \\ b_{21} & b_{22} & b_{23} \\ b_{31} & b_{32} & b_{33} \end{bmatrix}}
$$

$$
= \overset{2 \times 3}{\begin{bmatrix} (a_{11}b_{11} + a_{12}b_{21} + a_{13}b_{31}), (a_{11}b_{12} + a_{12}b_{22} + a_{13}b_{32}), \\ (a_{21}b_{11} + a_{22}b_{21} + a_{23}b_{31}), (a_{21}b_{12} + a_{22}b_{22} + a_{23}b_{32}), \end{bmatrix}}
$$

$$
\begin{bmatrix} (a_{11}b_{13} + a_{12}b_{23} + a_{13}b_{33}) \\ (a_{21}b_{13} + a_{22}b_{23} + a_{23}b_{33}) \end{bmatrix}
$$

In general, matrix multiplication does not follow all the rules for the multiplication of real numbers. When the matrix products are defined, the associative law and the distributive law hold, but the commutative law does not. Thus

$$([A][B])[C] = [A]([B][C]) = [A][B][C] \quad \text{(associative law)}$$

and

$$[A]([B] + [C]) = [A][B] + [A][C] \quad \text{(distributive law)}$$

but generally matrices do not commute, that is,

$$[A][B] \neq [B][A]$$

A.4   SPECIAL MATRIX PRODUCTS

A.4.1   Product of a Square Matrix and a Column Matrix

Suppose, for example, that we have a system of three simultaneous linear equations such as

$$
\begin{aligned}
a_{11}x_1 + a_{12}x_2 + a_{13}x_3 &= b_1 \\
a_{21}x_1 + a_{22}x_2 + a_{23}x_3 &= b_2 \\
a_{31}x_1 + a_{32}x_2 + a_{33}x_3 &= b_3
\end{aligned}
\tag{A.1}
$$

The a's and b's are known constants, while the x's are the unknowns. We have a convenient way to write equations A.1 if we recognize that a square matrix containing the a's premultiplying a column matrix containing the x's is equivalent to a column matrix containing the b's. In matrix notation equation A.1 is equivalent to the equation

$$
\begin{bmatrix}
a_{11} & a_{12} & a_{13} \\
a_{21} & a_{22} & a_{23} \\
a_{31} & a_{32} & a_{33}
\end{bmatrix}
\begin{Bmatrix} x_1 \\ x_2 \\ x_3 \end{Bmatrix}
=
\begin{Bmatrix} b_1 \\ b_2 \\ b_3 \end{Bmatrix}
$$

or

$$[A]\{x\} = \{b\} \tag{A.2}$$

Anyone would agree that equation A.2 is easier to write than equation A.1. Therefore the product of a square matrix and a column is a most useful way to symbolize simultaneous linear equations, and we use it often in this text.

A.4.2   Product of a Row Matrix and a Square Matrix

When a row matrix premultiplies a square matrix, the result is a row matrix of the same order as the initial row matrix. For example,

$$
\underset{1 \times n}{\lfloor A \rfloor} \; \underset{n \times n}{[B]} = \underset{1 \times n}{\lfloor C \rfloor}
$$

If $n = 2$,

$$\lfloor a_1 \quad a_2 \rfloor \begin{bmatrix} b_{11} & b_{12} \\ b_{21} & b_{22} \end{bmatrix} = \lfloor (a_1 b_{11} + a_2 b_{21}), (a_1 b_{12} + a_2 b_{22}) \rfloor$$

A.4.3   Product of a Row Matrix and a Column Matrix

When a row matrix premultiplies a column matrix, the result is a matrix of order 1 or, in other words, a scalar. For example,

$$\overset{1 \times n}{\lfloor A \rfloor} \overset{n \times 1}{\{B\}} = \overset{1 \times 1}{C} = \sum_{i=1}^{n} a_i b_i$$

If $n = 2$,

$$\lfloor a_1 \quad a_2 \rfloor \begin{Bmatrix} b_1 \\ b_2 \end{Bmatrix} = a_1 b_1 + a_2 b_2$$

A.4.4   Product of the Identity Matrix and Any Other Matrix

We can readily verify that for any matrix $[A]$

$$[I][A] = [A][I] = [A]$$

The dimensions of $[I]$ are adjusted so that the matrix products are defined.

A.5   MATRIX TRANSPOSE

The transpose of an $m \times n$ matrix $[A]$ is an $n \times m$ matrix $[A]^T$, whose rows are the columns of $[A]$. To form the transpose of a matrix we simply interchange its rows and columns so that row i of the matrix becomes column i of its transpose. Applying this definition, we find that the transpose of a row matrix is a column matrix, and that a symmetric matrix is its own transpose. Also, it is easy to show that the transpose of the product of two matrices is the product of their transposes in reverse order, that is, $([A][B])^T = [B]^T[A]^T$.

A.6   QUADRATIC FORMS

Although matrix notation is most often employed to deal with sets of linear equations, it is also useful in symbolizing special nonlinear expressions called quadratic forms. A function of n variables $x_1, \ldots, x_n$ in quadratic form is defined as

$$F(x_1, \ldots, x_n) = \sum_{i=1}^{n} \sum_{j=1}^{n} a_{ij} x_i x_j$$

$$= a_{11} x_1^2 + a_{12} x_1 x_2 + \cdots + a_{1n} x_1 x_n + \cdots + a_{21} x_2 x_1$$

$$+ \cdots + a_{2n} x_2 x_n + \cdots + a_{n1} x_n x_1 + \cdots + a_{nn} x_n^2$$

We encounter a quadratic form whenever we express the energy of a continuous system in a set of discretized coordinates of the system. A quadratic form in one variable, say x_1, is simply ax_1^2. In two variables x_1 and x_2, the most general quadratic form is

$$F(x_1, x_2) = a_{11} x_1^2 + 2a_{12} x_1 x_2 + a_{22} x_2^2$$

Using matrix notation, we can write this as

$$F(x_1, x_2) = \lfloor x_1 \quad x_2 \rfloor \begin{bmatrix} a_{11} & a_{12} \\ a_{21} & a_{22} \end{bmatrix} \begin{Bmatrix} x_1 \\ x_2 \end{Bmatrix}$$

or

$$F(x_1, x_2) = \{x\}^T [A]\{x\} = \lfloor x \rfloor [A]\{x\} \tag{A.3}$$

We wrote equation A.3 for a quadratic form of two variables, but the same matrix symbolism holds also for a quadratic form of n variables.

A quadratic form $F(x_1, \ldots, x_n)$ is positive definite if F is nonnegative for all possible combinations of real x_i, $i = 1, \ldots, n$, and if F is zero only when every x_i is zero. A useful property of a positive definite quadratic form is that the determinants of the coefficients a_{ij} and all of its principal minors are positive, that is,

$$a_{11} > 0, \begin{vmatrix} a_{11} & a_{12} \\ a_{21} & a_{22} \end{vmatrix} > 0, \ldots, \begin{vmatrix} a_{11} & a_{12} & \cdots & a_{1n} \\ a_{21} & a_{22} & \cdots & a_{2n} \\ \vdots & & & \\ a_{n1} & a_{n2} & \cdots & a_{nn} \end{vmatrix} > 0$$

This property may be used to check whether a quadratic form is positive definite. There are other types of quadratic forms including positive semidefinite, negative definite, negative semidefinite and indefinite. The type of a quadratic form is important in several applications; for example, a stable structural system is characterized by a positive definite potential energy function.[3]

A.7 MATRIX INVERSE

When we considered matrix multiplication, we avoided discussing *matrix division* because such an operation is *undefined*. However, there is a concept known as the *matrix inverse*, which is symbolized in much the same way as division is symbolized for real numbers. We know that for any real number $b \neq 0$ there exist an inverse b^{-1} such that $b^{-1}b = bb^{-1} = 1$. Under certain circumstances matrices also possess this inverse property. If for a given square matrix $[A]$ there exists another square matrix $[A]^{-1}$ such that $[A]^{-1}[A] = [A][A]^{-1} = [I]$, then $[A]^{-1}$ is the inverse of $[A]$. Matrix $[A]$ is *nonsingular* when its inverse $[A]^{-1}$ exists.

With this definition and symbolism we now have a convenient way to write the solution of a system of equations of the form $[A]\{x\} = \{b\}$. If we premultiply both sides of the equation by $[A]^{-1}$, we obtain

$$[A]^{-1}[A]\{x\} = [A]^{-1}\{b\}$$

or

$$[I]\{x\} = [A]^{-1}\{b\}$$

which reduces to

$$\{x\} = [A]^{-1}\{b\} \tag{A.4}$$

A.8 MATRIX PARTITIONING

We introduced a matrix as an array of numbers that obeys certain laws of manipulation. Sometimes one subset of an array is of greater interest than another, and special notation is useful to highlight the various subsets. For this purpose we use *matrix partitioning*. As an example of this technique let us consider a 4×5 matrix $[A]$ that has been divided into sections as shown

---

[3]A further discussion of quadratic form types and the identification of these types is beyond the scope of this brief discussion on matrix notation. Excellent discussions of the topic from an engineering mechanics viewpoint appear in *Energy Methods in Applied Mechanics* by H. L. Langhaar, Wiley, New York, 1962, and *Energy and Variational Methods in Applied Mechanics* by J. N. Reddy, Wiley, New York, 1984.

by the dashed lines:

$$[A] = \begin{bmatrix} a_{11} & a_{12} & a_{13} & \vdots & a_{14} & a_{15} \\ a_{21} & a_{22} & a_{23} & \vdots & a_{24} & a_{25} \\ a_{31} & a_{32} & a_{33} & \vdots & a_{34} & a_{35} \\ \hdashline a_{41} & a_{42} & a_{43} & \vdots & a_{44} & a_{45} \end{bmatrix}$$

We may now define submatrices as

$$[A_{11}] = \begin{bmatrix} a_{11} & a_{12} & a_{13} \\ a_{21} & a_{22} & a_{23} \\ a_{31} & a_{32} & a_{33} \end{bmatrix} \qquad [A_{12}] = \begin{bmatrix} a_{14} & a_{15} \\ a_{24} & a_{25} \\ a_{34} & a_{35} \end{bmatrix}$$

$$[A_{21}] = \begin{bmatrix} a_{41} & a_{42} & a_{43} \end{bmatrix} \qquad [A_{22}] = \begin{bmatrix} a_{44} & a_{45} \end{bmatrix}$$

and write $[A]$ in partitioned form as

$$[A] = \begin{bmatrix} [A_{11}] & [A_{12}] \\ [A_{21}] & [A_{22}] \end{bmatrix}$$

We could have partitioned $[A]$ in many other ways. The only requirement is that the partition lines are drawn straight and run the full length and width of the matrix.

The rules for the manipulation of partitioned matrices are the same as those for ordinary matrices, provided that all the submatrices involved are compatible for the particular operation to be performed. Consider matrix multiplication, for example. Suppose that we wish to postmultiply our partitioned 4×5 matrix $[A]$ by a 5×2 matrix $[B]$. Multiplication by submatrices, that is,

$$\overset{4 \times 5}{\begin{bmatrix} \overset{3 \times 3}{[A_{11}]} & \overset{3 \times 2}{[A_{12}]} \\ \underset{1 \times 3}{[A_{21}]} & \underset{1 \times 2}{[A_{22}]} \end{bmatrix}} \overset{5 \times 2}{\begin{bmatrix} [B_1] \\ [B_2] \end{bmatrix}} = \overset{4 \times 2}{\begin{bmatrix} [A_{11}][B_1] + [A_{12}][B_2] \\ [A_{21}][B_1] + [A_{22}][B_2] \end{bmatrix}}$$

is permissible only when $[B]$ is partitioned so that

$$[B] = \begin{bmatrix} b_{11} & b_{12} \\ b_{21} & b_{22} \\ b_{31} & b_{32} \\ \hdashline b_{41} & b_{42} \\ b_{51} & b_{52} \end{bmatrix} = \begin{bmatrix} [B_1] \\ [B_2] \end{bmatrix}$$

where

$$[B_1] = \begin{bmatrix} b_{11} & b_{12} \\ b_{21} & b_{22} \\ b_{31} & b_{32} \end{bmatrix} \qquad [B_2] = \begin{bmatrix} b_{41} & b_{42} \\ b_{51} & b_{52} \end{bmatrix}$$

We note that in the multiplication of partitioned matrices the submatrices are treated as though they were scalar matrix elements. The submatrices are treated in the same way for the other matrix operations as well.

A.9   THE CALCULUS OF MATRICES

A.9.1   Differentiation of a Matrix

If the elements of a matrix $[A]$ are themselves functions of n independent variables, $x_1, x_2, \ldots, x_n$, the matrix $[A]$ is called a matrix function of $x_1, x_2, \ldots, x_n$. The nth-order derivative of $[A]$ exists when the nth-order derivative of each of its elements exists. We calculate the derivative of $[A]$ by differentiating each of its elements. Hence the element in the ith row and jth column of $\partial[A]/\partial x_k$ is $\partial(a_{ij})/\partial x_k$.

A.9.2   Integration of a Matrix

The integral of a function matrix $[A]$ exists only when the integral of each element of the matrix exists. To calculate $\int[A]\,dx_k$ we must perform the integration on each of its elements. Hence the element in the ith row and jth column of $\int[A]\,dx_k$ is $\int a_{ij}\,dx_k$.

A.9.3   Differentiation of a Quadratic Functional

A quadratic functional $I(x_1, \ldots, x_x)$ is defined in matrix notation as

$$I(x_1, x_2, \ldots, x_n) = \tfrac{1}{2}\lfloor x \rfloor [A]\{x\} - \lfloor x \rfloor \{b\}$$

where $x_1, \ldots, x_n$ are n independent variables, $[A]$ is a positive definite matrix, and $\{b\}$ is a column vector. Both $[A]$ and $\{b\}$ have constant elements. To make the functional stationary we must compute the n derivatives of I with respect to each independent variable x_i and set the result equal to zero. Thus we require

$$\left\{ \frac{\partial I}{\partial x} \right\} = \frac{\partial I}{\partial x_i} = 0, \qquad i = 1, 2, \ldots, n$$

To perform these differentiations it is useful to note that

$$\frac{\partial}{\partial\{x\}}\lfloor x\rfloor[A]\{x\} = 2[A]\{x\}$$

and

$$\frac{\partial}{\partial\{x\}}\lfloor x\rfloor\{b\} = \{b\}$$

which the reader may verify by direct differentiation. Using these results, we may write

$$\frac{\partial I}{\partial\{x\}} = [A]\{x\} - \{b\} = 0 \tag{A.5}$$

which is a set of simultaneous equations in the form of equations A.2. We will use this result frequently in developing element equations.

A.10   NORMS

To ascertain convergence of finite element approximations, it is necessary to define some scalar measure of *magnitude* of vectors so that the "distance" between solution vectors from successive approximations may be determined. In one dimension we know the distance between two scalars is $|a - b|$, but in multidimensional spaces where vectors may have several thousand elements it is necessary to define some measure of vector magnitude to make sense of the distance between two vectors. Such a measure of magnitude in mathematics is called a *norm*, which is symbolically represented as $\|\{X\}\|$, where $\{X\}$ is a vector of n dimensions, and the distance (or magnitude of the difference) between two vectors $\{X\}$ and $\{Y\}$ with respect to some associated norm is $\|\{X\} - \{Y\}\|$. Though there are many specific norms, they all obey four axioms:

1. $\|\{X\}\| \geq 0$, that is, a norm (magnitude) cannot be negative.
2. $\|\{X\}\| = 0$ if and only if $\{X\}^T = \lfloor 0, 0, 0, \ldots, 0\rfloor$, that is, only a zero vector can have a zero norm.
3. $\|\alpha\{X\}\| = \alpha\|\{X\}\|$, where α is a scalar.
4. $\|\{X\} + \{Y\}\| \leq \|\{X\}\| + \|\{Y\}\|$, that is, the norm of a vector sum cannot be greater than the sum of the norms of each vector.

While many specific norms have been defined, we mention four that have proven useful in finite element computations. The first of these is the l_∞-*norm* (pronounced "ell-infinity") or *max-norm*, $\|\{X\}\|_\infty$ which is defined as

$$\|\{X\}\|_\infty \equiv \max_{1 \le i \le n} |x_i| \qquad (A.6)$$

This defines the norm as the absolute value of the largest element of the vector. Probably the most common norm is the l_2-*norm* (pronounced "ell-two") or *Euclidean-norm* defined as

$$\|\{X\}\|_2 \equiv \left[\sum_{i=1}^{n} x_i^2 \right]^{1/2} \qquad (A.7)$$

which we recognize as the Pythagorean theorem, the measure we use in our three-dimensional Euclidean world.

The above two norms are often adequate for measuring vector magnitudes and hence for measuring the 'distance' between successive solution approximations to ascertain convergence for certain problems. However, for measuring convergence of approximate solutions to differential equations other norms are often used. The distance between two vectors given by the above norms may be thought of as a linear distance, whereas often an area distance between two functions might be more meaningful. This leads to norms defined on inner products such as the L_2-*norm* (pronounced "ell-two") defined as

$$\|f(x)\|_0 \equiv \left[\int_a^b |f(x)|^2 \, dx \right]^{1/2} \qquad (A.8)$$

where $f(x)$ is a real-valued continuous function on the interval $a \le x \le b$. Then $\|f(x) - g(x)\|_0$ defines the distance between two functions $f(x)$ and $g(x)$ as the net area between the two curves when plotted from $x = a$ to $x = b$. Another quite useful norm in this class is the *energy-norm* given by

$$\|f(x)\|_m \equiv \left[\int_a^b \sum_{k=0}^{m} \left| \frac{d^k f(x)}{dx^k} \right|^2 \, dx \right]^{1/2} \qquad (A.9)$$

where $f(x)$ is real-valued with continuous derivatives through order m on the interval $a \le x \le b$, and the order of the differential equation being solved is $2m$. Many other norms are possible, but these will suffice for our purposes in this text.[4]

---

[4]The theory of norms, normed spaces, convergence, and error approximation are beyond the scope of this text, but there are many references available on this subject such as *Applied Functional Analysis and Variational Methods in Engineering* by J. N. Reddy, McGraw-Hill, New York, 1986.

APPENDIX B

VARIATIONAL CALCULUS

B.1 INTRODUCTION

The calculus of variations is in a sense an extension of calculus, although both were developed almost simultaneously. Here we review some of the concepts of differential calculus and variational calculus to show their parallel natures and to establish the terminology necessary for the variational approach to finite element analysis.

B.2 CALCULUS—THE MINIMA OF A FUNCTION

B.2.1 Definitions

Suppose we have a function F which depends on n independent variables, that is,

$$F = F(x_1, x_2, \ldots, x_n)$$

and suppose also that F has a minimum value at $(x_1^*, x_2^*, \ldots, x_n^*)$. Then, by

552

definition, F has an *absolute minimum* at $(x_1^*, x_2^*, \ldots, x_n^*)$ if

$$\Delta F = F(x_1^* + \Delta x_1, \ldots, x_n^* + \Delta x_n) - F(x_1^*, \ldots, x_n^*) > 0$$

for arbitrary $\Delta x_1, \ldots, \Delta x_n$ consistent with the domain of definition for the variables $x_1, x_2, \ldots, x_n$. Also we say that F has a *local minimum* at $(x_1^*, x_2^*, \ldots, x_n^*)$ if $\Delta F > 0$ for arbitrary infinitely small changes $\Delta x_1, \ldots, \Delta x_n$ consistent again with the domain of definition. We shall confine our discussion to local extremals (minima or maxima) of functions and functionals.

Functions and functionals are said to have a *stationary value* at the points in the space of independent variables or independent functions, respectively, where the necessary conditions for local extremals are met. In the following discussion we derive these necessary conditions for a number of different cases. Since stationary values are defined in terms of only necessary conditions for local extremals and not sufficient conditions, they may be neither maxima nor minima.

B.2.2 Functions of One Variable

As an example in one dimension consider the function $F = F(x_1)$ for $x_{1l} < x_1 < x_{1u}$. If we assume that F has a local minimum at x_1^*, then

$$\Delta F(x_1^*) = F(x_1^* + \Delta x_1) - F(x_1^*) > 0$$

Because Δx_1 is infinitely small, we can expand $F(x_1^* + \Delta x_1)$ about the value x_1^* and obtain

$$F(x_1^* + \Delta x_1) = F(x_1^*) + \frac{dF}{dx}\bigg|_{x_1 = x_1^*} \Delta x_1 + \frac{1}{2}\frac{d^2F}{dx_1^2}\bigg|_{x_1 = x_1^*} \Delta x_1^2 + \cdots$$

Thus

$$\Delta F(x_1^*) = \frac{dF}{dx_1}\bigg|_{x_1 = x_1^*} \Delta x_1 + \frac{d^2F}{dx_1^2}\bigg|_{x_1 = x_1^*} \frac{\Delta x_1^2}{2} + \cdots$$

Now it is obvious that to have $\Delta F(x_1^*) > 0$ we must have

$$\frac{dF}{dx_1}\bigg|_{x_1 = x_1^*} = 0 \tag{B.1}$$

Equation B.1 is the necessary condition for $F(x_1)$ to have a local minimum at x_1^*.

B.2.3   Functions of Two or More Variables

Now suppose that F is a function of two independent variables, that is, if F has a local minimum at (x_1^*, x_2^*) and $x_{1l} < x_1 < x_{1u}, x_{2l} < x_2 < x_{2u}$, then we must have

$$\Delta F = F(x_1^* + \Delta x_1, x_2^* + \Delta x_2) - F(x_1^*, x_2^*) > 0$$

Expanding $F(x_1^* + \Delta x_1, x_2^* + \Delta x_2)$ about x_1^*, x_2^*, we have, as before,

$$\Delta F = \frac{\partial F}{\partial x_1}\bigg|_{(x_1^*, x_2^*)} \Delta x_1 + \frac{\partial F}{\partial x_2}\bigg|_{(x_1^*, x_2^*)} \Delta x_2$$

$$+ \frac{1}{2}\left(\frac{\partial^2 F}{\partial x_1^2}\Delta x_1^2 + \frac{2\partial^2 F}{\partial x_1 \partial x_2}\Delta x_1 \Delta x_2 + \frac{\partial^2 F}{\partial x_2}\Delta x_2^2\right)_{x_1 = x_1^*, x_2 = x_2^*} + \cdots$$

A necessary condition for $\Delta F > 0$ is that at (x_1^*, x_2^*)

$$\frac{\partial F}{\partial x_1} = 0 \qquad \frac{\partial F}{\partial x_2} = 0 \tag{B.2}$$

Following the same reasoning, we can show that for a function of n independent variables, $F(x_1, x_2, \ldots, x_n)$, a necessary condition for a local minimum is that each of its first partials must vanish at $(x_1^*, x_2^*, \ldots, x_n^*)$:

$$\frac{\partial F}{\partial x_1}, \frac{\partial F}{\partial x_2}, \ldots, \frac{\partial F}{\partial x_n} = 0 \tag{B.3}$$

Equations B.1, B.2, and B.3 represent only necessary conditions for a minimum. Sufficient conditions involve inequalities among the second partial derivatives and are beyond the scope of this discussion.

B.3   VARIATIONAL CALCULUS—THE MINIMA OF FUNCTIONALS

B.3.1   Definitions

In contrast to the calculus, variational calculus is concerned primarily with the theory of maxima and minima, but the functions to be extremized are in general functions of functions, or *functionals*. A simple functional in terms of one independent variable would have the typical form

$$I(\phi) = \int_{x_1}^{x_2} F(x, \phi, \phi_x, \phi_{xx}) \, dx \tag{B.4}$$

where $\phi = \phi(x)$ and $\phi_x = d\phi/dx$, $\phi_{xx} = d^2\phi/dx^2$. Of course, more complicated functionals involving higher-order derivatives and more dependent variables are possible. The problem of variational calculus is to choose a function ϕ, called an extremal, so as to minimize or maximize (extremize) functionals like equation B.4. In other words, variational calculus is concerned with finding functions that extremize integrals whose integrands contain these functions.

Variational calculus has always been associated with realistic problems of continuum mechanics. Often the functionals whose extreme values are sought are expressions of some form of system energy. For example, many of the phenomena governing the elastic distortion of bodies can be deduced from the principle of minimum potential energy.

B.3.2   Functionals of One Variable

In the following we derive a necessary condition to be satisfied by $\phi(x)$ to give, say, a minimum[5] value to $I(\phi)$, equation B.4. Often values of ϕ at the end points of the interval $x_1 \le x \le x_2$ are given

$$\phi(x_1) = \phi_1 \qquad \phi(x_2) = \phi_2 \tag{B.5}$$

but, as we shall see, other end conditions or boundary conditions are permissible. We seek a function $\phi(x)$ that satisfies equation B.5 and minimizes the functional of equation B.4. Suppose that we postulate the existence of a comparison function $\tilde{\phi}(x, \epsilon)$ which differs slightly from the extremizing function $\phi(x)$ and depends on a small parameter ϵ. We require that

$$\lim_{\epsilon \to 0} \tilde{\phi}(x, \epsilon) = \phi(x)$$

The comparison function can approach the extremizing function in a number of ways. Two particular classes of comparison functions are known as weak and strong variations. These are classified as follows:

$$\text{Weak variation:} \quad \lim_{\epsilon \to 0} \frac{d\tilde{\phi}(x, \epsilon)}{dx} = \frac{d\phi(x)}{dx}$$

$$\text{Strong variation:} \quad \lim_{\epsilon \to 0} \frac{d\tilde{\phi}(x, \epsilon)}{dx} \ne \frac{d\phi(x)}{dx}$$

---

[5]For the purpose of discussion we restrict our attention to finding minima of a functional. The same arguments follow if conditions for maxima are sought, and the results are the same.

We shall consider only comparison functions characterized by weak variations. Since ϵ is a small parameter, we may expand $\tilde{\phi}(x, \epsilon)$ about ϵ to obtain

$$\tilde{\phi}(x, \epsilon) = \phi(x) + \left.\frac{\partial \tilde{\phi}}{\partial \epsilon}\right|_{\epsilon=0} \epsilon + (\text{terms of order } \epsilon^2)$$

We define the *first variation* in $\tilde{\phi}$ with respect to ϵ as

$$\delta \tilde{\phi} = \left.\frac{\partial \tilde{\phi}}{\partial \epsilon}\right|_{\epsilon=0} \epsilon$$

Let us consider, in the interval $x_1 \le x \le x_2$, a specific type of comparison function of the form

$$\tilde{\phi}(x, \epsilon) = \phi(x) + \epsilon \eta(x) \tag{B.6}$$

The function $\eta(x)$ is any continuously differentiable function with the property $\eta(x_1) = \eta(x_2) = 0$. The difference between the comparison function $\tilde{\phi}$ and the minimizing function ϕ is defined as the first variation in ϕ. Hence

$$\delta \phi \equiv \epsilon \eta(x)$$

The value of the integral, equation B.4, along the curve $\tilde{\phi}(x, \epsilon)$ now becomes a function of ϵ, that is,

$$I(\epsilon) = \int_{x_1}^{x_2} F\left(x, \tilde{\phi}, \tilde{\phi}_x, \tilde{\phi}_{xx}\right) dx$$

$$= \int_{x_1}^{x_2} F(x, \phi + \epsilon\eta, \phi_x + \epsilon\eta_x, \phi_{xx} + \epsilon\eta_{xx}) \, dx \tag{B.7}$$

Now, since $I(0)$ is the minimum value of I, we can be sure that ΔI, the change in I due to the change in the function ϕ, will be such that

$$\Delta I = I(\epsilon) - I(0) \ge 0 \tag{B.8}$$

Since ϵ is infinitely small, we may expand $I(\epsilon)$ about $\epsilon = 0$ to obtain

$$I(\epsilon) = I(0) + \left.\frac{dI}{d\epsilon}\right|_{\epsilon=0} \epsilon + \frac{1}{2} \left.\frac{d^2I}{d\epsilon^2}\right|_{\epsilon=0} \epsilon^2 + \cdots \tag{B.9}$$

To satisfy equation B.8 in view of equation B.9, it is necessary that

$$\left.\frac{dI}{d\epsilon}\right|_{\epsilon=0} = 0$$

since ϵ can be positive or negative. Thus $\delta I = 0$; that is, the first variation of I must vanish. From our previous definition we see that $\delta I = 0$ implies that I has a stationary value. The analogy between finding the minimum of a function via ordinary calculus and finding the minimum of a functional via variational calculus is now obvious. From equation B.7 we have

$$\frac{dI}{d\epsilon}\bigg|_{\epsilon=0} = \int_{x_1}^{x_2} \left(\frac{\partial F}{\partial \tilde{\phi}} \frac{\partial \tilde{\phi}}{\partial \epsilon} + \frac{\partial F}{\partial \tilde{\phi}_x} \frac{\partial \tilde{\phi}_x}{\partial \epsilon} + \frac{\partial F}{\partial \tilde{\phi}_{xx}} \frac{\partial \tilde{\phi}_{xx}}{\partial \epsilon} \right) dx = 0$$

$$= \int_{x_1}^{x_2} \left(\frac{\partial F}{\partial \phi} \eta + \frac{\partial F}{\partial \phi_x} \eta_x + \frac{\partial F}{\partial \phi_{xx}} \eta_{xx} \right) dx = 0 \qquad \text{(B.10)}$$

Frequently the δ notation is used, and multiplying equation B.10 by ϵ gives

$$\delta I = \int_{x_1}^{x_2} \left(\frac{\partial F}{\partial \phi} \delta\phi + \frac{\partial F}{\partial \phi_x} \delta\phi_x + \frac{\partial F}{\partial \phi_{xx}} \delta\phi_{xx} \right) dx = 0 \qquad \text{(B.11)}$$

This follows from the definition of the first variation of a functional. Before we proceed we shall establish several definitions. The change in F due to a change in the function ϕ is, as before,

$$\Delta F = F(x, \phi + \epsilon\eta, \phi_x + \epsilon\eta_x, \phi_{xx} + \epsilon\eta_{xx}) - F(x, \phi, \phi_x, \phi_{xx})$$

Expanding the right-hand side in powers of ϵ, we find

$$\Delta F = \frac{\partial F}{\partial \phi} \epsilon\eta + \frac{\partial F}{\partial \phi_x} \epsilon\eta_x + \frac{\partial F}{\partial \phi_{xx}} \epsilon\eta_{xx} + \text{(higher powers of } \epsilon)$$

By definition,

$$\delta F \equiv \frac{\partial F}{\partial \phi} \epsilon\eta + \frac{\partial F}{\partial \phi_x} \epsilon\eta_x + \frac{\partial F}{\partial \phi_{xx}} \epsilon\eta_{xx}$$

The analogy between the first variation of a functional and the first differential of a function now becomes obvious. In fact, all the laws of differentiation of sums, products, quotients, powers, and so forth carry over directly to become laws of the first variation. For example,

$$\delta(F_1 + F_2) = \delta F_1 + \delta F_2$$
$$\delta(F_1 F_2) = F_1 \delta F_2 + F_2 \delta F_1$$
$$\delta\left(\frac{F_1}{F_2}\right) = \frac{F_2 \delta F_1 - F_1 \delta F_2}{F_2^2}$$
$$\delta F_1^n = nF_1^{n-1} \delta F_1, \qquad \text{etc.}$$

The first variation is also a commutative operator with both differentiation and integration if the integration limits are not to be varied. Thus

$$\delta \int F \, dx = \int \delta F \, dx \qquad \delta \frac{d\phi}{dx} = \frac{d}{dx} \delta \phi$$

Obviously the first order variational operator, δ, like the differential operator is a *linear operator*, that is,

$$\delta(\alpha F_1 + \beta F_2) = \alpha \, \delta F_1 + \beta \, \delta F_2$$

Now we return our attention to equation B.11. The necessary conditions for $\phi(x)$ to be a minimizing function can be discovered if we integrate the second and third terms by parts. The result is

$$\delta I = \int_{x_1}^{x_2} \left[\frac{\partial F}{\partial \phi} - \frac{d}{dx} \left(\frac{\partial F}{\partial \phi_x} \right) + \frac{d^2}{dx^2} \left(\frac{\partial F}{\partial \phi_{xx}} \right) \right] \delta \phi \, dx$$

$$+ \left[\frac{\partial F}{\partial \phi_x} - \frac{d}{dx} \left(\frac{\partial F}{\partial \phi_{xx}} \right) \right] \delta \phi \Bigg|_{x_1}^{x_2} + \left[\left(\frac{\partial F}{\partial \phi_{xx}} \right) \delta \phi_x \right]_{x_1}^{x_2} = 0 \quad \text{(B.12)}$$

Because $\delta \phi$ and $\delta \phi_x$ are arbitrary admissible variations, equation B.12 can be satisfied only if the integrand and the remaining terms vanish. Thus the necessary conditions for $\phi(x)$ to minimize $I(\phi)$ are as follows:

$$\frac{\partial F}{\partial \phi} - \frac{d}{dx} \frac{\partial F}{\partial \phi_x} + \frac{d^2}{dx^2} \left(\frac{\partial F}{\partial \phi_{xx}} \right) = 0 \qquad \text{(B.13a)}$$

$$\left[\frac{\partial F}{\partial \phi_x} - \frac{d}{dx} \left(\frac{\partial F}{\partial \phi_{xx}} \right) \right] \delta \phi \Bigg|_{x_1}^{x_2} = 0 \qquad \text{(B.13b)}$$

$$\left[\left(\frac{\partial F}{\partial \phi_{xx}} \right) \delta \phi_x \right]_{x_1}^{x_2} = 0 \qquad \text{(B.13c)}$$

Equation B.13a is called the *Euler–Lagrange equation* and is the differential equation that $\phi(x)$ must satisfy. Equations B.13b and B.13c are satisfied by the variations of ϕ and ϕ_x which are zero on the boundaries, since the customary definition of η and its derivatives is

$$\eta(x_1) = \eta(x_2) = \eta_x(x_1) = \eta_x(x_x) = \cdots = \eta_{x^{n-1}}(x_1) = \eta_{x^{n-1}}(x_2) = 0$$

$$\text{(B.14)}$$

where n is the order of the functional. Although this derivation of the second-order Euler-Lagrange equation places no constraints on the boundary terms it is worth noting that the essential boundary conditions always appear as variations of ϕ as in equations B.13b and B.13c, in other words

$$\phi\Big|_{x_1}^{x_2} \qquad \text{and} \qquad \frac{\partial \phi}{\partial x}\Big|_{x_1}^{x_2}$$

are the essential boundary conditions. The other terms in equations B.13b and B.13c, namely

$$\left[\frac{\partial F}{\partial \phi_x} - \frac{d}{dx}\left(\frac{\partial F}{\partial \phi_{xx}}\right)\right]_{x_1}^{x_2} \qquad \text{and} \qquad \frac{\partial F}{\partial \phi_{xx}}\Big|_{x_1}^{x_2}$$

are the natural boundary conditions. There are other approaches to deriving the Euler–Lagrange equation using variational calculus that differ slightly from that presented here, and the interested reader encouraged to consult sources such references 1–5 for further details.

B.3.3 More General Functionals

Proceeding in a similar manner, we may derive Euler–Lagrange equations and boundary conditions for other functionals. For example, if we have extremal functions of two independent variables and derivatives up to and including second order, the functional has the form

$$I(\phi) = \iint_{\Omega} F(x, y, \phi, \phi_x, \phi_y, \phi_{xx}, \phi_{xy}, \phi_{yy})\, dx\, dy \qquad \text{(B.15)}$$

and the corresponding Euler–Lagrange equation is

$$\frac{\partial^2}{\partial x^2}\left(\frac{\partial F}{\partial \phi_{xx}}\right) + \frac{\partial^2}{\partial x\, \partial y}\left(\frac{\partial F}{\partial \phi_{xy}}\right) + \frac{\partial^2}{\partial y^2}\left(\frac{\partial F}{\partial \phi_{yy}}\right)$$

$$-\frac{\partial}{\partial x}\left(\frac{\partial F}{\partial \phi_x}\right) - \frac{\partial}{\partial y}\left(\frac{\partial F}{\partial \phi_y}\right) + \frac{\partial F}{\partial \phi} = 0 \qquad \text{(B.16)}$$

In terms of one independent variable and nth-order derivatives, the functional has the form

$$I(\phi) = \int_{x_1}^{x_2} F(x, \phi, \phi_x, \phi_{xx}, \dots, \phi_{x^n}, \dots)\, dx \qquad \text{(B.17)}$$

and the corresponding Euler–Lagrange equation is

$$\frac{\partial F}{\partial \phi} - \frac{d}{dx}\left(\frac{\partial F}{\partial \phi_x}\right) + \frac{d^2}{dx^2}\left(\frac{\partial F}{\partial \phi_{xx}}\right) - \cdots (-1)^n \frac{d^n}{dx^n}\left(\frac{\partial F}{\partial \phi_{x^n}}\right) = 0 \quad \text{(B.18)}$$

It is important to emphasize that the solution of an Euler–Lagrange equation may not yield a function ϕ that minimizes a given functional, because the Euler–Lagrange equation expresses only a necessary and not a sufficient condition for a minimum. In the general case we must verify whether the ϕ found by solving the Euler–Lagrange equation yields a minimum, and this involves computation of second variations. However, in many situations we can use some physical considerations to tell whether the ϕ so obtained actually minimizes the functional.

We have considered functionals of only one function and several independent variables, but this procedure may be used to derive necessary conditions for the minima of functionals of more than one function. For such a case several Euler–Lagrange equations would result—one for each independent function in the functional.

REFERENCES

1. H. L. Langhaar, *Energy Methods in Applied Mechanics*, John Wiley, New York, 1962.
2. I. H. Shames and C. L. Dim, *Energy and Finite Element Methods in Structural Mechanics*, Hemisphere, New York, 1985.
3. T. Mura and T. Koya, *Variational Methods in Mechanics*, Oxford, New York, 1992.
4. J. N. Reddy, *Applied Functional Analysis and Variational Methods in Engineering*, McGraw-Hill, New York, 1986.
5. R. Courant and D. Hilbert, *Methods of Mathematical Physics, Vol. 1*, John Wiley, 1953.

APPENDIX C

BASIC EQUATIONS FROM LINEAR ELASTICITY THEORY

C.1 INTRODUCTION

Here we summarize some of the concepts and basic equations of linear elasticity theory. Except for the general definitions of stress and strain, the equations given without derivation or proof apply to homogeneous isotropic materials unless otherwise noted. Because most finite element analyses are carried out in a Cartesian coordinate system, only this system is used in the following. Other special coordinate systems are sometimes employed in particular applications, and these are examined in the text where they appear.

This summary is intended only as a quick reference when the theory of finite element analysis is applied to solid mechanics problems in Part II of

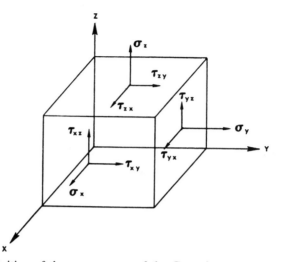

Figure C.1   Definition of the components of the Cartesian stress tensor.

this book. Hence the reader who questions the origin or limitations of a particular equation is encouraged to consult one of the many comprehensive texts on elasticity theory, for example, references 1–4.

C.2   STRESS COMPONENTS

A state of stress exists in a body acted upon by external forces. If these external forces act over some portion of the surface of the body, they are called *surface forces* or *surface tractions*; if they are distributed throughout the volume of the body, they are called *body forces*. Hydrostatic pressure over an area is an example of a surface traction, while gravitational force is an example of a body force.

Figure C.1 shows the notation used to define the state of stress in a three-dimensional body. The components of normal stresses are denoted as $\sigma_x, \sigma_y, \sigma_z$; the components of shear stress are denoted $\tau_{xy}, \tau_{yz}, \tau_{zx}$. Normal stress carries a single subscript to indicate that the stress acts on a plane normal to the axis in the subscript direction. The first letter of the double subscript on shear stress indicates that the plane on which the stress acts is normal to the axis in the subscript direction; the second letter designates the coordinate direction in which the stress acts.

We have designated only three components of shear stress because only three are independent. By considering the equilibrium of an element volume and neglecting the vanishing body force as the volume diminishes in size, we can show that $\tau_{xy} = \tau_{yx}, \tau_{zx} = \tau_{xz}, \tau_{yz} = \tau_{zy}$. Hence the six quantities $\sigma_x, \sigma_y, \sigma_z, \tau_{xy}, \tau_{xz}$, and τ_{yz} completely describe the state of stress at a point.

C.3   STRAIN COMPONENTS

When an elastic body is subjected to a state of stress, we shall assume that the particles of the body move only a small amount and the body in its deformed state remains perfectly elastic, so that its deformation disappears when the stress is removed. If the displacement field in the deformed body is represented by the three components u, v, and w parallel to the coordinate axes x, y, and z, respectively, the *strain* at a point in the body may be expressed as

$$\epsilon_x = \frac{\partial u}{\partial x} \qquad \epsilon_y = \frac{\partial v}{\partial y} \qquad \epsilon_z = \frac{\partial w}{\partial z}$$

$$\gamma_{xy} = \frac{\partial u}{\partial y} + \frac{\partial v}{\partial x} \qquad \gamma_{xz} = \frac{\partial u}{\partial z} + \frac{\partial w}{\partial x} \qquad \gamma_{yz} = \frac{\partial v}{\partial z} + \frac{\partial w}{\partial y} \qquad (C.1)$$

The strain ϵ_x, for example, is defined as the unit elongation of the body at a point, the shearing strain γ_{xy} as the distortion of the angle between the x–z and y–z planes, and so forth.

The relationships between strain and displacement expressed by equation C.1 are approximate but are consistent with the assumption of small deformations usually encountered in engineering situations. More exact strain–displacement relations contain higher-order terms [2].

A body acted upon by external forces may, in general, experience deformation characterized by strain, and it may translate and rotate. The small rotation of an elemental volume of the body can be expressed in terms of its displacements in the following manner:

$$\omega_x = \omega_{zy} = \frac{1}{2}\left(\frac{\partial w}{\partial y} - \frac{\partial v}{\partial z}\right)$$

$$\omega_y = \omega_{xz} = \frac{1}{2}\left(\frac{\partial u}{\partial z} - \frac{\partial w}{\partial x}\right)$$

$$\omega_z = \omega_{yx} = \frac{1}{2}\left(\frac{\partial v}{\partial x} - \frac{\partial u}{\partial y}\right) \qquad (C.2)$$

where ω_x is the small rotation about the x axis, and so forth. Sometimes it is convenient to define a quantity called the *dilatation* at a point in the body. This is related to strains and displacements as follows:

$$e = \epsilon_x + \epsilon_y + \epsilon_z = \frac{\partial u}{\partial x} + \frac{\partial v}{\partial y} + \frac{\partial w}{\partial z} \tag{C.3}$$

C.4 GENERALIZED HOOKE'S LAW (CONSTITUTIVE EQUATIONS)

The six components of stress are related to the six components of strain through a proportionality matrix $[C]$ containing 36 terms for a general anisotropic material:

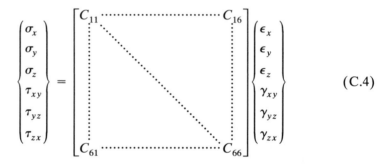

$$\begin{Bmatrix} \sigma_x \\ \sigma_y \\ \sigma_z \\ \tau_{xy} \\ \tau_{yz} \\ \tau_{zx} \end{Bmatrix} = \begin{bmatrix} C_{11} & \cdots & \cdots & C_{16} \\ & & & \\ & & & \\ & & & \\ & & & \\ C_{61} & \cdots & \cdots & C_{66} \end{bmatrix} \begin{Bmatrix} \epsilon_x \\ \epsilon_y \\ \epsilon_z \\ \gamma_{xy} \\ \gamma_{yz} \\ \gamma_{zx} \end{Bmatrix} \tag{C.4}$$

Because the matrix $[C]$ is symmetric, only 21 different coefficients can be identified. Using matrix notation, we may rewrite equation C.4 as

$$\{\sigma\} = [C]\{\epsilon\}$$

and by inversion the strains may be expressed as

$$\{\epsilon\} = [C]^{-1}\{\sigma\} = [D]\{\sigma\} \tag{C.5}$$

The matrix $[C]$ is the *material matrix* or *modulus matrix*, while its inverse $[D]$ is called the *material flexibility matrix* or *compliance matrix*. These relations between stress and strain take on less complex forms if we assume that the elastic body is *homogeneous* and *isotropic*. By homogeneous we mean that any elemental volume of the body possesses the same specific physical properties as any other elemental volume of the body; by isotropic we mean that the physical properties are the same in all directions. For homogeneous isotropic materials only two physical constants are required to express all the coefficients in Hooke's law; these are Young's modulus, E, and Poisson's ratio, ν. In terms of these constants the matrices in Hooke's law are as follows:

$$[C] = \frac{E}{(1 + \nu)(1 - 2\nu)}$$

$$\times \begin{bmatrix} 1 - \nu & \nu & \nu & 0 & 0 & 0 \\ \nu & 1 - \nu & \nu & 0 & 0 & 0 \\ \nu & \nu & 1 - \nu & 0 & 0 & 0 \\ 0 & 0 & 0 & \dfrac{1 - 2\nu}{2} & 0 & 0 \\ 0 & 0 & 0 & 0 & \dfrac{1 - 2\nu}{2} & 0 \\ 0 & 0 & 0 & 0 & 0 & \dfrac{1 - 2\nu}{2} \end{bmatrix} \quad (C.6)$$

$$[D] = [C]^{-1} = \frac{1}{E} \begin{bmatrix} 1 & -\nu & -\nu & 0 & 0 & 0 \\ -\nu & 1 & -\nu & 0 & 0 & 0 \\ -\nu & -\nu & 1 & 0 & 0 & 0 \\ 0 & 0 & 0 & 2(1 + \nu) & 0 & 0 \\ 0 & 0 & 0 & 0 & 2(1 + \nu) & 0 \\ 0 & 0 & 0 & 0 & 0 & 2(1 + \nu) \end{bmatrix}$$

$$(C.7)$$

Sometimes these matrices are given in terms of Lamé's constants, λ and μ, which are related to E and ν as follows:

$$E = \frac{\mu(3\lambda + 2\mu)}{\lambda + \mu} \quad (C.8a)$$

$$\nu = \frac{\lambda}{2(\lambda + \mu)} \quad (C.8b)$$

or

$$\lambda = \frac{E\nu}{(1 + \nu)(1 - 2\nu)} \quad (C.9a)$$

$$\mu = \frac{E}{2(1 + \nu)} \quad (C.9b)$$

In addition to these constants, two other constants characterizing a material's behavior under pure shear and pure volumetric distortion often appear in the engineering literature. The constant G, known as the *modulus of rigidity* or *shear modulus*, relates shearing stress and shearing strain and is defined in terms of E and ν as

$$G = \frac{E}{2(1 + \nu)} = \mu \qquad \text{(C.10a)}$$

The *modulus of volume expansion* or *bulk modulus*, K, is defined as

$$K = \frac{E}{3(1 - 2\nu)} \qquad \text{(C.10b)}$$

The constant K relates unit volume expansion (or contraction) to volumetric stresses such as hydrostatic pressure.

In general, the 21 individual coefficients in equation C.4 or any subset of these for special cases such as homogeneous isotropic solids must be determined by carefully controlled experiment. For general anisotropic materials, the constitutive equations may be found in reference 5.

When a body deforms, certain constraint conditions must be satisfied. These conditions, known as the equilibrium and compatibility conditions, impose force and geometric constraints, respectively, on the deformations. We now discuss the forms of these conditions.

C.5   STATIC EQUILIBRIUM EQUATIONS

The equilibrium of an elastic body in a state of stress is governed by three partial differential equations for the stress components. These equations may be derived by writing force balances for an elemental volume of the material acted upon by body forces X, Y, and Z. Forces acting on the element are calculated by assuming that the sides of the element have infinitesimal area and by multiplying the stresses at the centroid of the element by the area of the sides. Summing forces in the three coordinate directions gives

$$\frac{\partial \sigma_x}{\partial x} + \frac{\partial \tau_{xy}}{\partial y} + \frac{\partial \tau_{xz}}{\partial z} + X = 0$$

$$\frac{\partial \tau_{xy}}{\partial x} + \frac{\partial \sigma_y}{\partial y} + \frac{\partial \tau_{yz}}{\partial z} + Y = 0 \qquad \text{(C.11)}$$

$$\frac{\partial \tau_{xz}}{\partial x} + \frac{\partial \tau_{yz}}{\partial y} + \frac{\partial \sigma_z}{\partial z} + Z = 0$$

A balance of moments about the three coordinate directions shows that $\tau_{xy} = \tau_{yx}$, and so on. Boundary conditions for the equilibrium equations,

equations C.11, may be found by considering the external surface forces acting on the body. If the components of these unit surface forces are denoted as T_x, T_y, and T_z, we find that at any point *on the surface* of the body we have

$$T_x = \sigma_x l + \tau_{xy} m + \tau_{xz} n$$

$$T_y = \tau_{xy} l + \sigma_y m + \tau_{yz} n \qquad (C.12)$$

$$T_z = \tau_{xz} l + \tau_{yz} m + \sigma_z n$$

where l, m, and n are the direction cosines of the outward normal to the surface at the point of interest.

Given a set of boundary conditions in the form of equations C.12, it is impossible to obtain a unique solution to these equations for the six stress components because we have six unknown and only three equations. This apparent difficulty can be remedied by taking into account the displacement field u, v, w and its relation to strain in the body.

C.6 COMPATIBILITY CONDITIONS

The definition of strain in terms of derivatives of displacement (given in equation C.1) leads to other relations between strain and displacement that must be satisfied. These relations, known as *compatibility conditions*, can be derived by differentiation of equations C.1. For example, from the first of these equations we have

$$\frac{\partial^2 \epsilon_x}{\partial y^2} = \frac{\partial^3 u}{\partial x \, \partial y^2} \qquad \frac{\partial^2 \epsilon_y}{\partial x^2} = \frac{\partial^3 v}{\partial x^2 \, \partial y}$$

$$\frac{\partial^2 \gamma_{xy}}{\partial x \, \partial y} = \frac{\partial^3 u}{\partial x \, \partial y^2} + \frac{\partial^3 v}{\partial x^2 \, \partial y}$$

From these equations it is obvious that

$$\frac{\partial^2 \epsilon_x}{\partial y^2} + \frac{\partial^2 \epsilon_y}{\partial x^2} = \frac{\partial^2 \gamma_{xy}}{\partial x \, \partial y}$$

Continuing in this manner, we arrive at the following six differential relations, the compatibility conditions between strain components:

$$
\left.\begin{array}{l}
2\dfrac{\partial^2 \epsilon_x}{\partial y\,\partial z} = \dfrac{\partial}{\partial x}\left(-\dfrac{\partial \gamma_{yz}}{\partial x} + \dfrac{\partial \gamma_{zx}}{\partial y} + \dfrac{\partial \gamma_{xy}}{\partial z}\right) \\[3mm]
2\dfrac{\partial^2 \epsilon_y}{\partial z\,\partial x} = \dfrac{\partial}{\partial y}\left(-\dfrac{\partial \gamma_{zx}}{\partial y} + \dfrac{\partial \gamma_{yx}}{\partial z} + \dfrac{\partial \gamma_{yz}}{\partial x}\right) \\[3mm]
2\dfrac{\partial^2 \epsilon_z}{\partial x\,\partial y} = \dfrac{\partial}{\partial z}\left(-\dfrac{\partial \gamma_{xy}}{\partial z} + \dfrac{\partial \gamma_{yz}}{\partial x} + \dfrac{\partial \gamma_{zx}}{\partial y}\right) \\[3mm]
\dfrac{\partial^2 \gamma_{xy}}{\partial x\,\partial y} = \dfrac{\partial^2 \epsilon_x}{\partial y^2} + \dfrac{\partial^2 \epsilon_y}{\partial x^2} \\[3mm]
\dfrac{\partial^2 \gamma_{yz}}{\partial y\,\partial z} = \dfrac{\partial^2 \epsilon_y}{\partial z^2} + \dfrac{\partial^2 \epsilon_z}{\partial y^2} \\[3mm]
\dfrac{\partial^2 \gamma_{zx}}{\partial z\,\partial x} = \dfrac{\partial^2 \epsilon_z}{\partial x^2} + \dfrac{\partial^2 \epsilon_x}{\partial z^2}
\end{array}\right\} \qquad \text{(C.13)}
$$

The compatibility conditions may also be expressed in terms of stress components if we use Hooke's law. The results, known as the Beltrami–Michell equations, are as follows:

$$
\left.\begin{array}{l}
\nabla^2 \sigma_x + \dfrac{1}{1+\nu}\dfrac{\partial^2 I_1}{\partial x^2} = -\dfrac{\nu}{1-\nu}\nabla\cdot\mathbf{F} - 2\dfrac{\partial X}{\partial x} \\[3mm]
\nabla^2 \sigma_y + \dfrac{1}{1+\nu}\dfrac{\partial^2 I_1}{\partial y^2} = -\dfrac{\nu}{1-\nu}\nabla\cdot\mathbf{F} - 2\dfrac{\partial Y}{\partial y} \\[3mm]
\nabla^2 \sigma_z + \dfrac{1}{1+\nu}\dfrac{\partial^2 I_1}{\partial z^2} = -\dfrac{\nu}{1-\nu}\nabla\cdot\mathbf{F} - 2\dfrac{\partial Z}{\partial z} \\[3mm]
\nabla^2 \tau_{yz} + \dfrac{1}{1+\nu}\dfrac{\partial^2 I_1}{\partial y\,\partial z} = -\left(\dfrac{\partial Y}{\partial z} + \dfrac{\partial Z}{\partial y}\right) \\[3mm]
\nabla^2 \tau_{zx} + \dfrac{1}{1+\nu}\dfrac{\partial^2 I_1}{\partial z\,\partial x} = -\left(\dfrac{\partial Z}{\partial x} + \dfrac{\partial X}{\partial z}\right) \\[3mm]
\nabla^2 \tau_{xy} + \dfrac{1}{1+\nu}\dfrac{\partial^2 I_1}{\partial x\,\partial y} = -\left(\dfrac{\partial X}{\partial y} + \dfrac{\partial Y}{\partial x}\right)
\end{array}\right\} \qquad \text{(C.14)}
$$

where

$$
\nabla = \frac{\partial}{\partial x}\hat{i} + \frac{\partial}{\partial y}\hat{j} + \frac{\partial}{\partial z}\hat{k}
$$

and $\nabla^2 = \nabla\cdot\nabla$ is the Laplacian, $\nabla^2 = \partial^2/\partial x^2 + \partial^2/\partial y^2 + \partial^2/\partial z^2$; I_1 is the first stress invariant $I_1 = \sigma_x + \sigma_y + \sigma_z$; and $\mathbf{F} = X\hat{i} + Y\hat{j} + Z\hat{k}$. Equations

C.13 or C.14 are generally sufficient for determining the components of strain or stress, Respectively.

C.7 DIFFERENTIAL EQUATIONS FOR DISPLACEMENTS

If we return to the static equilibrium equations for stress, equations C.11, and their boundary conditions, equation C.13, we may derive a set of three differential equations for the three components of displacement. The derivation involves the use of Hooke's law to replace the stress components with strain components; then the strain components must be expressed in terms of derivatives of displacements via the definitions of equations C.1. The result is

$$\nabla^2 u + \frac{1}{1-2\nu}\frac{\partial}{\partial x}\left(\frac{\partial u}{\partial x} + \frac{\partial v}{\partial y} + \frac{\partial w}{\partial z}\right) + \frac{X}{\mu} = 0$$

$$\nabla^2 v + \frac{1}{1-2\nu}\frac{\partial}{\partial y}\left(\frac{\partial u}{\partial x} + \frac{\partial v}{\partial y} + \frac{\partial w}{\partial z}\right) + \frac{Y}{\mu} = 0 \qquad (C.15)$$

$$\nabla^2 w + \frac{1}{1-2\nu}\frac{\partial}{\partial z}\left(\frac{\partial u}{\partial x} + \frac{\partial v}{\partial y} + \frac{\partial w}{\partial z}\right) + \frac{Z}{\mu} = 0$$

On the surface the boundary conditions take the form

$$T_x = \lambda e\, l + \mu\left[2\frac{\partial u}{\partial x}l + \left(\frac{\partial u}{\partial y} + \frac{\partial v}{\partial x}\right)m + \left(\frac{\partial u}{\partial z} + \frac{\partial w}{\partial x}\right)n\right]$$

$$T_y = \lambda e\, m + \mu\left[2\frac{\partial v}{\partial y}m + \left(\frac{\partial v}{\partial z} + \frac{\partial w}{\partial y}\right)n + \left(\frac{\partial v}{\partial x} + \frac{\partial u}{\partial y}\right)l\right] \qquad (C.16)$$

$$T_z = \lambda e\, n + \mu\left[2\frac{\partial w}{\partial z}n + \left(\frac{\partial w}{\partial x} + \frac{\partial u}{\partial z}\right)l + \left(\frac{\partial w}{\partial y} + \frac{\partial v}{\partial z}\right)m\right]$$

where λ and μ are the previously defined Lamé constants (equation C.9), and e is the dilitational strain (equation C.3).

C.8 MINIMUM POTENTIAL ENERGY PRINCIPLE

In the preceding sections of this appendix we focused our attention on the differential equilibrium equations, the strain–displacement equations, the compatibility equations, and the constitutive equations as they apply to a point in a continuous material body. We now shift our attention to integral statements that hold throughout the entire system. The integral relations are the variational principles that apply to solid mechanics problems. We discuss only one such principle here, the *minimum potential energy principle*, but the interested reader may find an extensive treatment of several of these principles in Washizu [6].

Consider an elastic body of a given shape, deformed by the action of body forces and surface tractions. The potential energy of such a body is defined as the energy of deformation of the body (the strain energy) minus the work done on the body by the external forces. The theorem of minimum potential energy may be stated as follows [6]:

> The displacement (u, v, w) which satisfies the differential equations of equilibrium, as well as the conditions at the bounding surface, yields a smaller value for the potential energy than any other displacement which satisfies the same conditions at the bounding surface.

Hence, if $\Pi(u, v, w)$ is the potential energy, $U_P(u, v, w)$ is the strain energy, and $V_P(u, v, w)$ is the work done by the applied loads during displacement changes, then, according to the minimum principle, we have at equilibrium

$$\delta\Pi(u, v, w) = \delta[U_P(u, v, w) - V_P(u, v, w)]$$
$$= \delta U_P(u, v, w) - \delta V_P(u, v, w) = 0 \qquad (C.17)$$

In equation C.17 we tacitly assume that the variation is taken with respect to the displacements while all other parameters are held fixed.

The strain energy of a linear elastic body is defined as

$$U_P(u, v, w) = \frac{1}{2} \iiint_\Omega \lfloor \epsilon \rfloor \{\sigma\} \, dV$$

and with Hooke's law

$$U_P(u, v, w) = \frac{1}{2} \iiint_\Omega \lfloor \epsilon \rfloor [C] \{\epsilon\} \, dV \qquad (C.18)$$

where Ω is the volume of the body. From equation (C.1)

$$\{\epsilon\} = \begin{Bmatrix} \epsilon_x \\ \epsilon_y \\ \epsilon_z \\ \gamma_{xy} \\ \gamma_{xz} \\ \gamma_{yz} \end{Bmatrix} = \begin{bmatrix} \dfrac{\partial}{\partial x} & 0 & 0 \\ 0 & \dfrac{\partial}{\partial y} & 0 \\ 0 & 0 & \dfrac{\partial}{\partial z} \\ \dfrac{\partial}{\partial y} & \dfrac{\partial}{\partial x} & 0 \\ \dfrac{\partial}{\partial z} & 0 & \dfrac{\partial}{\partial x} \\ 0 & \dfrac{\partial}{\partial z} & \dfrac{\partial}{\partial y} \end{bmatrix} \begin{Bmatrix} u \\ v \\ w \end{Bmatrix}$$

or

$$\{\epsilon\} = [B]\{\tilde{\delta}\} \tag{C.19}$$

Substituting equation C.19 into equation C.18 gives the strain energy in terms of the displacement field, that is,

$$U_P(u, v, w) = \frac{1}{2} \iiint_\Omega \lfloor \tilde{\delta} \rfloor [B]^T [C][B]\{\tilde{\delta}\} \, dV$$

If initial strains $\{\epsilon_0\}$ are present, the strain energy becomes

$$U_P(u, v, w) = \frac{1}{2} \iiint_\Omega \left[\lfloor \tilde{\delta} \rfloor [B]^T [C][B]\{\tilde{\delta}\} - 2\lfloor \tilde{\delta} \rfloor [B]^T [C]\{\epsilon_0\} \right] dV$$

$$\tag{C.20a}$$

The work done by the external forces is given by

$$V_P(u, v, w) = \iiint_\Omega (Xu + Yv + Zw) \, dV + \iint_{S_1} (T_x u + T_y v + T_z w) \, dS$$

$$= \iiint_\Omega \lfloor F \rfloor \{\tilde{\delta}\} \, dV + \iint_{S_1} \lfloor T \rfloor \{\tilde{\delta}\} \, dS \tag{C.20b}$$

where

$$\lfloor F \rfloor = \lfloor X \quad Y \quad Z \rfloor \qquad \lfloor T \rfloor = \lfloor T_x \quad T_y \quad T_z \rfloor \qquad \lfloor \tilde{\delta} \rfloor = \lfloor u \quad v \quad w \rfloor$$

The body force components $\lfloor F \rfloor$ and the surface tractions $\lfloor T \rfloor$ are the given external forces and S_1 is the portion of the surface of the body on which the tractions are prescribed. Combining these equations, we can write the general potential energy functionals as

$$\Pi(u, v, w) = \frac{1}{2} \iiint_\Omega \left[\lfloor \tilde{\delta} \rfloor [B]^T [C][B]\{\tilde{\delta}\} - 2\lfloor \tilde{\delta} \rfloor [B]^T [C]\{\epsilon_0\} \right] dV$$

$$- \iiint_\Omega \lfloor F \rfloor \{\tilde{\delta}\} \, dV - \iint_{S_1} \lfloor T \rfloor \{\tilde{\delta}\} \, dS_1 \tag{C.21}$$

The displacement field u, v, w that minimizes Π and satisfies all the boundary conditions is the equilibrium displacement field. When we use the principle of minimum potential energy in finite element analysis, we assume the form of the displacement field in each element and then use the functional Π to derive the element equations as explained in Chapter 3. This approach is called the *displacement method* or the *stiffness method*, and the element equations that result are the approximate equilibrium equations. The compatibility conditions are identically satisfied.

C.9   PLANE STRAIN AND PLANE STRESS

Two of the many important specializations of three-dimensional linear elasticity theory are the cases of plane strain and plane stress. Both of these physically distinct situations can be described in terms of two independent coordinates, say x and y. In plane strain the component of displacement normal to the x–y plane is zero, while in plane stress the component of stress normal to the x–y plane is zero. For these two cases the foregoing equations of elasticity theory simplify to the following.

Plane strain

Stress components:

$$\lfloor \sigma \rfloor = \lfloor \sigma_x \quad \sigma_y \quad \tau_{xy} \rfloor$$
$$\sigma_z = \nu(\sigma_x + \sigma_y) \tag{C.22}$$

Strain components:

$$\lfloor \epsilon \rfloor = \lfloor \epsilon_x \quad \epsilon_y \quad \gamma_{xy} \rfloor \tag{C.23}$$

Hooke's law:

$$\{\sigma\} = [C]\{\epsilon\} \tag{C.24}$$

$$[C] = \frac{E}{(1 + \nu)(1 - 2\nu)} \begin{bmatrix} 1 - \nu & \nu & 0 \\ \nu & 1 - \nu & 0 \\ 0 & 0 & \dfrac{1 - 2\nu}{2} \end{bmatrix} \tag{C.25}$$

Static equilibrium:

$$\frac{\partial \sigma_x}{\partial x} + \frac{\partial \tau_{xy}}{\partial y} + X = 0$$

$$\frac{\partial \tau_{xy}}{\partial x} + \frac{\partial \sigma_y}{\partial y} + Y = 0 \tag{C.26}$$

$$\frac{\partial \sigma_z}{\partial z} + Z = 0$$

Compatibility:

$$\frac{\partial^2 \epsilon_x}{\partial y^2} + \frac{\partial^2 \epsilon_y}{\partial x^2} = \frac{\partial^2 \gamma_{xy}}{\partial x \, \partial y} \tag{C.27}$$

Differential equations for displacements:

$$\nabla^2 u + \frac{1}{1-2\nu} \frac{\partial}{\partial x} \left(\frac{\partial u}{\partial x} + \frac{\partial v}{\partial y} \right) + \frac{X}{\mu} = 0$$

$$\nabla^2 v + \frac{1}{1-2\nu} \frac{\partial}{\partial y} \left(\frac{\partial u}{\partial x} + \frac{\partial v}{\partial y} \right) + \frac{Y}{\mu} = 0$$

(C.28)

where

$$\nabla^2 = \frac{\partial^2}{\partial x^2} + \frac{\partial^2}{\partial y^2}$$

with boundary conditions

$$T_x = \lambda e\, l \; + \mu \left[2\frac{\partial u}{\partial x}l \; + \left(\frac{\partial u}{\partial y} + \frac{\partial v}{\partial x} \right)m \right]$$

$$T_y = \lambda e\, m + \mu \left[2\frac{\partial v}{\partial y}m + \left(\frac{\partial v}{\partial x} + \frac{\partial u}{\partial y} \right)l \right]$$

(C.29)

Plane stress
Stress components:

$$\lfloor \sigma \rfloor = \lfloor \sigma_x \quad \sigma_y \quad \tau_{xy} \rfloor$$

$$\sigma_z = 0$$

(C.30)

Strain components:

$$\lfloor \epsilon \rfloor = \lfloor \epsilon_x \quad \epsilon_y \quad \gamma_{xy} \rfloor$$

$$\epsilon_z = \frac{-\nu}{1-\nu}(\epsilon_x + \epsilon_y)$$

(C.31)

Hooke's law:

$$\{\sigma\} = [C]\{\epsilon\}$$

$$[C] = \frac{E}{1-\nu^2} \begin{bmatrix} 1 & \nu & 0 \\ \nu & 1 & 0 \\ 0 & 0 & \dfrac{1-\nu}{2} \end{bmatrix}$$

(C.32)

Static equilibrium:

$$\frac{\partial \sigma_x}{\partial x} + \frac{\partial \tau_{xy}}{\partial y} + X = 0$$

$$\frac{\partial \tau_{xy}}{\partial x} + \frac{\partial \sigma_y}{\partial y} + Y = 0 \tag{C.33}$$

Compatibility:

$$\frac{\partial^2 \epsilon_y}{\partial z^2} + \frac{\partial^2 \epsilon_z}{\partial y^2} = 0$$

$$\frac{\partial^2 \epsilon_z}{\partial x^2} + \frac{\partial^2 \epsilon_x}{\partial z^2} = 0$$

$$\frac{\partial^2 \epsilon_x}{\partial y^2} + \frac{\partial^2 \epsilon_y}{\partial x^2} = \frac{\partial^2 \gamma_{xy}}{\partial x \, \partial y} \tag{C.34}$$

$$2 \frac{\partial^2 \epsilon_x}{\partial y \, \partial z} = \frac{\partial^2 \gamma_{xy}}{\partial x \, \partial z}$$

Differential equations for displacements:

$$\nabla^2 u + \frac{1+\nu}{1-\nu} \frac{\partial}{\partial x} \left(\frac{\partial u}{\partial x} + \frac{\partial v}{\partial y} \right) + \frac{X}{\mu} = 0$$

$$\nabla^2 v + \frac{1+\nu}{1-\nu} \frac{\partial}{\partial y} \left(\frac{\partial u}{\partial x} + \frac{\partial v}{\partial y} \right) + \frac{Y}{\mu} = 0 \tag{C.35}$$

with the boundary conditions given by equations C.29.

C.10   THERMAL EFFECTS

Nonuniform temperature distributions affect the behavior of a linear elastic solid by altering the stress–strain relationship, equation C.5, and making it dependent on the absolute temperature $T(x, y, z)$. The total strain in the solid consists of two parts: the "mechanical" strains that depend on the stress state plus the thermal strains from thermal expansion. In an isotropic solid, the strains due to thermal expansion are uniform in all directions, hence, only normal strains are altered; the shearing strains are unaltered. The stress–strain equation for an isotropic solid including thermal effects may be written as

$$\{\epsilon\} = [D]\{\sigma\} + \{\epsilon_T\} \tag{C.36}$$

where $[D]$ is the material flexibility matrix, equation C.7, and $\{\epsilon_T\}$ is the vector of thermal strains. In three dimensions the thermal strains are given by

$$\lfloor \epsilon_T \rfloor = \lfloor \alpha(T - T_0), \alpha(T - T_0), \alpha(T - T_0), 0, 0, 0 \rfloor \qquad (C.37)$$

where α is the material coefficient of linear thermal expansion, and T_0 is a reference temperature for an unstrained state. The stresses in the solid, including thermal effects, may be computed by solving equation C.36 for the stress vector $\{\sigma\}$. The stress–strain law may then be written as

$$\{\sigma\} = [C]\{\epsilon\} + [C]\{\epsilon_T\} \qquad (C.38)$$

where $[C]$ is the material stiffness matrix, equation C.6.

For the plane strain and plane stress specializations of the three-dimensional elasticity equations with thermal effects, the stiffness matrix definitions given previously apply (equations C.25 and C.32, respectively), but the thermal strains are given by the following.

Plane stress:

$$\lfloor \epsilon_T \rfloor = \lfloor \alpha(T - T_0) \quad \alpha(T - T_0) \quad 0 \rfloor \qquad (C.39)$$

Plane strain:

$$\lfloor \epsilon_T \rfloor = \lfloor \alpha(1 + \nu)(T - T_0) \quad \alpha(1 + \nu)(T - T_0) \quad 0 \rfloor \quad (C.40)$$

In finite element analysis of elasticity problems thermal effects may be included conveniently by considering thermal strains as a vector of initial strains $\{\epsilon_0\}$, which was included, for example, in the elastic strain energy U_P given in equation C.20a. Thus, in formulating the elasticity finite element equations with thermal effects, we use one of the variational principles and replace the initial strains $\{\epsilon_0\}$ by the thermal strains, equation C.37. Total stresses in the body are, of course, calculated by equation C.38.

C.11 THIN-PLATE BENDING

We define a thin plate (see Figure 6.12) as a three-dimensional body of constant thickness h whose center plane is coincident with the x–y plane. The classical theory of thin-plate bending incorporates four ad hoc assumptions:

1. The deflection of the center plane is small compared to the thickness.
2. The center plane has no strain during bending.
3. The stress component normal to the center plane, σ_z, is small.
4. Normals to the center plane remain normal during bending.

According to these assumptions,

$$\epsilon_z \approx \gamma_{xz} \approx \gamma_{yz} \approx \sigma_z \approx 0 \tag{C.41}$$

and the in-plane displacements are related to the deflection by

$$u = -z\frac{\partial w}{\partial x} \quad \text{and} \quad v = -z\frac{\partial w}{\partial y} \tag{C.42}$$

These are the Kirchhoff constraints, which effectively reduce the plate-bending problem to that of finding only $w(x, y)$.

Instead of characterizing the plate-bending problem with the usual stress vector $\lfloor \sigma \rfloor = \lfloor \sigma_x \quad \sigma_y \quad \tau_{xy} \rfloor$ and the strain vector $\lfloor \epsilon \rfloor = \lfloor \epsilon_x \quad \epsilon_y \quad \gamma_{xy} \rfloor$, it is convenient to use other analogous parameters to play roles similar to those of stress and strain. In place of the stress vector we introduce the *moments* M_x, M_y, and M_{xy}, which are moments per unit length. The first subscript designates the axis normal to the plane on which the moment acts; the second subscript indicates the direction of the stress generating the moment. In place of the strain vector we shall use the *plate curvatures* $\kappa_x, \kappa_y, \kappa_{xy}$, defined as

$$\kappa_x = -\frac{\partial^2 w}{\partial x^2} \qquad \kappa_y = -\frac{\partial^2 w}{\partial y^2} \qquad \kappa_{xy} = -2\frac{\partial^2 w}{\partial x\,\partial y} \tag{C.43}$$

Replacing stress components by line moments and strain components by curvatures makes the problem formulation easier. In terms of line moments and curvatures, the constitutive equation for an orthotropic plate becomes

$$\begin{Bmatrix} M_x \\ M_y \\ M_{xy} \end{Bmatrix} = \begin{bmatrix} D_{K11} & D_{K12} & D_{K13} \\ D_{K21} & D_{K22} & D_{K23} \\ D_{K31} & D_{K32} & D_{K33} \end{bmatrix} \begin{Bmatrix} \kappa_x \\ \kappa_y \\ \kappa_{xy} \end{Bmatrix} \tag{C.44}$$

or

$$\{M\} = [D_K]\{\kappa\}$$

where $[D_K]$ is the flexural rigidity matrix. If the plate is composed of an isotropic material, we have

$$[D_K] = D\begin{bmatrix} 1 & \nu & 0 \\ \nu & 1 & 0 \\ 0 & 0 & \dfrac{1-\nu}{2} \end{bmatrix} \tag{C.45}$$

where D is called the flexural rigidity and is given by

$$D = \frac{Eh^3}{12(1 - \nu^2)}$$

When the transverse loading on the plate per unit area is $q(x, y)$, it can be shown that the differential equation governing the plate deflection for an orthotropic solid is

$$D_{K11}\frac{\partial^4 w}{\partial x^4} + 2(D_{K12} + 2D_{K33})\frac{\partial^4 w}{\partial x^2 \partial y^2} + D_{K22}\frac{\partial^4 w}{\partial y^4} = q(x, y) \quad \text{(C.46)}$$

and for an isotropic solid this reduces to

$$\nabla^4 w = \frac{\partial^4 w}{\partial x^4} + 2\frac{\partial^4 w}{\partial x^2 \partial y^2} + \frac{\partial^4 w}{\partial y^4} = \frac{q(x, y)}{D} \quad \text{(C.47)}$$

The boundary conditions for equations C.46 and C.47 involve w and its derivatives up to and inclusive of the third order, and they depend on the type of edge conditions, for example, simply supported, built-up, free, and specified deflection. Much more detailed information on elastic plate theory may be found in references 7 and 8.

REFERENCES

1. S. Timoshenko and J. N. Goodier, *Theory of Elasticity*, 3d ed., McGraw-Hill, New York, 1970.
2. Y. C. Fung, *Foundations of Solid Mechanics*, Prentice-Hall, Englewood Cliffs, NJ, 1965.
3. I. S. Sokolnikoff, *Mathematical Theory of Elasticity*, 2d. ed., McGraw-Hill, New York, 1956.
4. A. P. Boresi and K. P. Chong, *Elasticity in Engineering Mechanics*, Elsevier, New York, 1987.
5. J. M. Whitney, *Structural Analysis of Anisotropic Laminated Plates*, Technomic, Lancaster, PA, 1987.
6. K. Washizu, *Variational Methods in Elasticity and Plasticity*, 3d ed., Pergamon, Elmsford, NY, 1982.
7. S. Timoshenko and S. Woinowsky-Krieger, *Theory of Plates and Shells*, McGraw-Hill, New York, 1959.
8. A. C. Ugaral, *Stresses in Plates and Shells*, McGraw-Hill, New York, 1981.

APPENDIX D

BASIC EQUATIONS FROM FLUID MECHANICS

D.1 INTRODUCTION

Here we review introductory concepts and governing equations for the general flow of an isotropic single-species fluid. Again we assume that the continuum assumption holds. This means that the dimensions for a given problem are everywhere large compared to the mean-free molecular path of the fluid substance.[6] Hence a fluid particle is taken as the smallest lump of fluid having enough molecules to allow the continuum assumption.

Since most engineers and scientists use the Eulerian description to formulate their problems, this description is used here exclusively. The various equations for different kinds of flow are given in Cartesian coordinates

---

[6]For example, the mean-free molecular path for air at standard temperature and pressure is about 2×10^{-6} in.

578

without derivation or proof. Readers desiring more details or statements of the equations in different coordinate systems should consult the selected references given at the end of this appendix.

D.2   DEFINITIONS AND CONCEPTS [1–3]

A *fluid* is a substance (either a liquid or a gas) that continuously deforms under the action of applied surface stresses. This implies that shear stresses do not exist in a fluid unless it is deforming. As a continuum, a fluid possesses the following physical properties: density, ρ; viscosity, μ; thermal conductivity, k; thermal diffusivity, a; specific heats, c_v and c_p; electrical conductivity, σ; dielectric constant ϵ_R, magnetic permeability, ϵ_m; and expansivity factor, β.

The *state of stress* is characterized in the same way in a fluid as it is in a solid (see Section C.2). For the special case of a fluid at rest, the off-diagonal terms of the stress tensor are zero (no shear stresses), and the diagonal terms are all the same and are equal to the negative of the hydrostatic pressure. For a fluid in motion, all terms of the stress tensor may be nonzero, though the stress tensor remains symmetric.

A flow field is characterized by a velocity vector that is a continuous function of space and time. To represent a fluid motion graphically, it is convenient to introduce the concepts of streamlines, pathlines, and streak-lines. An imaginary line in a flow field along which flow velocity vectors are everywhere tangent at any instant is called a *streamline*. The trajectory or locus of points through which a fluid particle of fixed identity passes as it moves in a flow field is known as a *pathline*. A *streakline* is defined as a line that connects all fluid particles that at some time have passed through a fixed point in space. For an unsteady flow streamlines, pathlines, and streaklines are in general different; but for a steady flow they are identical.

In this book we classify a flow field as one-, two-, or three-dimensional, depending on the number of independent space coordinates needed to specify the velocity field. In addition to classifying flow fields, we may also classify the various types of flow. Admittedly such a classification is somewhat arbitrary because there can be much overlap between categories, but a given classification scheme can serve to highlight major divisions of the subject in spite of the hazards of oversimplification.

As Figure D.1 indicates, a flow may first be classified as either inviscid or viscous. Inviscid flows are frictionless flows characterized by zero viscosity. No real flows are inviscid, but there are numerous flow situations in which viscous effects can be neglected. The viscosity of the fluid and the magnitude of the gradient of the velocity field throughout the flow are the two primary factors that determine when viscous effects are insignificant. Often viscous effects are confined to a thin region or boundary layer near flow boundaries, and the rest of the flow can be considered frictionless.

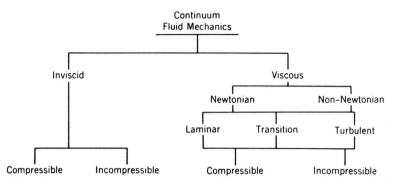

Figure D.1   Some major categories or types of flow in continuum fluid mechanics.

Inviscid flows may further be classified as either compressible or incompressible, depending on whether density variations are large or relatively unimportant. High-speed gas flows are an example of compressible flows, whereas liquid flows are *usually* assumed to be incompressible.

Since all real fluids have a finite nonzero viscosity, the subject of viscous flow is of paramount importance in fluid mechanics. A viscous flow is said to be Newtonian if Stokes'[7] law of friction prevails for the fluid in question, and non-Newtonian if some other law governs the relation between stress and rate of strain. Most fluids are assumed to be Newtonian or purely viscous, but some real fluids exhibit more complex behavior. We limit our review here to Newtonian fluids only.

The dynamic macroscopic behavior of a flow determines whether it is classified as laminar or turbulent, or whether it is in transition between the two. Fluid motion is said to be *laminar* if the fluid flows in imaginary laminas or layers and there is no macroscopic mixing of adjacent fluid layers. A flow is said to be *turbulent*, however, if such mixing occurs. The mixing of fluid particles from adjacent layers is caused by small velocity fluctuations superimposed on the mean flow velocity field. These velocity fluctuations have a random or irregular variation with the space and time coordinates, but they can be characterized by statistically distinct averages. Clearly, laminar flows are more amenable to detailed analysis than turbulent flows. *Transition* flow occurs when a laminar flow becomes unstable and approaches a turbulent flow. In other words, the type of flow that exists during the time span in which a flow changes from fully developed, laminar to fully developed, turbulent is known as transition flow. In such a flow there can be patches of turbulent flow where irregularity and disorderliness prevail, and regions of laminar flow characterized by orderliness. Essentially, the flow changes in time by alternating between being laminar and being turbulent.

---

[7]Stokes' law is presented in detail in the next section.

Any of these types of viscous flows can be compressible or incompressible. Furthermore, inviscid or viscous flows can be named according to the type of flow boundaries. We can have *internal flows* (i.e., pipe flows), *external flows* (i.e., flows over airfoils), or *internal–external flows* (i.e., flows around a center body in pipe). Flows may also be given a temporal description by labeling them as steady, periodic, quasisteady, or transient.

Here we have indicated only the major types of flow and the adjectives used to describe them. Obviously, there are many other special types of flow that we have not considered. For example, under the heading of compressible flow, we can have subsonic, transonic, supersonic, or hypersonic flow; and viscous flows may be cavitated, separated, or stratified.

D.3 LAWS OF MOTION [1–5]

Solving fluid mechanics problems ultimately involves solving in some form the equations expressing the three laws of motion of the fluid:

1. Conservation of mass
2. Conservation of momentum
3. Conservation of energy

In addition to these laws, it is necessary to include an equation of state and a viscosity equation for the fluid. These provide the pointwise relations among pressure, temperature, density, and viscosity regardless of whether the fluid is flowing or static. The conservation laws can be expressed in either integral or differential forms. Integral forms are most useful for determining the gross or global character of a flow as it interacts with a structure or other device. Differential forms provide the relations that hold at any point in the flow; hence they are most useful for studying the detailed character. In the following we summarize the differential form of the governing equations since finite element analyses of fluid flow problems rely on this form.

D.3.1 Differential Continuity Equation

By writing a mass balance for a differential control volume we can show that the conservation of mass principle is expressed by

$$\frac{\partial \rho}{\partial t} + \frac{\partial(\rho u)}{\partial x} + \frac{\partial(\rho v)}{\partial y} + \frac{\partial(\rho w)}{\partial z} = 0 \qquad \text{(D.1a)}$$

or, in vector form,

$$\frac{\partial \rho}{\partial t} + \nabla \cdot (\rho \mathbf{V}) = 0 \qquad \text{(D.1b)}$$

where ρ = mass density,

   u, v, w = velocity components in the x, y, and z directions, respectively,

   t = time,

   $\mathbf{V} = u\hat{i} + v\hat{j} + w\hat{k}$,

   $\nabla = \hat{i}(\partial/\partial x) + \hat{j}(\partial/\partial y) + \hat{k}(\partial/\partial z)$.

Equations D.1 hold for the motion of any fluid satisfying the continuum assumption.

D.3.2 Differential Momentum Equation (Navier–Stokes Equations)

Considering a differential volume (a fluid particle) and summing the forces acting on the particle results in the following differential equations of motion:

$$\rho\left(\frac{\partial u}{\partial t} + u\frac{\partial u}{\partial x} + v\frac{\partial u}{\partial y} + w\frac{\partial u}{\partial z}\right) = \rho B_x + \frac{\partial\sigma_{xx}}{\partial x} + \frac{\partial\tau_{xy}}{\partial y} + \frac{\partial\tau_{xz}}{\partial z} \qquad \text{(D.2a)}$$

$$\rho\left(\frac{\partial v}{\partial t} + u\frac{\partial v}{\partial x} + v\frac{\partial v}{\partial y} + w\frac{\partial v}{\partial z}\right) = \rho B_y + \frac{\partial\tau_{yx}}{\partial x} + \frac{\partial\sigma_{yy}}{\partial y} + \frac{\partial\tau_{yz}}{\partial z} \qquad \text{(D.2b)}$$

$$\rho\left(\frac{\partial w}{\partial t} + u\frac{\partial w}{\partial x} + v\frac{\partial w}{\partial y} + w\frac{\partial w}{\partial z}\right) = \rho B_z + \frac{\partial\tau_{zx}}{\partial x} + \frac{\partial\tau_{zy}}{\partial y} + \frac{\partial\sigma_{zz}}{\partial z} \qquad \text{(D.2c)}$$

Here B_x, B_y, and B_z are the x, y, z components of the body force vector per unit mass:

$$\mathbf{B} = B_x\hat{i} + B_y\hat{j} + B_z\hat{k}$$

The left-hand side of equations D.2 is the substantial or material derivative of the velocity components; for example, the substantial derivative of u is

$$\frac{Du}{Dt} = \frac{\partial u}{\partial t} + u\frac{\partial u}{\partial x} + v\frac{\partial u}{\partial y} + w\frac{\partial u}{\partial z}$$

The normal stresses σ_{xx}, σ_{yy}, and σ_{zz} include the fluid pressure P. Hence we can define new normal stresses of the form

$$\sigma_{xx} = -P + \sigma_x \qquad \sigma_{yy} = -P + \sigma_y \qquad \sigma_{zz} = -P + \sigma_z$$

Then, for instance, equation D.2a becomes

$$\rho\frac{Du}{Dt} = \rho B_x - \frac{\partial P}{\partial x} + \frac{\partial\sigma_x}{\partial x} + \frac{\partial\tau_{xy}}{\partial y} + \frac{\partial\tau_{xz}}{\partial z}$$

Equations D.2 are valid for any type of fluid. To use these equations to find velocity and pressure distributions for a particular flow and type of fluid, we must express the various stress components in terms of derivatives of velocity and the properties of the fluid. For Newtonian fluids, we have *Stokes' law of friction*, which is

$$
\left.
\begin{aligned}
\sigma_x &= 2\mu\frac{\partial u}{\partial x} - \frac{2}{3}\mu\nabla\cdot\mathbf{V} \\[4pt]
\sigma_y &= 2\mu\frac{\partial v}{\partial y} - \frac{2}{3}\mu\nabla\cdot\mathbf{V} \\[4pt]
\sigma_z &= 2\mu\frac{\partial w}{\partial z} - \frac{2}{3}\mu\nabla\cdot\mathbf{V} \\[4pt]
\tau_{xy} = \tau_{yx} &= \mu\left(\frac{\partial u}{\partial y} + \frac{\partial v}{\partial x}\right) \\[4pt]
\tau_{yz} = \tau_{zy} &= \mu\left(\frac{\partial v}{\partial z} + \frac{\partial w}{\partial y}\right) \\[4pt]
\tau_{zx} = \tau_{xz} &= \mu\left(\frac{\partial w}{\partial x} + \frac{\partial u}{\partial z}\right)
\end{aligned}
\right\}
\tag{D.3}
$$

Substitution of equations D.3 into equations D.2 gives the general form of the Navier–Stokes equations:

$$
\rho\frac{Du}{Dt} = \rho B_x - \frac{\partial P}{\partial x} + \frac{\partial}{\partial x}\left[2\mu\left(\frac{\partial u}{\partial x} - \frac{1}{3}\nabla\cdot\mathbf{V}\right)\right]
$$
$$
+ \frac{\partial}{\partial y}\left[\mu\left(\frac{\partial u}{\partial y} + \frac{\partial v}{\partial x}\right)\right] + \frac{\partial}{\partial z}\left[\mu\left(\frac{\partial w}{\partial x} + \frac{\partial u}{\partial z}\right)\right]
\tag{D.4a}
$$

$$
\rho\frac{Dv}{Dt} = \rho B_y - \frac{\partial P}{\partial y} + \frac{\partial}{\partial y}\left[2\mu\left(\frac{\partial v}{\partial y} - \frac{1}{3}\nabla\cdot\mathbf{V}\right)\right]
$$
$$
+ \frac{\partial}{\partial z}\left[\mu\left(\frac{\partial v}{\partial z} + \frac{\partial w}{\partial y}\right)\right] + \frac{\partial}{\partial x}\left[\mu\left(\frac{\partial u}{\partial z} + \frac{\partial v}{\partial x}\right)\right]
\tag{D.4b}
$$

$$
\underbrace{\rho\frac{Dw}{Dt}}_{\substack{\text{inertia}\\\text{force}\\\text{terms}}} = \underbrace{\rho B_z}_{\substack{\text{body}\\\text{force}\\\text{terms}}} - \underbrace{\frac{\partial P}{\partial z}}_{\substack{\text{pressure}\\\text{force}\\\text{terms}}}
$$

$$
\underbrace{+ \frac{\partial}{\partial z}\left[2\mu\left(\frac{\partial w}{\partial z} - \frac{1}{3}\nabla\cdot\mathbf{V}\right)\right] + \frac{\partial}{\partial x}\left[\mu\left(\frac{\partial w}{\partial x} + \frac{\partial u}{\partial z}\right)\right] + \frac{\partial}{\partial y}\left[\mu\left(\frac{\partial v}{\partial z} + \frac{\partial w}{\partial y}\right)\right]}_{\text{viscous force terms}}
$$
$$
\tag{D.4c}
$$

D.3.3   Thermal Energy Equation

For nonisothermal flows with temperature-dependent viscosity, the equations of continuity and momentum are coupled to the thermal energy equation, and all four equations must be solved simultaneously. In its most general form, the thermal energy equation is

$$\rho \frac{De}{Dt} + P\nabla \cdot V + \nabla \cdot \mathbf{q}_r - Q$$

$$= \frac{\partial}{\partial x}\left(k\frac{\partial T}{\partial x}\right) + \frac{\partial}{\partial y}\left(k\frac{\partial T}{\partial y}\right) + \frac{\partial}{\partial z}\left(k\frac{\partial T}{\partial z}\right) + \mu\tilde{\Phi}(x,y,z,t) \quad (D.5)$$

where e = internal energy per unit mass,
k = thermal conductivity,
T = temperature,
$\mathbf{q}_r$ = radiation heat flux vector,
Q = internal heat generation rate per unit volume,and
$\tilde{\Phi}$ = viscous dissipation function

$$= 2\left[\left(\frac{\partial u}{\partial x}\right)^2 + \left(\frac{\partial v}{\partial y}\right)^2 + \left(\frac{\partial w}{\partial z}\right)^2\right] + \left(\frac{\partial v}{\partial x} + \frac{\partial u}{\partial y}\right)^2$$

$$+ \left(\frac{\partial w}{\partial y} + \frac{\partial v}{\partial z}\right)^2 + \left(\frac{\partial u}{\partial z} + \frac{\partial w}{\partial x}\right)^2$$

$$- \frac{2}{3}\left(\frac{\partial u}{\partial x} + \frac{\partial v}{\partial y} + \frac{\partial w}{\partial z}\right)^2$$

D.3.4   Conservation Form of Equations

In computational fluid dynamics literature, the mass, momentum, and energy equations are often written in a more compact vector equation called the conservation form of the equations. The vector equation is called the conservation form because the equations are written exactly in the form that arises naturally from the integral conservation laws applied to a fixed control volume [4, 5]. Without body forces or external heat addition, the conservation form of the equations is

$$\frac{\partial}{\partial t}\{U\} + \frac{\partial}{\partial x}\{E_I - E_V\} + \frac{\partial}{\partial y}\{F_I - F_V\} + \frac{\partial}{\partial z}\{G_I - G_V\} = 0 \quad (D.6)$$

where the vector of conservation variables is

$$\{U\}^T = \begin{vmatrix} \rho & \rho u & \rho v & \rho w & \rho E_t \end{vmatrix} \qquad (D.7a)$$

where E_t is the total energy per unit mass given by

$$E_t = e + \tfrac{1}{2}(u^2 + v^2 + w^2) \qquad (D.7b)$$

and the inviscid flux components are

$$\{E_I\}^T = \begin{vmatrix} \rho u & \rho u^2 + P & \rho uv & \rho uw & (\rho E_t + P)u \end{vmatrix} \qquad (D.7c)$$

$$\{F_I\}^T = \begin{vmatrix} \rho v & \rho uv & \rho v^2 + P & \rho vw & (\rho E_t + P)v \end{vmatrix} \qquad (D.7d)$$

$$\{G_I\}^T = \begin{vmatrix} \rho w & \rho uw & \rho vw & \rho w^2 + P & (\rho E_t + P)w \end{vmatrix} \qquad (D.7e)$$

and the viscous flux components are given by

$$\{E_V\}^T = \begin{vmatrix} 0 & \sigma_x & \tau_{xy} & \tau_{xz} & u\sigma_x + v\tau_{xy} + w\tau_{xz} + q_x \end{vmatrix} \qquad (D.7f)$$

$$\{F_V\}^T = \begin{vmatrix} 0 & \tau_{xy} & \sigma_y & \tau_{yz} & u\tau_{xy} + v\sigma_y + w\tau_{yz} + q_y \end{vmatrix} \qquad (D.7g)$$

$$\{G_V\}^T = \begin{vmatrix} 0 & \tau_{xz} & \tau_{yz} & \sigma_z & u\tau_{xz} + v\tau_{yz} + w\sigma_z + q_z \end{vmatrix} \qquad (D.7h)$$

where the stress components are defined in equations D.3, and the conduction heat fluxes are defined by Fourier's law,

$$q_x = -k\frac{\partial T}{\partial x}$$

$$q_y = -k\frac{\partial T}{\partial y} \qquad (D.7i)$$

$$q_z = -k\frac{\partial T}{\partial z}$$

The vector equation D.6 represents the five conservation equations. The first row is the conservation of mass equation, the second through fourth rows are the three conservation of momentum equations, and the fifth row is the conservation of energy equation.

D.3.5   Supplementary Equations

The complete specification of a fluid flow problem requires three additional equations accompanying the conservation of mass, momentum, and energy equations. These are the equations of state, internal energy, and viscosity. Since the form of these equations is nonunique, we represent them generally as

$$\text{Equation of state:} \quad \rho = \rho(P, T) \qquad \text{(D.8a)}$$

$$\text{Internal energy:} \quad e = e(P, T) \qquad \text{(D.8b)}$$

$$\text{Viscosity equation:} \quad \mu = \mu(P, T) \qquad \text{(D.8c)}$$

D.3.6   Problem Statement

The solution of general flow problems may be summarized as follows: *Given* the geometry of the flow boundaries, physical properties of the fluid, the eight governing equations D.1, D.4, D.5 (or D.6), and D.8, and a complete set of boundary and initial conditions for the governing equations, *find* the velocity distribution u, v, w, pressure P, temperature T, density ρ, internal energy e, and viscosity μ.

General flow problems thus involve solving eight equations with eight unknown field variables. In many practical applications, however, explicit definitions of the density, internal energy, and viscosity in equations D.8 permit a reduction of the number of equations and unknowns from eight to five. The basic equations to be solved are the conservation of mass, the three momentum equations, and the energy equation; the basic unknowns are the velocity components u, v, w, the pressure P, and temperature T. The mix of knowns and unknowns is slightly different for *free boundary* problems. In these problems, the geometry of the free boundary of the flow is sought when flow conditions along the boundary are known.

In addition to the obvious complexity of the problem as stated thus far, other factors may cause difficulty. If the fluid forces or temperature acting on the bounding solids cause the shape of the boundary to deform, we must also simultaneously solve the elasticity equations for the solids. Thermal distortion of the boundaries may also alter the fluid flow. In this case, the flow and heat transfer solution would involve the simultaneous solution of the energy equation in the bounding solids (equation D.5 with $u = v = w = 0$).

The governing equations hold at any instant of time and apply to either laminar or turbulent flows. However, in a turbulent flow, the field variables are fluctuating randomly about their mean values, and the mathematical nature of the problem becomes hopelessly complex. Because the instantaneous equations cannot be solved for a turbulent flow, the standard approach is to time-average the equations and obtain new ones that describe the temporally averaged distributions of the field variables. Discussion of the theoretical methods for predicting the behavior of turbulent flows is far beyond our purpose here. The interested reader can find treatments of this topic in specialized textbooks such as references 6 and 7.

Even with attention restricted to laminar flows, a solution of the full set of equations has never been obtained. But for many problems of practical interest the governing equations simplify considerably and the mathematical difficulties become more tractable. In subsequent sections we present the governing equations for a number of special classes of problems. Often the governing equations and the form of a particular flow pattern can be described concisely in terms of a stream function. We consider this concept in the next section.

D.4   STREAM FUNCTIONS AND VORTICITY

Two-dimensional flows (both compressible and incompressible) can be conveniently characterized by introducing a mathematical artifice known as a stream function—$\psi(x, y, t)$. The stream function relates the concept of streamlines to the principle of mass conservation.

For a two-dimensional incompressible flow the continuity equation reduces to

$$\frac{\partial u}{\partial x} + \frac{\partial v}{\partial y} = 0$$

This equation is identically satisfied if we define

$$u = \frac{\partial \psi}{\partial y} \quad \text{and} \quad v = -\frac{\partial \psi}{\partial x} \tag{D.9}$$

since

$$\frac{\partial u}{\partial x} + \frac{\partial v}{\partial y} = \frac{\partial^2 \psi}{\partial x \, \partial y} - \frac{\partial^2 \psi}{\partial y \, \partial x} = 0$$

In the flow field, lines of constant ψ and streamlines are identical and the flow rate between two streamlines is proportional to the numerical difference between the two stream functions corresponding to the two streamlines.

For a two-dimensional steady compressible flow the continuity equation reduces to

$$\frac{\partial}{\partial x}(\rho u) + \frac{\partial}{\partial y}(\rho v) = 0$$

This equation is identically satisfied if we define

$$\rho u = \frac{\partial \psi}{\partial y} \quad \text{and} \quad \rho v = -\frac{\partial \psi}{\partial x} \tag{D.10}$$

where ρ is the local fluid density. A stream function formulation of two-dimensional flow problems is often advantageous since the continuity equation is automatically satisfied by the stream function definition. Clearly, if we can find the stream function for a given problem, we have essentially solved the problem. Equations governing the behavior of ψ for different types of flow problems will be given in the following sections as these problems are discussed.

Fluid rotation is defined as the average angular velocity of any two mutually perpendicular line elements of a fluid particle. For three-dimensional flow, the rotation vector is a three-component vector given by

$$\boldsymbol{\omega} = \omega_x \hat{i} + \omega_y \hat{j} + \omega_z \hat{k} \tag{D.11a}$$

where $\omega_x, \omega_y, \omega_z$ are the rotations[8] about the x, y, and z axes, respectively, and are defined by

$$\omega_x = \frac{1}{2} \left(\frac{\partial w}{\partial y} - \frac{\partial v}{\partial z} \right)$$

$$\omega_y = \frac{1}{2} \left(\frac{\partial u}{\partial z} - \frac{\partial w}{\partial x} \right) \tag{D.11b}$$

$$\omega_z = \frac{1}{2} \left(\frac{\partial v}{\partial x} - \frac{\partial u}{\partial y} \right)$$

In vector notation we have

$$2\boldsymbol{\omega} = \nabla \times \mathbf{V} \tag{D.11c}$$

The factor of two can be eliminated by defining a quantity called the vorticity, ζ, to be twice the rotation:

$$\zeta = 2\boldsymbol{\omega} = \nabla \times \mathbf{V} \tag{D.12}$$

The action of viscous forces in a flow field develops rotation. Hence any viscous flow is a rotational flow with $\boldsymbol{\omega} \neq 0$ [8]. In an initially irrotational flow field, rotation cannot be developed by the action of body forces or pressure forces. Only the action of shearing stress can produce rotationality. For this reason, inviscid ($\mu = 0$) flows initially irrotational can be assumed to remain irrotational, that is, $\boldsymbol{\omega} = 0$.

---

[8]By convention, positive rotation follows the right-hand screw rule.

D.5   POTENTIAL FLOW

Inviscid irrotational flow is called potential flow [8, 9] because the velocity field in the flow can be derived from a *potential function*. A well-known vector identity tells us that, for any function $\Phi(x, y, z, t)$ having continuous first and second derivatives,

$$\nabla \times (\nabla\Phi) = 0 \tag{D.13}$$

But the condition of irrotationality from equation D.12 states that $\nabla \times \mathbf{V} = 0$. Hence, for an irrotational flow, we must have $\mathbf{V} = \pm\nabla\Phi$. We choose the minus sign so that the positive direction of flow is in the direction of decreasing Φ. The potential function is Φ, and potential flow is characterized by the relation

$$\mathbf{V} = -\nabla\Phi \tag{D.14}$$

or

$$u = -\frac{\partial\Phi}{\partial x} \qquad v = -\frac{\partial\Phi}{\partial y} \qquad w = -\frac{\partial\Phi}{\partial z}$$

The definition of the stream function previously given applies to inviscid or viscous two-dimensional flows. But the potential function can be used for any *inviscid* (two- or three-dimensional) irrotational flow.

For an incompressible, irrotational flow if we substitute equation D.14 into the continuity equation, D.1,

$$\nabla \cdot \mathbf{V} = 0$$

we find that the velocity potential satisfies Laplace's equation,

$$\frac{\partial^2\Phi}{\partial x^2} + \frac{\partial^2\Phi}{\partial y^2} + \frac{\partial^2\Phi}{\partial z^2} = 0 \tag{D.15}$$

Also, the momentum conservation equations become, from equations D.4,

$$\rho\frac{Du}{Dt} = \rho B_x - \frac{\partial P}{\partial x} \tag{D.16a}$$

$$\rho\frac{Dv}{Dt} = \rho B_y - \frac{\partial P}{\partial y} \tag{D.16b}$$

$$\rho\frac{Dw}{Dt} = \rho B_z - \frac{\partial P}{\partial z} \tag{D.16c}$$

which are called Euler's equations. Note that because of the convective acceleration terms on the left-hand side, these momentum equations are nonlinear in the velocity components u, v, and w. But if the potential function formulation is used, the problem simply involves the solution of the linear Laplace's equation—equation D.15.

An incompressible, irrotational, two-dimensional flow can also be formulated in terms of a stream function. If we substitute the velocity components u and v from equation D.9 into the condition for zero rotation, $\omega_z = 0$, equation D.11b,

$$\frac{\partial v}{\partial x} - \frac{\partial u}{\partial y} = 0$$

we obtain

$$\frac{\partial^2 \psi}{\partial x^2} + \frac{\partial^2 \psi}{\partial y^2} = 0 \tag{D.17}$$

Hence the alternative stream function formulation also involves the solution of the linear Laplace's equation.

D.6   VISCOUS INCOMPRESSIBLE FLOW

D.6.1   Primitive Variable Formulation

In contrast to potential flow, where the convective inertia terms are predominant, in more general flow viscous effects are important [10]. However, in a number of practical applications the flow may be assumed incompressible, and the viscosity and thermal conductivity may be assumed constant. The continuity equation, D.1, and momentum equations, D.4, then simplify to

$$\frac{\partial u}{\partial x} + \frac{\partial v}{\partial y} + \frac{\partial w}{\partial z} = 0 \tag{D.18}$$

$$\rho \frac{Du}{Dt} = \rho B_x - \frac{\partial P}{\partial x} + \mu \nabla^2 u \tag{D.19a}$$

$$\rho \frac{Dv}{Dt} = \rho B_y - \frac{\partial P}{\partial y} + \mu \nabla^2 v \tag{D.19b}$$

$$\rho \frac{Dw}{Dt} = \rho B_z - \frac{\partial P}{\partial z} + \mu \nabla^2 w \tag{D.19c}$$

The energy equation also simplifies for incompressible flow because the internal energy of an incompressible fluid, equation D.8b, can be expressed as a function of temperature only, that is, $e = e(T)$. The energy differential then can be expressed as $de = c_v\, dT$, where c_v is the specific heat at constant volume. Utilizing this result and the continuity equation for incompressible flow we may reduce the energy equation, D.5, to the form

$$\rho c_v \frac{DT}{Dt} - Q = k\,\nabla^2 T + \mu\tilde{\Phi} \tag{D.20}$$

where we have taken the internal radiation q_r as zero. Thus for incompressible viscous flow the laws of motion reduce to five equations, D.18, D.19a, D.19b, D.19c, and D.20, for five unknowns, u, v, w, P, and T. The momentum equations D.21, are nonlinear in the velocity components u, v, and w due to the convective acceleration terms contained in the substantial derivatives.

D.6.2   Vorticity and Stream Function Formulation

The approach described above is known as a primitive variable formulation since the basic unknowns are the primitive flow variables. An alternative approach is to formulate the problem with vorticity and stream function as unknowns. The momentum equations may be written in terms of vorticity as follows. We write the momentum equations in vector form as

$$\rho\left[\frac{\partial \mathbf{V}}{\partial t} + (\mathbf{V}\cdot\nabla)\mathbf{V}\right] = \rho\mathbf{B} - \nabla P + \mu\nabla^2\mathbf{V} \tag{D.21}$$

and introduce the vorticity into both the acceleration and viscous terms by using two vector identities,

$$(\mathbf{V}\cdot\nabla)\mathbf{V} = \nabla\left(\frac{V^2}{2} - \mathbf{V}\times\boldsymbol{\zeta}\right) \tag{D.22}$$

$$\nabla^2\mathbf{V} = \nabla(\nabla\cdot\mathbf{V}) - \nabla\times\boldsymbol{\zeta}$$

and noting $\nabla\cdot\mathbf{V} = 0$ by the continuity equation. Then the momentum equations become

$$\rho\left[\frac{\partial \mathbf{V}}{\partial t} + \nabla\left(\frac{V^2}{2} - \mathbf{V}\times\boldsymbol{\zeta}\right)\right] = \rho\mathbf{B} - \nabla P - \mu\nabla\times\boldsymbol{\zeta}$$

The body force B is assumed to be conservative so that it can be derived from a scalar potential. By taking the curl of the above equation and noting that the curl of the gradient of a scalar is zero, we obtain after some manipulation

$$\frac{D\zeta}{Dt} = (\zeta \cdot \nabla)\mathbf{V} + \frac{\mu}{\rho} \nabla^2 \zeta \tag{D.23}$$

which is known as the vorticity transport equation. The continuity equation must also be satisfied along with the vorticity transport equation.

For two-dimensional flow the stream function can be used to satisfy the continuity equation. Also, the vorticity vector has only one component:

$$\zeta = \zeta \hat{k} = \left(\frac{\partial v}{\partial x} - \frac{\partial u}{\partial y} \right) \hat{k}$$

This definition of the vorticity and the stream function definition of the velocity components, equation D.10, can be used to express the vorticity transport equation in the form

$$\frac{\partial}{\partial t}(\nabla^2 \psi) + \frac{\partial \psi}{\partial y} \frac{\partial}{\partial x}(\nabla^2 \psi) - \frac{\partial \psi}{\partial x} \frac{\partial}{\partial y}(\nabla^2 \psi) = \frac{\mu}{\rho} \nabla^4 \psi \tag{D.24}$$

where

$$\nabla^4 = \frac{\partial^4}{\partial x^4} + 2\frac{\partial^4}{\partial x^2 \partial y^2} + \frac{\partial^4}{\partial y^4}$$

A limiting case of viscous incompressible flow occurs when the viscous forces are considerably greater than the inertia forces. When we have slow viscous or creeping flow the nonlinear terms in the momentum equations, D.18, may be neglected in comparison to the viscous terms. For this type of flow the momentum equations become

$$\rho \frac{\partial u}{\partial t} = \rho B_x - \frac{\partial P}{\partial x} + \mu \nabla^2 u \tag{D.25a}$$

$$\rho \frac{\partial v}{\partial t} = \rho B_y - \frac{\partial P}{\partial y} + \mu \nabla^2 v \tag{D.25b}$$

$$\rho \frac{\partial w}{\partial t} = \rho B_z - \frac{\partial P}{\partial z} + \mu \nabla^2 w \tag{D.25c}$$

which may be written succinctly as

$$\rho \frac{\partial}{\partial t}\mathbf{V} = \rho \mathbf{B} - \nabla P + \mu \nabla^2 \mathbf{V} \tag{D.26}$$

For an incompressible steady flow we have

$$\nabla P = \rho \mathbf{B} + \mu \nabla^2 \mathbf{V} \tag{D.27}$$

Taking the divergence of both sides of this equation gives

$$\nabla^2 P = \rho \nabla \cdot \mathbf{B} + \mu \nabla^2 (\nabla \cdot \mathbf{V})$$

But from the continuity equation $\nabla \cdot \mathbf{V} = 0$; hence

$$\nabla^2 P = \rho \nabla \cdot \mathbf{B} \tag{D.28}$$

and in the absence of body forces

$$\nabla^2 P = 0 \tag{D.29}$$

The pressure $P(x, y, z)$ for an incompressible creeping flow without body forces is seen to be a potential function. If we further simplify the problem to two dimensions, the stream function previously defined applies, and the momentum equations, when cross-differentiated to eliminate the pressure, may be written as

$$\nabla^4 \psi = 0 \tag{D.30}$$

Note that the general problem of slow viscous flow is linear in the pressure P and the velocity components u, v, and w.

D.7 BOUNDARY LAYER FLOW

When a fluid of low viscosity, such as air or water, flows past a streamlined boundary, the effect of the viscous forces is confined to a thin sublayer of flow adjacent to the boundary [10]. This thin sublayer is called a boundary layer. Outside of the boundary layer, viscous forces are small in comparison to the inertia forces and the flow may be treated as inviscid. Within the boundary layer viscous effects are most significant. The flow of a viscous fluid over a flat plate provides a convenient example of boundary layer flow. Determining the boundary layer thickness and the velocity distribution within the boundary layer become the central aspect of the problem. If the flow is in the $x-y$ plane and the free stream velocity is in the x direction, the governing boundary layer equations for a compressible fluid are as follows:

Continuity:

$$\frac{\partial \rho}{\partial t} + \frac{\partial(\rho u)}{\partial x} + \frac{\partial(\rho v)}{\partial y} = 0 \qquad \text{(D.31)}$$

Momentum:

$$\frac{\partial u}{\partial t} + \rho\left(u\frac{\partial u}{\partial x} + v\frac{\partial u}{\partial y}\right) = \rho g_x \beta(T - T_\infty) - \frac{dP}{dx} + \frac{\partial}{\partial y}\left(\mu\frac{\partial u}{\partial y}\right) \quad \text{(D.32)}$$

The y momentum equation within the boundary layer reduces to $\partial P/\partial y = 0$, which shows that the pressure is a function only of x. The first term on the right-hand side of equation D.32 represents a body force because of density variation due to temperature where g_x is the component of gravitational acceleration, β is the fluid coefficient of thermal expansion and T_∞ is a reference temperature.

Energy:

$$\rho c_p\left(u\frac{\partial T}{\partial x} + v\frac{\partial T}{\partial y}\right) = \frac{\partial}{\partial y}\left(k\frac{\partial T}{\partial y}\right) + \mu\left(\frac{\partial u}{\partial y}\right)^2 + u\frac{dP}{dx} \quad \text{(D.33)}$$

In the energy equation the fluid has been assumed to be a perfect gas.
Supplementary equations:

$$P = \rho RT \qquad \mu = \mu(T)$$

Since in the framework of boundary layer theory the pressure may be regarded as given, there are five equations for the five unknowns ρ, u, v, T, μ.

More complex equations apply for the boundary layers that form on curved walls. Note that even these simple boundary layer equations are nonlinear because of the presence of the convective inertia terms. Schlichting [10] provides a comprehensive discussion of numerous boundary layer problems and their solutions.

REFERENCES

1. F. M. White, *Viscous Fluid Flow*, McGraw-Hill, New York, 1974.

2. R. W. Fox and A. T. McDonald, *Introduction to Fluid Mechanics*, 2d ed., Wiley, New York, 1978.

3. W. F. Hughes, *An Introduction to Viscous Flow*, Hemisphere, Washington, DC, 1979.

4. J. D. Anderson, Jr., *Modern Compressible Flow, with Historical Perspective*, McGraw-Hill, New York, 1982.

5. D. A. Anderson, J. C. Tannehill, and R. H. Pletcher, *Computational Fluid Mechanics and Heat Transfer*, Hemisphere, Washington, DC, 1984.

6. J. O. Hinze, *Turbulence*, 2d ed., McGraw-Hill, New York, 1975.

7. A. A. Townsend, *The Structure of Turbulent Shear Flow*, 2d ed., Cambridge University Press, Cambridge, UK, 1976.

8. H. R. Vallentine, *Applied Hydrodynamics*, Butterworth, London, 1959.

9. L. M. Milne-Thomson, *Theoretical Hydrodynamics*, 6th ed., Macmillan, London, 1976.

10. H. Schlichting, *Boundary-Layer Theory*, 7th ed., McGraw-Hill, New York, 1979.

APPENDIX E

BASIC EQUATIONS
FROM HEAT TRANSFER

E.1   INTRODUCTION

Here we summarize concepts and fundamental equations of the three basic heat transfer modes: conduction, convection, and radiation. The governing equations appear without proof and serve as a convenient reference when we apply finite element analysis to heat transfer problems in Part II of this book. In the sections on conduction and convection the major emphasis is on summarizing basic heat transfer equations and boundary conditions often used in finite element heat transfer analysis. In the section on radiation we place additional emphasis on the physics and terminology of radiation since radiation concepts differ significantly from conduction and convection. The reader interested in further details of heat transfer should consult comprehensive texts on heat transfer, for example, references 1–5.

In heat transfer nondimensional parameters often characterize theory and results. For example, the well-known Reynolds number occurs in viscous flow theory and in convection heat transfer. Such parameters also appear in finite

596

element heat transfer literature. For convenient reference, we define some of the more frequently encountered nondimensional parameters in this appendix.

In many applications heat transfer is closely coupled to fluid flow. For example, conduction heat transfer in solids is strongly affected by the flow of an adjacent fluid. Convection heat transfer is intrinsically a flow phenomena. Because of the natural coupling of heat transfer and fluid mechanics, there is a close relationship between this appendix and Appendix D, *Basic Equations from Fluid Mechanics*. The reader interested in finite element analysis of heat transfer should review both appendices.

E.2   CONDUCTION

Conduction [6–8] is the transfer of thermal energy through a solid or fluid due to a temperature gradient. The transfer of thermal energy occurs at the molecular and atomic levels without net mass motion of the material. The rate equation describing this heat transfer mode is Fourier's law. For an isotropic medium Fourier's law is

$$q = -k\frac{\partial T}{\partial n} \qquad \text{(E.1)}$$

where q is the rate of heat flow per unit area in the n direction, k is the thermal conductivity that may be a function of the temperature T, and n indicates a normal direction. The minus sign appears because positive thermal energy transfer occurs from a warmer to a colder region; that is, the temperature gradient $\partial T/\partial n$ is negative in the direction of positive heat flow. In Cartesian coordinates the components of the heat flow vector are

$$q_x = -k\frac{\partial T}{\partial x} \qquad q_y = -k\frac{\partial T}{\partial y} \qquad q_z = -k\frac{\partial T}{\partial z}$$

or, in vector form,

$$\mathbf{q} = -k\nabla T \qquad \text{(E.2)}$$

where

$$\nabla = \frac{\partial}{\partial x}\hat{i} + \frac{\partial}{\partial y}\hat{j} + \frac{\partial}{\partial z}\hat{k}$$

For an anisotropic medium Fourier's law stated in Cartesian coordinates is

$$
\begin{aligned}
q_x &= -\left(k_{11}\frac{\partial T}{\partial x} + k_{12}\frac{\partial T}{\partial y} + k_{13}\frac{\partial T}{\partial z} \right) \\[2mm]
q_y &= -\left(k_{21}\frac{\partial T}{\partial x} + k_{22}\frac{\partial T}{\partial y} + k_{23}\frac{\partial T}{\partial z} \right) \\[2mm]
q_z &= -\left(k_{31}\frac{\partial T}{\partial x} + k_{32}\frac{\partial T}{\partial y} + k_{33}\frac{\partial T}{\partial z} \right)
\end{aligned}
\tag{E.3}
$$

where k_{ij} is the thermal conductivity tensor. The principles of irreversible thermodynamics show that the thermal conductivity tensor is symmetric, $k_{ij} = k_{ji}$.

E.2.1   Heat Conduction Equation

For an isotropic solid with temperature-dependent thermal conductivity the law of conservation of energy with Fourier's law yields the thermal energy equation. The law of conservation of energy is

$$
-\left(\frac{\partial q_x}{\partial x} + \frac{\partial q_y}{\partial y} + \frac{\partial q_z}{\partial z} \right) + Q = \rho c \frac{\partial T}{\partial t}
\tag{E.4}
$$

where Q is the internal heat generation rate per unit volume, ρ is the density, c is the specific heat, and t is time. No distinction is made between c_p and c_v, the specific heats at constant pressure and constant volume for a solid. The specific heat may in general be temperature dependent. When the thermal conductivity tensor and/or the specific heat c are temperature dependent, equation E.4 is nonlinear, but when these properties are constant the equation is linear. For constant thermal properties and an isotropic material, the *heat conduction* equation takes the form

$$
\nabla^2 T + \frac{Q}{k} = \frac{1}{\alpha}\frac{\partial T}{\partial t}
\tag{E.5}
$$

where α is the thermal diffusivity defined by

$$
\alpha = \frac{k}{\rho c}
\tag{E.6}
$$

With constant thermal properties and no internal heat generation, the heat conduction equation reduces to

$$
\nabla^2 T = \frac{1}{\alpha}\frac{\partial T}{\partial t}
\tag{E.7}
$$

which is the *diffusion equation*. With constant thermal properties and steady-state heat transfer the heat conduction equation reduces to

$$\nabla^2 T + \frac{Q}{k} = 0 \qquad (E.8)$$

which is *Poisson's equation*. For constant thermal properties, no internal heat generation, and steady-state heat transfer, the heat conduction equation reduces to

$$\nabla^2 T = 0 \qquad (E.9)$$

or *Laplace's equation*. The quantity $\nabla^2 T$ appearing in the different forms of the heat conduction equation, Equations E.5–E.9, takes different forms depending on the coordinate system. The two coordinate systems used in this book are Cartesian (x, y, z) where

$$\nabla^2 T = \frac{\partial^2 T}{\partial x^2} + \frac{\partial^2 T}{\partial y^2} + \frac{\partial^2 T}{\partial z^2} \qquad (E.10)$$

and cylindrical (r, θ, z) where

$$\nabla^2 T = \frac{\partial^2 T}{\partial r^2} + \frac{1}{r}\frac{\partial T}{\partial r} + \frac{1}{r^2}\frac{\partial^2 T}{\partial \theta^2} + \frac{\partial^2 T}{\partial z^2} \qquad (E.11)$$

E.2.2 Boundary Conditions

The heat conduction equation E.4 is solved subject to an initial condition and appropriate boundary conditions. The initial condition consists of specifying the temperature throughout the solid at an initial time. The boundary conditions take several forms and may be linear or nonlinear and must be specified on all boundaries of the solid. Typical boundary conditions encountered in finite element formulations are the following:

1. *Prescribed temperature.* The surface temperature of a boundary is specified to be constant or a function of a boundary coordinate and/or time. Prescribed temperature is an example of a *Dirichlet* boundary condition.

2. *Prescribed heat flow.* The rate of heat flow across a boundary is specified to be constant or a function of a boundary coordinate and/or time. For an isotropic solid Fourier's law, equation E.1, expresses surface heat flow as

$$-k\frac{\partial T}{\partial n} = q_s$$

where n is normal to the boundary, and q_s is the rate of the surface heat flow per area. Prescribed heat flow is an example of a *Neumann* boundary condition which specifies the normal derivative of the dependent variable.

3. *No heat flow (adiabatic boundary)*. When the rate of heat flow across a boundary is zero, we have, by Fourier's law,

$$\frac{\partial T}{\partial n} = 0$$

which is a special form of the previous case.

4. *Convective heat exchange*. When the rate of heat flow across a boundary is proportional to the difference between the surface temperature T_s and a convective exchange temperature T_e of an adjacent fluid, we have from Fourier's law the convective boundary condition,

$$-k\frac{\partial T}{\partial n} = h(T_s - T_e)$$

where h is a convection heat transfer coefficient. A convective boundary condition can be either linear or nonlinear since convection coefficients may be temperature dependent. The convective exchange temperature T_e may be a function of a boundary coordinate and/or time. Convection coefficients are discussed further in the section of this appendix on nondimensional parameters in convection heat transfer.

5. *Radiation heat exchange*. In this case the rate of heat flow across a boundary is specified in terms of the emitted energy from the surface and the incident radiant thermal energy, emitted and reflected from other solids and/or fluids. The boundary condition is

$$-k\frac{\partial T}{\partial n} = \sigma\epsilon T_s^4 - \alpha q_i$$

where σ is the Stefan–Boltzmann constant, ϵ is the surface emissivity, and T_s is the surface temperature. The coefficient α is the surface absorptivity, and q_i is the incident radiant thermal energy. The first term on the right-hand side is emitted energy from the surface, and the second term on the right-hand side is the absorbed incident radiant energy. The surface emissivity and absorptivity are normally a function of surface temperature. We discuss radiation boundary conditions further in the section of this appendix that describes radiation heat transfer.

6. *Other boundary conditions*. The boundary conditions previously described cover many cases of practical interest, but there are other important boundary conditions which arise in applications. Examples are the boundary conditions between two contacting solids not in perfect contact and the boundary conditions at a moving interface between a solid and a fluid due to a change of phase or ablation. A detailed discussion of these and other

boundary conditions is beyond the scope of this discussion, further details may be found, for example, in reference 7.

E.2.3 Nondimensional Parameters

Two nondimensional parameters associated with conduction heat transfer are the Fourier number, Fo, and the Biot number, Bi, defined as follows:

Fourier number:

$$Fo = \frac{\alpha t}{L^2}$$

Biot number:

$$Bi = \frac{hL}{k}$$

where α is the thermal diffusivity defined in equation E.6 and L is a characteristic length for a particular problem. The Fourier number is dimensionless time, and it characterizes the ratio of heat transfer by conduction to the rate of energy stored. The larger the Fourier number in a transient response, the larger the amount of heat conducted through the solid as compared with that stored in the solid, or the deeper the penetration of a temperature change into the solid in a given time. The Biot number characterizes the importance of the surface heat transfer compared to internal conduction heat transfer. High values of the Biot number correspond to specifying the surface temperature while low values correspond to insulated surfaces.

E.3 CONVECTION

Convection [9–11] is the transfer of thermal energy through a fluid due to motion of the fluid. The energy transfer from one fluid particle to another occurs by conduction, but thermal energy is transported by the motion of the fluid. The transfer of energy is called *forced convection* when the fluid motion is caused by external mechanical means (e.g., a fan or a pump). The transfer of energy is called *free* or *natural convection* when the fluid motion is caused by density differences in the fluid (buoyant effects). In convection heat transfer analysis, problems are often assumed to be either predominantly forced or free convection, but both heat transfer modes may occur simultaneously. To analyze convective heat transfer the concepts of conduction heat transfer are combined with the fundamental equations of fluid mechanics. Appendix D describes the basic types of fluid flow and presents the fundamental equations of fluid flow—conservation of mass, momentum, and energy. Here we present the basic equations of convection heat transfer for the specialized but important case of viscous incompressible flow.

E.3.1   Convection Equations

The *Boussinesq* approximation wherein the fluid is assumed to be Newtonian and incompressible means that the fluid density is taken as constant except for the effect of the density variation in producing buoyant forces. The density is assumed to vary only with temperature and only slightly from a reference value. With these assumptions the buoyant effects characteristic of free convection are included as body forces in the momentum equations. The fluid motion is assumed to be laminar, and for convenience the effects of radiation and viscous energy dissipation are neglected. With these assumptions the basic equations are

Conservation of mass:

$$\frac{\partial u}{\partial x} + \frac{\partial v}{\partial y} + \frac{\partial w}{\partial z} = 0 \tag{E.12}$$

Conservation of momentum:

$$\rho \frac{Du}{Dt} = -\frac{\partial p}{\partial x} - \rho_0 \beta (T - T_0) g_x + \frac{\partial \sigma_x}{\partial x} + \frac{\partial \tau_{xy}}{\partial y} + \frac{\partial \tau_{xz}}{\partial z} \tag{E.13a}$$

$$\rho \frac{Dv}{Dt} = -\frac{\partial p}{\partial y} - \rho_0 \beta (T - T_0) g_y + \frac{\partial \tau_{xy}}{\partial x} + \frac{\partial \sigma_y}{\partial y} + \frac{\partial \tau_{yz}}{\partial z} \tag{E.13b}$$

$$\rho \frac{Dw}{Dt} = -\frac{\partial p}{\partial z} - \rho_0 \beta (T - T_0) g_z + \frac{\partial \tau_{xz}}{\partial x} + \frac{\partial \tau_{yz}}{\partial y} + \frac{\partial \sigma_z}{\partial z} \tag{E.13c}$$

with the stress components

$$\sigma_x = 2\mu \frac{\partial u}{\partial x}$$

$$\sigma_y = 2\mu \frac{\partial v}{\partial y}$$

$$\sigma_z = 2\mu \frac{\partial w}{\partial z}$$

$$\tau_{xy} = \tau_{yx} = \mu \left(\frac{\partial u}{\partial y} + \frac{\partial v}{\partial x} \right)$$

$$\tau_{yz} = \tau_{zy} = \mu \left(\frac{\partial v}{\partial z} + \frac{\partial w}{\partial y} \right)$$

$$\tau_{zx} = \tau_{xz} = \mu \left(\frac{\partial w}{\partial x} + \frac{\partial u}{\partial z} \right)$$

Conservation of energy:

$$\rho c_v \frac{DT}{Dt} - Q = \frac{\partial}{\partial x}\left(k\frac{\partial T}{\partial x}\right) + \frac{\partial}{\partial y}\left(k\frac{\partial T}{\partial y}\right) + \frac{\partial}{\partial z}\left(k\frac{\partial T}{\partial z}\right) \qquad (E.14)$$

Supplementary equation:

$$\rho = \rho_0[1 - \beta(T - T_0)] \qquad (E.15)$$

where

$$\frac{D}{Dt} = \frac{\partial}{\partial t} + u\frac{\partial}{\partial x} + v\frac{\partial}{\partial y} + w\frac{\partial}{\partial z}$$

$$
\begin{aligned}
u, v, w &= \text{velocity components} \\
p &= \text{dynamic pressure} \\
\rho_0 &= \text{density at reference temperature } T_0 \\
\beta &= \text{coefficient of thermal expansion} \\
T_0 &= \text{reference temperature for buoyant forces} \\
g_x, g_y, g_z &= \text{components of gravitational acceleration} \\
c_v &= \text{specific heat at constant volume} \\
Q &= \text{internal heat generation rate} \\
k &= \text{thermal conductivity}
\end{aligned}
$$

The conservation of momentum equations, E.13, contain the dynamic pressure p. The total pressure P has been separated into hydrostatic and dynamic components, and the hydrostatic component has been canceled by the reference density component of the body force. The fluid properties such as the coefficient of thermal expansion β, viscosity μ, specific heat c_v, and thermal conductivity k are in general temperature dependent.

The momentum equations, E.13, and the energy equations, E.14, are coupled in general and must be solved simultaneously with the mass conservation equation, E.12. A simultaneous solution is also necessary for free convection since temperature appears in the momentum equations, or for forced convection with temperature-dependent fluid properties. For forced convection with constant fluid properties the flow problem uncouples from the thermal problem. The flow analysis that consists of solving the mass and momentum equations can be performed independently of the energy equation. Then the energy equation is solved with the velocity field from the flow analysis.

E.3.2 Boundary Conditions

The convection equations, E.12–E.15, for viscous incompressible flow are solved subject to initial conditions and appropriate boundary conditions. The initial conditions consist of specifying the velocity components, pressure, and temperature throughout the fluid at an initial time. The boundary conditions can be separated into boundary conditions for the flow variables (velocity components and pressure) and boundary conditions for temperature. The boundary conditions on the flow variables consist of specifying either the velocity components or the total surface tractions on the boundary of the fluid region. These boundary conditions are

$$u = u(s)$$
$$v = v(s) \qquad \text{on } S_v$$
$$w = w(s)$$

$$
\begin{Bmatrix} \bar{\sigma}_x \\ \bar{\sigma}_y \\ \bar{\sigma}_z \end{Bmatrix}
=
\begin{bmatrix}
-p + \sigma_x & \tau_{xy} & \tau_{xz} \\
\tau_{xy} & -p + \sigma_y & \tau_{yz} \\
\tau_{xz} & \tau_{yz} & -p + \sigma_z
\end{bmatrix}
\begin{Bmatrix} l \\ m \\ n \end{Bmatrix}
\qquad \text{on } S_T
$$

where s refers to a coordinate along the boundary, and S_v and S_T refer to portions of the boundary where velocity and tractions are specified, respectively. The components of surface tractions are denoted by $\bar{\sigma}_x$, $\bar{\sigma}_y$, and $\bar{\sigma}_z$, and l, m, and n are the direction cosines of the outward normal to the surface. The boundary conditions on temperature consist of specifying either the temperature or the heat flow on the boundary of the fluid region. The fluid thermal conditions may be any of the typical boundary conditions specified for heat conduction in a solid that appear in the previous section. For example, the types of heat flow boundary conditions include specified heating, an adiabatic (insulated) boundary or convective and radiative exchanges.

E.3.3   Nondimensional Parameters

Several nondimensional parameters characterize convective heat transfer. We list the most important ones here for later reference:

Reynolds number:

$$Re = \frac{\rho U L}{\mu}$$

Prandtl number:

$$Pr = \frac{\mu c_p}{k}$$

Peclet number:

$$Pe = Re\, Pr$$

Nusselt number:

$$Nu = \frac{hL}{k}$$

Grashof number:

$$Gr = \frac{g\beta\Delta T\rho^2 L^3}{\mu^2}$$

Rayleigh number:

$$Ra = Gr\,Pr$$

where ρ is density, U is a characteristic flow velocity, L is a characteristic length, μ is viscosity, c_p is specific heat at constant pressure, k is thermal conductivity, h is a convection coefficient, g is the acceleration of gravity, β is the coefficient of thermal expansion and ΔT is a characteristic temperature difference, for example $\Delta T = T_{wall} - T_0$.

The well-known *Reynolds* number characterizes the ratio of inertia forces to viscous forces in a viscous flow. The transition from laminar to turbulent flow in forced convection is typically expressed in terms of the Reynolds number. The *Prandtl* number is a measure of the ratio of viscous diffusion to thermal diffusion. For example, in hydrodynamic and thermally developing flows the Prandtl number is a measure of the rate at which velocity and temperature profiles develop [9]. The *Peclet* number characterizes forced convection; it represents the ratio of thermal energy transport by convection to thermal energy transfer by conduction. The *Nusselt* number is a dimensionless convection coefficient used in both free and forced convection. For forced convection low-speed flows the Nusselt number is expressed in terms of the Reynolds and Prandtl numbers. For forced convection high-speed flows the Nusselt number is also a function of the Mach number. For free convection the Nusselt number is expressed in terms of the Grashof and Prandtl numbers. Extensive analytical and/or experimental correlations for the Nusselt number are available for free and forced convection (see, e.g., references 1–5). The *Grashof* number characterizes free convection; it is interpreted as the ratio of buoyant to viscous forces. The *Rayleigh* number characterizes the transition from laminar to turbulent flow in free convection.

E.4   RADIATION

Radiation heat transfer [12–16] is the transfer of thermal energy between two locations by an electromagnetic wave. The electromagnetic radiation spectrum ranges from cosmic rays with wavelengths less than 10^{-12} m to radio waves with wavelengths of 10^4 m, and thermal radiation occupies the portion of the spectrum between 10^{-7} and 10^{-4} m that is detected as heat or light. The thermal radiation wavelength range includes the ultraviolet, visible, and infrared subranges. Radiation heat transfer differs from conduction and convection in two important ways. For conduction and convection, transfer of energy between two locations typically depends on the temperature difference between the two locations. Radiation heat transfer, however, typically depends on the differences of the absolute temperatures of the individual locations, each raised to the fourth power. A second distinguishing feature of radiation is that no medium is required for radiation heat transfer to occur. In conduction and convection heat transfer thermal energy is transported by a physical medium, but radiant energy will pass through a perfect vacuum.

The temperature of a medium governs the emission of thermal radiation. It can be demonstrated by the second law of thermodynamics that there is a maximum amount of radiant energy that can be emitted at a given temperature and wavelength. Such an ideal emitter is called a *blackbody*, and the blackbody concept is a standard for describing the radiant energy emitted by real surfaces.

When radiation is incident on the surface of a medium, some of the radiation is reflected and the remainder penetrates into the medium. The radiation that penetrates into the medium may be partially absorbed as it travels through the medium, and the unabsorbed portion may emerge and be transmitted from the medium. When all of the radiation that penetrates into a medium is absorbed internally, the medium is *opaque*. Most solids are opaque and do not transmit radiation. When all of the radiation that penetrates into a medium is transmitted the medium is *transparent*. Glass is an example of a solid that is transparent to radiation but only at certain wavelengths. Many liquids and gases are also transparent. Some gases, however, can emit, absorb, and scatter radiation. An example is the emission, absorption, and scattering of radiation in the earth's atmosphere.

E.4.1   Surface Radiation

The concept of a blackbody is basic to the study of radiation energy transfer. A blackbody is a perfect emitter at every wavelength and in each direction and a perfect absorber (no reflected energy). The total radiant energy emitted by a blackbody is a function only of its temperature. The emissive power $e_{b\lambda}$ of a blackbody at a specific wavelength λ is called the spectral or monochromatic power and is given by Planck's law [12]. The total emissive power of a blackbody is obtained by integrating $e_{b\lambda}$ over all wavelengths. The results of the integration is the total emitted energy rate e_b per unit area expressed as the *Stefan–Boltzmann* law. The Stefan–Boltzmann law states that the total emissive power of a blackbody is

$$e_b = \int_0^\infty e_{b\lambda}\, d\lambda = \sigma T^4 \qquad \text{(blackbody)} \qquad \text{(E.16)}$$

The monochromatic emissive power of a real (nonblack) surface is denoted by e_λ. The emissive power of a real surface is expressed in terms of the blackbody emissive power as

$$e_\lambda = \epsilon_\lambda e_b$$

where ϵ_λ is the monochromatic hemispherical emissivity of the surface. The *emissivity* is a measure of how well a body can radiate energy as compared to a blackbody. The emissivity of a real surface is a function of radiation wavelength, surface temperature, surface roughness, and surface material. Hemispherical means that the emissivity has been averaged with respect to all directions in a hemisphere above the surface. As an engineering approximation, emissivities of real surfaces are often assumed to be independent of radiation wavelength. A *gray body* is a surface for which the emissive power is independent of radiation wavelength. The emissive power for a gray body is given by

$$e = \epsilon(T)\sigma T^4 \qquad (\text{gray body}) \qquad (\text{E.17})$$

where the emissivity is a function of the surface temperature but independent of wavelength.

In addition to emitting energy, real surfaces absorb a portion of the energy incident on them. The monochromatic hemispherical *absorptivity* of a surface is defined as the fraction of the incident energy that is absorbed by the medium. *Kirchhoff's law* relates the emitting absorbing abilities of a surface. Kirchoff's law can have various conditions imposed on it [15], but in essence Kirchhoff's law states that for a specific wavelength the monochromatic absorptivity and emissivity of a surface at a given temperature are equal. Thus Kirchhoff's law states

$$\alpha_\lambda(T) = \epsilon_\lambda(T) \qquad (\text{E.18})$$

For a gray body α_λ and ϵ_λ are independent of wavelength and

$$\alpha(T) = \epsilon(T) \qquad (\text{gray body}) \qquad (\text{E.19})$$

The reflective properties of a surface are more complicated to specify than either the emissivity or absorptivity. Reflected energy depends not only on the angle at which the incident energy impinges on the surface but also on the direction of the reflected energy. There are two basic types of radiation reflections, *specular* and *diffuse*. The reflection is specular if the angle of reflection is equal to the angle of incidence, but diffuse if an incident beam is reflected uniformly in all directions. Reflections from highly polished smooth surfaces approach specular behavior, and reflections from rough surfaces approach diffuse behavior. For engineering calculations, reflections are often assumed to be diffuse. The monochromatic hemispherical *reflectivity* ρ_λ of a surface is defined as the fraction of the incident energy that is reflected back into the hemisphere space above the surface.

From the definitions of absorptivity and reflectivity as fractions of incident energy absorbed or reflected for an *opaque* medium,

$$\alpha_\lambda(T) + \rho_\lambda(T) = 1$$

Or using Kirchhoff's law, Equation E.18,

$$\epsilon_\lambda(T) + \rho_\lambda(T) = 1 \tag{E.20}$$

For a gray body the wavelength subscripts are omitted and

$$\epsilon(T) + \rho(T) = 1 \quad \text{(gray, opaque body)} \tag{E.21}$$

The relations above are for an opaque medium that does not transmit or scatter incident radiation. The study of energy transfer through media that absorb, emit, and scatter radiation is important in the study of various radiation phenomena, including radiation related to earth's and other planets' atmospheres, nuclear explosions, hypersonic shock layers, rocket propulsion, plasma generation for nuclear fusion, and ablating systems. Consideration of this topic, however, is beyond the scope of the present discussion. For a comprehensive discussion of the fundamentals of radiation in absorbing, emitting, and scattering media, the reader should see references 12–15.

E.4.2   Radiation Exchange Between Surfaces

In many engineering heat transfer problems the thermal radiation passing between surfaces is substantially unaffected by the intervening medium. Such a nonparticipating medium could be a gas at low temperatures, air, or a vacuum. For the purpose of computing radiation heat transfer between surfaces the intervening medium is often assumed nonparticipating, and the surfaces are typically idealized as gray diffuse emitters and reflectors of radiant energy.

Single Surface   A body at temperature T is within an enclosed space whose walls have a uniform temperature T_e. The walls emit and absorb energy perfectly with $\epsilon = \alpha = 1$ and reflectivity $\rho = 0$. The radiant energy emitted from the body per unit time and per unit area is $\epsilon\sigma T^4$, and the corresponding absorbed radiant energy from the walls is $\alpha\sigma T_e^4$. Consequently, the net rate of heat flow per unit area from the body surface is

$$q = \epsilon\sigma T^4 - \alpha\sigma T_e^4$$

By Kirchhoff's law $\epsilon = \alpha$, hence

$$q = \epsilon\sigma\left(T^4 - T_e^4\right) \tag{E.22}$$

The Radiation View Factor A concept fundamental to the computation of radiation heat exchange between surfaces is the radiation view factor (alternatively designated as the shape factor, configuration factor, geometrical factor, or angle factor). Consider the surface i with area A_i shown in Figure E.1. Radiant energy leaving surface i, if diffusely distributed, will flow into the hemispherical space above the surface. A fraction of this energy will impinge on surface j with area A_j. The *view factor* F_{ij} is the fraction of the radiation energy leaving i that arrives at j. View factors can be defined for radiation between infinitesimal areas, between infinitesimal and finite areas, and between finite areas [12–14]. If the radiation heat flow is assumed uniformly distributed over surfaces i and j, view factors based on finite areas are used. With the notation of Figure E.1, the view factor based on finite areas A_i and A_j is

$$F_{ij} = \frac{1}{A_i}\int_{A_i}\int_{A_j}\frac{\cos\theta_i\cos\theta_j\, dA_i\, dA_j}{\pi r^2} \tag{E.23}$$

A useful property of view factors is the reciprocity rule, which relates the view factor for radiant energy traveling from surface i to surface j to the view factor for radiant energy traveling from surface j to surface i. The reciprocity rule is

$$A_i F_{ij} = A_j F_{ji} \tag{E.24}$$

Another useful relationship between view factors is based on energy conservation. For radiation within an enclosure the radiant energy leaving any

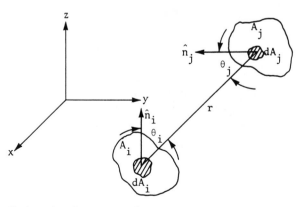

Figure E.1   Radiation view factor notation.

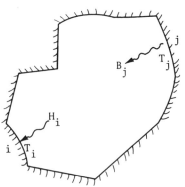

Figure E.2   Enclosure with N isothermal surfaces.

surface i in an enclosure must impinge on the surfaces of the enclosure. Energy conservation requires that all fractions of energy having surface i total to unity:

$$\sum_{j=1}^{N} F_{ij} = 1$$

where N denotes the number of surfaces within the enclosure. The summation includes F_{ii} because if surface i is concave, it will intercept a portion of its own emitted energy. Mathematical techniques for the evaluation of view factors and a catalogue of view factors for various configurations appear in reference 12. A comparative study of methods for computing diffuse radiation view factors for complex structures appears in reference 13.

Multiple Surfaces   An enclosure with N finite surfaces is shown in Figure E.2. We discuss the net heat flow out of surface i by making the following assumptions: (1) each surface of the enclosure is isothermal; (2) each surface is gray; (3) emitted and reflection radiation from surfaces are diffusely distributed; (4) radiant heat flow over each surface is uniformly distributed; and (5) the surface is opaque, $\rho = 1 - \epsilon$. The net heat flow out of a typical surface i is the difference between the emitted radiation and the absorbed portion of the radiation. Thus

$$q_i = \epsilon_i \sigma T_i^4 - \alpha_i H_i \tag{E.25}$$

where H_i is the radiation incident on surface i per unit time and unit area. By Kirchhoff's law, $\epsilon_i = \alpha_i$. The radiation incident on surface i consists of fractions of energy emitted and reflected from all surfaces in the enclosure. The rate of heat flow per unit area, B_j, leaving a typical surface j is called the radiosity. The radiosity B_j, consists of the energy emitted and reflected from surface j. Thus

$$B_j = \epsilon_j \sigma T_j^4 + \rho_j H_j \tag{E.26}$$

The energy incident on surface i is now expressed in terms of the sum of the contributions of the radiosity from all surfaces by making use of the view factors F_{ji}. Thus, radiation heat flow incident on surface i is

$$H_i A_i = \sum_{j=1}^{N} F_{ji} A_j B_j$$

Using the reciprocity rule, equation E.24, this last result becomes

$$H_i = \sum_{j=1}^{N} F_{ij} B_j \tag{E.27}$$

The radiosities are now eliminated by substituting equation E.26 into equation E.27 to yield a set of simultaneous equations for the incident radiation on each surface of the enclosure. For this purpose it is convenient to write the equations in matrix form. First the radiosity equation, E.26, is written

$$\{B\} = \{\epsilon \sigma T^4\} + [\searrow\rho_\searrow]\{H\}$$

where $[\searrow\rho_\searrow]$ denotes a diagonal matrix of reflectives. Next equation E.27 is written as

$$\{H\} = [F]\{B\}$$

Finally the vector of radiosities is eliminated to obtain the set of equations

$$[[I] - [F][\searrow\rho_\searrow]]\{H\} = [F]\{\epsilon \sigma T^4\} \tag{E.28}$$

Equation E.28 consists of N linear simultaneous equations for computing N unknown values of H_i. The N surface temperatures that appear in the vector on the right-hand side of the equation are assumed known. Alternative forms of these equations for other cases, such as prescribed heat flows and unknown temperatures, are discussed in references 14–16. The form presented in equation E.28 is the approach often employed in finite element analysis of radiation heat transfer between surfaces.

The calculation of radiation heat exchange between surfaces described above is based on the assumption of diffuse emission and reflection. A discussion of the radiant interchange among surfaces taking into account the directional distributions of emissions and reflections, that is, specular components of radiation, appears in reference 12.

E.5   HEAT TRANSFER UNITS

The following list provides examples of consistent heat transfer units for quantities encountered in the finite element thermal analysis:

Quantity	English	SI
Length	foot (ft)	meter (m)
Time	second (s)	second (s)
Mass	lb_m	kilogram (kg)
Energy	Btu	joules (J)
Temperature	Fahrenheit (°F) or Rankine (°R)	kelvin (K)
Density	lb_m/ft^3	kg/m^3
Specific heat	Btu/lb_m °F	$J/kg\,K$
Power	Btu/s	J/s (watt, W)
Heat flux	$Btu/ft^2\,s$	$J/m^2\,s\,(W/m^2)$
Convection heat transfer coefficient	$Btu/ft^2\,s\,°F$	$J/m^2\,s\,K\,(W/m^2\,K)$
Thermal conductivity	$Btu/ft\,s\,°F$	$J/m\,s\,K\,(W/m\,K)$
σ (Stefan–Boltzmann constant)	4.755×10^{-13} $\left(\dfrac{Btu}{s\,ft^2\,R^4}\right)$	5.6697×10^{-8} $\left(\dfrac{W}{m^2\,K^4}\right)$

The following are some relationships between the various temperature scales used in heat transfer:

$$T_R = T_F + 459.67$$

$$T_K = T_R/1.8$$

$$T_C = (T_F - 32)/1.8$$

$$T_K = T_C + 273.15$$

REFERENCES

1. W. M. Rohsenow and J. P. Hartnett (eds.), *Handbook of Heat Transfer*, McGraw-Hill, New York, 1973.
2. E. R. G. Eckert and R. M. Drake, Jr., *Analysis of Heat, and Mass Transfer*, McGraw-Hill, New York, 1972.
3. A. J. Chapman, *Heat Transfer*, 3d ed., Macmillan, New York, 1974.

4. F. Krieth, *Principles of Heat Transfer*, 3d ed., Intext, New York, 1973.

5. D. R. Pitts and L. E. Sissom, *Theory and Problems of Heat Transfer*, Schaum's Outline Series, McGraw-Hill, New York, 1977.

6. M. N. Ozisik, *Boundary Value Problems of Heat Conduction*, International, Scranton, PA, 1968.

7. V. S. Arpaci, *Conduction Heat Transfer*, Addison-Wesley, Reading, MA, 1966.

8. G. E. Myers, *Analytical Methods in Conduction Heat Transfer*, McGraw-Hill, New York, 1971.

9. W. M. Kays, *Convection Heat and Mass Transfer*, McGraw-Hill, New York, 1966.

10. R. B. Bird, W. E. Stewart, and E. N. Lightfoot, *Transport Phenomena*, Wiley, New York, 1960.

11. S. Ostrach, "Natural Convection in Enclosures," in *Advances in Heat Transfer*, Vol. 8, J. P. Hartnett and T. F. Irvine, Jr. (eds.), Academic, New York, 1972.

12. E. M. Sparrow and R. D. Cess, *Radiation Heat Transfer*, aug. ed., McGraw-Hill, New York, 1978.

13. A. F. Emery, O. Johnson, M. Lobo, and A. Abrous, "A Comparative Study of Methods for Computing the Diffuse Radiation Viewfactors for Complex Structures," *J. Heat Transfer*, Vol. 113, 1991, pp. 413–422.

14. T. J. Love, *Radiative Heat Transfer*, Merrill, Columbus, OH, 1968.

15. R. Siegel and J. R. Howell, *Thermal Radiation Heat Transfer*, Vol. I–III. NASA SP-164. 1968; also available in textbook form ass *Thermal Radiation Heat Transfer*, McGraw-Hill, New York, 1980.

16. M. N. Ozisik, *Radiative Transfer and Interactions With Conduction and Convection* Wiley–Interscience, New York, 1973.

AUTHOR INDEX

615

SUBJECT INDEX